TECHNIQUES IN PARTIAL DIFFERENTIAL EQUATIONS

INTERNATIONAL SERIES IN PURE AND APPLIED MATHEMATICS

William Ted Martin and E. H. Spanier, CONSULTING EDITORS

AHLFORS · Complex Analysis
BUCK · Advanced Calculus
BUSACKER AND SAATY · Finite Graphs and Networks
CHENEY · Introduction to Approximation Theory
CHESTER · Techniques in Partial Differential Equations
CODDINGTON AND LEVINSON · Theory of Ordinary Differential Equations
COHN · Conformal Mapping on Riemann Surfaces
DENNEMEYER · Introduction to Partial Differential Equations and Boundary Value Problems
DETTMAN · Mathematical Methods in Physics and Engineering
EPSTEIN · Partial Differential Equations
GOLOMB AND SHANKS · Elements of Ordinary Differential Equations
GRAVES · The Theory of Functions of Real Variables
GREENSPAN · Introduction to Partial Differential Equations
GRIFFIN · Elementary Theory of Numbers
HAMMING · Numerical Methods for Scientists and Engineers
HILDEBRAND · Introduction to Numerical Analysis
HOUSEHOLDER · Principles of Numerical Analysis
KALMAN, FALB, AND ARBIB · Topics in Mathematical System Theory
LASS · Elements of Pure and Applied Mathematics
LASS · Vector and Tensor Analysis
LEIGHTON · Introduction to the Theory of Differential Equations
LEPAGE · Complex Variables and the Laplace Transform for Engineers
MCCARTY · Topology: An Introduction with Applications to Topological Groups
MONK · Introduction to Set Theory
MOORE · Elements of Linear Algebra and Matrix Theory
MOSTOW AND SAMPSON · Linear Algebra
MOURSUND AND DURIS · Elementary Theory and Application of Numerical Analysis
NEF · Linear Algebra
NEHARI · Conformal Mapping
NEWELL · Vector Analysis
PIPES AND HARVILL · Applied Mathematics for Engineers and Physicists
RALSTON · A First Course in Numerical Analysis
RITGER AND ROSE · Differential Equations with Applications
RITT · Fourier Series
ROSSER · Logic for Mathematicians
RUDIN · Principles of Mathematical Analysis
SAATY AND BRAM · Nonlinear Mathematics
SAGAN · Introduction to the Calculus of Variations
SIMMONS · Introduction to Topology and Modern Analysis
SNEDDON · Elements of Partial Differential Equations
SNEDDON · Fourier Transforms
STRUBLE · Nonlinear Differential Equations
WEINSTOCK · Calculus of Variations
WEISS · Algebraic Number Theory
ZEMANIAN · Distribution Theory and Transform Analysis

TECHNIQUES IN PARTIAL DIFFERENTIAL EQUATIONS

Clive R. Chester
Formerly with the Department of Mathematics
Polytechnic Institute of Brooklyn

McGraw-Hill Book Company
New York
St. Louis
San Francisco
Düsseldorf
London
Mexico
Panama
Sydney
Toronto

Techniques in Partial Differential Equations

Library of Congress Catalog Card Number 71-114443
10740

1 2 3 4 5 6 7 8 9 0 M A M M 7 9 8 7 6 5 4 3 2 1 0

This book was set in Times New Roman, and printed on permanent paper and bound by The Maple Press Company. The designer was Merrill Haber; the drawings were done by B. Handelman Associates, Inc. The editors were Donald K. Prentiss and M. E. Margolies. Sally Ellyson supervised production.

Preface

The major part of this book (the first 14 chapters) stems from a year course on partial differential equations of mathematical physics given by the author for a number of years at the Polytechnic Institute of Brooklyn. The course was taken primarily by advanced engineering students in such fields as aerospace, electrical engineering, mechanical engineering, etc. It was also taken by students at General Electric in a cooperative program with the Polytechnic Institute.

Because the material is aimed primarily at applied scientists, including applied mathematicians, there is considerable emphasis on motivation, and the presentation is frequently inductive rather than deductive, the idea being to arrive, as far as possible, at each new concept in as natural way as possible. Of course, what is natural to one person may be unnatural to another, but most of the material here has been developed on the basis of experience in mathematics courses for advanced engineering and science students. Indeed, a number of the problems investigated were suggested to the author by such students, and the original material grew substantially in response to requests from various students for additional topics. One result of this is that the present book contains considerably more material than can be covered in the average one-year course. Some of the material has been used by the author in courses in applied techniques in boundary-value problems, in theoretical mechanics, in integral equations, in Laplace and Fourier transforms, and in calculus of variations, all given at the Polytechnic Institute. Thus there is ample material to allow a lecturer to choose according to his and/or his students' individual taste.

It is also possible to cover the material in an order different from that of the text. I have tried to arrange the material in order of increasing difficulty. Indeed, I found from experience that engineers and scientists take to the wave equation and waves traveling up and down strings much more readily than they do to Monge cones in space. Consequently, the general theory of first-order partial differential equations is not taken up until Chapter 8. However, anyone wishing to do so could follow Chapter 1 with Chapter 8. Similarly, Chapter 8 could be omitted entirely at the expense of omitting only a few sections in later chapters that depend on it.

Because the book starts off at about as elementary a level as possible for a course in partial differential equations, it is possible to use the first few chapters in an undergraduate course. The author has done so with sophomore honor students at the Polytechnic Institute with reasonable success.

As for prerequisites, a course in elementary differential equations and one in advanced calculus are probably advisable. Since there is a considerable variation in level of the material, some sections may require a deeper background than others if they are not to be omitted. Knowledge of complex variables is required for a number of parts, but there is ample material available here for a course to which complex variables is not a prerequisite. The chapter on integral transforms assumes at least cursory knowledge of the Laplace and Fourier transforms. An acquaintance with tensor analysis is necessary for parts of Chapters 10 and 11. Chapter 12 requires a little knowledge of linear algebra.

Although aimed primarily at applied scientists, the material may well be of benefit to pure mathematicians by forming an introduction to books of a more theoretical nature. Indeed, many an engineer and scientist has had his interest in seeing rigorous proofs whetted by an intuitive introduction to the subject, whereas an initial presentation that is too formal may have alienated him from the subject.

I hope that the book will also be useful as a reference book for applied mathematicians working both in academic and in industrial research.

Because too much motivation can occasionally obscure the logical structure of the final result, summaries of the salient facts developed in many of the chapters have been included at the ends of those chapters. Summaries are not included for all chapters because in some the structure stands out fairly clearly as it is.

The author wishes to express his gratitude to the many students whose suggestions and encouragement gave rise to this book. The people in the General Electric program at Schenectady were particularly helpful in this connection. Finally, he wishes to express his appreciation to Professor Bernard Epstein, of the University of New Mexico, for his very encouraging and helpful initial review of the manuscript.

CLIVE R. CHESTER

Contents

Chapter 8
SINGLE FIRST-ORDER EQUATIONS IN GENERAL 162

Chapter 9
SECOND-ORDER HYPERBOLIC EQUATIONS IN TWO INDEPENDENT VARIABLES 218

Chapter 10
LINEAR ELLIPTIC EQUATIONS 239

Chapter 11
MORE GENERAL LINEAR HYPERBOLIC EQUATIONS 247

Chapter 12
FIRST-ORDER SYSTEMS 266

Chapter 13

Chapter 14

Chapter 15

Chapter 16

1
Introduction to Linear First-order Equations

1-1. DIRECTIONAL DERIVATIVES; LINEAR FIRST-ORDER EQUATIONS

The principal purpose of this chapter is to introduce the notion of characteristics. The methods given here are more important for theoretical purposes and for an understanding of the nature of the solutions of certain types of partial differential equations (PDEs) than they are as practical methods for obtaining solutions. In Chap. 8 more efficient and more general methods for solving first-order partial differential equations will be discussed. The reader who is interested in the theory of first-order PDEs and in methods for solving them should study Chap. 8 in addition to the present chapter.

In the plane a direction can be specified by giving a single number, namely, the slope of any of the infinitely many parallel lines that go in that direction. In higher dimensions, a single number no longer suffices to determine direction. To generalize to higher dimensions, and for other purposes too, it is preferable to specify a direction in the plane by the ratio of a pair of numbers. Given a line, we call *direction numbers* for it any pair of numbers (a, b) whose ratio b/a gives the slope of the line. Thus, if (a, b) are direction numbers for a given line and λ is any number different from zero, then $(\lambda a, \lambda b)$ are also direction numbers for the same line. Therefore the pair of direction numbers for a given line is not unique. It is determined up to a scalar factor λ. We can take advantage of this fact to normalize the direction numbers for a given line by determining λ so that the sum of the squares of the direction numbers is 1. This requires

$$(\lambda a)^2 + (\lambda b)^2 = 1 \tag{1-1}$$

and so

$$\lambda = \frac{1}{\sqrt{a^2 + b^2}}$$

From (1-1) it follows that $| \lambda a | \leqslant 1$. Hence we can find an angle α such that $\lambda a = \cos \alpha$. From this and (1-1) again it follows that $\lambda b = \sin \alpha = \cos \beta$, where β is the complement of α. The pair of direction numbers $(\cos \alpha, \cos \beta)$ is called a pair of *direction cosines*. α is the angle between the given direction and the positive x axis, and β is the angle between the given direction and the positive y axis (see Fig. 1-1).

If we introduce the unit vectors **i** and **j** along the x and y axes respectively, then we can write a unit vector **e** in the given direction in the form

$$\mathbf{e} = (\cos \alpha)\,\mathbf{i} + (\cos \beta)\,\mathbf{j}$$

(Boldface symbols are used for vectors throughout the text.) A general vector in this direction has the form

$$\mathbf{v} = a\mathbf{i} + b\mathbf{j}$$

where $b/a = (\cos \beta)/(\cos \alpha)$.

As an example of the above, suppose we have a curve C given in parametric form by $x = a(t)$, $y = b(t)$. Then

$$\frac{b'(t)}{a'(t)} = \frac{dy/dt}{dx/dt} = \frac{dy}{dx}$$

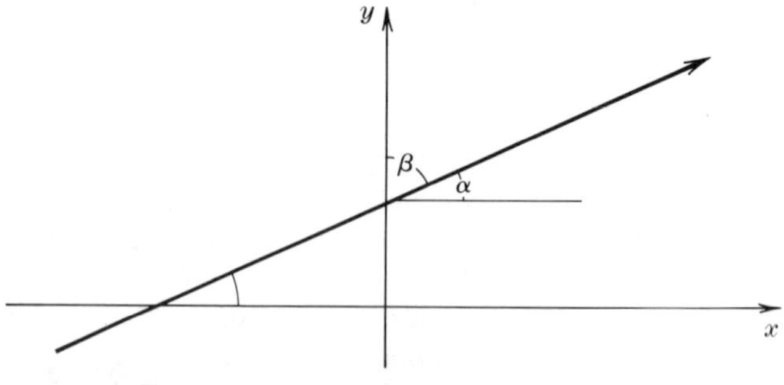

Figure 1-1

is the slope of the curve. Hence (a', b') are direction numbers for the curve. Moreover, if the arc length is chosen as parameter, then $(dx/ds)^2 + (dy/ds)^2 = 1$. Thus, $(\dot{a}, \dot{b})$, where the dot stands for d/ds, are direction cosines.

As a second example, consider the pair of ordinary differential equations

$$\begin{aligned} \frac{dx}{dt} &= a(x, y) \\ \frac{dy}{dt} &= b(x, y) \end{aligned} \tag{1-2}$$

Since t does not appear explicitly, we can divide the two equations and obtain a single first-order differential equation for y as a function of x. The solution of such an equation is a one-parameter family of curves. If we call this parameter τ and write the solution in parametric form, we will have

$$\begin{aligned} x &= \tilde{x}(t, \tau) \\ y &= \tilde{y}(t, \tau) \end{aligned} \tag{1-3}$$

as the parametric equations of a family of curves which, according to (1-2), have at every point (x, y) the direction $(a(x, y), b(x, y))$.

Suppose we have a function $u(x, y)$ defined in some region of the xy plane, and suppose we choose a point (x, y) and a direction $\gamma : (a, b)$ [which may depend upon (x, y)]. We find the direction cosines as before. Then we define the *directional derivative* u_γ of u in this direction to be†

$$u_\gamma = (\cos \alpha)\, u_x + (\cos \beta)\, u_y \tag{1-4}$$

Note that if C is a curve given parametrically by $x = \hat{a}(t)$, $y = \hat{b}(t)$, then, on C, $u(x, y) = u[\hat{a}(t), \hat{b}(t)] = \hat{u}(t)$ and therefore

$$\hat{u}'(t) = \hat{a}' u_x + \hat{b}' u_y \tag{1-5}$$

is proportional to the directional derivative of u along C.

1-2. CAUCHY PROBLEM

This last fact raises an interesting question. Suppose we wanted to solve the PDE

$$a(x, y)\, u_x + b(x, y)\, u_y = 0 \tag{1-6}$$

and suppose we can find the solutions (1-3) of the system of ordinary differential equations (1-2). Then, according to (1-5), (1-6) tells us that

$$\frac{du}{dt} = 0 \qquad \text{along each of the curves (1-3)} \tag{1-7}$$

† $u_x = \partial u/\partial x$; $u_y = \partial u/\partial y$.

Hence u is a constant along each of these curves (but not necessarily the same constant on each curve; it might be a different constant on each curve). If we knew the values of u on any curve Γ which crosses the curves (1-3), we would know the value of u everywhere. This suggests a natural problem to set up for the partial differential equation (1-6). Called the *Cauchy problem*, it is the following. Given (1) the PDE (1-6), (2) the initial curve

$$\Gamma : \begin{cases} x = \phi(\tau) \\ y = \psi(\tau) \end{cases}$$

and (3) the values $u = \omega(\tau)$ of u on Γ, find a function $u(x, y)$ which satisfies the PDE in a region containing Γ and which reduces to the given values $\omega(\tau)$ on Γ.

From the preceding discussion, it is clear how to solve this problem. We first write down the ordinary differential equations (1-2). We call these the *characteristic equations* of the PDE, and we call their solution curves the *characteristics* of the PDE. Let us assume that Γ is nowhere tangent to the characteristics (the consequences of this assumption will be investigated later). Under this assumption Γ crosses the characteristics. Its parameter τ can therefore be used as the arbitrary constant in (1-3). (Our notation has already anticipated this fact.) Since Eqs. (1-2) are invariant under a translation of the independent variable t, we can choose $t = 0$ wherever we wish. It is natural to take $t = 0$ on Γ. This means that we want the solutions (1-3) of (1-2) to satisfy the initial conditions (IC)

$$\begin{aligned} \tilde{x}(0, \tau) &= \phi(\tau) \\ \tilde{y}(0, \tau) &= \psi(\tau) \end{aligned} \tag{1-8}$$

The theory of ordinary differential equations tells us that under proper conditions on a and b, the differential equations (1-2) have a unique solution satisfying (1-8). t is the running variable along a characteristic, and τ is the variable that distinguishes one characteristic from another. If we pick a definite value of τ, then we are on a definite characteristic. On this characteristic t is the variable. According to (1-7), u does not depend on t. Hence we can evaluate u by choosing any value of t we please. We choose $t = 0$ because then we are on Γ and there we are given the value of u. According to the IC, $u = \omega(\tau)$ on Γ. Hence $u = \omega(\tau)$ everywhere. The only trouble is that we do not want u as a function of τ. We want it as a function of x and y. To obtain u as a function of x and y is easy, however. We have only to solve Eqs. (1-3) for τ as a function of x and y and insert the result into $u = \omega(\tau)$. The resulting function is the solution to the Cauchy problem.

Something needs to be checked, however. How do we know that we can solve Eqs. (1-3) for τ in terms of x and y? In order to be sure that we can, we need to know that the Jacobian, $j = x_t y_\tau - x_\tau y_t$, is different from zero. According to the differential equations (1-2), $x_t = a$ and $y_t = b$. Hence

$j = ay_\tau - bx_\tau$. On Γ [cf. (1-8)], $y_\tau = \psi'$ and $x_\tau = \phi'$. Hence on Γ, $j = a\psi' - b\phi'$. If $j = 0$, then $b/a = \psi'/\phi'$; that is, Γ is parallel to a characteristic. This shows how our previously vague assumption that Γ crosses the characteristic should be formulated analytically. We should assume that

$$a\psi' - b\phi' \neq 0 \tag{1-9}$$

This is a condition involving only known functions. (a and b are the given coefficients in the PDE, and ϕ and ψ are the given parametric representation of the given initial curve Γ.) If (1-9) holds, then Γ is not characteristic and Eqs. (1-3) can be inverted on Γ and therefore, by continuity, in a neighborhood of Γ. If (1-9) is violated, the solution may still exist, but further analysis is required. We leave this case aside.

EXAMPLE As an illustration of the above, let us try to solve the PDE

$$xu_x + yu_y = 0 \tag{1-10}$$

subject to the IC

$$u(x, 1) = F(x) \qquad F \text{ given but arbitrary} \tag{1-11}$$

As indicated by the notation, the initial curve Γ here is the line $y = 1$. A parametric representation for it is

$$\begin{aligned} x &= \tau \\ y &= 1 \end{aligned} \tag{1-12}$$

Then (1-11) becomes $u = F(\tau)$ on Γ.

The characteristic equations (1-2) are

$$\begin{aligned} \frac{dx}{dt} &= x \\ \frac{dy}{dt} &= y \end{aligned} \tag{1-13}$$

According to (1-8) and (1-12), the initial conditions are

$$\begin{aligned} \tilde{x}(0, \tau) &= \tau \\ \tilde{y}(0, \tau) &= 1 \end{aligned} \tag{1-14}$$

The solution to (1-13) satisfying (1-14) is

$$\begin{aligned} x &= \tau e^t \\ y &= e^t \end{aligned} \tag{1-15}$$

Equations (1-15) are parametric equations for the characteristics with t as the running parameter along the characteristics and τ as the "family" parameter that distinguishes one characteristic from another. If we eliminate t from (1-15), we obtain the equations of the characteristics in terms of x, y and the parameter τ: $x = \tau y$. The characteristics are therefore a one-parameter (τ) family of straight lines through the origin having slope τ^{-1}. Since $\tau = x/y$ and $u = F(\tau)$, we obtain $u = F(x/y)$ as the solution of our problem (see Fig. 1-2).

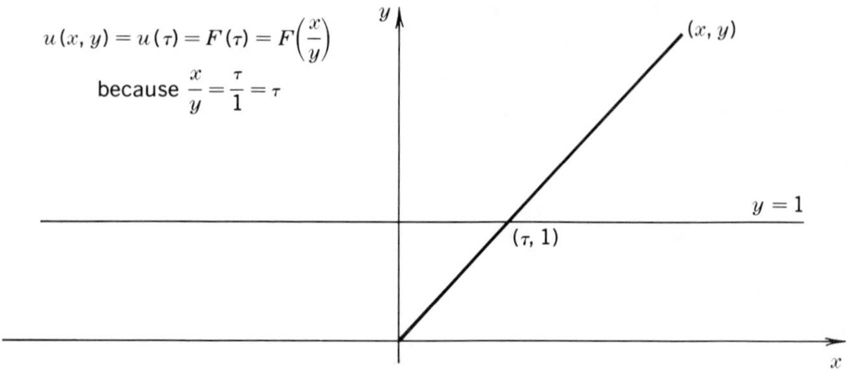

Figure 1-2

The constancy of u along the characteristics is due to the fact that the PDE we have treated is homogeneous and does not contain u explicitly. If the PDE contains u explicitly, then u *will no longer be constant along the characteristics.* Nevertheless, the characteristics can still be used to simplify the problem. To see how, consider the same Cauchy problem as before, only this time for the PDE

$$a(x, y)\, u_x + b(x, y)\, u_y = c(x, y)\, u \tag{1-16}$$

This time the PDE no longer reduces to $du/dt = 0$ along the characteristics but rather to

$$\frac{du}{dt} = c(x, y)\, u \tag{1-17}$$

on them. However, if we take the parametric equations (1-3) of the characteristics and substitute into (1-17) the values they give for x and y in terms of t and τ, we see that (1-17) becomes an ordinary differential equation (ODE) for u as a function of t with τ as parameter. Moreover, just as before, u satisfies the initial condition $\tilde{u}(0, \tau) = \omega(\tau)$. We therefore solve this initial-value problem for u which now depends upon t as well as τ. We obtain a function $u = u(t, \tau)$. We then solve (1-3) for t and τ in terms of x and y and thus obtain the desired solution, $u(x, y)$.

Later on (Chap. 8) by modifying the above procedure we shall show how the characteristics can be used to solve more general types of PDEs, including those which contain u in a nonlinear way.

EXAMPLE To illustrate the difference between (1-6) and (1-16) we replace the 0 on the right side of (1-10) by ku and consider the PDE

$$xu_x + uy_y = ku \qquad k = \text{const} \tag{1-18}$$

with the same IC (1-11) as before. The previous analysis all carries over up to the point where we used the fact that u in that problem was constant on the characteristics. In the case of (1-18) the PDE tells us that

$$\frac{du}{dt} = ku \tag{1-19}$$

The IC for u is still

$$u(0, \tau) = F(\tau) \tag{1-20}$$

The solution to (1-19) satisfying (1-20) is

$$u = F(\tau)e^{kt} \tag{1-21}$$

Using (1-15) in this, we obtain the solution in terms of x and y as

$$u(x, y) = y^k F\left(\frac{x}{y}\right)$$

1-3. GENERAL SOLUTION; DEPENDENCE ON ARBITRARY FUNCTIONS

Since the function $F(x)$ in (1-11) is arbitrary, the solution $u = F(x/y)$ to the PDE (1-10) depends upon an arbitrary function. As the example $u_x(x, y) = 0$ shows, it is to be expected that the general solution to a first-order PDE will contain one arbitrary function. In Chap. 8 we shall examine the nature of that dependence in greater detail, but for the present we restrict ourselves to deriving the dependence for the special case of the PDE (1-6).

In the case of (1-6) we have already seen that the equations of the characteristics can be written in the form

$$\begin{aligned} x &= f(t, \tau) \\ y &= g(t, \tau) \end{aligned} \tag{1-22}$$

where t is the running parameter along the characteristics and τ is the running parameter along any noncharacteristic initial curve Γ. Since Γ is noncharacteristic, (1-22) can be inverted to give

$$\begin{aligned} t &= \alpha(x, y) \\ \tau &= \beta(x, y) \end{aligned} \tag{1-23}$$

We have seen that the solution of (1-6) is constant along a characteristic; that is,

$$u(t, \tau) = u(0, \tau) = F(\tau) \tag{1-24}$$

where $F(\tau)$ represents the arbitrary initial values of u on Γ.

Substituting (1-23) into (1-24), we have

$$u(x, y) = F[\beta(x, y)] \tag{1-25}$$

where F is an arbitrary function of one variable and $\beta(x, y)$ is a determined function of x and y.

Equation (1-25) is called the *general solution* to (1-6). Just as for ODEs, the general solution does not necessarily contain all solutions of the equation. A few "singular" solutions may get left out. We shall say more on this topic in Chap. 8.

1-4. SEMILINEAR EQUATIONS

We can generalize (1-16) somewhat further and still carry through all the previous analysis. A perusal of the analysis in Sec. 1-2 shows that everything would still work for the PDE

$$a(x, y)\, u_x + b(x, y)\, u_y = c(x, y, u) \tag{1-26}$$

where now c, the right-hand side, can depend upon u in any way at all, even nonlinearly. The only change that this creates in the method of solution is that the differential equation for u

$$\frac{du}{dt} = c(x, y, u) \tag{1-27}$$

is now nonlinear and therefore much harder to solve in practice. The characteristics are still defined by (1-2), however, and once they have been determined, (1-27) is an ODE for u along a characteristic. Notice that if a or b, the coefficients of u_x and u_y in (1-26), depended upon u, then the methods of the present chapter would fail, since (1-2) would no longer determine the characteristics independently of the solution u.

An equation of the type (1-26) which is linear in the first derivatives of u and in which the coefficients of these first derivatives are independent of u is called *semilinear*. Notice that a semilinear PDE has characteristics which are independent of the particular solution u considered. The characteristics of a semilinear PDE can be determined in advance of the solution. It is this property that makes a semilinear PDE little harder to solve than a linear one.

A first-order PDE which is linear in the derivatives of u but whose coefficients may depend upon u is called *quasi-linear*. First-order quasi-linear equations will be considered in Chap. 8.

SUMMARY

DEFINITION S1-1

A function $f(x)$ is said to belong to class C^n in an interval if $f^{(n)}(x)$ exists and is continuous in that interval. If we wish to indicate the interval under consideration, we write $C^{(n)}(a, b)$ or $C^{(n)}[a, b)$, etc., using parentheses to exclude an end point and brackets to include it.

The Cauchy problem Let $a(x, y)$, $b(x, y)$ and $c(x, y)$ be continuous in a region $\mathscr{R}$ of the xy plane. Let Γ be an arc of a simple, i.e., non-self-intersecting, curve in $\mathscr{R}$ whose parametric equations are $x = \phi(\tau)$, $y = \psi(\tau)$, $0 \leqslant \tau \leqslant \tau_0$, where ϕ and ψ both belong to $C^{(1)}[0, \tau_0]$. Suppose that $a[\phi(\tau), \psi(\tau)]\, \psi'(\tau) - b[\phi(\tau), \psi(\tau)]\, \phi'(\tau) \neq 0$ for $0 \leqslant \tau \leqslant \tau_0$. Then there exists a neighborhood of Γ and a continuously differentiable function

$u(x, y)$ defined therein such that $a(x, y)\,u_x(x, y) + b(x, y)\,u_y(x, y) = c(x, y)\,u(x, y)$ and $u[\phi(\tau), \psi(\tau)] = \omega(\tau)$, ω given, $0 \leqslant \tau \leqslant \tau_0$.

DEFINITION S1-2

The ODEs

$$\frac{dx}{dt} = a(x, y) \qquad \frac{dy}{dt} = b(x, y) \tag{S1-1}$$

are called the *characteristic equations* of the PDE $a(x, y)\,u_x + b(x, y)\,u_y = c(x, y)u$. Their solution curves are called the *characteristic curves* of the PDE.

Method for solving Cauchy problem Write down the characteristic equations (S1-1). Solve them subject to the ICs $x = \phi(\tau)$, $y = \psi(\tau)$, when $t = 0$. Let the solutions so obtained be

$$x = f(t, \tau) \qquad y = g(t, \tau) \tag{S1-2}$$

Invert Eqs. (S1-2) to obtain

$$t = \hat{f}(x, y) \qquad \tau = \hat{g}(x, y) \tag{S1-3}$$

Solve the ODE $du/dt = c[f(t, \tau), g(t, \tau)]$ subject to the IC $u = \omega(\tau)$ when $t = 0$. Let the result be

$$u = h(t, \tau) \tag{S1-4}$$

Substitute (S1-3) into (S1-4). The result is the solution u of the Cauchy problem.

PROBLEMS

1. Solve the following Cauchy problem for $u(x, y)$:

PDE:

$$(x + 2)\,u_x + 2yu_y = 2u$$

IC:

$$u(-1, y) = \sqrt{y}$$

2. Solve the Cauchy problem

PDE:

$$u_x + u_y = 1$$

IC:

$$u = \phi(x) \text{ on the curve } y = 2x$$

3. Solve the Cauchy problem

PDE:

$$x^2u_x - y^2u_y = 0$$

IC:

$$u(1, y) = F(y)$$

4. Solve the Cauchy problem

PDE:

$$xu_x + yu_y = 3$$

IC:

$$u(1, y) = \log y$$

5. Find the general solution of $3u_x + 4u_y + 5u = 0$.

6. Find the general solution of $au_x + bu_y + cu = 0$ when a, b, and c are all constants.

2
One-dimensional Wave Equations

2-1. DERIVATION OF THE ONE-DIMENSIONAL LINEAR WAVE EQUATION

Suppose we have a string of infinite length lying along the entire x axis, and suppose that it is set into motion in such a way that every point of the string moves only in the vertical direction. Further assumptions will be added shortly.

How can we describe the motion of such a string? Since each point moves only in the vertical direction, a given point always retains the same x coordinate. Its distance above the x axis varies, however. Let u represent the distance of a given point above the x axis. (If the point is below the x axis, then u is negative.) Clearly u is a function of the given point x and the time $t: u = u(x, t)$. We wish to derive a partial differential equation (PDE) satisfied by u. Let $\rho(x)$ be the linear density, i.e., mass per unit length, and let $\mathbf{T}(x) = H(x)\,\mathbf{i} + V(x)\,\mathbf{j}$ be the tension in the string. (Boldface type is used for vectors.) Let $\boldsymbol{\tau}$ be a tangent vector to the string. The equation of the string is $u = u(x, t)$. From elementary calculus we know that its tangent $\boldsymbol{\tau}$ has slope

u_x, hence direction numbers 1 and u_x. Therefore $\boldsymbol{\tau} = \mathbf{i} + u_x\mathbf{j}$. In a string the tension is always tangential. Hence the cross product, $\mathbf{T} \times \boldsymbol{\tau} = 0$; that is,

$$H(x)\, u_x - V(x) = 0$$

and so

$$V(x) = H(x)\, u_x$$

Consider a segment of string extending from a to b. Since we have assumed no horizontal motion, there can be no net horizontal force on any segment. Hence $H(a) = H(b) = H$ (say) ($H =$ const). The net force F on the segment in the vertical direction is therefore

$$F = V(b) - V(a) = H[u_x(b, t) - u_x(a, t)]$$

By Newton's second law,

$$F = \int_a^b \rho u_{tt}\, dx$$

Therefore,

$$\int_a^b \rho u_{tt}\, dx = H[u_x(b, t) - u_x(a, t)]$$

We can rewrite this in the form

$$\int_a^b \rho u_{tt}\, dx = H \int_a^b u_{xx}(x, t)\, dx \tag{2-1}$$

Since (2-1) holds for arbitrary (a, b), we conclude that

$$\rho(x)\, u_{tt} = H u_{xx} \tag{2-2}$$

or

$$u_{xx} = \frac{\rho}{H} u_{tt}$$

Now

$$\rho = \frac{\text{mass}}{\text{length}} = \frac{M}{L}$$

and

$$H = \text{force} = \text{mass} \times \text{acceleration} = \frac{\text{mass} \times \text{length}}{\text{time}^2} = \frac{ML}{T^2}$$

Therefore,

$$\frac{H}{\rho} = \frac{ML}{T^2}\frac{L}{M} = \frac{L^2}{T^2}$$

Thus, H/ρ has the dimensions of a velocity squared. Since H and ρ are both inherently positive, we can put

$$\frac{\rho}{H} = c^{-2} \tag{2-3}$$

and be sure that c is real and represents a velocity. Just what it is the velocity of will appear later.

In any event we have arrived at the *equation of the vibrating string*:

$$u_{xx} - c^{-2}u_{tt} = 0 \tag{2-4}$$

In the present chapter we shall assume that c is a *constant*. For a case in which c is a function of u_x, $c = c(u_x)$, see Sec. 13-9.

Equation (2-4) arises in many other physical problems. For example, it is satisfied by the velocity of a particle of gas in a one-dimensional sound wave (see Sec. 2-9), the density of the gas in the wave, the components of the electric or magnetic field vectors in certain special problems in electromagnetic wave propagation, and many other quantities of physical importance. We will formulate most of our problems in terms of strings, but the reader should be able to carry over the results to other phenomena of interest to him.

Notice that our derivation of (2-4) has assumed that the only forces acting on the string are internal ones due to the tension in the string. External forces, such as gravity, have been neglected in deriving (2-4). The results of the next few sections should therefore not be misinterpreted by mistakenly assuming that the strings considered in them are under the influence of gravity. They are *not*. The case wherein external forces, such as gravity, act is taken up in Sec. 2-7.

2-2. SOLUTION OF THE WAVE EQUATION IN ONE DIMENSION

In a self-explanatory operational notation, Eq. (2-4) can be written

$$(D_x^2 - c^{-2}D_t^2)u = 0$$

or

$$(D_x - c^{-1}D_t)(D_x + c^{-1}D_t)u = 0$$

This is the product of two directional derivatives in the xt plane, the first in the direction with direction numbers $(1, -c^{-1})$ and the second in the direction with direction numbers $(1, c^{-1})$. It suggests that we introduce these two directions as new coordinate directions. Accordingly we set

$$\begin{aligned} \xi &= x - ct \\ \eta &= x + ct \end{aligned} \tag{2-5}$$

By the chain rule we have

$$u_x = u_\xi \xi_x + u_\eta \eta_x$$

But from (2-5) we have $\xi_x = 1$ and $\eta_x = 1$. Hence

$$u_x = u_\xi + u_\eta \tag{2-5a}$$

To compute u_{xx} we observe that (2-5*a*) can be written in operational form as

$$\frac{\partial}{\partial x} = \frac{\partial}{\partial \xi} + \frac{\partial}{\partial \eta}$$

Hence

$$\frac{\partial^2}{\partial x^2} = \left(\frac{\partial}{\partial \xi} + \frac{\partial}{\partial \eta}\right)^2 = \frac{\partial^2}{\partial \xi^2} + 2\frac{\partial^2}{\partial \xi\, \partial \eta} + \frac{\partial^2}{\partial \eta^2}$$

and thus

$$u_{xx} = u_{\xi\xi} + 2u_{\xi\eta} + u_{\eta\eta} \tag{2-5b}$$

Similarly we find that

$$u_{tt} = c^2(u_{\xi\xi} - 2u_{\xi\eta} + u_{\eta\eta}) \tag{2-5c}$$

Substituting (2-5*b*) and (2-5*c*) into (2-4), we then get

$$4u_{\xi\eta} = 0 \tag{2-6}$$

Writing (2-6) as $(\partial/\partial\eta)\, u_\xi = 0$, we see that u_ξ is independent of η. It can, however, depend upon ξ in an arbitrary way. Thus,

$$u_\xi = \hat{\phi}(\xi) \tag{2-6a}$$

where $\hat{\phi}$ is an arbitrary function of ξ.

Integrating (2-6*a*) and letting $\int \hat{\phi}(\xi)\, d\xi = \phi(\xi)$, we have

$$u(\xi, \eta) = \phi(\xi) + \psi(\eta) \tag{2-7}$$

where ϕ and ψ are arbitrary functions of their arguments. Replacing ξ and η by x and t [cf. (2-5)] we have

$$u(x, t) = \phi(x - ct) + \psi(x + ct) \tag{2-8}$$

as the general solution to (2-4).

Notice that the functions ϕ and ψ occurring in (2-8) are each functions of a single variable. In forming the general solution (2-8) to the wave equation, we choose two arbitrary functions ϕ and ψ, each of which is a function of a single variable, say $\phi(\xi)$ and $\psi(\eta)$. Then we replace the single variable ξ by the special combination $x - ct$ and similarly replace η by $x + ct$. Thus the variables x and t cannot enter ϕ or ψ in an arbitrary way but only in the special combination $x - ct$ in ϕ and $x + ct$ in ψ.

To illustrate the foregoing remarks we compute $\partial\phi/\partial t$. By the chain rule this is

$$\frac{\partial \phi}{\partial t} = \frac{d\phi}{d\xi}\frac{\partial \xi}{\partial t} = \frac{d\phi}{d\xi}(-c) = -c\,\frac{d\phi}{d\xi} \tag{2-8a}$$

Since ϕ is, as just emphasized, a function of the single variable ξ [which is why we have written $d\phi/d\xi$ instead of $\partial\phi/\partial\xi$ in (2-8*a*)], we can write ϕ' for $d\phi/d\xi$ without danger of ambiguity. Thus,

$$\frac{\partial \phi}{\partial t} = -c\phi'$$

We shall make use of these facts shortly (Sec. 2-3).

The two terms of (2-8) can be interpreted as waves traveling to the right and left respectively. To see this, first take $t = 0$ in the argument of ϕ and then plot ϕ (Fig. 2-1). Select any point $x = a$. At time $t = 0$, ϕ has the value $\phi(a)$ at that point. $\phi(x - ct)$ will have the same value $\phi(a)$ at any space-time point (x, t) such that $x - ct = a$. In particular, at $t = 1$, ϕ will

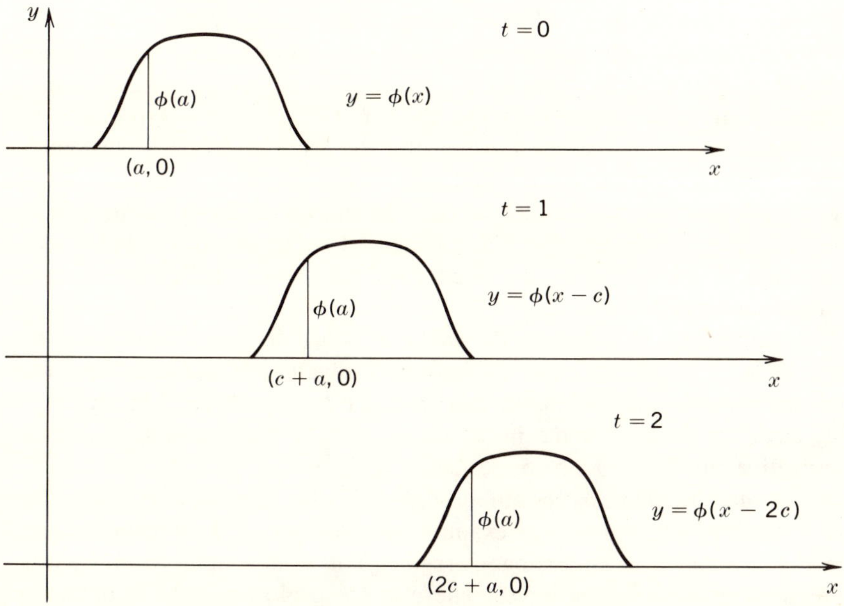

Figure 2-1

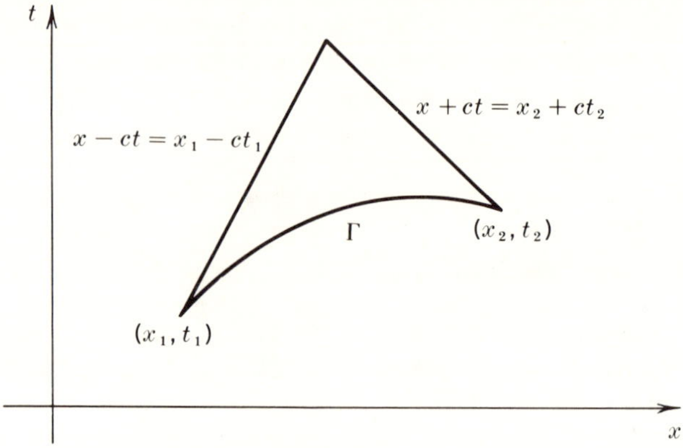

Figure 2-2

have the value $\phi(a)$ at the point $x = a + c$; at $t = 2$, $\phi = \phi(a)$ at the point $x = a + 2c$, etc. Thus, the point at which ϕ takes the value $\phi(a)$ moves to the right with velocity c.

In Chap. 1 we saw that a first-order linear PDE could be interpreted as a directional derivative along a given curve. In the present chapter we have just seen that the wave equation is the product of *two* directional derivatives, one in the direction of the lines $x - ct =$ const and the other in the direction of the lines $x + ct =$ const. Again we call these significant directions *characteristic directions* and the corresponding curves *characteristic curves.* We shall denote the characteristics $x - ct =$ const, which have positive slopes, by C^+ and the others by C^-. From (2-8) we see that the solution to the wave equation can be written as the sum of two functions, one of which is constant on the C^+ characteristics and the other of which is constant on the C^- characteristics. As a result, knowledge of both these parts of the solution along any curve which crosses the characteristics will determine the solution in a certain triangle. For example, if we know both ϕ and ψ along the curve Γ shown in Fig. 2-2, then u will be known inside the curvilinear triangle indicated.

In practice, it is usually the values of u and its normal derivative that are given on the initial curve rather than the values of ϕ and ψ. The latter values, however, can be determined from the former. Much later (Chap. 9) we shall see that for some initial curves only u itself or only its normal derivative, not both, can be prescribed.

Note, too, that in a certain sense, "half" of u, namely, ϕ or ψ, is known in a much larger region. For example, ψ is known in the region indicated in Fig. 2-3 and ϕ in an analogous region. This is in keeping with the interpretation of ϕ and ψ as waves. They, so to speak, carry the initial values outward.

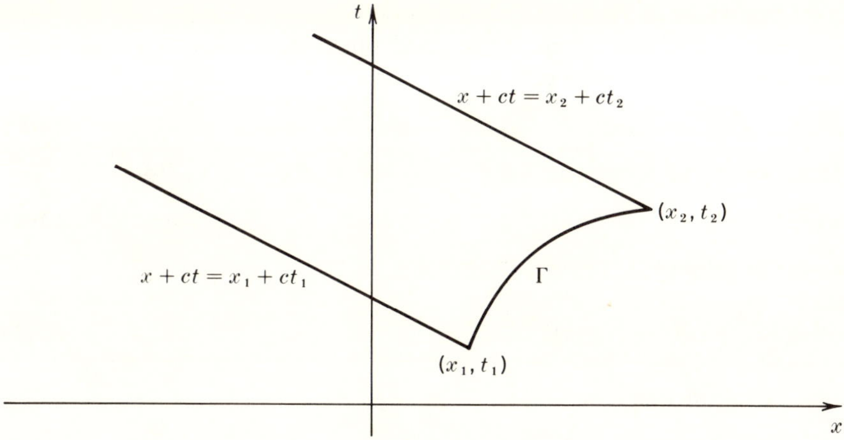

Figure 2-3

2-3. D'ALEMBERT'S SOLUTION TO THE INITIAL-VALUE PROBLEM

In Chap. 9 we shall go into the ramifications of these facts in much greater detail. For the present we return to the case of the string. According to newtonian mechanics, the natural conditions to impose are the prescription of the initial displacement and the initial velocity. This means that we prescribe the shape of the string at $t = 0$ to be that of a given function $f(x)$ and the velocity at $t = 0$ to be another given function $g(x)$. Then we have the following.

Initial-value problem

PDE:

$$u_{xx} - c^{-2}u_{tt} = 0 \qquad \begin{array}{l} -\infty < x < \infty \\ 0 < t < \infty \end{array}$$

ICs:

$$u(x, 0) = f(x) \qquad -\infty < x < \infty \tag{2-9}$$

$$u_t(x, 0) = g(x) \tag{2-10}$$

To solve it, we try to determine the functions ϕ and ψ in the general solution (2-8) so as to satisfy (2-9) and (2-10). Insertion of (2-8) into (2-9) yields

$$\phi(x) + \psi(x) = f(x) \tag{2-10a}$$

while substituting (2-8) into (2-10) gives [recall the remarks about ϕ' following Eq. (2-8*a*)]

$$-c\phi'(x) + c\psi'(x) = g(x) \tag{2-10b}$$

Equations (2-10a) and (2-10b) are two equations for the two unknown functions ϕ and ψ in terms of the given functions f and g. To solve them, we differentiate (2-10a) to get

$$\phi'(x) + \psi'(x) = f'(x) \tag{2-10c}$$

Then, writing (2-10b) in the form

$$\phi'(x) - \psi'(x) = -c^{-1}g(x) \tag{2-10d}$$

we see by adding (2-10c) and (2-10d) that

$$\phi'(x) = \frac{1}{2}f'(x) - \frac{1}{2c}\,g(x) \tag{2-11}$$

and, by subtracting them, that

$$\psi'(x) = \frac{1}{2}f'(x) + \frac{1}{2c}\,g(x) \tag{2-12}$$

To obtain ϕ and ψ we now merely have to integrate (2-11) and (2-12). We integrate both sides of these equations from 0 to x and add an arbitrary constant to the right side of each resulting equation, lumping $\phi(0)$, $f(0)$, etc., into these constants. Thus we obtain

$$\phi(x) = \frac{1}{2}f(x) - \frac{1}{2c}\int_0^x g(z)\,dz + k_1 \tag{2-13}$$

$$\psi(x) = \frac{1}{2}f(x) + \frac{1}{2c}\int_0^x g(z)\,dz + k_2 \tag{2-14}$$

However, $\phi + \psi = f$. Therefore $k_1 + k_2 = 0$. Hence $k_1 = -k_2 = k$ (say), and therefore

$$\phi(x - ct) = \frac{1}{2}f(x - ct) - \frac{1}{2c}\int_0^{x-ct} g(z)\,dz + k \tag{2-15}$$

or

$$\phi(x - ct) = \frac{1}{2}f(x - ct) + \frac{1}{2c}\int_{x-ct}^{0} g(z)\,dz + k$$

Also

$$\psi(x + ct) = \frac{1}{2}f(x + ct) + \frac{1}{2c}\int_0^{x+ct} g(z)\,dz - k \tag{2-16}$$

Finally [see (2-8)]

$$u(x, t) = \frac{1}{2}\,[f(x - ct) + f(x + ct)] + \frac{1}{2c}\int_{x-ct}^{x+ct} g(z)\,dz \tag{2-17}$$

Formula (2-17) is frequently referred to as D'Alembert's solution to the wave equation. It has a very interesting interpretation. In order to present this interpretation, we choose a point (x_0, t_0) in the xt plane and draw the two characteristics through this point backward until they intersect the initial line, i.e., the x axis (Fig. 2-4*a*). The equations of these two characteristics are obviously $x \pm ct = x_0 \pm ct_0$. They hit the x axis at the points $x_0 \pm ct_0$. Equation (2-17) therefore tells us that the solution u at (x_0, t_0) is obtained by averaging the values of f at these two points and integrating g along the x axis between them.

Among other things, this shows that the value of the solution at a given point $P(x, t)$ in the xt plane depends only upon the values of the initial data on the segment of the initial line obtained by drawing the characteristics backward from the point P to the initial line. The segment of the initial line on which the values of u at a point $P(x, t)$ depend is called the *domain of dependence* of P.

Another result of the above is this: the initial values on a segment $\mathscr{I}$ of the initial line can influence the solution only in the region bounded by the two characteristics which diverge from the end points of the interval. The resulting region in the xt plane is called the *range of influence* of the interval (Fig. 2-4*b*). The triangle formed by the two converging characteristics from the end points of the interval is called the *domain of determinancy* of the interval (Fig. 2-4*c*) because the solution is completely determined in it by the initial data on the interval.

In particular, suppose the initial values are identically zero for $x > b$. Then the solution must be identically zero for $x > b + ct$ (Fig. 2-5). Moreover, since ψ is constant on the C^- characteristics, $x + ct = \text{const}$, and is zero in the region $x > b + ct$, it follows that $\psi = 0$ for $b - ct < x < b + ct$ also; i.e., the C^- characteristics carry the zero values of ψ into the adjacent region (see Fig. 2-5). Thus the solution consists only of a wave going to the right in the region adjacent to that in which it is zero.

EXAMPLE

DE:

$$u_{xx} - c^{-2}u_{tt} = 0 \qquad \begin{matrix} -\infty < x < \infty \\ 0 < t < \infty \end{matrix}$$

ICs

$$u(x, 0) = \begin{cases} h & \text{if } |x| < a \\ 0 & \text{if } |x| > a \end{cases}$$

$$u_t(x, 0) = 0 \qquad -\infty < x < \infty$$

The solution is best illustrated by an xt diagram (Fig. 2-6) obtained by using the remarks below formula (2-17).

We can interpret the results in the xt diagram (Fig. 2-6) in two different ways. First of all we can look at the situation in the physical (xu) plane at different times t. If we do, we get the sequence of diagrams in Fig. 2-7. They show that the original rectangular bump of height h splits into two smaller bumps, each of height $h/2$ and each moving outward from the origin with velocity c.

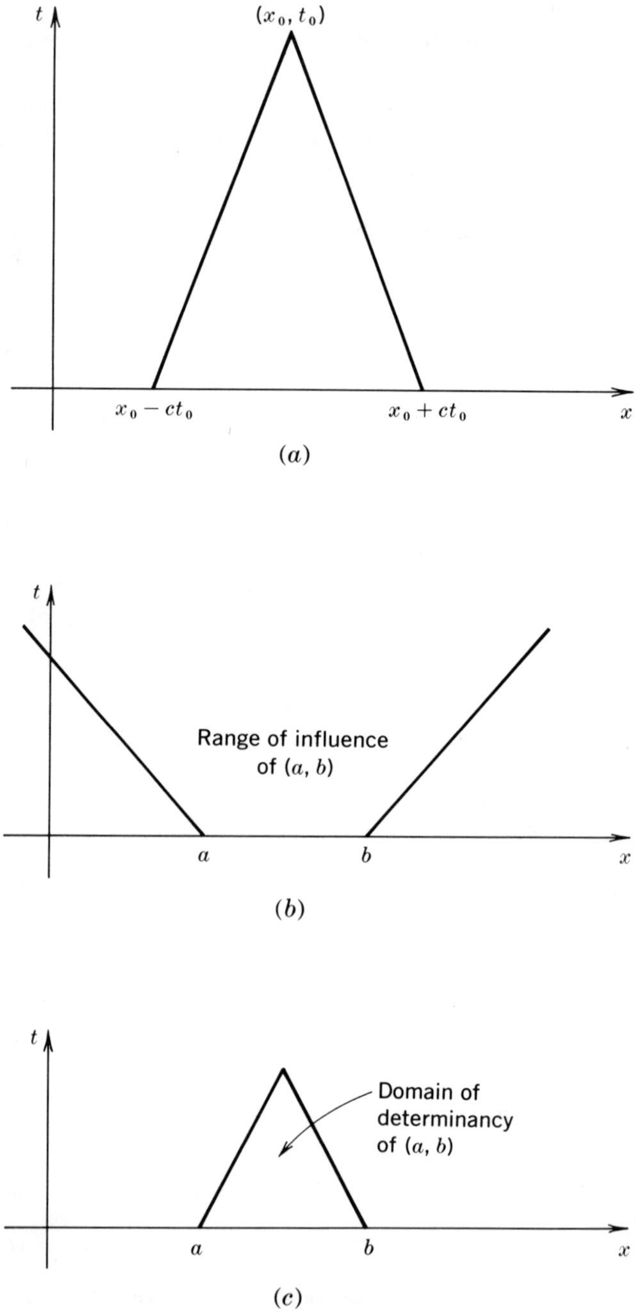
t
(x0, t0)
x0 − ct0
x0 + ct0
x
(a)
t
Range of influence
of (a, b)
a
b
x
(b)
t
Domain of
determinancy
of (a, b)
a
b
x
(c)

Figure 2-4

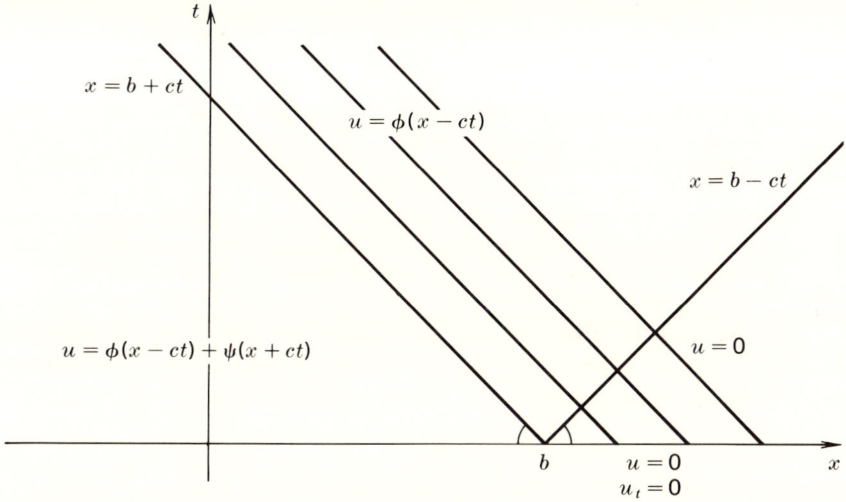

Figure 2-5

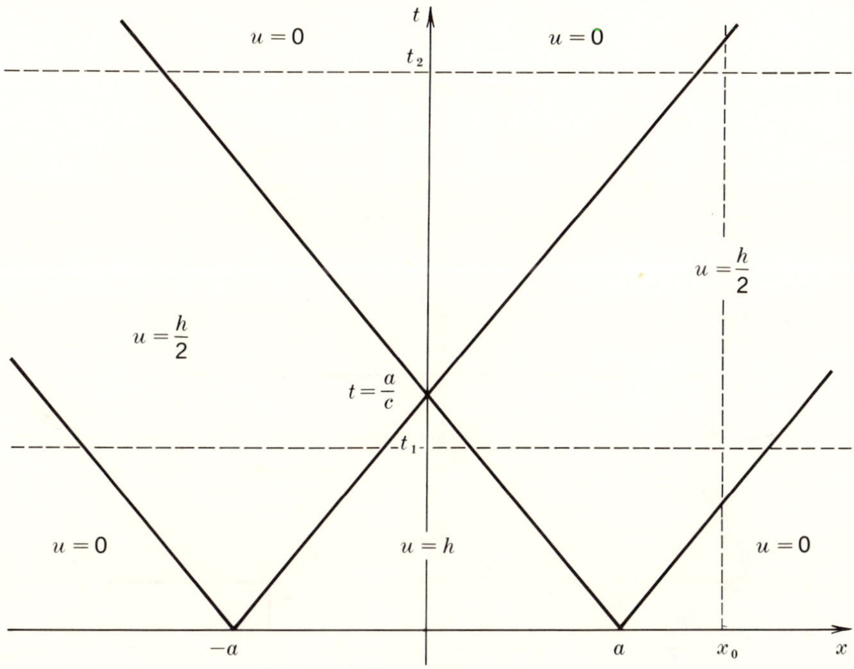

Figure 2-6

Another insight is gained by fixing a point $x = x_0 > a$. An observer stationed at x_0 will initially see zero displacement and will continue to see zero until the time $t = (x_0 - a)/c$. At that time he will see the wave arrive. It will continue to pass through until the time $t = (x_0 + a)/c$, after which he will again see zero. Thus, there is a sharp arrival time and a sharp departure time. Note, however, that this is due to the fact that we have taken $g(x) = 0$ in the present example. If we had assumed that $g(x)$ was different from zero in some finite interval, say again the interval $|x| < a$, then the observer would see the displacement $G = \int_{-a}^{a} g(x)\,dx$ from the time $(x_0 + a)/c$ ever afterward. In other words, the wave would have a sharp arrival time but no sharp departure time. We shall investigate this phenomenon in greater detail when we deal with wave propagation in higher dimensions (Chap. 7).

Notice that the discontinuity in the initial conditions in this example does not disappear as time increases but rather propagates outward along the characteristics. This propagation of initial discontinuities is a characteristic

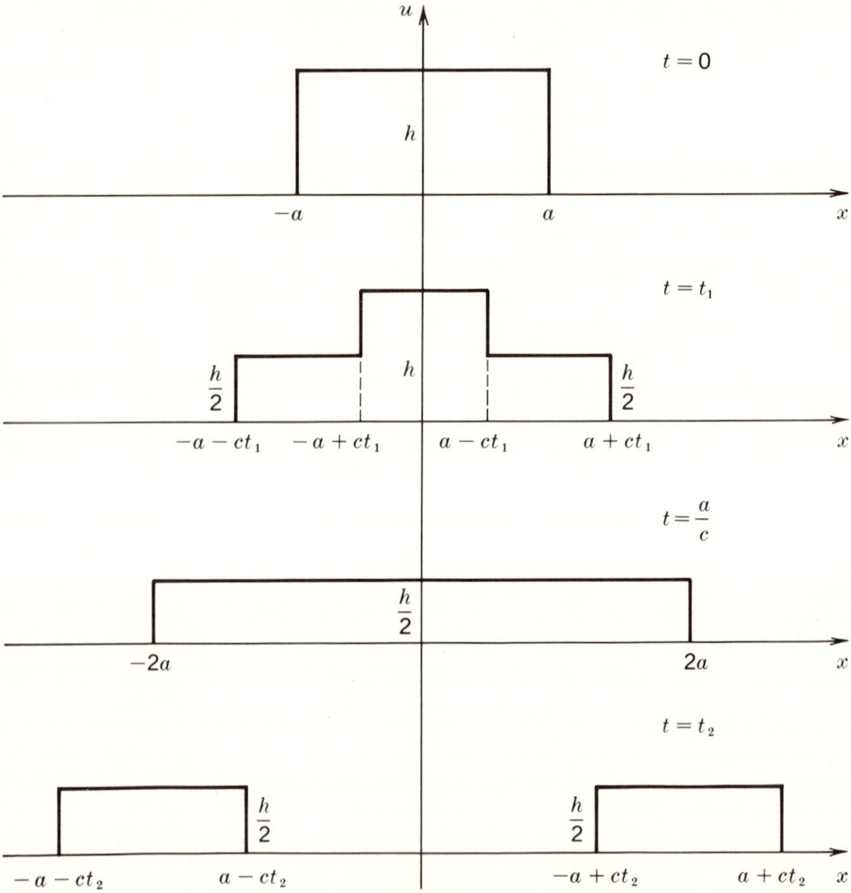

Figure 2-7

feature of a type of PDEs known as *hyperbolic equations* (see Chap. 6 for the definition of hyperbolic). It stands in sharp contrast to the situation for elliptic equations, as will be pointed out in Chap. 3.

2-4. SEMI-INFINITE STRING

If the string extends only from the origin to $x = +\infty$, the problem becomes more complicated. On physical grounds it is obvious that we have to specify what is to be done with the end point at $x = 0$. Various things can be done with it. For simplicity, we assume first that the end point is held fixed. This means that we add to the initial-value problem (IVP) at the outset of Sec. 2-3 the additional condition $u(0, t) = 0$, for $0 < t < \infty$. Because this condition is imposed at the end of the string, it is called a *boundary condition* (BC). We also have to make one alteration in the conditions in the previous problem, where the differential equation and the initial conditions were required to hold for all x. Now we want them to hold only for positive x. This seemingly slight alteration will cause a considerable change in the solution. First, however, let us formulate the problem precisely.

Boundary-initial-value problem for the semi-infinite string

DE:

$$u_{xx} - c^{-2}u_{tt} = 0 \qquad \begin{array}{l} 0 < x < \infty \\ 0 < t < \infty \end{array} \tag{2-18}$$

ICs:

$$u(x, 0) = f(x) \qquad 0 < x < \infty \tag{2-19}$$

$$u_t(x, 0) = g(x) \qquad 0 < x < \infty \tag{2-20}$$

BC:

$$u(0, t) = 0 \qquad 0 < t < \infty \tag{2-21}$$

In order to ensure the compatibility of the ICs and the BC, we assume $f(0) = 0$.

At first sight it appears that we do not need the BC (2-21) because our previous method of solution (see Sec. 2-3) seems to produce the solution without ever needing (2-21). This appearance is illusory, however, for look at the representation (2-13) for the function ϕ. Formula (2-13) expresses ϕ in terms of the given functions f and g. In the present case of a semi-infinite string, f and g are known only for positive values of their arguments. Hence (2-13) defines ϕ only for positive values of its argument. In order to get a solution of the wave equation out of ϕ, however, we have to replace its argument by $x - ct$. Since both x and t range from 0 to ∞, $x - ct$ ranges from $-\infty$ to ∞. Hence we need to know ϕ for negative as well as for positive values of its argument. In other words, the expression (2-15) for $\phi(x - ct)$ is good only if the argument of all the functions involved is positive, i.e., only if $x > ct$. For $x < ct$, $\phi(x - ct)$ remains to be determined. Before attempting to determine it, let us first see what the situation is with regard to ψ. If x and t are both positive, then so is $x + ct$. Hence we need to know $\psi(x + ct)$ only for positive values of the argument. Formula (2-14) gives the values of ψ for positive argument. Hence (2-14) completely determines ψ for all values for which we need it in the present problem. Therefore, to complete the solution we have only to determine ϕ for negative values of its argument. In doing so, we can assume that ψ is already known, because it is given by (2-16).

It goes without saying that the BC (2-21) must be the condition that enables us to determine ϕ for negative argument. From (2-8) and (2-21) we have

$$\phi(-ct) + \psi(ct) = 0 \tag{2-22}$$

or

$$\phi(-ct) = -\psi(ct)$$

If we let $w = -ct$, then we see that

$$\phi(w) = -\psi(-w) \tag{2-23}$$

Since ψ is known for *positive* argument, this determines ϕ for *negative* argument. Replacing w by $x - ct$ in (2-23) and using (2-14), we have

$$\phi(x - ct) = -\frac{1}{2}f(ct - x) - \frac{1}{2c}\int_0^{ct-x} g(z)\,dz + k \qquad \text{for } 0 < x < ct \tag{2-24}$$

As remarked above, $\phi(x - ct)$ is still given by (2-15) for $x > ct$, and $\psi(x + ct)$ is given by (2-16) for all positive x and t. Thus [cf. (2-8)], we find the solution

$$u(x, t) = \begin{cases} \dfrac{1}{2}[f(x - ct) + f(x + ct)] + \dfrac{1}{2c}\displaystyle\int_{x-ct}^{x+ct} g(z)\,dz & \text{for } x > ct \\ \dfrac{1}{2}[f(ct + x) - f(ct - x)] + \dfrac{1}{2c}\displaystyle\int_{ct-x}^{ct+x} g(z)\,dz & \text{for } 0 < x < ct \end{cases} \tag{2-25}$$

This solution, too, can be interpreted in an illuminating way in the xt plane (see Fig. 2-8). Since x and t are both positive, the solution exists only in the first quadrant. The characteristic $x = ct$ plays a decisive role because [cf. (2-25)] the representation for the solution changes as we cross this characteristic. To the right of $x = ct$, the formula for the solution is the same as for the full infinite string. Hence the solution at any point (x_0, t_0), with $x_0 > ct_0$, is obtained in the same way as previously explained. If (x_0, t_0) lies in the region $x < ct$, then the method for obtaining the solution is slightly

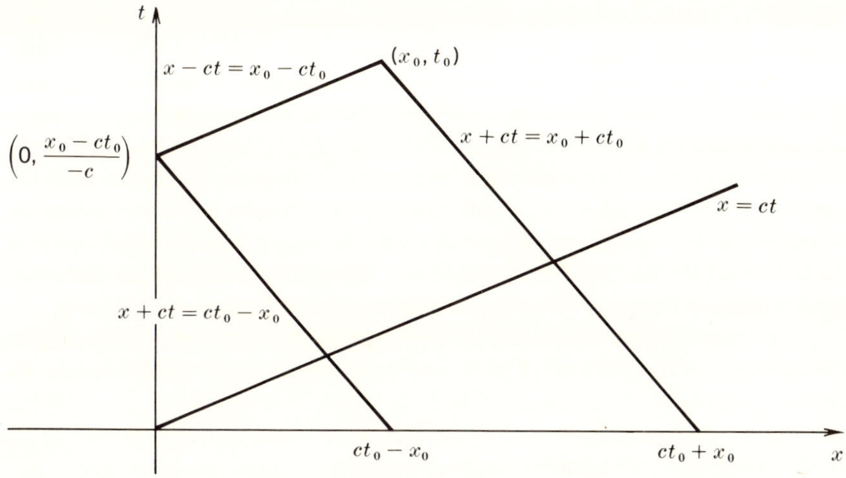

Figure 2-8

more complicated. From (x_0, t_0) we draw backward the two characteristics, $x - ct = x_0 - ct_0$ and $x + ct = x_0 + ct_0$, as before. The characteristic $x + ct = x_0 + ct_0$ still hits the initial line, i.e., the x axis, at the point $x = x_0 + ct_0$ as before. However, the characteristic $x - ct = x_0 - ct_0$ bumps into the boundary, $x = 0$, before it ever reaches the initial line. It hits the boundary at the point $t = (x_0 - ct_0)/-c$. Unable to follow this characteristic past this point, we follow it up to this point and then follow the characteristic $x + ct = ct_0 - x_0$ from this point down to the x axis. We then arrive at the x axis at the point $x = ct_0 - x_0$, which is precisely one of the points at which the function f is to be evaluated in (2-25). Thus, we see that in the case $x_0 < ct_0$, the solution at (x_0, t_0) is obtained as follows: draw the characteristic $x + ct = x_0 + ct_0$ backward until it hits the x axis at a point P; draw the characteristic $x - ct = x_0 - ct_0$ backward until it hits the t axis; then reflect it from the t axis down to the x axis, where it will hit at a point Q. The solution at (x_0, t_0) is obtained by taking $\frac{1}{2}[f(P) - f(Q)]$ and integrating $(2c)^{-1}g$ between P and Q.

The situation here is very much in keeping with the physical interpretation of solutions of the wave equation. We have already seen that disturbances satisfying the wave equation propagate with velocity c. Applying this principle to disturbances due to the end point $x = 0$ in the present problem, we see that such disturbances must propagate down the string with velocity c. This fact is in perfect agreement with the fact found above that the boundary influences the solution only behind the characteristic $x = ct$; that is, the effect of the boundary propagates outward with velocity c.

Caution: The preceding depends upon having the BC $u(0, t) = 0$. If the BC is changed, then the above prescription for obtaining the solution must be changed.

Notice that while we prescribed two initial conditions, we have prescribed only one boundary condition. Physically, this seems reasonable, but mathematically, it is more than reasonable: it is necessary. Mathematically it is not necessary that we prescribe two conditions on the x axis and only one on the t axis. We could just as well have prescribed only one condition on the x axis and two on the t axis. More will be said on this subject later on (Chap. 9). For the present, suffice it to say that we cannot prescribe two conditions on *both* the x and t axes, the reason being that the C^- characteristics connect points on these two axes and, so to speak, carry information from one axis to the other. For example, prescribing two conditions on the x axis is tantamount to prescribing both ϕ and ψ there. Since ψ is constant on the C^- characteristics, its known values on the x axis are carried to the t axis by the C^- characteristics, and so only ϕ is unknown on the t axis. Hence we can prescribe only one datum, not two, there.

The foregoing analysis shows that, in general, it would not be possible to prescribe any data at all on a line $t = t_0 > 0$ if data have already been prescribed on $t = 0$. Thus a boundary-value problem for the wave equation in which the solution is prescribed on the entire boundary of some closed, bounded region will not in general be solvable. This situation stands in marked contrast to that for the potential equation (see Chap. 3).

A generalization of the BC (2-21) is

$$u(0, t) = h(t) \qquad 0 < t < \infty \tag{2-26}$$

where $h(t)$ is a given function, with $h(0) = f(0)$ [cf. (2-19)]. Physically, it corresponds to moving the end point in such a way that its displacement is $h(t)$ at time t.

In order for (2-26) to be compatible with (2-19) and (2-20) we require that $h(0) = f(0)$ and $h'(0) = g(0)$.

If we look through the analysis in Sec. 2-4, we see that for (2-26), (2-22) becomes

$$\phi(-ct) + \psi(ct) = h(t)$$

and therefore (2-23) becomes

$$\phi(x) = h\left(-\frac{x}{c}\right) - \psi(-x)$$

Hence (2-24) will be

$$\phi(x - ct) = h\left(t - \frac{x}{c}\right) - \frac{1}{2}f(ct - x) - \cdots \qquad \text{for } x < ct \tag{2-27}$$

This makes the alteration in the solution due to h evident.

We leave it to the reader as an exercise to investigate the case in which the BC (2-21) is replaced by

$$u_x(0, t) = 0 \qquad 0 < t < \infty \tag{2-28}$$

This BC corresponds to a string with its end point free to move in a vertical direction.

EXAMPLE

DE:

$$u_{xx} - c^{-2}u_{tt} = 0 \qquad \begin{array}{l} 0 < x < \infty \\ 0 < t < \infty \end{array}$$

IC:

$$u(x, 0) = \begin{cases} 0 & \text{if } 0 < x < a \\ & \text{or if } a + 2l < x < \infty \\ h & \text{if } a < x < a + 2l \end{cases}$$

$$u_t(x, 0) = 0 \qquad 0 < x < \infty$$

BC:

$$u(0, t) = 0 \qquad 0 < t < \infty$$

The solution can be found by specializing (2-25) to the present case. The results are again best illustrated by an xt diagram (see Fig. 2-9).

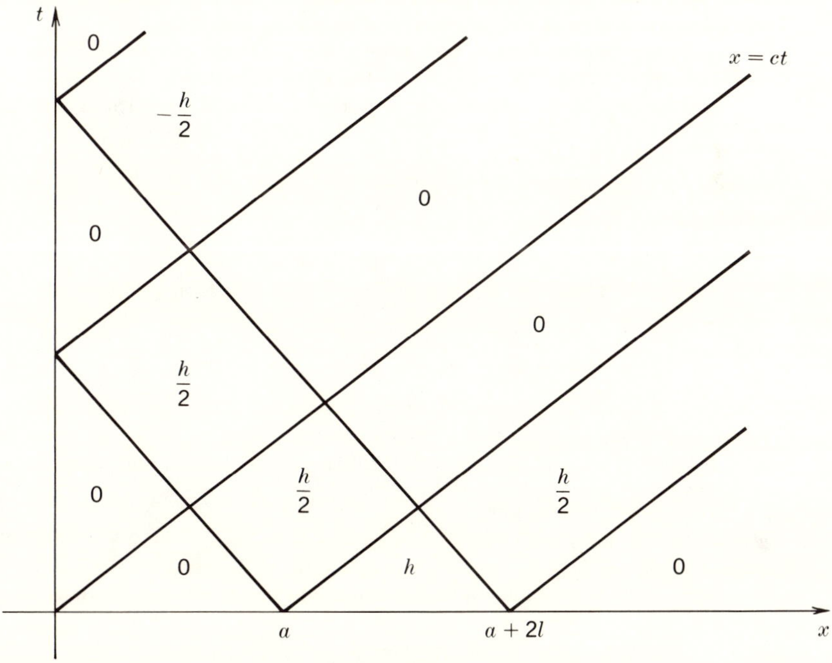

Figure 2-9

An observer stationed at a point $x < a$ thus sees zero initially, then sees a wave of amplitude $h/2$ going to the left, then zero for a while again, then a wave of amplitude $-h/2$ going to the right, and then zero again forevermore. An observer stationed at a point $x > a + 2l$ first sees zero, then sees a wave going to the right with amplitude $h/2$, then zero, then another wave of amplitude $-h/2$ going to the right, then zero.

Thus, the initial disturbance of amplitude h splits into two waves of equal amplitude, numerically, one going to the right and the other to the left, each with speed c. The rightward-going wave continues to move to the right, unmolested. The leftward-moving wave, however, bumps into the boundary at $x = 0$ and is reflected therefrom. Because of the BC $u = 0$ there, this wave is reflected with reversed amplitude. If we had taken the BC $u_x(0, t) = 0$ instead, then the wave would have been reflected with no change in the sign of the amplitude. After reflection, this wave moves to the right and is henceforth undisturbed.

We proceed to generalize the BC still further. We assume that the end point is attached to the origin by a linear spring with spring constant γ. In Sec. 2-1 we saw that the vertical component of the tension force is Hu_x. Since we assume only vertical motion, and since no net force can act at a point, the spring force must balance the tension force. Hence $Hu_x(0, t) = \gamma u(0, t)$. Thus, we are led to consider the BC

$$u_x(0, t) - ku(0, t) = 0 \qquad \begin{matrix} 0 < t < \infty \\ k = \text{const} \end{matrix} \tag{2-29}$$

Notice that (2-29) contains (2-21) and (2-28) as special cases. If we let $k = 0$, then the spring at the end point has no resistance and we get the *free boundary condition* (2-28). On the other hand, if we let $k \to \infty$, then the spring becomes rigid, i.e., the end point cannot move, and we get the BC (2-21).

In order to ensure the compatibility of the BC (2-29) with the IC (2-19) and (2-20) we assume that $f'(0) - kf(0) = 0$ and $g'(0) - kg(0) = 0$.

As an illustration, suppose $g(x) \equiv 0$. Then (2-15) and (2-16) tell us that

$$\phi(x - ct) = \tfrac{1}{2} f(x - ct) \qquad \text{for} \quad x > ct$$

and

$$\psi(x + ct) = \tfrac{1}{2} f(x + ct) \qquad \text{for} \quad x > ct$$

We have already remarked, however, that ψ is constant on the characteristics $x + ct = \text{const}$. Hence $\psi(x + ct) = \frac{1}{2} f(x + ct)$ for $x < ct$, too. Therefore

$$u(x, t) = \begin{cases} \frac{1}{2}[f(x - ct) + f(x + ct)] & \text{for } x > ct \\ \frac{1}{2} f(x + ct) + \phi(x - ct) & \text{for } x < ct \end{cases} \tag{2-30}$$

where ϕ still has to be determined. The BC (2-29) will do this for us. In order to avoid unnecessary complications, we observe that a function of $x - ct$ is equally well a function of $t - x/c$ [because $t - x/c = -(x - ct)/c$], and therefore we can write

$$u(x, t) = \frac{1}{2} f(x + ct) + \tilde{\phi}\left(t - \frac{x}{c}\right) \qquad \text{for } x < ct \tag{2-31}$$

We need $\tilde{\phi}$ for positive values of its argument. For brevity we drop the tilde. Then, applying (2-29), we have

$$\frac{1}{2} f'(ct) - \frac{1}{c}\phi'(t) - \frac{k}{2} f(ct) - k\phi(t) = 0$$

or,

$$\phi'(t) + kc\phi(t) = \frac{c}{2}[f'(ct) - kf(ct)]$$

This is a first-order linear ordinary differential equation for ϕ as a function of t. Its solution is

$$\phi(t) = \frac{c}{2}e^{-kct}\int_0^t [f'(c\tau) - kf(c\tau)]\,e^{kc\tau}\,d\tau + \kappa e^{-kct} \tag{2-32}$$

where $\kappa = \phi(0)$. To determine κ, we use the fact that the two different expressions for u in the two different regions $x > ct$ and $x < ct$ must match on their common boundary, $x = ct$. From the first part of (2-30) and from (2-31) we therefore have

$$\tfrac{1}{2}f(0) = \phi(0)$$

Hence $\kappa = \frac{1}{2}f(0)$. With this value inserted for κ, (2-32) then gives the solution to our example.

2-5. SEMI-INFINITE STRING WITH MOVING END POINT

Suppose that the end point of a semi-infinite string is not held fixed at the origin but moved with constant velocity U. We take U as positive if the end point moves to the right and negative if the end point moves to the left. Anticipating later developments (Chaps. 13 and 14), we call the motion of the end point *supersonic* if $|U| > c$ and *subsonic* if $|U| < c$. Notice that it is only the end point of the medium that is being permitted to move with velocity U, not the waves in the medium. The PDE therefore remains unchanged and stays hyperbolic (see Chap. 6 for the definition of a hyperbolic PDE). In gas dynamics, subsonic flows are usually governed by elliptic PDEs and supersonic flows by hyperbolic ones. (Again, see Chap. 6 for the definition of ellipticity.) The present case of a string with a moving end point is therefore not an exact analog of compressible-flow problems. Nevertheless, it does give some preliminary insight into phenomena which will be investigated in greater detail in later chapters, particularly Chaps. 13 and 14.

Formulated analytically, our problem is the following:

PDE:

$$u_{xx} - c^{-2}u_{tt} = 0 \qquad \begin{array}{l}\text{for } 0 < t < \infty \\ \text{and } Ut < x < \infty\end{array} \tag{2-33}$$

ICs:

$$\begin{array}{ll} u(x, 0) = f(x) & 0 < x < \infty \\ u_t(x, 0) = 0 & f(0) = 0 \end{array} \tag{2-34}$$

BC:

$$u(Ut, t) = 0 \qquad 0 < t < \infty \tag{2-35}$$

We have chosen a zero BC and one zero IC in order to simplify the analysis. The more complicated case of a *finite* string with nonzero conditions has been treated by Balazs [7].†

Physically, it is clear that three cases should be distinguished: I: $U > c$, II: $-c < U < c$, and III: $U < -c$. In case I the end point moves up the string at a speed faster than that at which waves propagate up and down the string. In case II the end point is moving at a speed slower than the wave propagation speed, and in the case III the end point is receding at a speed faster than the wave propagation speed.

First we remark that the preceding section has shown that for $x > ct$ only the ICs play a role and that the solution is $u = \frac{1}{2}[f(x - ct) + f(x + ct)]$ in that region.

Case I $U > c$ According to (2-33), $x > Ut$ always. Hence in the present case, $x > ct$ always. Therefore

$$u(x, t) = \tfrac{1}{2}[f(x - ct) + f(x + ct)] \qquad x > Ut \tag{2-36}$$

is the solution in case I (see Fig. 2-10).

Case II $-c < U < c$ As mentioned above, the solution is given by (2-36) for $x > ct$. For $Ut < x < ct$, we have to proceed as in Sec. 2-4. Writing $u = \phi(x - ct) + \psi(x + ct)$, we find, as there, that $\psi(x) = \frac{1}{2}f(x)$. Then applying the BC (2-35), we get

$$0 = \phi[(U - c)t] + \tfrac{1}{2}f[(U + c)t] \tag{2-37}$$

Let

$$(U - c)t = z \tag{2-38}$$

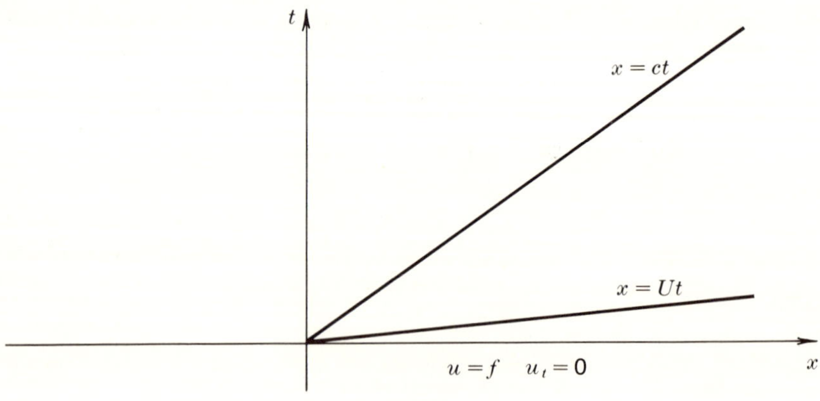

Figure 2-10

† Numbered references will be found at the end of the book.

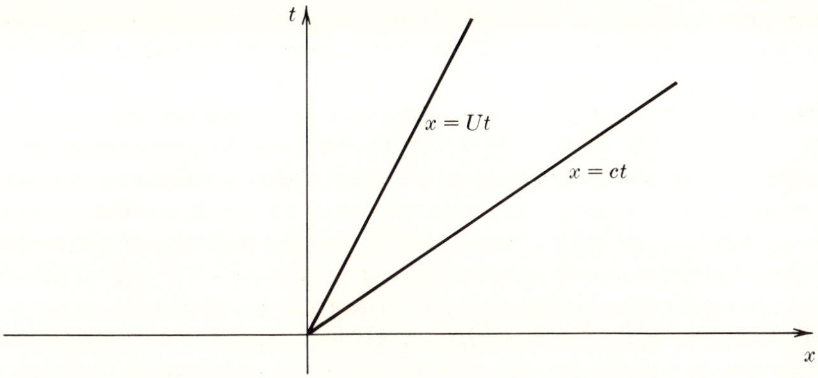

Figure 2-11

and

$$\alpha = \frac{U + c}{U - c}$$

Then (2-37) tells us that

$$\phi(z) = -\tfrac{1}{2}f(\alpha z) \tag{2-39}$$

Since f is defined for positive argument, and since α is negative for U in the range considered in the present case, (2-39) defines ϕ for negative argument. Hence we obtain the solution (Fig. 2-11)

$$u(x, t) = -\tfrac{1}{2}\{f(x + ct) - f[\alpha(x - ct)]\} \qquad \text{for } Ut < x < ct \tag{2-40}$$

Case III $U < -c$ In this case (see Fig. 2-12) the solution is given by (2-36) for $x > ct$ and by (2-40) for $-ct < x < ct$. For $Ut < x < -ct$,

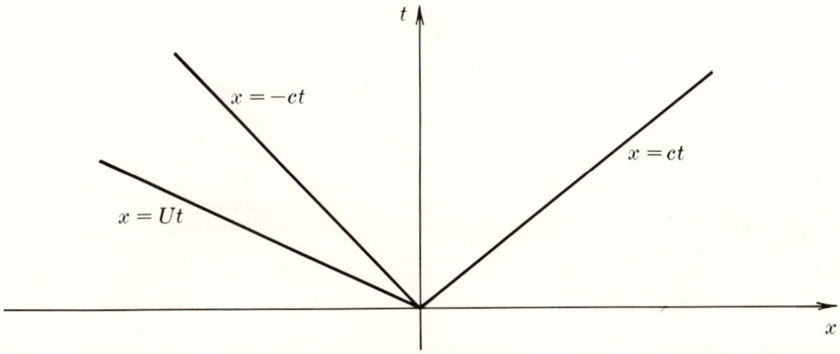

Figure 2-12

the solution should, on physical grounds, be zero since the waves generated by the initial conditions cannot reach that region. Mathematically, however, the solution is not determined in this region because both characteristics from a point in the region will hit the line $x = Ut$. It is therefore necessary to specify an additional condition on the line $x = Ut$ (for example, u_t) in order to determine the solution in the region under consideration. An initial line having this property, as, for example, the x axis does, is called *spacelike*. More precisely, an initial curve is called spacelike with respect to a second-order hyperbolic equation in two independent variables if both characteristics when traversed in the direction of increasing t point to the same side of the curve. For a second-order hyperbolic equation in two independent variables, two initial data must be prescribed on any spacelike initial curve (see Fig. 2-13). A fuller discussion of space- and timelike initial curves is given in Sec. 9-6.

The qualitative differences among the three cases are pretty much what were to be expected. If $U > c$, the end point moves up the string at a speed faster than that at which any waves it might create can move. Thus the end point arrives at each point unannounced by any waves it might generate. Qualitatively, this is a feature of the shock waves to be considered in Chap. 14.

If $0 < U < c$, the waves generated by the end point arrive at any point of the string ahead of the end point itself. If $-c < U < 0$, the end point is receding but at a speed slow enough for its effect to propagate down the string. If $U < -c$, the end point is receding so fast that it carries its effects away faster than they can propagate up the string.

Of course, all this may seem rather artificial, but, as we shall see in Sec. 2-9, the wave equation also describes the motion of gas in a tube if certain simplifying assumptions are made. If we consider an infinite tube with

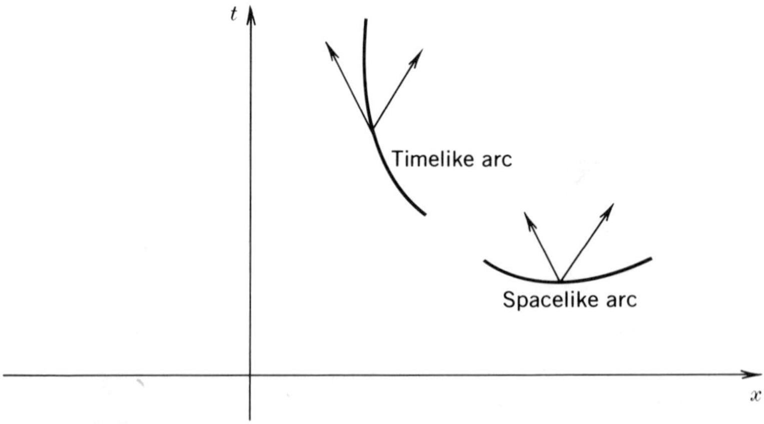

Figure 2-13

a piston inserted from the left having its face initially at $x = 0$ and with gas occupying the semi-infinite half of the tube to the right of $x = 0$, then the results of this section explain what happens when the piston is moved with a constant velocity U. Unfortunately, however, the wave equation is only an approximation to this problem of gas dynamics. When a more exact description is necessary, the situation becomes more complicated, as we shall see in Chap. 13. Nevertheless, the results of the present section give some initial insight into the situation. A deeper investigation will have to be postponed till later.

2-6. SUPERPOSITION PRINCIPLE

Suppose $u(x, t)$ is a solution to the following problem:

PDE:

$$u_{xx} - c^{-2}u_{tt} = 0 \qquad \begin{matrix} -\infty < x < \infty \\ 0 < t < \infty \end{matrix}$$

ICs:

$$\begin{aligned} u(x, 0) &= f(x) \\ u_t(x, 0) &= 0 \end{aligned} \qquad -\infty < x < \infty$$

and suppose $v(x, t)$ is a solution to

PDE:

$$v_{xx} - c^{-2}v_{tt} = 0 \qquad \begin{matrix} -\infty < x < \infty \\ 0 < t < \infty \end{matrix}$$

ICs:

$$\begin{aligned} v(x, 0) &= 0 \\ v_t(x, 0) &= g(x) \end{aligned} \qquad -\infty < x < \infty$$

Put $w = u + v$. Then clearly w is a solution to

PDE:

$$w_{xx} - c^{-2}w_{tt} = 0$$

ICs:

$$\begin{aligned} w(x, 0) &= f(x) \\ w_t(x, 0) &= g(x) \end{aligned} \qquad -\infty < x < \infty$$

This illustrates a general fact which is due to the linearity of all the problems with which we have dealt so far. Suppose we have a number, say n, of nonhomogeneous conditions in a problem. If we wish, we can select, say, any k of them and replace these k by homogeneous conditions, leaving the other $n - k$ conditions alone. We solve the resulting problem and then set up another problem in which we now keep the k nonhomogeneous conditions previously replaced by homogeneous ones and replace the other $n - k$

nonhomogeneous conditions by homogeneous ones. We solve this problem. The solution to the original problem is then the sum of the solutions to these two problems. Obviously, we can, if we wish, break the solution of the original problem into the superposition of the solution of several instead of just two subproblems by taking only a few nonhomogeneous conditions at a time.

2-7. NONHOMOGENEOUS WAVE EQUATION

Next we consider an initial-value problem for a nonhomogeneous equation:

PDE:

$$u_{xx} - c^{-2}u_{tt} = F(x, t) \qquad \begin{array}{c} -\infty < x < \infty \\ 0 < t < \infty \end{array} \tag{2-41}$$

ICs:

$$\begin{aligned} u(x, 0) &= f(x) \\ u_t(x, 0) &= g(x) \end{aligned} \qquad -\infty < x < \infty \tag{2-42}$$

Here $F(x, t)$ is a given function representing some impressed force, e.g., gravity.

Suppose $v(x, t)$ is a solution of the nonhomogeneous equation (2-41) satisfying homogeneous ICs, $f = g = 0$, and suppose w is a solution of the homogeneous equation $w_{xx} - c^{-2}w_{tt} = 0$ satisfying the nonhomogeneous ICs (2-42). Then $u = v + w$ is the desired solution of the nonhomogeneous equation (2-41) satisfying the nonhomogeneous ICs (2-42). Since the problem of finding w has been solved previously, it therefore suffices to find v, that is, to solve (2-41) with the homogeneous ICs

$$u(x, 0) = u_t(x, 0) = 0 \qquad -\infty < x < \infty \tag{2-43}$$

To solve (2-41) with the ICs (2-43) we transform to the characteristic coordinates ξ and η previously introduced in (2-5). In terms of ξ and η (2-41) becomes

$$4u_{\xi\eta} = \hat{F}(\xi, \eta) \tag{2-44}$$

where

$$\hat{F}(\xi, \eta) = F\left(\frac{\xi + \eta}{2}, \frac{\eta - \xi}{2}\right)$$

Naturally, the next step is to integrate (2-44), first with respect to η and then with respect to ξ. These integrations will introduce certain arbitrary functions of ξ and η into the solution. It is easy to see, however, that the solution of (2-41) that satisfies (2-43) or, more generally, (2-42), is unique. To see this, suppose that u_1 and u_2 are two different solutions of (2-41) both satisfying the same initial conditions (2-42) with the same f and g. Then the difference, $u = u_1 - u_2$, satisfies the homogeneous wave equation with homogeneous initial conditions. Hence $u \equiv 0$. So $u_1 \equiv u_2$.

Since the solution to (2-41) satisfying (2-42) is unique, it must be possible to determine the arbitrary functions that will enter into the integration of (2-44). It is clear that the ICs must be used to determine them. We need to transfer the ICs from the xt to the $\xi\eta$ plane so that they can be used in the integration of (2-44). We also need to determine the domain in the $\xi\eta$ plane in which (2-44) holds so that we know where the integration is to be carried out.

In order to do this, we have to examine the effects of the mapping (2-5) on the upper half of the xt plane. Under (2-5) $x = 0$ goes into $\xi + \eta = 0$, and $t = 0$ goes into $\xi = \eta$. Since $\eta - \xi = 2ct$, $t > 0$ implies $n > \xi$. Hence the upper half of the xt plane maps into the part of the $\xi\eta$ plane above the line $\xi = \eta$.

As already mentioned, the initial line, $t = 0$, maps into the line $\xi = \eta$. The ICs in the $\xi\eta$ plane must therefore be imposed on this line. One IC is $u = 0$ on $t = 0$. This becomes

$$u = 0 \qquad \text{for } \xi = \eta \tag{2-45}$$

In order to transfer the IC $u_t = 0$ for $t = 0$, we observe that $u(x, 0) = 0$ for all x implies that $u_x(x, 0) = 0$ for all x because the variable t is held constant during differentiation with respect to x anyway. Thus, we can add to the ICs the IC $u_x = 0$ for $t = 0$ because this additional IC is a consequence of the IC $u = 0$ for $t = 0$ and is therefore no additional restriction. It is, however, an additional advantage for our purposes because from the knowledge that $u_x = u_t = 0$ for $t = 0$ we can conclude that

$$u_\xi = u_\eta = 0 \qquad \text{for } \xi = \eta \tag{2-46}$$

because u_ξ and u_η are linear combinations of u_x and u_t.

Again not all three conditions in (2-45) and (2-46) are independent. Either of the two conditions in (2-46) follows from the other in conjunction with (2-45).

Now that we have the IC expressed in terms of ξ and η, we can integrate (2-44). To do this, we choose a point P: (ξ_0, η_0) (see Fig. 2-14) above the line $\xi = \eta$ and integrate first with respect to η from a point P_1 on the initial line, $\xi = \eta$, to P. Since integration with respect to η means keeping ξ constant, the integration in question must be along a vertical line. Hence P_1 must have ξ_0 for its ξ coordinate. Since P_1 lies on the line $\xi = \eta$, its η coordinate is therefore also equal to ξ_0 (see Fig. 2-14). So the indicated integration gives

$$4[u_\xi(\xi_0, \eta_0) - u_\xi(\xi_0, \xi_0)] = \int_{\xi_0}^{\eta_0} \hat{F}(\xi_0, \hat{\eta})\, d\hat{\eta} \tag{2-47}$$

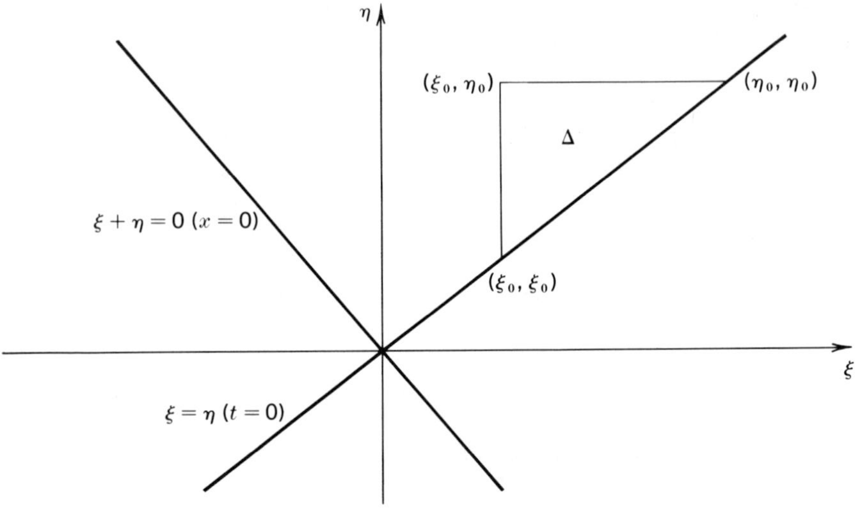

Figure 2-14

The IC (2-46) implies that $u_\xi(\xi_0, \xi_0) = 0$ and so (2-47) reduces to

$$4u_\xi(\xi_0, \eta_0) = \int_{\xi_0}^{\eta_0} \hat{F}(\xi_0, \hat{\eta})\, d\hat{\eta} \tag{2-48}$$

or, dropping the subscripts [because (ξ_0, η_0) is arbitrary],

$$4u_\xi(\xi, \eta) = \int_{\xi}^{\eta} \hat{F}(\xi, \hat{\eta})\, d\hat{\eta} \tag{2-49}$$

Now, as indicated in Fig. 2-14, we integrate (2-49) with respect to ξ from $\xi = \xi_0$ to the point where this horizontal line hits the initial line, namely, to $\xi = \eta_0$. Since u vanishes on the initial line, we get

$$-4u(\xi_0, \eta_0) = \int_{\xi_0}^{\eta_0} \int_{\hat{\xi}}^{\eta_0} \hat{F}(\hat{\xi}, \hat{\eta})\, d\hat{\eta}\, d\hat{\xi} \tag{2-50}$$

Notice that the iterated integral on the right of (2-50) is just the double integral of $\hat{F}$ over the triangle $\triangle$ formed by the characteristics through (ξ_0, η_0) and the initial line. Thus,

$$u(\xi, \eta) = -\frac{1}{4} \iint_{\triangle} \hat{F}(\xi, \eta)\, d\xi\, d\eta \tag{2-51}$$

If we transform back to xt coordinates, $\triangle$ becomes the triangle with

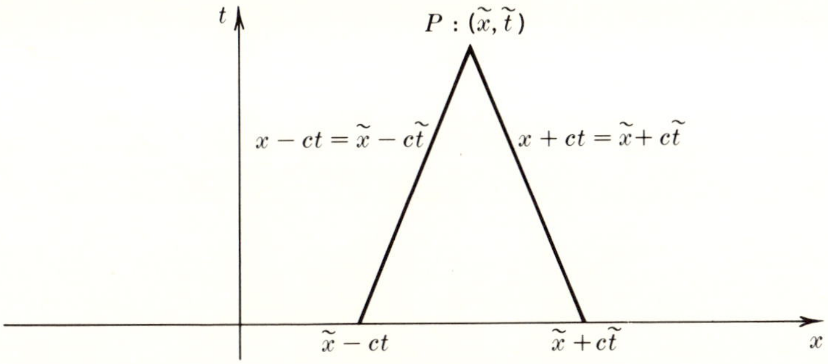

Figure 2-15

vertex at (x, t), base on the initial line $(t = 0)$, and sides the characteristics through (x, t). The Jacobian of the transformation is

$$\frac{\partial(\xi, \eta)}{\partial(x, t)} = \begin{vmatrix} 1 & -c \\ 1 & c \end{vmatrix} = 2c$$

Hence

$$u(x, t) = -\frac{c}{2} \iint_{\triangle} F(x, t)\, dx\, dt \tag{2-52}$$

that is (see Fig. 2-15),

$$u(\tilde{x}, \tilde{t}) = -\frac{c}{2} \int_0^{\tilde{t}} dt \int_{\tilde{x}-c\tilde{t}+ct}^{\tilde{x}+c\tilde{t}-ct} F(x, t)\, dx \tag{2-53}$$

For a semi-infinite string, formula (2-52) has to be modified in the region $x < ct$, for if $(\tilde{x}, \tilde{t})$ lies in that region, then the characteristic $x - ct = \tilde{x} - c\tilde{t}$ intersects the boundary, $x = 0$, and part of the characteristic triangle lies outside the region in which the problem is defined (see Figs. 2-8 and 2-16).

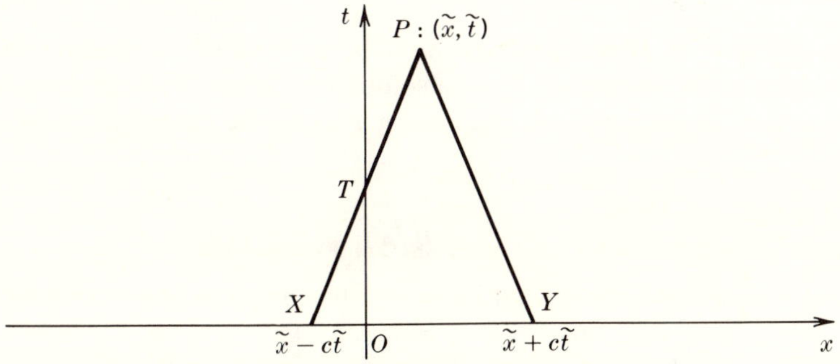

Figure 2-16

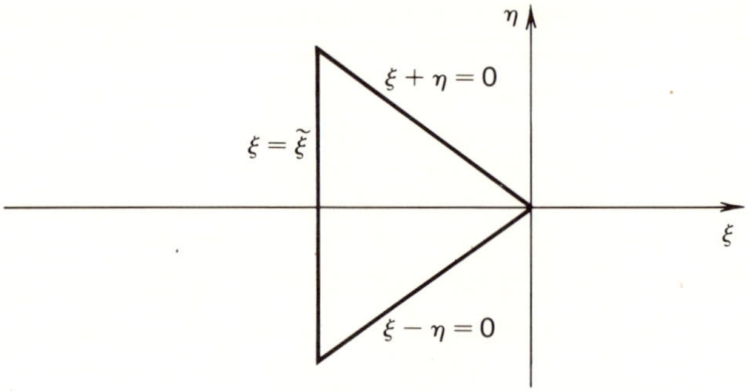

Figure 2-17

It would be nice if we could just neglect the part of the integral taken over the right triangle XOT and merely use the integral of F over the trapezoid $OTPY$. We have already seen, however, that, as a rule, we cannot arbitrarily throw away parts of a region merely because we have no information on the solution there. For example, we could not arbitrarily define the initial values to be zero for $x < 0$ because the solution satisfying zero initial conditions for $x < 0$ will not generally satisfy whatever BCs have been imposed on $x = 0$. It is therefore amazing that in the present case we can indeed neglect the integral of F over the right triangle XOT. The reason, as will appear below, is that the neglected integral is a solution of the homogeneous wave equation and therefore is not needed in finding a particular solution of the nonhomogeneous equation.

In order to show that the integral in question can be neglected, we return to characteristic coordinates (ξ, η). In the $\xi\eta$ plane, triangle XOT maps into the triangle shown in Fig. 2-17.

Apart from a constant factor, the integral in question is

$$I = \int_{\tilde{\xi}}^{0} d\xi \int_{\xi}^{-\xi} F(\xi, \eta)\, d\eta$$

Obviously, $I_{\tilde{\eta}} = 0$. Hence $I_{\tilde{\xi}\tilde{\eta}} = 0$. So I is, as claimed, a solution to the homogeneous wave equation. Thus, a solution to the nonhomogeneous wave equation can be obtained by integrating F over the trapezoid mentioned above.

2-8. INCOMING WAVES; TRANSMISSION PROBLEMS

Next we wish to investigate what happens when an incoming wave impinges upon some sort of discontinuity in a medium. First, however, we have to be able to say precisely what we mean by an incoming wave. Roughly speaking it is a function $\Phi(x - ct)$. As on a previous occasion, we shall find it slightly more convenient to write this in the form $\Phi(t - x/c)$. Since we want our incoming wave to have a sharp "head," we want it to be zero ahead of some characteristic, say the characteristic $x = ct$. Thus, by an *incoming wave* we shall mean a function $u(x, t)$ defined as follows:

$$u(x, t) = \begin{cases} \Phi\left(t - \dfrac{x}{c}\right) & \text{for } x < ct \\ 0 & \text{for } x > ct \end{cases} \tag{2-54}$$

We notice that an incoming wave, i.e., (2-54), could also be defined as a solution of the wave equation satisfying the following ICs:

$$\begin{aligned} u(x, 0) &= \Phi\left(-\frac{x}{c}\right) \\ & \qquad\qquad x < 0 \\ u_t(x, 0) &= \Phi'\left(-\frac{x}{c}\right) \\ u(x, 0) &= u_t(x, 0) = 0 \qquad x > 0 \end{aligned} \tag{2-55}$$

where $\Phi(\xi)$ is a given function of a single variable. Specifying Φ means specifying the shape of the incoming wave.

Now we are ready to consider a transmission problem. Suppose a string has a jump in its density at some point, say at the origin; i.e., suppose it has density ρ_1 to the left of the origin and density ρ_2 to the right. Then [cf. (2-3)] there will be a propagation speed c_1 for waves moving through the left half of the string and a propagation speed c_2 for waves moving through the right half. Thus, we will have a wave equation

$$u_{xx} - c_1^{-2}u_{tt} = 0 \qquad \text{for } x < 0 \tag{2-55a}$$

and a wave equation

$$u_{xx} - c_2^{-1}u_{tt} = 0 \qquad \text{for } x > 0 \tag{2-55b}$$

For $x < 0$ the general solution will be

$$u(x, t) = \phi_1(x - c_1t) + \psi_1(x + c_1t)$$

and for $x > 0$ it will be

$$u(x, t) = \phi_2(x - c_2t) + \psi_2(x + c_2t)$$

The transmission problem we wish to investigate is one in which an incoming wave impinges upon the "interface" $x = 0$. More precisely, we wish to assume that we have an incoming wave in the part of the string to the left of the origin, a wave that arrives at the origin at time $t = 0$. We wish the part of the string to the right of the origin to be at rest at $t = 0$.

According to (2-55), this means that we are given the ICs

$$\begin{aligned} u(x,0) &= \Phi\left(-\frac{x}{c_1}\right) \\ u_t(x,0) &= \Phi'\left(-\frac{x}{c_1}\right) \end{aligned} \qquad -\infty < x < 0 \tag{2-56}$$
$$u(x,0) = u_t(x,0) = 0 \qquad 0 < x < \infty$$

Let us look at the xt plane (Fig. 2-18) and see what we can expect the solution to look like. Since the ICs are zero for $x > 0$, $u = 0$ for $x > c_2 t$. Since ψ is constant on the C^- characteristics, $\psi_2 = 0$ for $0 < x < c_2 t$. These characteristics end at $x = 0$ because c changes there. Hence we cannot conclude that $\psi_1 = 0$ for $x < 0$. We can, however, conclude that $\psi_1 = 0$ for $x < -c_1 t$ because of the ICs (2-56). In fact u is completely determined in that region. It is the incoming wave Φ. Because of the constancy of ϕ on the C^+ characteristics, it follows that $\phi_1 = \Phi$ for $-c_1 t < x < 0$ too, but $\psi_1 \neq 0$ in this region.

To summarize, we have found the solution must look like

$$u(x,t) = \begin{cases} \Phi\left(t - \dfrac{x}{c_1}\right) & -\infty < x < -c_1 t \\ \Phi\left(t - \dfrac{x}{c_1}\right) + \psi_1\left(t + \dfrac{x}{c_1}\right) & -c_1 t < x < 0 \\ \phi_2\left(t - \dfrac{x}{c_2}\right) & 0 < x < c_2 t \\ 0 & c_2 t < x < \infty \end{cases} \tag{2-57}$$

In (2-57) Φ is the given incoming wave. ψ_1 and ϕ_2 are as yet unknown.

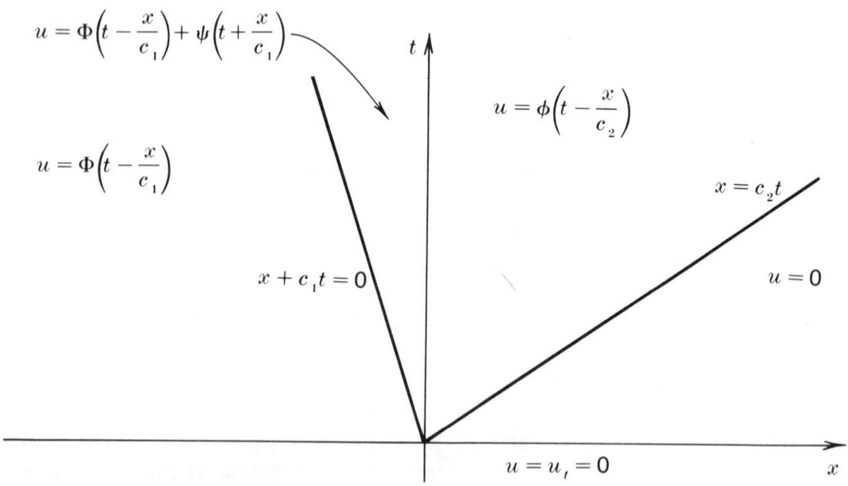

Figure 2-18

The form of the solution (2-57) was to be expected. Since the incoming wave arrives at the interface at $t = 0$, the effects of its interaction with the interface propagate outward from the interface with velocity c_1 to the left of the interface and velocity c_2 to the right of it. Points in the region $x < -c_1t$ cannot be reached by disturbances which propagate outward from $x = 0$ with velocity c_1. Hence only the original incoming wave is present in this region. Similarly, points in the region $x > c_2t$ have not yet been reached by the effects of the wave's interaction with the interface. Hence the solution is zero in that region.

In the region $-c_1t < x < 0$, (2-57) shows that the solution consists of the incoming wave going to the right and a wave ψ_1 going to the left. Since ψ_1 is obviously the result of the interaction of the incoming wave with the interface, and since ψ_1 is receding from the interface, it is natural to call ψ_1 the *reflected wave*.

Finally, in the region $0 < x < c_2t$, we have only a wave ϕ_2 going to the right. Since ϕ_2 can be thought of as the part of Φ that gets through the interface, we call it the *transmitted wave*.

Using the terminology just introduced, we can say that the transmission problem consists of determining the reflected and transmitted waves in terms of the incoming one. Since we have already made use of all the information in (2-56), it is clear that we will need additional information to determine ψ_1 and ϕ_2 because any function $u(x, t)$ given by (2-57) will satisfy (2-55*a*), (2-55*b*), and (2-56) no matter what we choose for ψ_1 and ϕ_2. This is not surprising on mathematical grounds because (2-55*a*) and (2-55*b*) are really two different wave equations. The solutions of one have no mathematical connection with those of the other unless we impose some additional conditions connecting them; i.e., mathematically, we have to say how the solution for $x < 0$ is to be related to that for $x > 0$.

There is relatively little mathematical restriction on what conditions we impose at $x = 0$, but since we want the resulting solution to have physical relevance, we turn to physics to find some natural conditions to impose.

Since we don't want our string to break, we require that

$$u^{(1)}(0, t) = u^{(2)}(0, t) \tag{2-58}$$

the meaning of the notation being obvious. In addition, there can be no net vertical force at the origin. Hence Hu_x must be continuous there. Since there can be no net horizontal force there either, H must be continuous. Hence

$$u_x^{(1)}(0, t) = u_x^{(2)}(0, t) \tag{2-59}$$

Thus, the correct formulation of our problem is not "solve (2-55*a*) and (2-55*b*) with the ICs (2-56)" but rather "solve (2-55*a*) and (2-55*b*) with the ICs (2-56) and subject to the *interface conditions* (2-58) and (2-59)."

Having finally arrived at the correct formulation of the transmission problem, we proceed to solve it. We have already seen that the solution to (2-55*a*) and (2-55*b*) satisfying (2-56) is (2-57). All that remains, then, is to use (2-58) and (2-59) to determine ψ_1 and ϕ_2. For brevity, we drop the subscripts on ψ_1 and ϕ_2 and write them as merely ψ and ϕ. From (2-57) to (2-59) we have

$$\Phi(t) + \psi(t) = \phi(t)$$

and

$$-\frac{1}{c_1}\,\Phi'(t) + \frac{1}{c_1}\,\psi'(t) = -\frac{1}{c_2}\,\phi'(t)$$

or

$$\Phi' + \psi' = \phi'$$

and

$$\Phi' - \psi' = \frac{c_1}{c_2}\,\phi'$$

Therefore,

$$2\Phi' = \left(1 + \frac{c_1}{c_2}\right)\phi$$

or

$$\phi = \frac{2c_2}{c_1 + c_2}\,\Phi \tag{2-60}$$

Also

$$2\psi = \left(1 - \frac{c_1}{c_2}\right)\phi$$

or

$$\psi = \frac{c_2 - c_1}{2c_2} \qquad \phi = \frac{c_2 - c_1}{c_1 + c_2}\,\Phi$$

For obvious reasons, $T = 2c_2/(c_1 + c_2)$ is called the *transmission coefficient* and $R = (c_2 - c_1)/(c_1 + c_2)$ the *reflection coefficient*. Note that $T - R = 1$.

Note, too, that if $c_1 = c_2$, then $R = 0$ and $T = 1$. This is not surprising because $c_1 = c_2$ means that the medium has no discontinuity at the origin and $R = 0$, $T = 1$ means the entire wave is then transmitted with none of it reflected. In such a case we speak of *total transmission*.

It is important to remember that the transmission and reflection coefficients just obtained depend upon having interface conditions which require u and u_x to be continuous at the interface. In the next section we will illustrate a different pair of interface conditions; still other types of interface conditions exist, but we will not discuss them. The reader who is familiar with another relationship between R and T should bear in mind that he may be thinking of a problem in which the interface conditions are different from those we have taken in our illustrations.

2-9. LAGRANGIAN REPRESENTATION OF ONE-DIMENSIONAL GAS FLOW

Suppose we have a gas filling a tube and performing a motion that is one-dimensional. "One-dimensional" means that all particles on any plane perpendicular to the axis of the tube perform the same motion as the particle on the axis where the plane intersects the axis. This being the case, the motion is completely described by describing the motion of the particles on the axis.

First we need a way of identifying the different particles on the axis. This can be done in numerous ways. We choose this one: let α denote the initial position of each particle. Then the position x of this particle at any later time t will be a function of α and t: $x(\alpha, t)$. Since our particles are not permitted to devour each other, we will have $x(\alpha_1, t) \neq x(\alpha_2, t)$ if $\alpha_1 \neq \alpha_2$ and

$$x(\alpha, 0) = \alpha \tag{2-61}$$

In addition to the position of each particle it is desirable to know the density $\rho(\alpha, t)$ and the pressure $p(\alpha, t)$. To determine the quantities x, ρ, and p, we need three equations. The first equation is the equation of state, an equation which is determined by the properties of the particular medium at hand. Neglecting entropy, we assume the equation of state to be given in the form

$$p = f(\rho) \tag{2-62}$$

where f is a given, i.e., known, function.

Having (2-62), we still need two more equations. The first of these comes from the law of conservation of mass. Consider a section of particles extending at time $t = 0$ from α_1 to α_2. The mass of this section is $\int_{\alpha_1}^{\alpha_2} \rho(\alpha, 0)\, d\alpha$. At a later time t, the mass of the same collection will be $\int_{x_1}^{x_2} \rho(\alpha, t)\, dx$. The volume or length of the section may change in time, but the mass must remain constant. Hence

$$\int_{\alpha_1}^{\alpha_2} \rho(\alpha, 0)\, d\alpha = \int_{x_1}^{x_2} \rho(\alpha, t)\, dx$$

or

$$\int_{\alpha_1}^{\alpha_2} \rho(\alpha, 0)\, d\alpha = \int_{\alpha_1}^{\alpha_2} \rho(\alpha, t)\, x_\alpha(\alpha, t)\, d\alpha$$

Divide by $\alpha_2 - \alpha_1$ and let $\alpha_2 \to \alpha_1 = \alpha$ (say). Then

$$p(\alpha, 0) = x_\alpha(\alpha, t)\, \rho(\alpha, t) \tag{2-63}$$

We assume henceforth that the initial density $\rho(\alpha, 0)$ is a constant, ρ_0:

$$\rho(\alpha, 0) = \rho_0 = \text{const} \tag{2-64}$$

Then

$$x_\alpha(\alpha, t)\, \rho(\alpha, t) = \rho_0 \tag{2-65}$$

Next we apply Newton's second law to this segment. We assume the only forces acting are due to the pressure. At the right-hand end point of the segment the pressure is $p(\alpha_2, t)$. At the left-hand end point it is $p(\alpha_1, t)$. Since pressure is a compressive force, $p(\alpha_2, t)$ acts in the negative direction and $p(\alpha_1, t)$ in the positive. Hence the net force is $-[p(\alpha_2, t) - p(\alpha_1, t)]$. Newton's second law therefore reads

$$\int_{x_1}^{x_2} \rho(\alpha, t)\, x_{tt}(\alpha, t)\, dx = -[p(\alpha_2, t) - p(\alpha_1, t)]$$

or

$$\int_{\alpha_1}^{\alpha_2} \rho(\alpha, t)\, x_{tt}(\alpha, t)\, x_\alpha(\alpha, t)\, d\alpha = -[p(\alpha_2, t) - p(\alpha_1, t)]$$

Use of (2-65) reduces this to

$$\int_{\alpha_1}^{\alpha_2} \rho_0 x_{tt}(\alpha, t)\, d\alpha = -[p(\alpha_2, t) - p(\alpha_1, t)]$$

Divide by $\alpha_2 - \alpha_1$ and let $\alpha_2 \to \alpha_1 = \alpha$. Then

$$\rho_0 x_{tt}(\alpha, t) = -p_\alpha(\alpha, t) \tag{2-66}$$

Thus, we have found the following set of equations for x, ρ, and p:

$$\rho x_\alpha = \rho_0 \tag{2-65}$$

$$\rho_0 x_{tt} = -p_\alpha \tag{2-66}$$

$$p = f(\rho) \tag{2-62}$$

In them α, the initial position of a particle, and t, the time, are the independent variables, and x, the position of a particle at any time, p, the pressure, and ρ, the density, are the dependent ones. Such a representation of the flow is called the *Lagrangian representation.*

If we differentiate (2-62) with respect to α, we get $p_\alpha = f'(\rho)\, \rho_\alpha$, which enables us to eliminate p from (2-66) and obtain

$$\rho_0 x_{tt} = -f'(\rho)\, \rho_\alpha$$

Then we use (2-65) to eliminate ρ from this equation to obtain an equation for x alone

$$\rho_0 x_{tt} = -f'\left(\frac{\rho_0}{x_\alpha}\right)\frac{-\rho_0 x_{\alpha\alpha}}{{x_\alpha}^2}$$

or

$$x_{tt} = \frac{f'(\rho_0/x_\alpha)}{{x_\alpha}^2}\, x_{\alpha\alpha} \tag{2-67}$$

a highly nonlinear equation for x.

Because (2-67) is complicated, it is natural to look for means of simplifying it. As an example of a frequently useful approximation to (2-67), let us assume that the motion of the gas is relatively peaceful, so that a given particle will not move very far from its original position in a short time. Since $x(\alpha, t) = \alpha$ when $t = 0$, this assumption is tantamount to saying that x does not differ too much from α for small t. We cannot assume that x stays equal to α because then there would be no motion at all. So we do the next best thing; we assume that $x_\alpha \sim 1$, because this expresses our assumption that x does not change greatly from α for small t.

The assumption $x_\alpha \sim 1$, when used in (2-67), linearizes (2-67) to

$$x_{tt} = f'(\rho_0)\, x_{\alpha\alpha} \tag{2-68}$$

Since it is natural to assume that pressure increases with density, we can assume that $f'(\rho) > 0$ and therefore put $f'(\rho_0) = {c_0}^2$. Then (2-68) becomes the wave equation

$$x_{tt} = {c_0}^2 x_{\alpha\alpha} \tag{2-69}$$

We can also obtain a nonlinear wave equation for p and then assume $x_\alpha \sim 1$ to linearize it. First we eliminate x from (2-65) and (2-66). From (2-65) we have

$$x_\alpha = \rho_0 \rho^{-1}$$

Therefore,

$$x_{\alpha tt} = \rho_0 (\rho^{-1})_{tt}$$

From (2-66) we get

$$\rho_0 x_{tt\alpha} = -p_{\alpha\alpha}$$

Then $x_{\alpha tt} = x_{tt\alpha}$ implies that

$$\rho_0^2 (\rho^{-1})_{tt} = -p_{\alpha\alpha} \tag{2-70}$$

To eliminate ρ from this, use (2-62):

$$p = f(\rho)$$

whence

$$p_t = f'(\rho)\, \rho_t$$

Therefore,

$$(\rho^{-1})_t = -\rho^{-2}\rho_t = -[\rho^2 f'(\rho)]^{-1}\, p_t$$
$$(\rho^{-1})_{tt} = -\{[\rho^2 f'(\rho)]^{-1}\, p_t\}_t$$

Hence [cf. (2-70)]

$$\rho_0^{2}\{[\rho^2 f'(\rho)]^{-1}\, p_t\}_t = p_{\alpha\alpha}$$

If we again assume that $x_\alpha \sim 1$, this, too, reduces to the wave equation

$$c_0^{-2} p_{tt} = p_{\alpha\alpha} \tag{2-71}$$

2-10. INTERFACE CONDITIONS FOR ONE-DIMENSIONAL GAS FLOW

Next suppose we have an infinite tube filled with two different gases, one of density ρ_1 filling the part of the tube to the left of the origin and the other of density ρ_2 filling the part to the right of the origin. Then there will be a propagation speed $c_1 = \sqrt{f'(\rho_1)}$ for $x < 0$ and a propagation speed $c_2 = \sqrt{f'(\rho_2)}$ for $x > 0$. The obvious conditions at the origin are

$$x^{(1)} = x^{(2)} \qquad \text{and} \qquad p^{(1)} = p^{(2)} \tag{2-72}$$

The only trouble with them is the fact that one of them applies to x and the other to p. We need two conditions on the same function, not two conditions on two different functions. However, it is easy to get a second condition on p from the above two; (2-66) tells us that $x_{tt} = -(1/\rho)\, p_\alpha$. Since $x^{(1)}(0, t) = x^{(2)}(0, t)$ implies that $x^{(1)}_{tt}(0, t) = x^{(2)}_{tt}(0, t)$, we find immediately

$$\frac{1}{\rho_1}\, p^{(1)}_\alpha = \frac{1}{\rho_2}\, p^{(2)}_\alpha \tag{2-72a}$$

Condition (2-72*a*) differs from the second condition for the vibrating string in the presence of the factor $1/\rho$. We can remove this difference by introducing the new variable $\mu = \rho\alpha$, representing the mass of gas between the origin and the point α in the equilibrium state. In terms of μ the differential equation (2-71) becomes

$$p_{tt} = (\rho c)^2\, p_{\mu\mu} \tag{2-73}$$

The interface conditions become

$$p^{(1)} = p^{(2)} \qquad \text{and} \qquad p^{(1)}_\mu = p^{(2)}_\mu$$

Hence (see page 66) the transmission and reflection coefficients are

$$T = \frac{2\rho_2 c_2}{\rho_1 c_1 + \rho_2 c_2} \qquad R = \frac{\rho_2 c_2 - \rho_1 c_1}{\rho_1 c_1 + \rho_2 c_2}$$

Thus, it is not necessary to have $c_1 = c_2$ in order to obtain total transmission. It is sufficient that the *impedances* ρc of both media be the same.

2-11. EULERIAN DESCRIPTION OF ONE-DIMENSIONAL GAS FLOW

In practice, it is usually more advantageous to use the so-called Eulerian description of fluid flow rather than the Lagrangian one. In the Eulerian representation x and t are used as independent variables and u, the velocity, p, the pressure, and ρ, the density, are the dependent variables. Conceptually the difference between the two representations is that in the Lagrangian representation the observer follows a particle and watches its motion. In the Eulerian the observer stations himself at a particular point in the gas and watches how the motion changes at that point.

We shall show that the equations of the Lagrangian representation, namely,

$$\rho x_\alpha = \rho_0 \tag{2-74}$$

$$\rho_0 x_{tt} = -p_\alpha \tag{2-75}$$

are equivalent to the following equations of the Eulerian representation:

$$\rho_t + (\rho u)_x = 0$$

$$u_t + uu_x + \frac{1}{\rho} p_x = 0$$

To see this, first observe that $x = x(\alpha, t)$ and, by definition, $u = x_t$. Thus,

$$x_t = u(x, t) = u[x(\alpha, t), t]$$

Hence

$$x_{tt} = u_x x_t + u_t = uu_x + u_t$$

Thus,

$$x_{tt} = u_t + uu_x$$

Also

$$p = p(x, t) = p[x(\alpha, t), t]$$

Therefore,

$$p_\alpha = p_x x_\alpha = \frac{\rho_0}{\rho} p_x \qquad \text{by (2-74)}$$

Hence (2-75), which says

$$\rho_0 x_{tt} = -p_\alpha$$

becomes

$$\rho_0(u_t + uu_x) = -\frac{\rho_0}{\rho} p_x$$

or

$$u_t + uu_x + \frac{1}{\rho} p_x = 0$$

In addition, (2-74), which says

$$\rho x_\alpha = \rho_0$$

implies

$$(\rho x_\alpha)_t = 0$$

that is,

$$(\rho_t + u\rho_x)\, x_\alpha + \rho x_{\alpha t} = 0$$

But

$$x_{\alpha t} = x_{t\alpha} = u_\alpha = u_x x_\alpha$$

Hence

$$\rho_t + u\rho_x + \rho u_x = 0$$

or

$$\rho_t + (\rho u)_x = 0$$

If the entropy S is taken into account, then a third equation, $S_t + uS_x = 0$, must be added.

To summarize, we can say that the Eulerian equations of one-dimensional compressible isentropic flow are

$$\rho_t + (\rho u)_x = 0 \tag{2-76}$$

$$u_t + uu_x + \frac{1}{\rho} p_x = 0 \tag{2-77}$$

$$p = f(\rho) \qquad f \text{ a given function with } f' > 0 \tag{2-78}$$

Since $f(\rho)$ is known, we can use (2-78) to eliminate p from (2-77). From (2-78) we have

$$p_x = f'(\rho)\,\rho_x \tag{2-79}$$

Since $f'(\rho) > 0$, we can put

$$c^2(\rho) = f'(\rho) \tag{2-80}$$

and write (2-79) in the form

$$p_x = c^2(\rho)\,\rho_x \tag{2-81}$$

Using (2-81) in (2-77), we obtain

$$u_t + uu_x + \frac{c^2(\rho)}{\rho}\,\rho_x = 0 \tag{2-82}$$

Equations (2-76) and (2-82) constitute two PDEs for the two unknown functions $u(x, t)$ and $\rho(x, t)$. $c^2(\rho)$ is a known function of ρ defined by (2-80). Equations (2-76) and (2-82) are nonlinear, and their general treatment will be postponed to Chap. 13. In the present chapter we restrict ourselves to indicating a method for obtaining a linear approximation to (2-76) and (2-82). To do this, we imagine that we have a known flow, i.e., a known pair of functions, $\hat{u}(x, t)$ and $\hat{\rho}(x, t)$, satisfying (2-76) and (2-82) and certain BCs and ICs. Then we imagine that the boundary and/or initial conditions are changed slightly. "Changed slightly" means that there is some parameter ϵ in the boundary and/or initial conditions, that the problem with the known solution corresponds to $\epsilon = 0$, and that the problem we are interested in has ϵ very small. In practice, ϵ can be the thickness of a thin airfoil in a flow, the height of a little bump on the boundary of some region, or any one of numerous appropriate small parameters in the problem.

Since ϵ is small, it is natural to attempt to expand the solution to the perturbed problem in powers of ϵ. It is natural to do this, but it is by no means clear that it is legitimate, because there is no guarantee that such a series will converge. It is not even guaranteed that the series is asymptotic. In fact, at present, there is no general theory to decide this question for PDEs. Hence in practice it is necessary to investigate whether the perturbed solution found by the method given below is really a good approximation to the true solution. We shall not enter into such an investigation in the present case but rather confine ourselves to obtaining a formal first-order approximation.

We do this by writing the solution to the perturbed problem in the form

$$\begin{aligned} u(x, t) &= \hat{u}(x, t) + \epsilon\tilde{u}(x, t) + \epsilon^2\tilde{\tilde{u}}(x, t) + \cdots \\ \rho(x, t) &= \hat{\rho}(x, t) + \epsilon\tilde{\rho}(x, t) + \cdots \end{aligned} \tag{2-83}$$

Then we insert (2-83) into (2-76) and (2-82), use the fact that

$$(\hat{\rho} + \epsilon\tilde{\rho} + \cdots)^{-1} = \hat{\rho}^{-1}\left(1 + \epsilon\,\frac{\tilde{\rho}}{\hat{\rho}} + \cdots\right)^{-1} = \hat{\rho}^{-1} - \epsilon\,\frac{\tilde{\rho}}{\hat{\rho}^2} + \cdots$$

and

$$c^2(\hat{\rho} + \epsilon\tilde{\rho} + \cdots) = c^2(\hat{\rho}) + \epsilon\,\frac{dc^2(\hat{\rho})}{d\rho}\,\tilde{\rho} + \cdots \qquad \text{by Taylor's theorem}$$

and equate coefficients of ϵ.

We get

$$\tilde{\rho}_t + (\hat{\rho}\tilde{u})_x + (\tilde{\rho}\hat{u})_x = 0 \tag{2-84}$$

$$\tilde{u}_t + \hat{u}\tilde{u}_x + \hat{u}_x\tilde{u} + \frac{c^2(\hat{\rho})}{\hat{\rho}}\,\tilde{\rho}_x + \frac{1}{\hat{\rho}}\,\frac{dc^2(\hat{\rho})}{d\rho}\,\hat{\rho}_x - \frac{\tilde{\rho}}{\hat{\rho}^2}\,c^2(\hat{\rho})\,\hat{\rho}_x = 0 \tag{2-85}$$

Notice that these are *linear* equations for the perturbation functions $\tilde{u}$ and $\tilde{\rho}$. The variable coefficients depend upon the solution $\hat{u}$, $\hat{\rho}$ around which the perturbation is made. If we assume that the unperturbed flow is constant, i.e., that

$$\hat{u}(x, t) = u_0 = \text{const} \qquad \text{and} \qquad \hat{\rho}(x, t) = \rho_0 = \text{const} \tag{2-86}$$

then (2-84) and (2-85) simplify further to

$$\tilde{\rho}_t + \rho_0\tilde{u}_x + u_0\tilde{\rho}_x = 0 \tag{2-87}$$

and

$$\tilde{u}_t + u_0\tilde{u}_x + \frac{{c_0}^2}{\rho_0}\,\tilde{\rho}_x = 0 \tag{2-88}$$

By writing (2-87) and (2-88) in operational form,

$$\left(\frac{\partial}{\partial t} + u_0\,\frac{\partial}{\partial x}\right)\tilde{\rho} + \rho_0\,\frac{\partial}{\partial x}\,\tilde{u} = 0 \tag{2-89}$$

$$\frac{{c_0}^2}{\rho_0}\,\frac{\partial}{\partial x}\,\tilde{\rho} + \left(\frac{\partial}{\partial t} + u_0\,\frac{\partial}{\partial x}\right)\tilde{u} = 0 \tag{2-90}$$

we see that both $\tilde{u}$ and $\tilde{\rho}$ are solutions of the PDE

$$\left[\left(\frac{\partial}{\partial t} + u_0\,\frac{\partial}{\partial x}\right)^2 - {c_0}^2\,\frac{\partial^2}{\partial x^2}\right]u = 0 \tag{2-91}$$

If we make the change of variables

$$\xi = x - u_0 t \qquad \tau = t \tag{2-92}$$

then (2-91) becomes the wave equation

$$u_{\tau\tau} - {c_0}^2 u_{\xi\xi} = 0 \tag{2-93}$$

Notice that the change of variables (2-92) corresponds to a change to a coordinate system moving with velocity u_0, that is, with the velocity of the unperturbed flow. Notice, too, that if the unperturbed flow is the state of rest, then (2-91) is the wave equation in its usual form with no change of variables necessary.

2-12. FINITE STRING

The problem of the finite string turns out to have a more complicated solution than that of the infinite string. This is due to the repeated reflection of the waves from the boundaries.

For simplicity we first assume that the initial velocity $g(x) = 0$ and that both end points are held fixed. Specifically, we consider the problem:

DE:

$$u_{xx} - c^{-2}u_{tt} = 0 \qquad \begin{matrix} 0 < x < l \\ 0 < t < \infty \end{matrix}$$

ICs:

$$\begin{matrix} u(x, 0) = 2f(x) \\ u_t(x, 0) = 0 \end{matrix} \qquad 0 \leqslant x \leqslant l$$

BCs:

$$u(0, t) = u(l, t) = 0 \qquad 0 \leqslant t < \infty \tag{2-94}$$

To ensure compatibility of the BCs and ICs we assume that $f(0) = f(l) = 0$. As usual we write

$$u = \phi(x - ct) + \psi(x + ct) \tag{2-95}$$

and find that

$$\begin{matrix} \phi(x) = f(x) \\ \psi(x) = f(x) \end{matrix} \qquad 0 \leqslant x \leqslant l \tag{2-96}$$

However, as indicated, f is defined only for $0 \leqslant x \leqslant l$. Since t ranges from 0 to ∞, and since u is given by (2-95), we need $\phi(x)$ for $-\infty < x < \infty$ and $\psi(x)$ for $0 \leqslant x < \infty$. Equation (2-96) gives ϕ and ψ only for $0 \leqslant x \leqslant l$. In order to extend their definition outside this interval we use the BCs. From (2-94) we have

$$\phi(-ct) + \psi(ct) = 0 \tag{2-97}$$

and

$$\phi(l - ct) + \psi(l + ct) = 0 \tag{2-98}$$

If we let $-ct = x$, then (2-97) tells us that

$$\phi(x) = -\psi(-x) = -f(x) \qquad \text{by (2-96)} \tag{2-99}$$

Since $f(x)$ is defined for $0 \leqslant x \leqslant l$, (2-99) extends the definition of $\phi(x)$ to the range $-l \leqslant x \leqslant 0$ as

$$\phi(x) = -f(-x) \qquad -l \leqslant x \leqslant 0 \tag{2-100}$$

Next put $x = l + ct$ in (2-98) to obtain

$$\psi(x) = -\phi(2l - x) \tag{2-101}$$

Since $\phi(x)$ has thus far been determined in the range $-l \leqslant x \leqslant l$, (2-101) determines $\psi(x)$ in the range $-l \leqslant 2l - x \leqslant l$, that is,

$$\psi(x) = -\phi(2l - x) \qquad l \leqslant x \leqslant 3l \tag{2-102}$$

Now knowing $\psi(x)$ for $l \leqslant x \leqslant 3l$, we can return to (2-99) to find

$$\phi(x) = -\psi(x) \qquad -3l \leqslant x \leqslant -l \tag{2-103}$$

Then we feed this information back into (2-101) to get (note that $-3l \leqslant 2l - x \leqslant -l$ implies $3l \leqslant x \leqslant 5l$)

$$\psi(x) = -\phi(2l - x) \qquad 3l \leqslant x \leqslant 5l \tag{2-104}$$

It is obvious how to continue the process. Let us obtain the first few terms explicitly in terms of the known function f. From (2-96), (2-100), and (2-102) we have

$$\psi(x) = \begin{cases} -f(2l - x) & l \leqslant x \leqslant 2l \\ f(x - 2l) & 2l \leqslant x \leqslant 3l \end{cases}$$

Therefore, by (2-103),

$$\phi(x) = \begin{cases} -f(2l - x) & -3l \leqslant x \leqslant -2l \\ f(x - 2l) & -2l \leqslant x \leqslant -l \end{cases}$$

and so forth.

2-13. SEPARATION OF VARIABLES

As can be seen from the above, the solution of the BVP for the finite string is complicated if we use the general representation, (2-95), of the solution. If the BCs are more involved than the ones assumed above, then the above method is even more difficult to apply. We therefore abandon it in favor of another method, one which has wide applicability.

Instead of finding the general solution to the PDE we try to find special solutions and then put them together somehow in order to satisfy the ICs and BCs. In particular, we look for solutions of

$$u_{xx} - c^{-2}u_{tt} = 0 \tag{2-105}$$

which are of the special form

$$u(x, t) = \alpha(x)\,\beta(t) \tag{2-106}$$

i.e., a product of a function of x alone and a function of t alone. We hope to be able to find so many special solutions of the form (2-106) that we can put them together to satisfy all the ICs and BCs.

Substituting (2-106) into (2-105), we get

$$\alpha''\beta - c^{-2}\alpha\beta'' = 0$$

or

$$\frac{\alpha''}{\alpha} = \frac{\beta''}{c^2\beta} \tag{2-107}$$

The left side of (2-107) does not vary with t. Therefore, the right side, which is equal to it, cannot vary with t. Similarly, the right side does not vary with x. Hence neither can the left. The result is that both sides must be constant. Calling this constant λ, we have

$$\frac{\alpha''}{\alpha} = \frac{\beta''}{\beta} = \lambda$$

whence

$$\alpha'' - \lambda\alpha = 0 \tag{2-108}$$

and

$$\beta'' - \lambda c^2\beta = 0 \tag{2-109}$$

The IC $u(x, 0) = f(x)$ applied to (2-106) requires $\alpha(x)\,\beta(0) = f(x)$. Since α is a solution of (2-108), hence a sine, cosine, or exponential, it is impossible to satisfy this condition unless the given initial displacement $f(x)$ is a sine, cosine, or exponential. In general it will not be such a function. Therefore we cannot in general determine α so as to satisfy it. At this point our method therefore seems doomed to failure. Nevertheless it will turn out to be possible to circumvent this difficulty. For the time being, we merely avoid trying to satisfy this troublesome IC and try instead to satisfy the other IC and the two BCs. Because all three of these conditions are homogeneous, it will turn out to be possible to satisfy them. Having satisfied them we will then turn to the task of satisfying the nonhomogeneous condition $u(x, 0) = f(x)$.

The reader might well ask at this point: What if more than one of the BCs and ICs are nonhomogeneous? The answer is simple. If several of the BCs and ICs are nonhomogeneous, then replace all but one of them by homogeneous conditions; do this in all possible ways; solve the problems thus obtained; and then use the superposition principle to obtain the solution to the original problem.

Returning to our problem, we note that the IC $u_t(x, 0) = 0$ implies that $\alpha(x)\,\beta'(0) = 0$, $0 \leqslant x \leqslant l$, hence that

$$\beta'(0) = 0 \tag{2-110}$$

Similarly, $u(0, t) = u(l, t) = 0$ implies

$$\alpha(0) = \alpha(l) = 0 \tag{2-111}$$

If λ is positive, say $\lambda = \mu^2 \neq 0$, then the solution to (2-108) is

$$\alpha(x) = Ae^{\mu x} + Be^{-\mu x}$$

Equations (2-111) imply

$$A + B = 0$$

and

$$Ae^{\mu l} + Be^{-\mu l} = 0$$

The first of these tells us that $B = -A$, and the second then says $A(e^{\mu l} - e^{-\mu l}) = 0$, that is, $2A \sinh \mu l = 0$. Since the only real root of the sinh is 0, it follows that $A = 0$ and we get the useless solution $u = 0$.

If $\lambda = 0$, then $\alpha = Ax + B$ and conditions (2-111) imply $B = 0$ and $Al + B = 0$; that is, $A = B = 0$ again.

Hence we assume that $\lambda = -\mu^2 \neq 0$. Then the general solution to (2-108) is

$$\alpha(x) = A \sin \mu x + B \cos \mu x$$

$\alpha(0) = 0$ implies $B = 0$. Then $\alpha(l) = 0$ implies

$$A \sin \mu l = 0$$

Since we wish $A \neq 0$, we need to have

$$\sin \mu l = 0 \tag{2-112}$$

and therefore

$$\mu l = n\pi, \qquad n = \pm 1, \pm 2, \ldots$$

Thus μ must be restricted to the values $n\pi/l$ if we wish to satisfy the BCs.

With $\lambda = -\mu^2 = -n^2\pi^2/l^2$, (2-109) becomes

$$\beta'' + \frac{n^2\pi^2c^2}{l^2}\,\beta = 0$$

the general solution to which is

$$\beta(t) = C \sin \frac{n\pi ct}{l} + D \cos \frac{n\pi ct}{l}$$

Since $u_t(x, 0) = 0$, we must have $C = 0$. If we let $AD = a$ and combine the expressions we have found for α and β into an expression for u, then we have

$$u(x, t) = a \sin \frac{n\pi x}{l} \cos \frac{n\pi ct}{l} \tag{2-113}$$

If we take different values of n, we get different solutions, each of which can be multiplied by its own arbitrary constant a_n. In this way we get an infinite number of solutions

$$u_n(x, t) = a_n \sin \frac{n\pi x}{l} \cos \frac{n\pi ct}{l} \qquad n = 1, 2, \ldots \tag{2-114}$$

each of which satisfies both BCs and the homogeneous IC.

The solutions (2-114) are called *standing waves*, as is any solution of the form (2-106). The reason for this is the fact that such solutions can be looked upon as having the fixed shape $\alpha(x)$ with varying amplitude $\beta(t)$. The zeros of a standing wave, i.e., of $\alpha(x)$, are called *nodes*. Note that the number of nodes of the solutions (2-114) depends upon n. The wave for $n = 1$ has no nodes, that for $n = 2$ has one node, etc. The solutions (2-114) are also called *harmonics* or *modes*, and so we can say that the lowest harmonic has no nodes, the next, one node, etc.

The representation (2-95) of the solution to the wave equation is a representation in terms of progressing waves. In the present problem we have found standing instead of progressing waves. The two types are connected, however. From elementary trigonometry we know that

$$a_n \sin \frac{n\pi x}{l} \cos \frac{n\pi ct}{l} = \frac{a_n}{2}\left[\sin \frac{n\pi}{l}(x - ct) + \sin \frac{n\pi}{l}(x + ct)\right]$$

which expresses a standing wave as a sum of two progressing waves of equal amplitudes but going in opposite directions.

The u_n's do not, in general, satisfy the nonhomogeneous IC. Suppose, however, we add several of them together. Since the problem is linear, the resulting sum will also be a solution that satisfies the homogeneous conditions. If it contains only a finite number of terms, it still will not satisfy the nonhomogeneous condition. Suppose, though, we take an infinite sum,

$$u(x, t) = \sum_{n=1}^{\infty} a_n \sin \frac{n\pi x}{l} \cos \frac{n\pi ct}{l} \tag{2-115}$$

Then to satisfy the condition $u(x, 0) = f(x)$ we would have to determine the a_n's so that

$$f(x) = \sum_{n=1}^{\infty} a_n \sin \frac{n\pi x}{l} \tag{2-116}$$

This can be done, because it merely requires the a_n's to be the Fourier coefficients of $f(x)$, extended as an odd function into the interval $-l \leqslant x \leqslant 0$. The a_n's are given by

$$a_n = \frac{2}{l}\int_0^l f(x) \sin \frac{n\pi x}{l}\, dx \tag{2-117}$$

2-14. MORE GENERAL BOUNDARY CONDITIONS

As a generalization we replace the BC at $x = l$ by the more general third BC

$$u_x(l, t) - ku(l, t) = 0 \tag{2-118}$$

retaining the other BC and both of the ICs.

It is again easy to show that λ must be negative, $\lambda = -\mu^2$, and that $\alpha(x) = A \sin \mu x$ satisfies the BC at the origin for any μ. To satisfy (2-118) we need

$$\alpha'(l) - k\alpha(l) = 0$$

that is,

$$\mu \cos \mu l - k \sin \mu l = 0$$

or

$$\tan \mu l = \frac{\mu}{k} \tag{2-119}$$

This is a transcendental equation for μ. A look at the graphs of $y = \tan \mu l$ and the straight line $y = \mu/k$ shows that they have infinitely many intersections. Hence there is an infinite sequence of roots, $\mu_1, \mu_2, \ldots,$ with $\mu_n \to \infty$. For each root μ_n we get a solution

$$u_n(x, t) = a_n \sin \mu_n x \cos \mu_n ct \tag{2-120}$$

We then need to be able to expand $f(x)$ into a series of the form

$$f(x) = \sum_{n=1}^{\infty} a_n \sin \mu_n x \tag{2-121}$$

To do this we need to know that the functions $\sin \mu_n x$ are orthogonal over the interval $-l \leqslant x \leqslant l$, that is, that

$$\int_{-l}^{l} \sin \mu_n x \sin \mu_m x\, dx = 0 \qquad \text{if } n \neq m$$

We also need to know the normalizing factor

$$N_n = \int_{-l}^{l} \sin^2 \mu_n x\, dx$$

We leave these details to the reader as an exercise.

2-15. TRANSMISSION-LINE EQUATIONS; THE TELEGRAPH EQUATION

Suppose a pair of transmission lines has a voltage $u(x, t)$ across them and a current $\pm j(x, t)$ respectively. Let C, L, R, and G denote respectively the capacitance, self-inductance, resistance, and leakage conductance per unit length. Here $R/2$ is the resistance per unit length of each line. Then Ohm's and Kirchhoff's laws lead to the PDE

$$(Lj)_t + Rj + u_x = 0 \tag{2-122}$$

while the law of conservation of charge leads to the equation

$$(Cu)_t + Gu + j_x = 0 \tag{2-123}$$

Equations (2-122) and (2-123) are two differential equations for the two unknowns u and j. If we eliminate either u or j, we can obtain a single PDE for the other one. If L and C are constant, it turns out to be the same PDE no matter which one we eliminate, namely,

$$LCu_{tt} + (RC + LG)\, u_t + RGu - u_{xx} = 0 \tag{2-124}$$

To simplify the notation put

$$c = \frac{1}{\sqrt{LC}} \qquad a = RC + LG \qquad b = RG \tag{2-125}$$

Then (2-124) reads

$$u_{xx} - c^{-2}u_{tt} - au_t - bu = 0 \tag{2-126}$$

Equation (2-126) is called the *telegraph equation.* If $a = b = 0$, it reduces to the wave equation. If either a or b is different from zero, then (2-126) represents a generalization of the wave equation. The term au_t, as we shall see below, is a dissipative term. It corresponds to a damping force. The term bu, as we shall also see below, produces dispersion in the waves.

The term au_t is the easiest to deal with. The method for dealing with it will at the same time reveal the effect it has on the solution. Strangely enough, the method consists of removing au_t from the PDE. More precisely, we shall show that we can always change variables in (2-126) in such a way that the new equation has no first-derivative terms. To do this we put

$$u(x, t) = \omega(t)\, v(x, t) \tag{2-127}$$

where $v(x, t)$ is the new unknown function, and we try to determine $\omega(t)$ so that the new equation for v has no first derivatives of v in it.

From (2-127) we have

$$\begin{aligned} u_t &= \omega v_t + \omega' v \\ u_{tt} &= \omega v_{tt} + 2\omega' v_t + \omega'' v \\ u_{xx} &= \omega v_{xx} \end{aligned}$$

Hence (2-126) becomes

$$\omega v_{xx} - c^{-2}\omega v_{tt} - (2c^{-2}\omega' + a\omega)\, v_t - (c^{-2}\omega'' + a\omega' + b\omega)\, v = 0 \qquad (2\text{-}128)$$

Thus, if we choose ω so that $2c^{-2}\omega' + a\omega = 0$, that is, $\omega = e^{-ac^2t/2}$, then (2-128) contains no first derivatives of v and reduces to

$$v_{xx} - c^{-2}v_{tt} + kv = 0 \qquad (2\text{-}129)$$

where

$$k = b - \frac{a^2c^2}{4} \qquad (2\text{-}130)$$

Summarizing, we can say that the solution $u(x, t)$ of the telegraph equation (2-126) can be expressed in terms of the solution $v(x, t)$ of the special telegraph equation (2-129) by means of the formula

$$u(x, t) = e^{-ac^2t/2}v(x, t) \qquad (2\text{-}131)$$

An interesting case arises when $k = 0$. In this case (2-129) is the wave equation and (2-131) then implies that the general solution to (2-126) *in this case* is

$$u(x, t) = e^{-ac^2t/2}[\phi(x - ct) + \psi(x + ct)] \qquad (2\text{-}132)$$

The interesting thing about (2-132) is that it shows that the waves retain their shape except for an amplitude-attenuation factor that depends on time. The waves are therefore called *relatively undistorted.*

Returning briefly to our transmission-line problem, we note that in order to have discernible signals, it is essential that the waves propagated over cables be relatively undistorted, i.e., that $k = 0$. Using (2-125) and (2-130), we see that this means that the circuit parameters have to be adjusted so that $RC = LG$. Usually this is done by distributing inductances in each cable at appropriate intervals, which, in practice, can be as long as $\frac{1}{4}$ or $\frac{1}{2}$ mile. The idea is due to Pupin.

If $k \neq 0$, then (2-129) is not the wave equation and its general solution is more complicated. Later (Chap. 9) we shall give what amounts to a general solution of (2-129), but it will be given in terms of arbitrary initial values on an arbitrary noncharacteristic curve, somewhat the way the general solution to the first-order equation considered in Chap. 1 was given by the solution to a general Cauchy problem. For the present we confine ourselves to investigating the existence of some simple types of solutions to (2-129).

First we look for progressing waves of the same form as those which satisfy the ordinary wave equation; i.e., we look for solutions of (2-129) of the form

$$v(x, t) = \phi(x - ct) \qquad (2\text{-}133)$$

Inserting (2-133) into (2-129), we find that

$$\phi'' - \phi'' + k\phi = 0$$

and thus, if $k \neq 0$, then $\phi = 0$.

This shows that if $k \neq 0$, then (2-129) has no solutions of the form (2-133) at all. The same is true for solutions of the form $\psi(x + ct)$. However, we notice something about the above computation. Suppose, instead of solutions of the form $\phi(x - ct)$, we try solutions of the form

$$v = \phi(x - \gamma t) \tag{2-134}$$

where γ is some constant different from c. Then we find

$$\left[1 - \left(\frac{\gamma}{c}\right)^2\right] \phi'' + k\phi = 0$$

or

$$\phi'' + \frac{kc^2}{c^2 - \gamma^2}\,\phi = 0 \tag{2-135}$$

and this equation does have solutions. Of course, the solutions are only sines, cosines, or exponentials or some combination of these, but at least we do find some solutions of the form (2-134) as long as $\gamma \neq c$.

Thus there are progressing-wave solutions of (2-129), but none of them propagate with speed c. In fact, they propagate with every speed but c. Since some of them may have exponential factors, some of them may be attenuated.

The fact that the waves do not all propagate with the same velocity is referred to as *dispersion*. For this reason (2-129) is a special case of what is known as a *dispersive hyperbolic equation*. Notice that the dispersion is due to the presence of the term ku in the equation.

Although the only solutions we have thus far found to (2-129) are essentially of the form $v(x, t) = e^{\alpha i(x - \gamma t)}$, they are far from useless because they depend upon a parameter γ. If we multiply them by an arbitrary function $A(\gamma)$ and integrate over γ, then we obtain a large class of solutions. Since this is essentially equivalent to using the Fourier transform, and since transform methods are taken up much later in the book, we shall not pursue these ideas further here.

The methods of Chap. 9 will show that the results on domains of dependence and ranges of influence for the wave equation *do* carry over to the telegraph equation in spite of the dispersion term.

2-16. NONLINEAR WAVE EQUATIONS

The argument given in the preceding section to show that progressing-wave solutions of the telegraph equation cannot possess arbitrary forms can be carried over to more general wave equations. If the nonlinear equation

$$u_{xx} - c^{-2}u_{tt} = f(u, u_x, u_t) \tag{2-136}$$

is to have solutions of the form

$$u = \phi(x - \gamma t) \tag{2-137}$$

then ϕ must be a solution of the ODE

$$(1 - \gamma^2 c^{-2})\, \phi'' = f(\phi, \phi' - \gamma\phi') \tag{2-138}$$

If $\gamma \neq c$, then (2-138) is a second-order ODE for ϕ as a function of its argument, say ξ. Since the independent variable is missing, (2-138) can be reduced to a first-order ODE. Notice that if $\gamma = c$, the order of (2-138) is one already. Readers interested in the further analysis of the problem should consult Fleishman [106, 107].

SUMMARY

DEFINITION S2-1

The lines $x \pm ct$ = const are called the *characteristics* of the wave equation $u_{xx} - c^{-2}u_{tt} = 0$. The variables $\xi = x - ct$, $\eta = x + ct$ are called *characteristic variables* for that equation.

Theorem S2-1 If $\phi(z)$ and $\psi(z)$ belong to $C^{(2)}$, then

$$u(x, t) = \phi(x - ct) + \psi(x + ct)$$

is a solution to the wave equation. Conversely, if $u(x, t)$ is a twice continuously differentiable solution of the wave equation, then there exist two functions, $\phi(z)$ and $\psi(z)$, belonging to $C^{(2)}$ such that $u(x, t) = \phi(x - ct) + \psi(x + ct)$.

DEFINITION S2-2

If $\phi(z)$ and $\psi(z)$ belong to $C^{(0)}$, then $u(x, t) = \phi(x - ct) + \psi(x + ct)$ is called a *weak solution* of the wave equation.

Theorem S2-2 Let $f(x)$ belong to $C^{(2)}(-\infty, \infty)$ and $g(x)$ belong to $C^{(1)}(-\infty, \infty)$. Then

$$u(x, t) = \frac{1}{2}[f(x - ct) + f(x + ct)] + \frac{1}{2c}\int_{x-ct}^{x+ct} g(z)\, dz \tag{S2-1}$$

is the unique solution to the wave equation satisfying $u(x, 0) = f(x)$ and $u_t(x, 0) = g(x)$, $-\infty < x < \infty$.

Theorem S2-3 Let $f(x)$ and $h(x)$ belong to $C^{(2)}[0, \infty)$. Let $g(x)$ belong to $C^{(1)}[0, \infty)$, and let $f(0) = h(0)$ and $g(0) = h'(0)$. Then the unique solution of the wave equation satisfying $u(x, 0) = f(x)$, $0 \leqslant x < \infty$,

$u_t(x, 0) = g(x)$, $0 \leqslant x < \infty$, and $u(0, t) = h(t)$, $0 \leqslant t < \infty$, is given by (S2-1) for $x \geqslant ct$ and by

$$u(x, t) = h\left(t - \frac{x}{c}\right) + \frac{1}{2}[f(ct - x) - f(ct + x)] + \frac{1}{2c}\int_{ct-x}^{ct+x} g(z)\,dz$$

for $x \leqslant ct$.

Theorem S2-4 Let $F(x, t)$ be continuous in (x, t) for $-\infty < x < \infty$, $0 \leqslant t < \infty$. Then the (unique) solution to $u_{tt} - c^2 u_{xx} = F(x, t)$ in the indicated upper half of the xt plane satisfying $u(x, 0) = u_t(x, 0) = 0$ is

$$u(x, t) = \frac{1}{2c}\int_0^t d\tau \int_{x-ct+c\tau}^{x+ct-c\tau} F(\xi, \tau)\,d\xi \tag{S2-2}$$

Theorem S2-5 If $f(x)$ belongs to $C^{(2)}[0, 1]$ and $f(0) = f(l) = 0$, then the unique solution to $u_{xx} - c^{-2}u_{tt} = 0$, $0 < t < \infty$, satisfying

$$\left.\begin{aligned} u(x, 0) &= f(x) \\ u_t(x, 0) &= 0 \end{aligned}\right\} \quad 0 \leqslant x \leqslant l$$

$$u(0, t) = u(l, t) = 0 \qquad 0 \leqslant t < \infty$$

is

$$u(x, t) = \sum_{n=1}^{\infty} a_n \sin\frac{n\pi x}{l}\cos\frac{n\pi ct}{l}$$

where

$$a_n = \frac{2}{l}\int_0^l f(x)\sin\frac{n\pi x}{l}\,dx$$

Theorem S2-6 The general solution of the PDE

$$u_{tt} + pu_t + qu - c^2 u_{xx} = 0$$

p, q, c constants, is given by $u(x, t) = e^{-pt/2}v(x, t)$, where $v(x, t)$ is the general solution of the telegraph equation $v_{tt} - c^2 v_{xx} - hv = 0$, with $h = p^2/4 - q$.

Theorem S2-7 The telegraph equation $u_{tt} - c^2 v_{xx} - hu = 0$ or, more generally, the equation $u_{tt} - c^2 u_{xx} - f(u, u_x, u_t) = 0$, possesses solutions of the form $u(x, t) = \phi(x - \gamma t)$, $\gamma = \text{const}$, if and only if $\phi(z)$ satisfies the ODE $(\gamma^2 - c^2)\,\phi'' - f(\phi, \phi', -\gamma\phi') = 0$. Here ϕ' means $d\phi/dz$.

PROBLEMS

1. Verify that (2-8) satisfies the wave equation.

2. Find $u(x, t)$ if

$$u_{xx} - u_{tt} = 0$$

and

$$u(x, 0) = \begin{cases} 1 & \text{if } 0 \leqslant x \leqslant 1 \\ & \text{or if } 2 \leqslant x \leqslant 3 \\ 0 & \text{otherwise} \end{cases}$$

$$u_t(x, 0) = 0 \qquad -\infty < x < \infty$$

(You may give the answer by means of a diagram in the xt plane.)

3. Find $u(x,t)$ if

$$u_{xx} = u_{tt} \qquad \begin{matrix} 0 \leqslant t < \infty \\ -t \leqslant x \leqslant t \end{matrix}$$

and

$$u(x, -x) = g(x) \qquad -\infty < x \leqslant 0$$
$$u(x, x) = f(x) \qquad 0 \leqslant x < \infty$$

Assume $f(0) = g(0)$, where f and g are given functions.

4. Given that $u(x, t)$ satisfies the differential equation $u_{xx} - 4u_{tt} = 0$ for all positive t and for all x, with ICs $u(x, 0) = x$ and $u_t(x, 0) = 3x^2$. Find $u(3, 2)$.

5. Find and discuss the solution to:

DE:

$$u_{xx} - c^{-2}u_{tt} = 0 \qquad \begin{matrix} 0 < t < \infty \\ 0 < x < \infty \end{matrix}$$

ICs:

$$u(x, 0) = f(x) \qquad x > 0$$
$$u_t(x, 0) = g(x) \qquad x > 0$$

BC:

$$u_x(0, t) = 0 \qquad t > 0$$

6. Find and discuss (include an xt diagram) the solution of

DE:

$$u_{xx} - c^{-2}u_{tt} = 0 \qquad \begin{matrix} 0 < t < \infty \\ 0 < x < \infty \end{matrix}$$

ICs:

$$u(x, 0) = 0 \qquad x > 0$$

$$u_t(x, 0) = \begin{cases} 0 & x \leqslant a \\ 1 & x \geqslant a \end{cases}$$

BCs:

$$u_x(0, t) - ku(0, t) = 0 \qquad t > 0$$

Also find the behavior of the solution as $t \to \infty$.

7. The equation $\mu = k \tan \mu l$ has infinitely many roots μ_1, μ_2, etc. Let μ_n and μ_m be two of them. Evaluate

$$\int_0^l \sin \mu_n x \sin \mu_m x \, dx$$

8. Find the general solution of $u_{xx} - c^{-2}u_{tt} = x$.

9. Find the general solution of $u_{xx} - c^{-2}u_{tt} = xt + x$.

10. Find $u(x, t)$ if

$$u_{xx} - u_{tt} = x \qquad \begin{matrix} -\infty < x < \infty \\ 0 < t < \infty \end{matrix}$$

and

$$u(x, 0) = \begin{cases} 1 & \text{if } 0 \leqslant x \leqslant 1 \\ 0 & \text{otherwise} \end{cases}$$

and

$$u_t(x, 0) = 0 \qquad -\infty < x < \infty$$

11. Solve:

DE:

$$c^2u_{xx} - u_{tt} = 16x^2 \qquad \begin{matrix} 0 < x < L \\ 0 < t < \infty \end{matrix}$$

ICs:

$$u(x, 0) = x(L - x)$$
$$u_t(x, 0) = 0$$

BCs:

$$u(0, t) = u(L, t) = 0$$

Hint: Note that

$$x^2 = \frac{[(x + ct) + (x - ct)]^2}{4}$$

and transform to characteristic coordinates.

12. Solve $u_{xx} - c^{-2}u_{tt} = xt$, $-\infty < x < \infty, 0 < t < \infty$, with the ICs $u(x, 0) = u_t(x, 0) = 0$.

13. Find the solution of the following BVP:

DE:

$$u_{tt} - c^2u_{xx} = 0 \qquad \begin{matrix} 0 < x < x_0 \\ x_0 < x < l \\ 0 < t < \infty \end{matrix}$$

ICs:

$$u(x, 0) = u_t(x, 0) = 0 \qquad 0 < x < l$$

BCs:

$$u(0, t) = u(l, t) = 0$$
$$u(x_0 - 0, t) = u(x_0 + 0, t) \qquad 0 < t < \infty$$
$$T[u_x(x_0 + 0, t) - u_x(x_0 - 0, t)] = A \sin \omega t$$

3
Potential Theory

3-1. THE POTENTIAL EQUATION; DIRICHLET AND NEUMANN PROBLEMS

As we have seen, the wave equation describes phenomena that change with time. Next we take up an equation which describes steady-state phenomena, i.e., phenomena whose state depends only on position and not on time.

One example of such a phenomenon is the distribution of temperature in some body. If the boundary of a body is kept at some fixed temperature, then, on physical grounds, we expect that the temperature is uniquely determined at every interior point of the body. When we say "the boundary is kept at some fixed temperature," we mean that the temperature of each point on the boundary remains the same for all time. The temperature may, however, vary from point to point on the boundary. It is just not permitted to vary from time to time. Of course, this is somewhat unrealistic, because at some time in the perhaps distant past, the boundary of the body had to be brought to the given temperature. In a steady-state problem, however, it is always assumed that the boundary has been at the given temperature for so long a time that the interior temperature has reached its final state and is no longer changing in time.

Suppose that the body occupies the region $\mathscr{R}$ in xyz space and that its boundary is some surface which we denote by $\partial\mathscr{R}$. Then the temperature u at any point (x, y, z) in $\mathscr{R}$ is obviously a function of (x, y, z): $u(x, y, z)$. It turns out (cf. [48]) that u satisfies the following partial differential equation (PDE):

$$u_{xx} + u_{yy} + u_{zz} = 0 \qquad \text{in } \mathscr{R} \tag{3-1}$$

Using the standard notation of vector analysis, we can write this in the abbreviated form

$$\nabla^2 u = 0 \qquad \text{in } \mathscr{R} \tag{3-2}$$

Thus, our problem is to solve (3-1) in $\mathscr{R}$ given the values of u on $\partial\mathscr{R}$.

Before discussing this mathematical problem, we first mention some other physical situations that give rise to the same or similar problems. One is the distribution of electric charge inside a body due to a given charge distribution on its boundary. Another is the determination of the temperature inside a body when instead of keeping the boundary at a fixed temperature, we steadily, i.e., in a time-independent manner, supply or remove heat through the boundary. The latter problem leads to the same PDE (3-1) but with a different boundary condition (BC). Instead of the function u being given on $\partial\mathscr{R}$, now $\partial u/\partial n$, the normal derivative of u, is given there. The latter BC also arises in the problem of torsion of a cylindrical bar (see [258]).

There are many other physical situations which lead to (3-1). Rather than add to the list, we shall turn to the mathematics of the problem.

DEFINITION 3-1

The PDE (3-1), that is, $\nabla^2 u = 0$, is called *Laplace's equation* or *the potential equation* or *the harmonic equation*. Its solutions are called *potential functions* or *harmonic functions*.

DEFINITION 3-2

The problem of finding a function harmonic in a region $\mathscr{R}$ and taking given values on the boundary $\partial\mathscr{R}$ of $\mathscr{R}$ is called the *Dirichlet* or *first boundary-value problem* for $\mathscr{R}$.

DEFINITION 3-3

The problem of finding a function harmonic in a region $\mathscr{R}$ and whose normal derivative takes given values on $\partial\mathscr{R}$ is called the *Neumann* or *second boundary-value problem* for $\mathscr{R}$.

There is also a *third boundary-value problem* (BVP), sometimes called the *Robbin problem* [86] or the *Churchill problem* [258], in which a linear combination of function and normal derivative are given on $\partial\mathscr{R}$, but we shall not treat it until much later (Chap. 18).

Usually the region $\mathscr{R}$ will be bounded, but it need not always be. However, since bounded regions are easier to treat than unbounded ones, we shall take them up first.

If $\mathscr{R}$ is bounded, the first thing we note is that the Dirichlet and Neumann problems are quite different from any of the problems we considered for the wave equation. We never considered the wave equation in a bounded region of the xt plane, and we never gave only one condition on the entire boundary of any region in which we considered the wave equation. For the finite string, for example, we gave one BC on $x = 0$ and on $x = l$, but two BCs [called initial conditions (ICs) there because one variable was called "time"] on $t = 0$. So the potential equation differs from the wave equation in that the physics of the situation it describes leads to problems quite different from those for the wave equation.

3-2. CONNECTION WITH COMPLEX VARIABLES

The methods for solving these problems are also quite different from those for the wave equation. In most cases the methods we shall use in the present chapter work as well in three dimensions as in two, and so we shall usually work in three dimensions. Right now, however, in order to facilitate comparison with the wave equation, we consider the two-dimensional potential equation

$$u_{xx} + u_{yy} = 0 \tag{3-2a}$$

It arises from the three-dimensional one by assuming the phenomena under investigation to be independent of z. Then $\mathscr{R}$ is a region in the xy plane, and $\partial\mathscr{R}$ is a curve.

If we factor the operator (3-2a), then we obtain complex factors. This shows that (3-2a) has no real characteristics. Since virtually our entire analysis in Chap. 2 was built on the theory of characteristics, we have two choices: either to use complex variables or to develop a new theory. In the present chapter we shall do the latter, but before doing so, we briefly sketch how complex variables could be used.

As suggested above, we first factor the operator ∇^2 into

$$\nabla^2 = \left(\frac{\partial}{\partial x} + i\frac{\partial}{\partial y}\right)\left(\frac{\partial}{\partial x} - i\frac{\partial}{\partial y}\right)$$

By analogy to the wave equation, this suggests putting $\xi = x + iy$ and $\eta = x - iy$. We notice that $\eta = \bar{\xi}$. Therefore we put $z = x + iy$ and $\bar{z} = x - iy$ and have $\xi = z$ and $\eta = \bar{z}$. In terms of z and $\bar{z}$ the potential equation becomes $u_{z\bar{z}} = 0$. Hence $u = \phi(z) + \psi(\bar{z})$; that is, $u = \phi(x + iy) + \psi(x - iy)$. Since ψ is arbitrary, we can take $\psi(z) = \bar{\phi}(z)$. Note that $\bar{\phi}(z)$ does *not* mean $\overline{\phi(z)}$. It is defined by $\bar{\phi}(z) = \overline{\phi(\bar{z})}$; that is, it is obtained from ϕ by taking the complex conjugates of the coefficients in ϕ. Then $\psi(\bar{z}) = \bar{\phi}(\bar{z}) = \overline{\phi(z)}$, and $u = 2 \operatorname{Re} [\phi(z)]$. If we take ϕ as analytic and replace it by $\phi/2$, we see that the real part of any analytic function of a complex variable is a harmonic function of x and y. The converse is also true. Any harmonic function is the real part of some analytic function. The Cauchy-Riemann equations show this. Thus, just as for the one-dimensional wave equation, we have a formula for the general solution of the potential equation. The general solution is the real part of an arbitrary analytic function of a complex variable. Solving the Dirichlet problem for a plane domain $\mathscr{R}$ is therefore equivalent to finding in that domain an analytic function whose real part takes the given boundary values. For some examples of how to do this, see [47].

3-3. GREEN'S IDENTITIES

The complex-variable method requires more than a knowledge of complex variables. It also requires a *two-dimensional* potential problem. In three or higher dimensions, complex-variable methods do not work without some special symmetry assumptions. It therefore pays to develop some methods that do not rely on complex variables.

In order to develop these methods we shall need some well-known integral identities which are more or less integration by parts in higher dimensions. We shall state most things in three dimensions, but, with the obvious alterations, they hold in any number of dimensions. Some of the immediate sequel may look rather abstract, but the concepts introduced are very useful in stating results about the solutions of the potential equation. It turns out that the mere fact that a function satisfies Laplace's equation, i.e., is harmonic in a region, enables us to derive quite a number of interesting properties of that function with very little information about the nature of the region. Likewise, we can derive a number of interesting properties of the solutions of the Dirichlet and Neumann problems without knowing too much about the region.

In order to state things clearly we need a number of definitions. Let $\mathscr{R}$ be a region in xyz space, and let P: (ξ, η, ζ) be a point in $\mathscr{R}$. P is called an *interior point* of $\mathscr{R}$ if it is possible to draw about P a sphere

$$(x - \xi)^2 + (y - \eta)^2 + (z - \zeta)^2 = r^2$$

such that all points (x, y, z) inside the sphere, i.e., with

$$(x - \xi)^2 + (y - \eta)^2 + (z - \zeta)^2 < r^2$$

belong to $\mathscr{R}$. P is called an *exterior point* of $\mathscr{R}$ if there exists a sphere centered at P none of whose points belong to $\mathscr{R}$. If every sphere about P contains both points in $\mathscr{R}$ and points not in $\mathscr{R}$, then P is called a *boundary point* of $\mathscr{R}$.

$\mathscr{R}$ is called *open* if all its points are interior points. The set of all points in space not belonging to $\mathscr{R}$ is called the *complement* of $\mathscr{R}$. $\mathscr{R}$ is called *closed* if its complement is open.

A set can be either open or closed or both open and closed or neither open nor closed, depending upon whether it contains none, some, or all of its boundary points. By definition, the empty set (consisting of no points) is taken to be both open and closed. Its complement, the whole space, is therefore also both open and closed.

We shall denote the set of all boundary points of $\mathscr{R}$ by $\partial\mathscr{R}$. By the *closure* of $\mathscr{R}$ is meant the set of all interior points of $\mathscr{R}$ together with all boundary points of $\mathscr{R}$. We shall write $\overline{\mathscr{R}}$ for the closure of $\mathscr{R}$ and indicate its definition symbolically by writing $\overline{\mathscr{R}} = \mathscr{R} \cup \partial\mathscr{R}$. If $\mathscr{R}$ is closed, then $\overline{\mathscr{R}} = \mathscr{R}$.

A region $\mathscr{R}$ is called *connected*† if any two points of $\mathscr{R}$ can be joined by a continuous curve all of whose points belong to $\mathscr{R}$. A continuous curve means one whose parametric representation is given by a pair of continuous functions of the parameter. An open connected region is called a *domain*.

A region is said to be *bounded* if there exists some sphere completely containing the region in its interior. An unbounded region is said to contain a *full neighborhood of infinity* if there exists some sphere such that all points outside that sphere belong to the region.

A region is called *simply connected* if every closed curve lying in the region can be shrunk continuously to a point without leaving the region in the process.

A function is said to belong to C^n if all its derivatives of order n are continuous. If it belongs to C^0, then it is merely continuous.

An important tool in the derivation of many of the results which follow is the divergence theorem.‡

Theorem 3-1 Divergence theorem

$$\iiint_{\mathscr{R}} \nabla \cdot \boldsymbol{\omega}\, dV = \iint_{\partial\mathscr{R}} \mathbf{n} \cdot \boldsymbol{\omega}\, dS \tag{3-2b}$$

where $\mathscr{R}$ = *bounded* region in which $\boldsymbol{\omega}$ is a vector belonging to C^1 in $\mathscr{R}$ and continuous on $\overline{\mathscr{R}}$

$\partial\mathscr{R}$ = boundary of $\mathscr{R}$
dV = element of volume
dS = element of surface area
$\mathbf{n}$ = *outward*-drawn normal

It is extremely important to remember that $\boldsymbol{\omega}$ cannot have any singularities in the region $\mathscr{R}$. If $\boldsymbol{\omega}$ is singular at even one point inside $\mathscr{R}$, then in

† What we have called *connected* should really be called *arcwise connected*.
‡ Recall that boldface type is used for vectors.

order to apply the divergence theorem, we must exclude that point from $\mathscr{R}$ by encircling it with a little sphere or the like and then apply the divergence theorem to the new region $\mathscr{R}'$ consisting of $\mathscr{R}$ with the points in the interior of the sphere deleted. It should also be noted that there is no requirement that $\mathscr{R}$ be simply connected or, for that matter, connected at all. $\mathscr{R}$ can consist of many disjoint parts and be full of holes. All that matters is that $\boldsymbol{\omega}$ belong to C^1 in $\mathscr{R}$ and that $\partial\mathscr{R}$ consist of the entire boundary of $\mathscr{R}$. Note, too, that $\mathscr{R}$ must be *bounded.*

A number of variants of the divergence theorem, known as *Green's theorems*, will also be useful. Among them are the following:

$$\iiint\limits_{\mathscr{R}} \boldsymbol{\omega} \cdot \nabla\phi \, dV = \iint\limits_{\partial\mathscr{R}} \mathbf{n} \cdot \boldsymbol{\omega}\phi \, dS - \iiint\limits_{\mathscr{R}} \phi \, \nabla \cdot \boldsymbol{\omega} \, dV \tag{3-2c}$$

$$\iiint\limits_{\mathscr{R}} \nabla\phi \cdot \nabla\psi \, dV = \iint\limits_{\partial\mathscr{R}} \phi \frac{\partial\psi}{\partial n} \, dS - \iiint\limits_{\mathscr{R}} \phi \, \nabla^2\psi \, dV \tag{3-3}$$

$$\iiint\limits_{\mathscr{R}} (\phi \, \nabla^2\psi - \psi \, \nabla^2\phi) \, dV = \iint\limits_{\partial\mathscr{R}} \left(\phi \frac{\partial\psi}{\partial n} - \psi \frac{\partial\phi}{\partial n}\right) dS \tag{3-4}$$

In all three of these formulas all functions involved must be continuous on $\overline{\mathscr{R}}$. In (3-2c) they must belong to C^1 in $\mathscr{R}$. In (3-3) ψ must belong to C^2 and ϕ to C^1 in $\mathscr{R}$. In (3-4) both must belong to C^2 in $\mathscr{R}$. In all cases $\mathscr{R}$ must be bounded.

For future reference we note the special case of (3-3) in which $\phi = \psi$:

$$\iiint\limits_{\mathscr{R}} (\nabla\phi)^2 \, dV = \iint\limits_{\partial\mathscr{R}} \phi \frac{\partial\phi}{\partial n} \, dS - \iiint\limits_{\mathscr{R}} \phi \, \nabla^2\phi \, dV \tag{3-4a}$$

3-4. PROPERTIES OF HARMONIC FUNCTIONS IN BOUNDED REGIONS

As mentioned above, we can use these integral identities to derive a number of interesting properties enjoyed by harmonic functions in general. For example, if ϕ is harmonic in $\mathscr{R}$, (3-4a) implies

$$\iiint\limits_{\mathscr{R}} (\nabla\phi)^2 \, dV = \iint\limits_{\partial\mathscr{R}} \phi \frac{\partial\phi}{\partial n} \, dS \tag{3-5}$$

Formula (3-5) holds for any function ϕ harmonic in a region $\mathscr{R}$. An immediate consequence of it is the following theorem.

Theorem 3-2 If a harmonic function vanishes identically on the boundary of a bounded region throughout which it is harmonic, then that harmonic function vanishes identically throughout that region.

Proof If $\phi = 0$ on $\partial\mathscr{R}$, then (3-5) implies $\nabla\phi \equiv 0$ in $\mathscr{R}$. Hence $\phi = \text{const}$ in $\mathscr{R}$. But $\phi = 0$ on $\partial\mathscr{R}$. Therefore $\phi \equiv 0$ in $\mathscr{R}$; Q.E.D.

Notice that Theorem 3-2 is false if the function fails to be harmonic at even one point in $\mathscr{R}$. For example, the function $\phi(x, y) = \log | z |^2 = \log (x^2 + y^2)$, being the real part of an analytic function, is obviously a two-dimensional potential function. It also obviously vanishes everywhere on the circle $| z | = 1$. Yet it does not vanish identically *inside* the circle. Theorem 3-2 is not applicable here because $\log | z |$ has a singularity at the point $z = 0$ inside the circle in question and the Green's identity from which the proof of Theorem 3-2 stems demands that the functions involved be everywhere regular in $\mathscr{R}$.

Theorem 3-2 also is false without further conditions if $\mathscr{R}$ is unbounded. Again $\phi(x, y) = \log | z |$ shows this if we take $\mathscr{R}$ to be the exterior of the circle $| z | = 1$. In this region $\log | z |$ is harmonic everywhere and vanishes on the boundary yet is not identically zero. The extra condition needed to guarantee the validity of Theorem 3-2 for unbounded regions will be given in Sec. 3-8.

Theorem 3-3 If the normal derivative of a harmonic function vanishes everywhere on the boundary of a bounded region in which the function is harmonic, then that function is a constant in that region.

Proof See (3-5) and the proof of Theorem 3-2.

Theorem 3-4 If the Dirichlet problem for a bounded region has a solution, then that solution is unique.

Proof Suppose ϕ and ψ are both harmonic in a region $\mathscr{R}$ and that $\phi = \psi$ on $\partial\mathscr{R}$. Put $u = \phi - \psi$. Then u is harmonic in $\mathscr{R}$, and $u = 0$ on $\partial\mathscr{R}$. Hence, by Theorem 3-2, $u \equiv 0$ in $\mathscr{R}$. Therefore $\phi \equiv \psi$ in $\mathscr{R}$; Q.E.D.

Notice that Theorem 3-4 says that if the Dirichlet problem has a solution, that solution is unique. Theorem 3-4 makes no statement about the *existence* of a solution to the Dirichlet problem. We will take for granted the existence of a solution to the Dirichlet problem. There are a number of different proofs of this fact, but none of them are easy and we will omit them (see [65, vol. II], [88], [95], [115]).

Theorem 3-5 If the Neumann problem for a bounded region has a solution, then that solution is unique to within an additive constant.

Proof Analogous to that of Theorem 3-4.

Like Theorem 3-4, Theorem 3-5 makes no assertion about the *existence* of the solution of the Neumann problem. In fact, the Neumann problem does not always have a solution. This follows from the next theorem.

Theorem 3-6 If ϕ is harmonic in a bounded region $\mathscr{R}$, then

$$\iint_{\partial\mathscr{R}} \frac{\partial\phi}{\partial n}\, dS = 0$$

Proof In (3-4) take ϕ harmonic in $\mathscr{R}$ and take $\psi \equiv 1$ there.

Theorem 3-6 implies the following.

Theorem 3-7 The Neumann problem for a bounded region $\mathscr{R}$ can have a solution only if

$$\iint_{\partial\mathscr{R}} \frac{\partial\phi}{\partial n}\, dS = 0$$

Proof Corollary of Theorem 3-6.

Again we emphasize that Theorem 3-7 does not prove *existence* of the solution to the Neumann problem. It gives a necessary condition for existence, not a sufficient one. It does not claim that if the average value of the normal derivative on the boundary is zero, then there is a solution to the Neumann problem. It states rather that if there is a solution to the Neumann problem, then the average value of the normal derivative of that solution on the boundary is zero.

3-5. THE MEAN-VALUE PROPERTY OF HARMONIC FUNCTIONS

As another application of Theorem 3-7, let $\phi(x, y)$ be a two-dimensional potential function, and let $\mathscr{R}$ be a circle completely contained in the domain of harmonicity of ϕ. Suppose the center of $\mathscr{R}$ is at (ξ, η) and its radius is r. Thus, $r^2 = (x - \xi)^2 + (y - \eta)^2$. Then according to Theorem 3-6

$$\int_{\partial\mathscr{R}} \frac{\partial\phi}{\partial n}\, ds = 0$$

that is,

$$\int_0^{2\pi} \left[\frac{\partial}{\partial r}\, \phi(\xi + r\cos\theta, \eta + r\sin\theta)\right] r\, d\theta = 0$$

Since the r next to the $d\theta$ is constant on the circle, we can cancel it from the equation and then write the $\partial/\partial r$ in front of the integral to obtain

$$\frac{\partial}{\partial r}\int_0^{2\pi} \phi(\xi + r\cos\theta, \eta + r\sin\theta)\, d\theta = 0$$

Hence the integral does not vary with r. Its value for any value of r is therefore the same as for any other value of r, say for $r = 0$. Thus,

$$\int_0^{2\pi} \phi(\xi + r\cos\theta, \eta + r\sin\theta)\, d\theta = \int_0^{2\pi} \phi\Big|_{r=0} d\theta$$

But

$$\int_0^{2\pi} \phi\Big|_{r=0} d\theta = \int_0^{2\pi} \phi(\xi, \eta)\, d\theta = 2\pi\phi(\xi, \eta)$$

(since ξ and η are not involved in the integration).

Theorem 3-8 Let $\mathscr{R}$ be a circle centered at (ξ, η) inside which ϕ is harmonic. Then ϕ has the *mean-value property*

$$\phi(\xi, \eta) = \frac{1}{2\pi} \int_0^{2\pi} \phi(\xi + r\cos\theta, \eta + r\sin\theta)\, d\theta \tag{3-6}$$

The analogous result holds in any number of dimensions. For example, in three dimensions

$$\phi(\xi, \eta, \zeta) = \frac{1}{4\pi} \iint_{\mathscr{S}} \phi(S)\, dS$$

where $\mathscr{S}$ is a sphere.

The converse of Theorem 3-8 is also true. This is the content of the next theorem.

Theorem 3-9 Let $\phi(x, y)$ be continuous and have the mean-value property for *every* circle in a domain $\mathscr{D}$. Then ϕ is harmonic in $\mathscr{D}$.

Proof We omit the proof.

Theorems 3-8 and 3-9 show that the mean-value property is an intrinsic property of harmonic functions. It characterizes them completely. This fact can be exploited to great advantage in finite-difference schemes for solving the potential equation.

Theorem 3-10 The maximum principle The maximum of a nonconstant harmonic function occurs on the boundary. More precisely, if a harmonic function takes on a maximum at an interior point of a bounded connected domain of harmonicity, then that harmonic function must be a constant.

Proof Suppose ϕ is a nonconstant harmonic function, and suppose that the maximum M of ϕ occurs at some interior point P. Because M

is the maximum of ϕ, and because ϕ is not a constant, there must exist a circle γ about P such that some of the values of ϕ on γ are less than M. By the mean-value principle the value of ϕ at P is the average of the values of ϕ on γ, hence less than M. This contradicts the assumption that $\phi = M$ at P.

Thus, ϕ must be constant over the entire circle γ. If we choose any other point Q in the domain which can be connected to P by an arc lying entirely within the domain, then by covering this arc with circles, using the Heine-Borel theorem to choose a finite number of covering circles and continuing the foregoing argument stepwise, we arrive at the conclusion that ϕ equals the same constant at Q as at P.

Corollary If a harmonic function takes on its minimum in a bounded domain of harmonicity, then it is a constant in that domain.

Proof Apply Theorem 3-10 to $-\phi$.

Note that Theorem 3-10 and its corollary again imply uniqueness for the solution to the Dirichlet problem.

The maximum principle for harmonic functions serves another useful purpose. In a physical problem there is never perfect accuracy in the given data. If the boundary data are measured to be $f(s)$, where s is some parameter on the boundary, then there may be a small error of magnitude ϵ in $f(s)$; that is, the actual boundary data may be $f(s) + \epsilon$. If the solution to the corresponding mathematical problem with boundary data $f(s) + \epsilon$ differs greatly from the solution to the problem with boundary data $f(s)$, that solution is virtually useless physically because, due to the possibility of experimental error, it is manifestly necessary that small changes in the given data produce small changes in the solution. A mathematical problem in which the latter is the case is called *well posed*, a notion first introduced by Hadamard.

The maximum principle shows that the Dirichlet problem is well posed for the potential equation because if the boundary values of two harmonic functions differ by less than ϵ, then the maximum principle implies that the difference of these two harmonic functions (which is also harmonic) must be less than ϵ throughout the domain.

The reader should verify for himself that all the problems considered in Chaps. 2 and 3 are well posed. As an example of a problem that is *not* well posed, consider the following *initial-value* problem (IVP) for the potential equation:

PDE:

$$u_{xx} + u_{yy} = 0 \qquad \begin{matrix} 0 < x < \infty \\ -\infty < y < \infty \end{matrix} \tag{3-7a}$$

ICs:

$$u(0, y) = 0 \qquad u_x(0, y) = \frac{1}{n} \sin ny \qquad -\infty < y < \infty \tag{3-7b}$$

One solution of this problem is

$$u(x, y) = \frac{1}{n^2} \sinh nx \sin ny \tag{3-8}$$

As $n \to \infty$, the Cauchy data (3-7b) tend to zero but the solution (3-8) does not. In fact, very small initial data give rise to a very large solution in this case. This is one reason why IVPs are not usually meaningful for elliptic PDEs.

3-6. SOLUTION OF THE POTENTIAL EQUATION FOR A CIRCLE; POISSON'S INTEGRAL

Formula (3-6) gives the values of a harmonic function ϕ at the center of a circle in terms of its values on the boundary of the circle. What about the values of ϕ at the interior points of the circle other than the center? How can we express the value of ϕ at a general point inside a circle in terms of its values on the circle? That is, how do we solve the Dirichlet problem for the circle:

DE:

$$\nabla^2 u = 0 \qquad \text{in } 0 \leqslant r \leqslant R,\ 0 \leqslant \theta \leqslant 2\pi$$

BC:

$$u = f(\theta) \qquad \text{on } \partial\mathscr{R} \qquad \begin{array}{c} r = R \\ 0 \leqslant \theta \leqslant 2\pi \\ f \text{ given} \end{array} \tag{3-9}$$

This problem is easy to solve if we introduce polar coordinates and then separate variables. In polar coordinates $\nabla^2 u = 0$ becomes

$$u_{rr} + \frac{1}{r} u_r + \frac{1}{r^2} u_{\theta\theta} = 0$$

If

$$u(r, \theta) = \alpha(r)\,\beta(\theta)$$

then

$$\alpha''\beta + \frac{1}{r}\alpha'\beta + \frac{1}{r^2}\alpha\beta'' = 0$$

or

$$\frac{r^2\alpha'' + r\alpha'}{\alpha} = -\frac{\beta''}{\beta} = \lambda$$

Hence

$$r^2\alpha'' + r\alpha' - \lambda\alpha = 0$$
$$\beta'' + \lambda\beta = 0$$

Since we want our solutions to be periodic in θ with period 2π, the equation for β implies that $\lambda = n^2$ (n an integer). Therefore the equation for α reads

$$r^2\alpha'' + r\alpha' - n^2\alpha = 0 \tag{3-10}$$

This is a Euler-Cauchy equation. Its solutions are of the form

$$\alpha = r^{\mu} \tag{3-11}$$

Substituting (3-11) into (3-10), we find

$$\mu(\mu - 1) + \mu - n^2 = 0$$

or

$$\mu = \pm n$$

We reject $\mu = -n$ because the resulting solution would have a singularity at the origin. Therefore

$$u_n(r, \theta) = r^n(a_n \cos n\theta + b_n \sin n\theta)$$

are the resulting solutions. So we put

$$u(r, \theta) = \sum_{n=0}^{\infty} r^n(a_n \cos n\theta + b_n \sin n\theta)$$

We need to determine the a_n and b_n so that $u(R, \theta) = f(\theta)$; that is,

$$f(\theta) = \sum_{n=0}^{\infty} R^n(a_n \cos n\theta + b_n \sin n\theta)$$

Hence

$$a_0 = \frac{1}{2\pi}\int_0^{2\pi} f(\theta)\, d\theta$$
$$a_n = \frac{1}{\pi R^n}\int_0^{2\pi} f(\theta)\cos n\theta\, d\theta$$
$$b_n = \frac{1}{\pi R^n}\int_0^{2\pi} f(\theta)\sin n\theta\, d\theta$$

Therefore

$$u(r, \theta) = \frac{1}{2\pi}\int_0^{2\pi} f(\phi)\, d\phi + \frac{1}{\pi}\sum_{n=0}^{\infty}\left(\frac{r}{R}\right)^n \left[\cos n\theta \int_0^{2\pi} f(\phi)\cos n\phi\, d\phi + \sin n\theta \int_0^{2\pi} f(\phi)\sin n\phi\, d\phi\right]$$

If $r < R$, the series converges uniformly in θ (by comparison with a geometric series with terms independent of θ). Hence we can interchange the order of summation and integration to obtain

$$u(r, \theta) = \frac{1}{\pi}\int_0^{2\pi} f(\phi)\left[\frac{1}{2} + \sum_{n=1}^{\infty}\left(\frac{r}{R}\right)^n (\cos n\theta \cos n\phi + \sin n\theta \sin n\phi)\right] d\phi$$

or

$$u(r, \theta) = \frac{1}{\pi}\int_0^{2\pi} f(\phi)\left[\frac{1}{2} + \sum_{n=1}^{\infty}\left(\frac{r}{R}\right)^n \cos n(\theta - \phi)\right] d\phi$$

Now

$$\cos z = \frac{e^{iz} + e^{-iz}}{2}$$

Therefore ($z = \theta - \phi$, $\rho = r/R$),

$$\begin{aligned}\sum_{n=1}^{\infty} \rho^n \cos nz &= \frac{1}{2}\sum_{n=1}^{\infty} \rho^n(e^{inz} + e^{-inz})\\ &= \frac{1}{2}\sum_{n=1}^{\infty} [(\rho e^{iz})^n + (\rho e^{-iz})^n]\\ &= \frac{1}{2}\left(\frac{1}{1 - \rho e^{iz}} + \frac{1}{1 - \rho e^{-iz}} - 2\right)\\ &= \frac{1}{2}\left[\frac{2 - \rho(e^{iz} + e^{-iz})}{1 - \rho(e^{iz} + e^{-iz}) + \rho^2} - 2\right]\\ &= \frac{1}{2}\frac{\rho(e^{iz} + e^{-iz}) - 2\rho^2}{1 - \rho(e^{iz} + e^{-iz}) + \rho^2}\\ &= \frac{1}{2}\frac{2\rho \cos z - 2\rho^2}{1 - 2\rho \cos z + \rho^2}\end{aligned}$$

Hence

$$\frac{1}{2} + \sum \rho^n \cos nz = \frac{1}{2}\left(1 + \frac{2\rho \cos z - 2\rho^2}{1 - 2\rho \cos z + \rho^2}\right) = \frac{1}{2}\frac{1 - \rho^2}{1 - 2\rho \cos z + \rho^2}$$

Hence

$$u(r, \theta) = \frac{1}{2\pi}\int_0^{2\pi} \frac{1 - (r^2/R)}{1 - 2(r/R)\cos(\theta - \phi) + (r^2/R)} f(\phi)\, d\phi$$

or

$$u(r, \theta) = \frac{R^2 - r^2}{2\pi}\int_0^{2\pi} \frac{f(\phi)\, d\phi}{R^2 + r^2 - 2rR\cos(\theta - \phi)} \tag{3-12}$$

The integral in formula (3-12) is known as *Poisson's integral.* The coefficient of $f(\phi)$, namely,

$$\frac{R^2 - r^2}{2\pi[R^2 + r^2 - 2rR\cos(\theta - \phi)]}$$

is called the *Poisson kernel.*

If we take $f(\phi) \equiv 1$, then we can solve the Dirichlet problem by inspection. $u \equiv 1$ is obviously a harmonic function which takes the boundary value 1. By Theorem 3-4, $u = 1$ must be the unique solution taking these boundary values. Thus, if $f \equiv 1$ in (3-12), then u must be, too. This shows indirectly that the integral of the Poisson kernel must be one.

Note, too, that

$$u(0, 0) = \frac{1}{2\pi}\int_0^{2\pi} f(\phi)\,d\phi$$

in agreement with the mean-value principle.

Because of its method of derivation, formula (3-12) should satisfy the BC $u(R, \theta) = f(\theta)$. However, if we put $r = R$ in (3-12), the factor in front of the integral is zero and the integral is infinite. Therefore, in order for (3-12) to satisfy the BC (3-9), (3-9) has to be reformulated. The proper reformulation turns out to be

$$\lim_{r\to R} u(r, \theta) = f(\theta) \tag{3-13}$$

Thus, instead of substituting R into (3-12), we have to take the limit of (3-12) as $r \to R$.

The proof that (3-12) satisfies (3-13) depends upon the fact that the Poisson kernel acts like a delta function. As (r, θ) approaches the boundary of the circle, only those values of ϕ close to θ turn out to contribute significantly to the integral (3-12). The rigorous proof of (3-13) is an excellent exercise in advanced calculus. We refer to [65, vol. II] for it.

Notice that a discontinuous but integrable $f(\phi)$ still gives rise to a continuous $u(r, \theta)$ inside the circle because Poisson's integral is a continuous function of r and θ. Discontinuities in the boundary data f of a harmonic function are immediately smoothed out in the interior of the domain. This situation stands in sharp contrast to that for the wave equation, where any discontinuities in the initial data are propagated outward along the characteristics (cf. Sec. 2-3).

We mention in passing that if $f(\phi)$ has a jump at a point ϕ_0, then the limiting value of (3-12) as (r, θ) approaches (R, ϕ_0) depends upon the direction of approach. At points where f is continuous the limit is still unique and equal to the value of f at the point approached.

3-7. FUNDAMENTAL SOLUTIONS; GREEN'S FUNCTIONS

Next we look for potential functions with spherical symmetry, i.e., which depend only on r. In n dimensions

$$\nabla^2 u = u_{rr} + \frac{n-1}{r} u_r + \text{angular terms}$$

Therefore if we look for functions $u(r, \ldots) = \psi(r)$, then

$$\psi'' + \frac{n-1}{r} \psi' = 0$$

or

$$\frac{\psi''}{\psi'} + \frac{n-1}{r} = 0$$

and so

$$\log \psi' + (n-1) \log r = \log c$$

$$\psi' = \frac{c}{r^{n-1}}$$

Therefore, if $n = 2$,

$$\psi = c \log r + k \tag{3-14}$$

If $n > 2$, then

$$\psi = \frac{c_1}{r^{n-2}} + c_2 \tag{3-15}$$

For the next few sections we shall assume that $n = 3$.

Among other things, we see from (3-15) that there are no nonconstant spherically symmetric potential functions which do not have singularities. Since the origin can be taken anywhere (the potential equation is invariant under translation), we could always choose it to be outside the region we are interested in and then $1/r$ would be regular in the region. It is more fruitful, however, to take the origin inside the region of interest. Suppose, for example, that $Q = (\xi, \eta, \zeta)$ is a point inside $\mathscr{R}$ and

$$r^2 = (x-\xi)^2 + (y-\eta)^2 + (z-\zeta)^2 \tag{3-16}$$

Looked at as a function of $P = (x, y, z)$, $1/r$ is harmonic except at $P = Q$, that is,

$$\nabla^2 \frac{1}{r} = 0 \qquad \text{if } P \neq Q$$

and therefore

$$\iiint\limits_{\mathscr{R}} u \, \nabla^2 \frac{1}{r} \, dV = 0$$

for any regular function u since one point cannot affect the value of a Riemann integral.

Let u be harmonic in $\mathscr{R}$ and apply (3-4) to u and $1/r$. To do this it is necessary to delete from $\mathscr{R}$ a sphere centered at Q and contained in $\mathscr{R}$. Let σ be such a sphere, with ϵ its radius. Then (see Fig. 3-1)

$$0 = \iint\limits_{\partial\mathscr{R}} \left(u \frac{\partial(1/r)}{\partial n} - \frac{1}{r}\frac{\partial u}{\partial n}\right) dS + \iint\limits_{\partial\sigma} \left(u \frac{\partial(1/r)}{\partial n} - \frac{1}{r}\frac{\partial u}{\partial n}\right) dS$$

On $\partial\sigma$, $\partial/\partial n = -\partial/\partial r$. Therefore the above equation reduces to

$$\iint\limits_{\partial\sigma} \left(u \frac{\partial(1/r)}{\partial r} - \frac{1}{r}\frac{\partial u}{\partial r}\right) dS = \iint\limits_{\partial\mathscr{R}} \left(u \frac{\partial(1/r)}{\partial n} - \frac{1}{r}\frac{\partial u}{\partial n}\right) dS \tag{3-17}$$

On $\partial\sigma$, $r^2 = \epsilon^2$. Therefore

$$\frac{\partial(1/r)}{\partial r} = -\frac{1}{r^2} = -\frac{1}{\epsilon^2} \qquad \text{and} \qquad \frac{1}{r} = \frac{1}{\epsilon}$$

Hence

$$\iint\limits_{\partial\sigma} \left(u \frac{\partial(1/r)}{\partial r} - \frac{1}{r}\frac{\partial u}{\partial r}\right) dS = -\iint\limits_{\partial\sigma} \left(\frac{u}{\epsilon^2} + \frac{1}{\epsilon}\frac{\partial u}{\partial r}\right) dS$$

If we let $d\Omega$ be the element of area of the unit sphere, then $dS = \epsilon^2 \, d\Omega$ and the integral on the right can be converted into an integral over the unit sphere Ω. Then

$$-\iint\limits_{\partial\sigma} \left(\frac{u}{\epsilon^2} + \frac{1}{\epsilon}\frac{\partial u}{\partial r}\right) dS = -\iint\limits_{\Omega} \left(u + \epsilon \frac{\partial u}{\partial r}\right) d\Omega$$

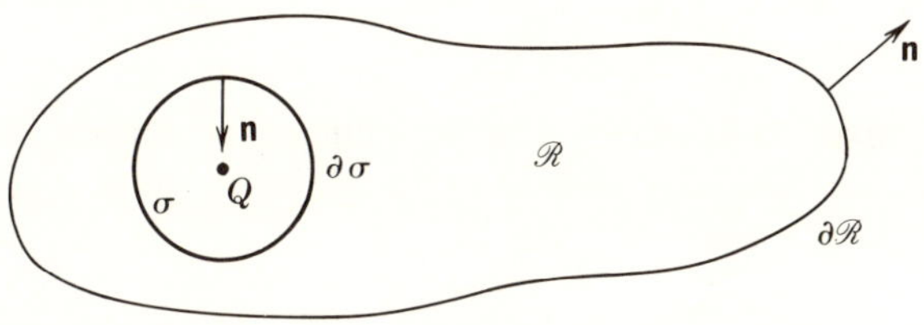

Figure 3-1

Inserting this back into (3-17), we have

$$-\iint_{\Omega} \left(u + \epsilon \frac{\partial u}{\partial r}\right) d\Omega = \iint_{\partial\mathscr{R}} \left(u \frac{\partial(1/r)}{\partial n} - \frac{1}{r}\frac{\partial u}{\partial n}\right) dS$$

This holds for all ϵ. Let $\epsilon \to 0$. Then

$$-\lim_{\epsilon\to 0} \iint_{\Omega} u \, d\Omega = \iint_{\partial\mathscr{R}} \left(u \frac{\partial(1/r)}{\partial n} - \frac{1}{r}\frac{\partial u}{\partial n}\right) dS$$

Recall that the u in the integral on the left side of (3-17) is to be evaluated on $\partial\sigma$. This is still true for u in $\iint_\Omega u\, d\Omega$ because a mere change of variables in an integral cannot change the value of the integral. As $\epsilon \to 0$, σ shrinks to Q. Hence only $u(Q) = u(\xi, \eta, \zeta)$ remains in the limit. Since $\iint_\Omega d\Omega = 4\pi$, we have found

$$-4\pi u(\xi, \eta, \zeta) = \iint_{\partial\mathscr{R}} \left(u \frac{\partial(1/r)}{\partial n} - \frac{1}{r}\frac{\partial u}{\partial n}\right) dS$$

or

$$u(\xi, \eta, \zeta) = -\frac{1}{4\pi} \iint_{\partial\mathscr{R}} \left(u \frac{\partial(1/r)}{\partial n} - \frac{1}{r}\frac{\partial u}{\partial n}\right) dS \tag{3-18}$$

This is a very interesting formula. It expresses the value of a potential function u at any point Q inside a domain $\mathscr{R}$ in terms of the values of u and $\partial u/\partial n$ on $\partial\mathscr{R}$. From Theorems 3-4 and 3-5, however, we know that knowledge of the boundary values of u alone is sufficient to determine u in the interior. The boundary values of either u or $\partial u/\partial n$ can be prescribed but not both. Therefore we need a way of eliminating either u or $\partial u/\partial n$ from (3-18). Before developing this method, we wish to make a few more comments on (3-18) and its derivation. The key to (3-18) is the behavior of $1/r$ at its singularity. $1/r$ blows up at just the right rate in this sense; as $r \to 0$, the surface integral taken over a sphere surrounding the singularity approaches the value of u at the center. In other words, $1/r$ picks out the value of u at the singularity (of $1/r$). In symbolic notation we can write

$$\frac{1}{4\pi} \iiint_{r\leqslant\epsilon} u \, \nabla^2 \frac{1}{r} \, dV = u(Q) \tag{3-19}$$

where this formula really means that the value of the surface integrals when Green's theorem is correctly applied to the sphere $r \leqslant \epsilon$ is $u(Q)$. Even more symbolically, we can write

$$\nabla^2 \frac{1}{4\pi r} = \delta(x - \xi, y - \eta, z - \zeta) \tag{3-20}$$

where (3-20) is to be regarded as an abbreviation for (3-19), which, in turn, is an abbreviation for the derivation above.

If we add to $1/r$ any regular harmonic function h, then the resulting function $g = 1/r + h$ is also a solution of

$$\nabla^2 g = 4\pi\delta(x - \xi, y - \eta, z - \zeta) \tag{3-21}$$

Any solution of (3-21) is called a *fundamental solution* of the potential equation. Fundamental solutions possess a number of properties any one of which can be used as their definition. For example, the fundamental solution is the solution which has a singularity of the lowest order possible at a given point.

As remarked above, the use of the function $1/r$ in Green's identity leads to an interesting result but still leaves us short of our goal because (3-18) contains the boundary values of both u and $\partial u/\partial n$. We would like to get rid of one of these two quantities, but we cannot hope to accomplish this by using just $1/r$ because $1/r$ is too simple a function. It has the right singularity for our purposes, but it is not general enough. For one thing, it is spherically symmetric whereas most harmonic functions are not. We therefore add to $1/r$ a regular (but not necessarily symmetric) harmonic function $h(x, y, z)$. If we put $g = 1/r + h$, then the previous analysis is virtually unaltered and we obtain

$$u(\xi, \eta, \zeta) = -\frac{1}{4\pi} \iint_{\partial\mathscr{R}} \left(u \frac{\partial g}{\partial n} - g \frac{\partial u}{\partial n}\right) dS \tag{3-22}$$

Since h is an arbitrary regular harmonic function, we are free to choose it any way we like. Suppose we choose h so that $h = -1/r$ on $\partial\mathscr{R}$. Then $g = 0$ on $\partial\mathscr{R}$, and the boundary values of $\partial u/\partial n$ will be eliminated from (3-22), leaving us with

$$u(\xi, \eta, \zeta) = -\frac{1}{4\pi} \iint_{\partial\mathscr{R}} u \frac{\partial g}{\partial n}\, dS \tag{3-23}$$

as the explicit solution to the Dirichlet problem for u. Of course, finding h also means solving a Dirichlet problem for h, but if we can solve this one Dirichlet problem for h, then we can solve all Dirichlet problems for the given domain. Moreover, there are occasions where such an h can be found easily.

Thus we see that we can solve the Dirichlet problem provided we can find the function g. To find g we have to solve another Dirichlet problem, but finding g means solving only *one* Dirichlet problem. Once we have solved it, we can solve any Dirichlet problem for the given region, no matter what the prescribed boundary values. Note that r is a function of six variables, $x, y, z, \xi, \eta, \zeta$, so that g is, too.

g is called the *Green's function* for the domain. It has the following properties:

1. g is harmonic in (x, y, z) except when $(x, y, z) = (\xi, \eta, \zeta)$.
2. $g - 1/r$ is harmonic everywhere in $\mathscr{R}$.
3. $g = 0$ on $\partial\mathscr{R}$.

Originally the term "Green's function" meant only a function satisfying the above conditions. Some people still use the term in this restricted sense. Most, however, use the term in a much more general sense. Nowadays, a Green's function need not be harmonic. Solutions of equations other than the potential equation are frequently called Green's functions provided they have the "picking-out" or delta-function property and provided they satisfy appropriate boundary conditions. Later we shall give a rather general definition of a Green's function. For the present we confine ourselves to emphasizing some of their important features and discussing some specific problems in which they are of use.

It is important to bear in mind that any time a Green's function has been found, it is a solution to a particular PDE in a particular domain and suited to solving a particular type of BVP. If the DE or the domain or the type of BC is changed, then the Green's function will change. For example, if we change the BC from one in which the function is prescribed on the boundary to one in which the normal derivative of the function is prescribed on the boundary, then the Green's function defined above is no good. The first two conditions have to be retained, but the third one now must be replaced by the condition $\partial g/\partial n = 0$ on $\partial\mathscr{R}$. Similarly, if we change the domain, we have to find a new Green's function for that new domain.

Formula (3-18) gives us still other information about a harmonic function. In it (ξ, η, ζ) is a point in the interior of $\mathscr{R}$, and (x, y, z) is a point on the boundary. As long as (ξ, η, ζ) remains in the interior, $1/r$ is an analytic function of (ξ, η, ζ). From this fact and (3-14) it can be proved that u is also an analytic function of (ξ, η, ζ). As such it cannot vanish to fractional order. Its zeros must be of integral order. Similarly, it can be proved that a harmonic function cannot become infinite of order less than $1/r$ at a singularity.

In Sec. 3-9 we shall give some examples of Green's functions. To do this we need some of the results of the next section.

3-8. UNBOUNDED DOMAINS

All of the foregoing has assumed the regions under consideration to be bounded. If a region $\mathscr{R}$ extends to infinity, some modifications are necessary. This is due to the fact that our integral theorems do not apply directly. In order to be able to adapt them to an unbounded region $\mathscr{R}$ we shall assume that $\mathscr{R}$ is the region exterior to a closed surface or surfaces; i.e., we assume that $\mathscr{R}$ contains a full neighborhood of infinity. Then we apply our integral theorems to the region $\mathscr{R} \cap \sigma$, where σ is some huge sphere of radius R containing all the boundaries of $\mathscr{R}$ in its interior. Each formula which has a term involving an integral over $\partial\mathscr{R}$ will now also have an integral with the same integrand integrated over $\partial\sigma$. For example, (3-5) now reads

$$\iiint_{\mathscr{R}\cap\sigma} (\nabla\phi)^2\, dV = \iint_{\partial\mathscr{R}} \phi \frac{\partial\phi}{\partial n}\, dS + \iint_{\partial\sigma} \phi \frac{\partial\phi}{\partial n}\, dS \tag{3-24}$$

In order to conclude that $\phi = 0$ on $\partial\mathscr{R}$ implies $\phi = 0$ in $\mathscr{R}$ we now need to know that

$$\lim_{R\to\infty} \iint_{\partial\sigma} \phi \frac{\partial\phi}{\partial n}\, dS = 0 \tag{3-25}$$

For (3-25) to hold it is sufficient that $\phi \to 0$, at infinity, because, as remarked above, potential functions are analytic. Therefore, if $\phi \to 0$ at infinity, then $\phi = O(1/r)$ there.† Since $\partial/\partial n = \partial/\partial r$ on σ, $\partial\phi/\partial n = O(1/r^2)$. Therefore, $\phi(\partial\phi/\partial n) = O(1/r^3)$. Since $dS = O(r^2)$, $\iint_{\partial\sigma} \to 0$.

In two dimensions, it is sufficient that ϕ merely be bounded at infinity because then $\phi = c + k/r + O(1/r^2)$ there and therefore, $\partial\phi/\partial n = O(1/r^2)$. Since $ds = O(r)$, $\int_{\partial\sigma} \to 0$ in this case, too.

From this discussion it is clear how Theorems 3-2 to 3-5 must be modified if $\mathscr{R}$ is unbounded. In three dimensions, it is sufficient to add the condition $\phi \to c =$ const at infinity. The value of the constant c here has to be *prescribed* as part of the problem. Usually we shall take $c = 0$. In two dimensions, the weaker condition, ϕ bounded at infinity, is sufficient. Then the Dirichlet problem has a unique solution if it has one at all, and the Neumann problem now has a unique solution, too.

The nonuniqueness for an unbounded domain of the solution to the Dirichlet problem as originally formulated is illustrated by the fact that $1/r - 1$ vanishes on the unit sphere, is regular and harmonic everywhere outside that sphere, yet is not identically zero outside the sphere. Of course, it does not tend to zero at infinity.

3-9. CONSTRUCTION OF GREEN'S FUNCTIONS; METHOD OF REFLECTION

Because it is harmonic everywhere except at one point, and because it vanishes at infinity, $1/r$ is often called the *free-space Green's function*. It has the physical interpretation of being the potential of the electrostatic field due

† A function $\phi(r)$ is said to be $O(1/r)$ (read "large oh of $1/r$") at infinity if $|r\phi(r)| < K$, for all sufficiently large r. K is a constant.

to a point charge at the point (ξ, η, ζ). This physical interpretation is of use in constructing Green's functions for other domains. Consider, for example, the upper half-space. We would like to construct a Green's function for it. First, however, we have to realize that the problem as stated, "construct a Green's function for the upper half-space," does not make sense. As remarked previously, such a problem is incomplete until we have stated the BCs. There are many possible BCs. We shall consider two of them. First we assume $g = 0$ on the xy plane. This means that we are looking for a Green's function which will enable us to solve the Dirichlet problem for the upper half-space. Still the conditions stated are not sufficient. Since the domain extends to infinity, we have to give a "BC at infinity." The most natural one is that $g \to 0$ there. This will ensure that the condition $u \to 0$ at infinity is sufficient for uniqueness.

To summarize, we want a function, $g(x, y, z, \xi, \eta, \zeta)$, which is harmonic in (x, y, z) in the upper half-space, which vanishes for $z = 0$, which is such that

$$g = \frac{1}{[(x-\xi)^2 + (y-\eta)^2 + (z-\zeta)^2]^{1/2}} + \text{a regular harmonic function}$$

in the upper half-space, and which goes to zero at infinity.

If we look at g as the potential of a field due to a unit charge at (ξ, η, ζ), plus other charges somewhere, we see that these other charges must be located in the lower half-space or else g would have singularities at points other than (ξ, η, ζ) in the upper half-space. Since the lower half-space is not in the region we are interested in, g can have all the singularities it likes there. All that is necessary is that in the upper half-space g has exactly one singularity at (ξ, η, ζ). We see, however, that if g has a singularity at the mirror image of (ξ, η, ζ) in the xy plane, namely, at $(\xi, \eta, -\zeta)$, and if the charge there is a negative unit charge, then the resulting field will be zero on the xy plane as desired. Thus, if

$$\tilde{r} = [(x-\xi)^2 + (y-\eta)^2 + (z+\zeta)^2]^{1/2} \tag{3-26}$$

then

$$g = \frac{1}{r} - \frac{1}{\tilde{r}} \tag{3-27}$$

will be harmonic everywhere except at (ξ, η, ζ) and $(\xi, \eta, -\zeta)$, will vanish on the xy plane and at infinity, and will therefore be the desired Green's function.

If, instead of the BC $g = 0$, we take the BC $\partial g/\partial n = 0$ on $z = 0$, then it is obvious that $g = 1/r + 1/\tilde{r}$.

As an explicit example of the use of a Green's function we use (3-27) to solve the Dirichlet problem for the upper half-space:

DE:

$$\nabla^2 u = 0 \qquad \begin{array}{l} -\infty < x < \infty \\ -\infty < y < \infty \\ 0 < z < \infty \end{array}$$

BCs:

$$u(x, y, 0) = f(x, y) \qquad \begin{array}{l} -\infty < x < \infty \\ -\infty < y < \infty \end{array}$$

$$u \to 0 \text{ at } \infty$$

From (3-16), (3-26), and (3-27) we have

$$-g_z\,|_{z=0} = \frac{-2\zeta}{[(x - \xi)^2 + (y - \eta)^2 + \zeta^2]^{3/2}}$$

Noting that $g_n = -g_z$ (n is the *outward* normal), and inserting this into (3-23), we get

$$u(\xi, \eta, \zeta) = \frac{\zeta}{2\pi} \iint_{-\infty}^{\infty} \frac{f(x, y)\, dx\, dy}{[(x - \xi)^2 + (y - \eta)^2 + \zeta^2]^{3/2}} \tag{3-28}$$

as the solution to the problem. Formula (3-28) is called *Poisson's integral for the half-space.* Note that again the boundary values are taken on in the sense that

$$\lim_{\zeta \to 0} u(\xi, \eta, \zeta) = f(\xi, \eta)$$

The method of reflection for finding Green's functions can be extended to more complicated domains. For example, suppose we want a Green's function for the Dirichlet problem for a wedge of angle π/n, where n is an integer. Then we can place a singularity at a point (ξ, η, ζ) in the wedge, reflect it over one side, reflect the image over the image of that side, etc., until we return, after an even number of reflections to the original domain. The Green's function is then the sum of the $1/r$'s thus obtained, each with its proper sign. This method works for either the first or second boundary-value problem.

Similarly, the Green's function for an infinite strip can be found as an infinite series by repeated reflection across the boundaries. In this case, however, it is necessary to combine the terms properly in order to obtain a convergent series.

All the regions considered so far have straight boundaries. If the boundaries are curved, the reflection method cannot usually be carried out successfully. For a sphere, however, it can be shown that the image should be placed at the point $(R^2\xi/r_0^2, R^2\eta/r_0^2, R^2\zeta/r_0^2)$, where $r_0^2 = \xi^2 + \eta^2 + \zeta^2$ and where R is the radius of the sphere. We leave the details to the reader.

3-10. GREEN'S FUNCTION VIA CONFORMAL MAPPING

In the plane the problem of finding a Green's function can be reduced to a problem in conformal mapping. For the sake of illustration, let $\mathscr{R}$ be a simply connected region with boundary $\partial\mathscr{R}$. Riemann's mapping theorem says that $\mathscr{R}$ can be mapped conformally onto the unit circle.

Moreover, the mapping can be so chosen that a prescribed point in $\mathscr{R}$ maps into the origin. Let $z = x + iy$ be the coordinates in the plane in which $\mathscr{R}$ lies, let $\hat{z} = \hat{x} + i\hat{y}$ be the coordinates in the image plane, and let $\zeta = \xi + i\eta$ be the prescribed point which maps into $\hat{z} = 0$. The mapping function then depends upon z and the parameter ζ. Let it be given by $\hat{z} = F(z, \zeta)$.

The Green's function for the Dirichlet problem for the unit circle with singularity at the origin is obviously $g(\hat{z}, 0) = \log \hat{r} = \log |\hat{z}|$. Since harmonicity is invariant under conformal mapping, the Green's function for $\mathscr{R}$ is

$$g(z, \zeta) = \log |F(z, \zeta)| \tag{3-29}$$

Thus if the Riemann mapping function for $\mathscr{R}$, $F(z, \zeta)$, is known, formula (3-29) gives the Green's function for $\mathscr{R}$. Hence the Dirichlet problem for $\mathscr{R}$ can be solved. Conversely, if the Green's function g for $\mathscr{R}$ is known, then, since $\log |z| = \text{Re}(\log z)$, formula (3-29) tells us that g is the real part of $\log F$. The Cauchy-Riemann equations then enable us to determine $\text{Im}(\log F)$ to within a constant. Hence $\log F$, and therefore F, can be found.

For multiply connected domains the results are more complicated. For example, it is not possible to map all *doubly* connected domains conformally onto a single canonical domain. Rather, there is a whole one-parameter family of canonical domains for doubly connected domains. Usually these canonical domains are taken as annuli with fixed inner radius and variable outer radius, the latter being the one parameter mentioned above. It then follows that the Green's function can be found for a doubly connected domain if the conformal mapping of that domain onto its canonical annulus is known and if the Green's function can be found for the annulus. The Green's function for the annulus is known, however. It involves a non-elementary type of function, called a *theta function*. The formula for the Green's function for the circular annulus can be found in Courant and Hilbert [65, vol. I, p. 386].

For domains of connectivity greater than two, the canonical domain is usually taken to be the complex plane with a certain number of slits. Naturally, there is a whole family of canonical domains depending on a number of parameters, which turns out to be equal to 0 for simply connected domains, 1 for doubly connected domains, and $3n - 6$ for $n$tuply connected domains if $n > 2$.

3-11. POISSON'S EQUATION

The nonhomogeneous potential equation,

$$\nabla^2 u = F(x, y, z) \tag{3-30}$$

is called *Poisson's equation*. We have already seen that a nonhomogeneous equation can be solved completely if we can solve the corresponding homogeneous equation completely and then find any particular solution to the nonhomogeneous equation. Using delta functions, we can easily obtain a particular solution of (3-30). We know that

$$\iiint_{\mathscr{R}} F(x, y, z)\, \delta(x - \xi, y - \eta, z - \zeta)\, dx\, dy\, dz = F(\xi, \eta, \zeta)$$

But from (3-20) we know that $\nabla^2(1/4\pi r) = -\delta(x - \xi, y - \eta, z - \zeta)$. Hence we expect that

$$\iiint_{\mathscr{R}} F(x, y, z)\, \nabla^2 \frac{1}{4\pi r}\, dx\, dy\, dz = F(\xi, \eta, \zeta) \tag{3-31}$$

Now $\nabla^2 1/r$ is unchanged if we interchange the roles of (x, y, z) and (ξ, η, ζ). Therefore, if it is legitimate to interchange the order of integration and differentiation, then (3-31) suggests that

$$u(\xi, \eta, \zeta) = \frac{1}{4\pi} \iiint_{\mathscr{R}} \frac{F(x, y, z)}{r}\, dx\, dy\, dz \tag{3-32}$$

is a solution to Poisson's equation.

Instead of attempting to justify its derivation, we will prove directly from (3-32) that (3-30) holds.

To do this we have to differentiate (3-32) twice with respect to ξ, etc. If we attempt to differentiate twice under the integral sign, we produce a divergent integral. We can, however, differentiate once to obtain

$$4\pi u_\xi = \iiint F \left(\frac{1}{r}\right)_\xi dx\, dy\, dz \tag{3-33}$$

Since

$$\left(\frac{1}{r}\right)_\xi = \frac{-r_\xi}{r^2}$$

the integral in (3-33) converges. If we were to differentiate once more, we would get a term containing r^{-3} and the integral would diverge. In order to be able to differentiate once more, then, we have to get rid of the derivative on the r^{-1}. We do this by first observing that $(r^{-1})_\xi = -(r^{-1})_x$, so that

$$4\pi u_\xi = -\iiint \left(\frac{1}{r}\right)_x F\, dx\, dy\, dz$$

Since x is an integration variable, we can get it off the $1/r$ by integrating by parts, i.e., using the divergence theorem. Since $1/r$ is singular, we have to cut out the point (ξ, η, ζ) by a little sphere σ of radius ϵ and then let $\epsilon \to 0$. This gives

$$4\pi u_\xi = \lim_{\epsilon\to 0}\left(-\iint_{\partial\mathscr{R}} \frac{F}{r}\cos\alpha\, dS + \iint_{r=\epsilon} \frac{F}{r}\cos\alpha\, dS + \iiint_{\mathscr{R}-\sigma} \frac{F_x}{r}\, dx\, dy\, dz\right)$$

But $dS = O(\epsilon^2)$, and therefore $\lim_{\epsilon\to 0}\iint_{r=\epsilon} = 0$. Hence

$$4\pi u_\xi = -\iint_{\partial\mathscr{R}} \frac{F}{r}\cos\alpha\, dS + \iiint_{\mathscr{R}} \frac{F_x}{r}\, dx\, dy\, dz \tag{3-34}$$

which shows that the singularity had no influence on the integral after all.

Since the singularity of $1/r$ is inside $\mathscr{R}$, not on $\partial\mathscr{R}$, the surface integral in (3-34) can be differentiated without difficulty. The volume integral in (3-34) can also be differentiated, since the $1/r$ in it has no derivatives on it. Thus, we can obtain

$$4\pi u_{\xi\xi} = -\iint_{\partial\mathscr{R}} \left(\frac{1}{r}\right)_\xi F\cos\alpha\, dS + \iiint_{\mathscr{R}} F_x\left(\frac{1}{r}\right)_\xi dx\, dy\, dz \tag{3-35}$$

The volume integral in (3-35) exists because $1/r$ has only one derivative on it. Nevertheless, we would like to use the fact that $1/r$ is harmonic. In order to do this, we need to put another derivative on $1/r$. So we apply the divergence theorem again to get

$$\begin{aligned}4\pi u_{\xi\xi} = \lim_{\epsilon\to 0}\Bigg[&-\iint_{\partial\mathscr{R}} F\left(\frac{1}{r}\right)_\xi \cos\alpha\, dS + \iint_{\partial\mathscr{R}} F\left(\frac{1}{r}\right)_\xi \cos\alpha\, dS \\ &- \iint_{r=\epsilon} F\left(\frac{1}{r}\right)_\xi \cos\alpha\, dS - \iiint_{\mathscr{R}-\sigma} F\left(\frac{1}{r}\right)_{x\xi} dx\, dy\, dz\Bigg]\end{aligned}$$

Canceling two terms and again using the fact that $(1/r)_x = -(1/r)_\xi$, we have

$$4\pi u_{\xi\xi} = \lim_{\epsilon\to 0}\left[-\iint_{r=\epsilon} F\left(\frac{1}{r}\right)_\xi \cos\alpha\, dS + \iiint_{\mathscr{R}-\sigma} F\left(\frac{1}{r}\right)_{\xi\xi} dV\right]$$

Similar expressions obviously obtain for $u_{\eta\eta}$ and $u_{\zeta\zeta}$. Combining them, and using the fact that $1/r$ is harmonic, we find

$$4\pi\,\nabla^2 u = \lim_{\epsilon\to 0} -\iint_{r=\epsilon} F\frac{\partial(1/r)}{\partial n}\, dS$$

We have already seen (Sec. 3-7) that

$$\lim_{\epsilon\to 0} -\iint_{r=\epsilon} F\frac{\partial(1/r)}{\partial n}\, dS = F$$

Hence the proof is complete. Notice that is necessary to assume the existence of first partial derivatives for F in order to carry out the above proof.

If we replace $1/r$ in (3-32) by g, where g is the Green's function for the Dirichlet problem (so that $g = 0$ on $\mathscr{R}$), we obtain a solution of Poisson's equation which vanishes on the boundary. We state this important result as a theorem.

Theorem 3-11 The solution to the BVP $\nabla^2 u = F$ in $\mathscr{R}$, $u = 0$ on $\partial\mathscr{R}$, is

$$u(\xi, \eta, \zeta) = -\frac{1}{4\pi} \iiint g(x, y, z, \xi, \eta, \zeta)\, F(x, y, z)\, dx\, dy\, dz \qquad (3\text{-}36)$$

where g is the Green's function for the Dirichlet problem for $\mathscr{R}$.

The concept of a Green's function arises in another way, one which is similar to, but slightly different from, the way in which Green's functions have been used so far in this chapter. With apologies to those readers who are unfamiliar with delta functions (whom we refer to Friedman [113, chap. 3]), we note that finding a solution $g(x, y, z, \xi, \eta, \zeta)$ of the special Poisson equation

$$\nabla^2_{(x,y,z)}\, g(x, y, z, \xi, \eta, \zeta) = \delta(x - \xi, y - \eta, z - \zeta)$$

satisfying the BC $g = 0$ on the boundary leads, by virtue of (3-33), to a solution u of Poisson's equation with a general function F on the right-hand side.

This method of introducing the Green's function for the purpose of solving a nonhomogeneous equation with homogeneous boundary conditions is probably more familiar to those readers who have previously come into contact with Green's functions than the method of Sec. 3-7, in which Green's functions were introduced for the purpose of solving a homogeneous equation with nonhomogeneous boundary conditions. They both lead to the same Green's function, however, because the nonhomogeneity can be transferred from the BCs to the PDE or from the PDE to the BCs at will. For example, if we wish to solve the DE $\nabla^2 u = 0$ in $\mathscr{R}$ with the BC $u = f$ on $\partial\mathscr{R}$, we can do this by first finding any function v (not necessarily a harmonic one) which takes the boundary values f. Then $w = v - u$ satisfies the DE $\nabla^2 w = v$ in $\mathscr{R}$ with $w = 0$ on $\partial\mathscr{R}$.

Conversely, to find a function u satisfying $\nabla^2 u = F$ in $\mathscr{R}$ with $u = 0$ on $\partial\mathscr{R}$, we have only to find any function v such that $\nabla^2 v = F$ in $\mathscr{R}$ and then put $w = v - u$. Then w satisfies $\nabla^2 w = 0$ in $\mathscr{R}$ and $w = v$ on $\partial\mathscr{R}$.

If both the DE and the BCs are nonhomogeneous, it is clear from the above how to throw the entire nonhomogeneity into either the DE or the BCs. Nevertheless, it is possible to use symbolic functions in such a way that a generalized Green's function can be used in the case where both DE and BCs are nonhomogeneous. An exposition of this method can be found in Friedman [113].

3-12. FURTHER REMARKS ON COMPLEX-VARIABLE METHODS

In Sec. 3-10 we pointed out that conformal mapping can be used to find the Green's function for the Dirichlet problem in two dimensions. It is clear that the Green's function for the Neumann problem could be found in an analogous way. In fact, the Dirichlet and Neumann problems for particular cases can be solved directly by conformal mapping if the given domain can be mapped conformally onto any domain for which these problems can be solved easily. For a detailed discussion of these techniques, see Churchill [47].

The use of complex variables in potential theory requires some care if the domain involved is multiply connected. In Sec. 3-2 we pointed out that a harmonic function is the real part of an analytic function. The harmonic function that solves a given problem—Dirichlet, Neumann, or otherwise—is necessarily single-valued. Hence the real part of the corresponding analytic function is single-valued. Its imaginary part may turn out to be multiple-valued, however. To see this, let $f(z) = \phi(x, y) + i\psi(x, y)$ and suppose that ϕ is known and single-valued in some domain $\mathscr{D}$. Pick an arbitrary point (x_0, y_0) in $\mathscr{D}$. By elementary calculus and the Cauchy-Riemann equations, we have

$$\psi(x, y) = \int_{(x_0,y_0)}^{(x,y)} d\psi = \int_{(x_0,y_0)}^{(x,y)} (\psi_x\, dx + \psi_y\, dy) = \int_{(x_0,y_0)}^{(x,y)} (-\phi_y\, dx + \phi_x\, dy) \tag{3-37}$$

The line integral defining ψ will be zero around a closed curve (and hence ψ will be single-valued) if the interior of the curve lies entirely inside the domain of harmonicity of ϕ. Otherwise ψ may be multiple-valued. Its multiple-valuedness is highly restricted, however. It can increase only by an additive constant and that only on curves that surround the "holes" in $\mathscr{D}$. Moreover, the constant must be the same on any two curves which are homotopic to each other, i.e., continuously deformable into each other without leaving $\mathscr{D}$.

Thus, the problem "find a harmonic function in $\mathscr{D}$ given its values on $\partial\mathscr{D}$" and the problem "find an analytic function in $\mathscr{D}$ given the values of its real part on $\partial\mathscr{D}$" are *not* equivalent unless $\mathscr{D}$ is simply connected. Here "analytic function" means "single-valued analytic function," just as "harmonic function" means "single-valued harmonic function." If $\mathscr{D}$ is multiply connected, we have seen above that there may not be a single-valued analytic function in $\mathscr{D}$ whose real part takes given boundary values on $\partial\mathscr{D}$. In this case it is necessary to consider a *modified Dirichlet problem*, in which the real part is prescribed only to within an additive constant on each of the connected components of $\partial\mathscr{D}$ (the separate closed curves which comprise $\partial\mathscr{D}$). The additive constants then have to be determined so that the imaginary part of the analytic function and hence the whole analytic function is single-valued in $\mathscr{D}$. For a discussion of the modified Dirichlet problem, see Mikhlin [213] (cf. also the first paragraph of this section).

Since it contains a logarithm, the analytic function of which the Green's function is the real part will, in general, be multiple-valued. The normal derivative of the complex Green's function is, however, single-vlaued because any two branches of the complex Green's function differ at most by a constant. The normal derivative of the complex Green's function is called the *Schwartz kernel.*

As an illustration of the connection between harmonic and analytic functions we mention the *Schwartz formula.* This tells us that the function $F(z)$ analytic in the unit circle whose real part is $f(\zeta)$ on the boundary is given by

$$F(z) = \frac{1}{2\pi i}\int_{|\zeta|=1} f(\zeta)\,\frac{\zeta + z}{\zeta - z}\,\frac{d\zeta}{\zeta} + ic \tag{3-38}$$

where c is an arbitrary real constant.

EXERCISE Rederive Poisson's formula by taking the real part of (3-38). Conversely by starting with Poisson's formula, derive (3-38).

The Cauchy-Riemann equations are also of use in relating the solutions of the Dirichlet and Neumann problems for a given two-dimensional domain. We leave it to the reader as an exercise to show that the Cauchy-Riemann equations hold in any orthogonal coordinate system, not just in rectangular coordinates. More precisely, if n and τ are orthogonal curvilinear coordinates whose relative orientation is the same as that of the x and y axes, and if ϕ and ψ are conjugate harmonic functions of x and y, then, as functions of n and τ, ϕ and ψ satisfy

$$\begin{aligned} \phi_n &= \psi_\tau \\ \phi_\tau &= -\psi_n \end{aligned} \tag{3-39}$$

Equations (3-39) enable us to transform a Dirichlet problem for a given domain into a Neumann one for that domain and vice versa. For, suppose that ϕ is harmonic in a domain $\mathscr{R}$ and satisfies $\phi_n = g(\tau)$ on $\partial\mathscr{R}$, where τ is the running parameter on $\partial\mathscr{R}$. Then ψ, the harmonic conjugate of ϕ in $\mathscr{R}$, is harmonic in $\mathscr{R}$ and satisfies $\psi_\tau = g(\tau)$ on $\partial\mathscr{R}$. Integrating this BC with respect to τ, we find that ψ satisfies $\psi = G(\tau)$ on $\partial\mathscr{R}$, where $G(\tau)$ is any indefinite integral of $g(\tau)$. The arbitrary constant in $G(\tau)$ can be determined later. Thus, the Neumann problem for ϕ is equivalent to a Dirichlet problem for its conjugate ψ and, vice versa, the Dirichlet problem for ψ can be solved by solving instead a Neumann problem for its conjugate ϕ.

As pointed out earlier in this section, the conjugate harmonic function of a given harmonic function may be multiple-valued if the domain is multiply connected. The preceding paragraph gives us a chance to see how such a situation can arise in practice. If $\mathscr{R}$ is multiply connected, when we integrate $g(\tau)$ along a particular connected component of $\partial\mathscr{R}$, we need not always get the same additive constant in the indefinite integral. In fact, on each connected component $\partial\mathscr{R}_j$ of $\partial\mathscr{R}$ we can add a different constant c_j. If $G_0(\tau)$ is some definite integral of $g(\tau)$, we will have the BC $\psi = G_0(\tau) + c_j$ on $\partial\mathscr{R}_j$ and it will be necessary to determine the constants c_j so that the resulting ψ is single-valued in $\mathscr{R}$. As mentioned above, this can be done. It turns out that one of the c_j's can be chosen arbitrarily. The rest are then determined by the requirement of single-valuedness for ψ.

3-13. THE INVERSE METHOD FOR THE DIRICHLET PROBLEM

In the application of the linear theory of elasticity to the problem of torsion of cylindrical beams the following Dirichlet problem arises. Find a function $\phi(x, y)$, harmonic in the region $\mathscr{R}$, which represents a typical cross section of the beam, and satisfying the BC $\phi(x, y) = \frac{1}{2}(x^2 + y^2)$ on $\partial\mathscr{R}$.

Naturally this problem cannot be solved explicitly for arbitrary cross sections $\mathscr{R}$. Given an explicit cross section, it may be very difficult to solve the above Dirichlet problem for this cross section. In engineering, however, it is rare that only one particular cross section is admissible for a beam. Therefore, instead of first choosing a cross section and then solving the Dirichlet problem for it, workers in the field of elasticity sometimes use an inverse method. They first choose a harmonic function $\phi(x, y)$ and then see on what curves ϕ satisfies the given BC. Such curves are given by the equation $\phi(x, y) = \frac{1}{2}(x^2 + y^2)$. If one of these curves bounds a region $\mathscr{R}$, then ϕ obviously solves the given Dirichlet problem for that region. By choosing different harmonic functions it is possible to compile a list of cross sections for which the problem can be solved. For examples of this method, see Sokolnikoff [258, pp. 120–127].

SUMMARY

Theorem S3-1 If $\mathscr{R}$ is a bounded domain in n-dimensional euclidean space to which Green's theorem is applicable, if $\nabla^2 u = 0$ in $\mathscr{R}$, and if $u = 0$ on $\partial\mathscr{R}$, then $u \equiv 0$ in $\mathscr{R}$.

Theorem S3-2 If $\mathscr{R}$ is a bounded domain in n-dimensional euclidean space to which Green's theorem applies, if $\nabla^2 u = 0$ in $\mathscr{R}$, and if $\partial u/\partial n = 0$ on $\partial\mathscr{R}$, then $u \equiv \text{const}$ in $\mathscr{R}$.

Theorem S3-3 If the domain $\mathscr{R}$ in Theorems S3-1 and S3-2 is unbounded, and if, in addition, the hypothesis $\lim_{P\to\infty} u(P) = 0$ is added, then the conclusions of Theorems S3-1 and S3-2 are still valid.

Theorem S3-4 If $\mathscr{R}$ is a bounded region in n-dimensional euclidean space to which Green's theorem is applicable, if $\nabla^2 u = \nabla^2 v = 0$ in $\mathscr{R}$, and if $u = v$ on $\partial\mathscr{R}$, then $u \equiv v$ in $\mathscr{R}$. If $\mathscr{R}$ is unbounded, then the condition $\lim_{P\to\infty} u(P) = \lim_{P\to\infty} v(P)$ must be appended in order to guarantee the conclusion.

Theorem S3-5 If $\mathscr{R}$ is a bounded region in n-dimensional euclidean space, and if $\nabla^2 u = 0$ in $\mathscr{R}$, then $\iint_{\partial\mathscr{R}} \partial u/\partial n\, dS = 0$. If $\mathscr{R}$ is unbounded, then it is necessary to add the condition $\lim_{P\to\infty} u(P) = 0$ in order to guarantee this result.

Theorem S3-6 If $\nabla^2 u = 0$ in a sphere $\mathscr{S}$, then

$$u(Q) = \frac{1}{\sigma_n} \iint\limits_{\partial\mathscr{S}} u(S)\, dS$$

where Q is the center of $\mathscr{S}$ and σ_n is the surface area of the unit sphere in n dimensions.

Theorem S3-7 (See Theorem S3-6 for notation.) If

$$u(Q) = \frac{1}{\sigma_n} \iint\limits_{\partial\mathscr{S}} u(S)\, dS$$

for every sphere $\mathscr{S}$ in a region $\mathscr{R}$, then $\nabla^2 u = 0$ in $\mathscr{R}$.

Theorem S3-8 If $\nabla^2 u = 0$ in a bounded region $\mathscr{R}$, then both the maximum and minimum of u occur on $\partial\mathscr{R}$.

Theorem S3-9 If $f(\theta)$ is continuous on $0 \leqslant \theta \leqslant 2\pi$, then the unique solution of $\nabla^2 u = 0$, $0 \leqslant r < R$, $0 \leqslant \theta \leqslant 2\pi$, satisfying $\lim_{r\to R} u(r, \theta) = f(\theta)$, is given by *Poisson's integral*

$$u(r, \theta) = \frac{R^2 - r^2}{2\pi} \int_0^{2\pi} \frac{f(\phi)\, d\phi}{R^2 + r^2 - 2rR\cos(\theta - \phi)}$$

DEFINITION

$g(x, y, z, \xi, \eta, \zeta)$ is called a *Green's function for the Dirichlet problem for* $\nabla^2 u = 0$ *in* $\mathscr{R}$ if

1. $\nabla^2_{(x,y,z)} g = 0$ in $\mathscr{R}$ except when $(x, y, z) = (\xi, \eta, \zeta)$.
2. $\nabla^2_{(x,y,z)}(g - 1/r) = 0$ everywhere in $\mathscr{R}$.
3. $g = 0$ on $\partial\mathscr{R}$.

If $\mathscr{R}$ is unbounded, then

4. $g \to 0$ at infinity

must be added. g is a *Green's function for the Neumann problem* if item 3 is replaced by $\partial g/\partial n = 0$ on $\partial\mathscr{R}$.

Theorem S3-10 If $f(x, y)$ is continuous in (x, y) and tends to zero at infinity, then the unique solution to $\nabla^2 u = 0$, $-\infty < x < \infty$, $0 < y < \infty$, satisfying

$$\lim_{z \to 0} u(x, y, z) = f(x, y) \qquad \text{and} \qquad \lim_{(x,y,z)\to\infty} u = 0$$

is given by *Poisson's integral for the half-space*

$$u(x, y, z) = \frac{z}{2\pi} \iint_{-\infty}^{\infty} \frac{f(\xi, \eta)\, d\xi\, d\eta}{[(x - \xi)^2 + (y - \eta)^2 + z^2]^{3/2}}$$

Theorem S3-11 Let $\mathscr{R}$ be a simply connected domain in the complex z plane with at least two boundary points. Let $F(z, \zeta)$ be the analytic function of z which maps $\mathscr{R}$ conformally onto the unit circle, taking the arbitrary point ζ in $\mathscr{R}$ into the origin. Then the Green's function for $\mathscr{R}$ with singularity at ζ is given by $g(z, \zeta) = \log |F(z, \zeta)|$.

Theorem S3-12 If $F(x, y, z)$ has continuous first partial derivatives in a bounded region $\mathscr{R}$, then

$$u(x, y, z) = \frac{1}{4\pi} \iiint_{\mathscr{R}} \frac{F(\xi, \eta, \zeta)}{r}\, d\xi\, d\eta\, d\zeta$$

satisfies *Poisson's equation*

$$\nabla^2 u = F \qquad \text{in } \mathscr{R}$$

Theorem S3-13 If $g(x, y, z, \xi, \eta, \zeta)$ is the Green's function for the Dirichlet problem for $\nabla^2 u = 0$ in $\mathscr{R}$, then

$$u(x, y, z) = \frac{1}{4\pi} \iiint_{\mathscr{R}} F(\xi, \eta, \zeta)\, g(x, y, z, \xi, \eta, \zeta)\, d\xi\, d\eta\, d\zeta$$

satisfies Poisson's equation, $\nabla^2 u = F$, in $\mathscr{R}$, and $u = 0$ on $\partial\mathscr{R}$.

PROBLEMS

1. Find $u(x, y)$ if $\nabla^2 u = 0, 0 \leqslant x \leqslant 1, 0 \leqslant y \leqslant 1$, and $u(x, 0) = x(1-x)$, $u(0, y) = y(1-y)$, $u(x, 1) = u(1, y) = 0$.

2. Prove that the Green's function for a region, if it exists, is unique.

3. Use the maximum principle to prove uniqueness for the solution of the Dirichlet problem for the potential equation.

4. Prove or disprove that the product of two harmonic functions is harmonic.

5. (*a*) Prove that the following problem has no solution:

$$\nabla^2 u(r, \theta) = 0 \qquad \begin{array}{l} 0 \leqslant r \leqslant 3 \\ 0 \leqslant \theta \leqslant 2\pi \end{array}$$

$$u_r(3, \theta) = 2 \qquad 0 \leqslant \theta \leqslant 2\pi$$

(*b*) Find a solution of

$$\nabla^2 u(r, \theta) = 0 \qquad \begin{array}{l} 3 < r < \infty \\ 0 < \theta < 2\pi \end{array}$$

$$u_r(3, \theta) = 2 \qquad 0 \leqslant \theta \leqslant 2\pi$$

6. Find the Green's function that would be appropriate for solving the following BVP:

$$\nabla^2 u(x, y) = 0 \qquad \begin{array}{l} 0 \leqslant x < \infty \\ 0 \leqslant y < \infty \end{array}$$

$$u(0, y) = g(y) \qquad 0 \leqslant y < \infty$$
$$u_y(x, 0) = f(x) \qquad 0 \leqslant x < \infty$$

7. Solve the following BVP for the potential equation on the square of width π:

$$\nabla^2 u(x, y) = 0 \qquad \begin{array}{ll} u(x, 0) = 0 & u(\pi, y) = 1 \\ u(x, \pi) = 0 & u(0, y) = 0 \end{array}$$

8. Poisson's integral gives us the solution to the potential problem for a circle. If $f(\theta) \equiv 1$, we know that $u \equiv 1$ throughout the circle. Consequently, show by direct calculation that

$$\frac{l^2 - r^2}{2\pi} \int_0^{2\pi} \frac{d\varphi}{l^2 + r^2 - 2rl \cos(\theta - \varphi)} \equiv 1$$

9. Explain the following paradoxes:

(*a*) The solution to the Dirichlet problem is unique. Yet $u = 1$ and $u = (x^2+y^2+z^2)^{-1/2}$ are both solutions of $\nabla^2 u = 0$ in the unit sphere, and both take the boundary values 1.

(*b*) The maximum of a nonconstant harmonic function occurs on the boundary; $-\log(x^2 + y^2)$ is harmonic on the unit circle; yet it is zero on the boundary and positive everywhere inside.

10. Use the reflection method to find the Green's function for the first boundary-value problem for a circle of radius R. Using the Green's function, obtain Poisson's integral for the circle.

11. Show that in three dimensions if $f(x, y, z)$ is harmonic in a bounded region not containing the origin, then

$$F(x, y, z) = \frac{1}{r} f\left(\frac{x}{r^2}, \frac{y}{r^2}, \frac{z}{r^2}\right)$$

is also harmonic. Here $r^2 = x^2 + y^2 + z^2$.

12. Find the electrostatic potential $f(r, \theta)$ in an infinitely long hollow right circular cylinder of inner radius a and outer radius b if

$$f(a, \theta) = 0 \qquad f(b, \theta) = \begin{cases} 1 & 0 < \theta < \pi \\ -1 & \pi < \theta < 2\pi \end{cases}$$

Note: $f(r, \theta)$ satisfies $\nabla^2 f = 0$ inside the cylinder.

13. Let $\alpha(x, y)$ and $\beta(x, y)$ be conjugate harmonic functions, i.e., satisfy the Cauchy-Riemann equations $\alpha_x = \beta_y$, $\alpha_y = -\beta_x$. Let

$$\xi = \alpha(x, y) \qquad \eta = \beta(x, y) \qquad \text{and} \qquad u_{xx} + u_{yy} = 0$$

Show that $u_{\xi\xi} + u_{\eta\eta} = 0$.

14. Derive Poisson's formula (3-12) from Schwarz's formula (3-38) and, vice versa, derive Schwarz's formula (3-38) from Poisson's formula (3-12).

15. Solve the PDE $\nabla^2 u = 0$ in the sphere of radius R about the origin with the BC $u(R, \theta, \phi) = R \cos \theta$, where θ is the angle between the radius vector and the positive z axis.

16. Find the Green's function for the Dirichlet problem for the potential equation in the region $a < r < \infty$, $0 < \theta < \pi$.

17. (*a*) Find the solution of Poisson's equation $\nabla^2 u = 1$ inside a circle of radius $r = a$ if $u|_{r=a} = 0$.

(*b*) Find the solution of Poisson's equation $\nabla^2 u = A + B/r$ inside the spherical layer $a < r < b$ for the boundary condition $u|_{r=a} = 0$ and $u|_{r=b} = 0$.

4
The Reduced Wave Equation

4-1. DERIVATION OF THE REDUCED WAVE EQUATION

In Chap. 2 we considered the one-dimensional wave equation. The generalization of this equation to two dimensions is

$$u_{xx} + u_{yy} = c^{-2}u_{tt} \tag{4-1}$$

One interpretation of the solution u of (4-1) is as the vertical displacement of some two-dimensional object each point of which moves only in the vertical direction. In addition, the object must be incapable of supporting any shear stresses; otherwise a more complicated equation than (4-1) is necessary to describe the motion. An object whose motion is described by (4-1) is called a *membrane*. If the object resists shearing stresses, it is called a *plate* and the PDE describing its motion is a more complicated one than (4-1).

We have already mentioned that the one-dimensional wave equation is one of the few partial differential equations (PDEs) of any importance whose general solution is both known and useful. Equation (4-1) is not the one-dimensional wave equation, and its general solution is not known. However, as also pointed out before, general solutions of PDEs are of little interest per se in applied mathematics. The key interest lies in the solutions to initial-value problems (IVPs) or boundary-value problems (BVPs), or both, for PDEs. A natural BIVP for (4-1) arises if we consider the vibrations of a membrane which at time $t = 0$ covers some region $\mathscr{R}$ in the xy plane. As in the one-dimensional case, we have to prescribe the initial deflection $u(x, y, 0)$ and the initial velocity $u_t(x, y, 0)$ of the membrane over $\mathscr{R}$ and to say how the boundary of the membrane is constrained during the motion; i.e., we have to prescribe u on $\partial\mathscr{R}$ or $\partial u/\partial n$ on $\partial\mathscr{R}$, or the like. For simplicity, we assume that u is given as a prescribed function of t on $\partial\mathscr{R}$. Thus we consider the following BIVP:

DE:

$$u_{xx} + u_{yy} = c^{-2}u_{tt} \qquad \text{for } (x, y) \text{ in } \mathscr{R}, \quad t \geqslant 0 \tag{4-1}$$

ICs:

$$\begin{aligned} u(x, y, 0) &= f(x, y) \\ u_t(x, y, 0) &= g(x, y) \end{aligned} \qquad \text{for } (x, y) \text{ in } \mathscr{R} \tag{4-2}$$

BC:

$$u = h \text{ on } \partial\mathscr{R} \qquad \text{for } 0 \leqslant t < \infty$$

Since we do not have the general solution of (4-1), it is natural to try to solve this problem by separation of variables. From Chap. 2 we recall that the success of this method depends on having some homogeneity in the ICs and/or BCs. We also recall (cf. Sec. 2-6) that it is always possible to replace a nonhomogeneous BC for a homogeneous DE by a homogeneous BC for a nonhomogeneous DE. Later we shall show how to solve the nonhomogeneous version of the DE (4-1) (see Duhamel's method, Sec. 7-3). Therefore, in the present chapter, we shall make the key assumption that the BC we are dealing with is

$$u = 0 \text{ on } \partial\mathscr{R} \qquad \text{for } 0 \leqslant t < \infty \tag{4-2a}$$

The importance of having a homogeneous BC will be evident in the following analysis. Incidentally, we remark that the IVP for an infinite membrane covering the whole xy plane, i.e., when $\mathscr{R}$ is the whole xy plane, so that no BC need be imposed, can be solved in a manner quite different from that for the finite membrane. The solution to the IVP for (4-1) in the whole plane is given in Chap. 7.

Just as for the potential equation, it is of interest to obtain some rather general results before specializing. In separating variables in (4-1) we first separate out only the t dependence, writing

$$u(x, y, t) = v(x, y)\, w(t) \tag{4-2b}$$

Substituting (4-2*b*) into (4-1) and separating terms appropriately, we obtain

$$c^2 w \, \nabla^2 v = v w''$$

or

$$\frac{\nabla^2 v}{v} = \frac{w''}{c^2 w} = \lambda$$

Hence

$$w'' - \lambda c^2 w = 0 \tag{4-2c}$$

and

$$\nabla^2 v - \lambda v = 0 \tag{4-2d}$$

Equation (4-2*d*) is called the *reduced wave equation* or the *Helmholtz equation*. In order for (4-2*b*) to satisfy (4-2*a*) v must vanish on $\partial\mathscr{R}$. Thus, we need to find a solution of (4-2*d*), the reduced wave equation, in $\mathscr{R}$ which vanishes on $\partial\mathscr{R}$. In general, the only solution of (4-2*d*) in $\mathscr{R}$ which vanishes on $\partial\mathscr{R}$ will be $v \equiv 0$. In particular, for $\lambda = 0$, this follows from our results in Chap. 3 on the potential equation. However, just as in the case of the finite string, there may be certain values of λ for which (4-2*d*) has a solution which vanishes on $\partial\mathscr{R}$ but which does not vanish identically in $\mathscr{R}$. Such values of λ are called *eigenvalues* of the problem. Notice that in order to have an eigenvalue problem we have to have a DE and BCs, not just a DE.

It is interesting that it is possible to show in general that if there are eigenvalues of this problem, then they cannot be positive. To show this, we use the two-dimensional analog of Green's identity (3-4*a*), which tells us that

$$\iint_{\mathscr{R}} v \, \nabla^2 v \, dA = \int_{\partial\mathscr{R}} v \frac{\partial v}{\partial n} \, ds - \iint_{\mathscr{R}} (\nabla v)^2 \, dA$$

Since $v = 0$ on $\partial\mathscr{R}$ and $\nabla^2 v = \lambda v$ in $\mathscr{R}$, this reduces to

$$\lambda \iint_{\mathscr{R}} v^2 \, dA = - \iint_{\mathscr{R}} (\nabla v)^2 \, dA$$

or

$$\lambda = - \frac{\iint_{\mathscr{R}} (\nabla v)^2 \, dA}{\iint_{\mathscr{R}} v^2 \, dA}$$

which shows that $\lambda < 0$.

Knowing that if there are any eigenvalues, they must be negative, we can put $\lambda = -k^2$. Then the solution to (4-2*c*) is

$$w = A \cos kct + B \sin kct \tag{4-3}$$

and (4-2d) reads

$$\nabla^2 v + k^2 v = 0 \tag{4-4}$$

The problem now is to find values of k for which (4-4) has a solution that vanishes on $\partial\mathscr{R}$ but does not vanish identically in $\mathscr{R}$. The eigenvalues, i.e., the values of k^2, and the corresponding eigenfunctions obviously depend upon the nature of the domain. We therefore abandon further generality at this point in favor of the consideration of specific problems. We will find the eigenvalues and eigenfunctions of (4-4) for a number of simple domains. In so doing we will encounter some of the standard special functions of mathematical physics.

The simplest domain to begin with is the rectangular membrane.

4-2. THE RECTANGULAR MEMBRANE

Here the domain $\mathscr{R}$ is $0 < x < a$, $0 < y < b$. In this case, we set $v(x, y) = \alpha(x)\,\beta(y)$. Then we find

$$\alpha''\beta + \alpha\beta'' + k^2\alpha\beta = 0$$

or

$$\frac{\alpha''}{\alpha} + k^2 = -\frac{\beta''}{\beta} = \mu^2$$

We have chosen the separation constant to be positive because we want β to vanish both at $y = 0$ and at $y = b$. Hence we need the trigonometric solutions to the equation for β. This equation is

$$\beta'' + \mu^2\beta = 0$$

In order for $\beta(0) = \beta(b) = 0$, we must have $\mu = m\pi/b$ (m an integer). Then, since

$$\alpha'' + (k^2 - \mu^2)\alpha = 0$$

and since we need $\alpha(0) = \alpha(a) = 0$, we must have

$$k^2 - \frac{m^2\pi^2}{b^2} = \frac{n^2\pi^2}{a^2} \qquad n \text{ an integer}$$

Thus,

$$k = \pi\sqrt{\frac{n^2}{a^2} + \frac{m^2}{b^2}} = k_{mn} \tag{4-5}$$

gives the eigenvalues for this case. The corresponding eigenfunctions are

$$v_{mn}(x, y) = A_{mn} \sin\frac{n\pi x}{a} \sin\frac{m\pi y}{b} \tag{4-6}$$

Note that sometimes more than one eigenfunction can correspond to a given eigenvalue k. From (4-3) and (4-6) we have

$$u(x, y, t) = \sum_{m,n} (a_{mn} \cos k_{mn}ct + b_{mn} \sin k_{mn}ct) \sin \frac{n\pi x}{a} \sin \frac{m\pi y}{b}$$

Therefore the ICs (4-2) require that

$$f(x, y) = \sum_{m,n} a_{mn} \sin \frac{n\pi x}{a} \sin \frac{m\pi y}{b}$$

and

$$g(x, y) = c \sum_{m,n} k_{mn} b_{mn} \sin \frac{n\pi x}{a} \sin \frac{m\pi y}{b}$$

which shows that the a_{mn} and $ck_{mn}b_{mn}$ are the Fourier coefficients of f and g respectively.

4-3. CIRCULAR MEMBRANE; BESSEL FUNCTIONS

As a second example, consider a circular membrane of radius R with the boundary held fixed. Then it is natural to use polar coordinates, and we have

DE:

$$u_{rr} + \frac{1}{r} u_r + \frac{1}{r^2} u_{\theta\theta} = \frac{1}{c^2} u_{tt} \qquad \begin{matrix} 0 \leqslant r < R \\ 0 \leqslant \theta \leqslant 2\pi \\ 0 < t < \infty \end{matrix} \tag{4-7}$$

BC:

$$u(R, \theta, t) = 0 \qquad \begin{matrix} 0 \leqslant \theta \leqslant 2\pi \\ 0 < t < \infty \end{matrix}$$

ICs:

$$\begin{matrix} u(r, \theta, 0) = f(r, \theta) \\ u_t(r, \theta, 0) = g(r, \theta) \end{matrix} \qquad 0 \leqslant r \leqslant R, 0 \leqslant \theta \leqslant 2\pi$$

Separate variables

$$u(r, \theta, t) = v(r, \theta)\, w(t) \tag{4-8}$$

Then

$$\left(v_{rr} + \frac{1}{r} v_r + \frac{1}{r^2} v_{\theta\theta}\right) w = \frac{1}{c^2} v w''$$

We have already seen (Sec. 4-1) that the separation constant must be negative. Therefore,

$$\frac{v_{rr} + (1/r)\, v_r + (1/r^2)\, v_{\theta\theta}}{v} = \frac{w''}{c^2 w} = -k^2$$

Hence

$$w = a \cos kct + b \sin kct \tag{4-9}$$

and

$$v_{rr} + \frac{1}{r} v_r + \frac{1}{r^2} v_{\theta\theta} + k^2 v = 0 \tag{4-10}$$

Separate variables again

$$v(r, \theta) = \alpha(r)\,\beta(\theta)$$

Then

$$\alpha''\beta + \frac{1}{r}\alpha'\beta + \frac{1}{r^2}\alpha\beta'' + k^2\alpha\beta = 0$$

or

$$\frac{r^2}{\alpha}\left(\alpha'' + \frac{1}{r}\alpha' + k^2\alpha\right) = -\frac{\beta''}{\beta} = \lambda$$

At first sight it seems as if there were no restriction on λ. There is a restriction, though, because of a hidden condition on the solution. Because $\theta = 0$ and $\theta = 2\pi$ represent the same physical point, u and hence β must be periodic in θ with period divisible by 2π. This necessitates taking $\lambda = n^2$ (n an integer). Then

$$\beta = A \cos n\theta + B \sin n\theta$$

and

$$\alpha'' + \frac{1}{r}\alpha' + \left(k^2 - \frac{n^2}{r^2}\right)\alpha = 0 \tag{4-11}$$

Put

$$kr = \rho \tag{4-12}$$

Then

$$\alpha'' + \frac{1}{\rho}\alpha' + \left(1 - \frac{n^2}{\rho^2}\right)\alpha = 0 \tag{4-13}$$

Equation (4-13) is called *Bessel's equation of index n*. Since the leading coefficient is 1, and since the other coefficients are singular at the origin, it is to be expected that some of the solutions of (4-13) may be singular at the origin. It turns out that for n an integer there is exactly one linearly independent Bessel function that is not singular at the origin. When properly normalized, it is denoted by $J_n(\rho)$. The properties of J_n and of all other Bessel functions are well known and can be found in many books (see, for example, Churchill [48]). In the present problem, the singular Bessel functions cannot be used because we want a solution that is regular at the origin. The reader is cautioned not to throw away the other Bessel functions without good cause. For example, if the region in question were an annulus instead of a circle, the singular Bessel functions would be needed because boundary conditions have to be satisfied on both circles. Since the origin is not in the ring, there is no need to exclude functions which are singular there.

Thus, we take

$$\alpha_n(\rho) = J_n(\rho) = J_n(kr) \tag{4-14}$$

and then

$$v_n(r, \theta) = J_n(kr)(a_n \cos n\theta + b_n \sin n\theta)$$

Therefore in order to satisfy the BC $v(R, \theta) = 0$, we need

$$J_n(kR)(a_n \cos n\theta + b_n \sin n\theta) = 0 \qquad \text{for all } \theta \text{ between 0 and } 2\pi$$

This requires us to choose k so that

$$J_n(kR) = 0 \tag{4-15}$$

Thus, we need to know something about the zeros of J_n. Again we must refer the reader to other books for the details, but we can obtain a correct conjecture by observing that for large ρ the DE (4-13) is approximately $\alpha'' + \alpha = 0$, the solutions to which are sines and cosines. Since the solutions of this approximating DE have an infinite sequence of real zeros, one is led to the dangerous (but in the present case correct) conclusion that so too does J_n. Since one cannot blandly throw away terms in a DE without strict justification, we recommend that the reader consult Tricomi [275] for the details of the proof that J_n has an infinite sequence of zeros, whose consecutive distance approaches π as the argument of J_n tends to infinity.

In any event, it is known that J_n has an infinite number of real zeros, $z_{n1}, z_{n2}, \ldots, z_{nm}, \ldots$ with the property that $\lim_{m\to\infty} z_{nm} = \infty$. Note that, as indicated by the notation, the zeros depend upon n, the index of the Bessel function. m indexes the set of zeros for a given J_n. Thus, for every n we get an infinite set of k's satisfying (4-15)

$$J_n(k_{nm}R) = 0 \tag{4-16}$$

If z_{nm} is a typical root of J_n, that is, if $J_n(z_{nm}) = 0$, then $k_{nm} = R^{-1}z_{nm}$. For each k_{nm} we get a solution

$$v_{nm}(r, \theta) = J_n(k_{nm}r)(A_{nm} \cos n\theta + B_{nm} \sin n\theta)$$

Therefore

$$u(r, \theta, t) = \sum_{n,m} J_n(k_{nm}r)(A_{nm} \cos n\theta + B_{nm} \sin n\theta) \times (c_{nm} \cos k_{nm}ct + d_{nm} \sin k_{nm}ct)$$

Renaming the coefficients, we therefore need to determine new coefficients A_{nm} and B_{nm} so that

$$f(r, \theta) = \sum_{n,m} J_n(k_{nm}r)(A_{nm} \cos n\theta + B_{nm} \sin n\theta) \tag{4-17}$$

and

$$g(r, \theta) = c \sum_{n,m} k_{nm} J_n(k_{nm}r)(\tilde{A}_{nm} \cos n\theta + \tilde{B}_{nm} \sin n\theta)$$

In other words, we need to know how to expand a function in a series of the type (4-17). We solve this problem as follows. Multiply (4-17) by $1/\pi \cos \tilde{n}\theta$ and integrate with respect to θ from 0 to 2π. If we let

$$F_{\tilde{n}}(r) = \frac{1}{\pi} \int_0^{2\pi} f(r, \theta) \cos \tilde{n}\theta \, d\theta \tag{4-18}$$

and then, for simplicity, drop the tilde everywhere, we get

$$F_n(r) = \sum_{m=1}^{\infty} A_{nm} J_n(k_{nm}r) \tag{4-19}$$

Since f is given, F_n is known. We therefore have to be able to expand a given function in a series like (4-19). Notice that n is fixed in (4-19). The sum is over the different zeros k_{nm} of the same Bessel function J_n.

In order to determine the A_{nm}, we need the right orthogonality relation among the Bessel functions. The one to use here is

$$\int_0^R rJ_n(k_{nm}r)\, J_n(k_{n\tilde{m}}r)\, dr = \tfrac{1}{2}\delta_{m\tilde{m}} R^2 J_{n+1}^2(k_{nm}R) \tag{4-20}$$

$\delta_{m\tilde{m}}$ here is the Kronecker delta:

$$\delta_{m\tilde{m}} = \begin{cases} 1 & \text{if } m = \tilde{m} \\ 0 & \text{if } m \neq \tilde{m} \end{cases}$$

Equation (4-20) can be deduced merely by using the DE (4-13) satisfied by the J_n (see, for example, Churchill [48]). The Bessel functions should not be looked upon as anything mysterious. They are well-tabulated functions like trigonometric functions and satisfy many different identities just as the trigonometric functions do.

If we let

$$N_{nm} = \tfrac{1}{2} R^2 J_{n+1}^2(k_{nm}R)$$

we see that we can solve (4-19) for the A_{nm} by multiplying it by $rJ_n(k_{m\tilde{m}}r)$ and integrating with respect to r from 0 and R. This gives

$$A_{nm} = \frac{1}{N_{nm}} \int_0^R rF_n(r)\, J_n(k_{nm}r)\, dr$$

and it is now obvious how to conclude the solution to the original problem.

Bessel functions also arise in problems in potential theory. Suppose, for example, we wish to find the steady-state temperature in a right circular cylinder of radius R and length L with base in the xy plane and axis along the z axis. We assume that the lateral surface is insulated, the top is kept at a temperature $f(r)$, and the bottom at temperature 0°.

Since the BCs are all independent of θ, the solution will be too, and we get the following BVP:

DE:

$$\nabla^2 u(r, z) = 0 \qquad \begin{array}{l} 0 \leqslant r \leqslant R \\ 0 \leqslant z \leqslant L \end{array}$$

BCs:

$$\begin{array}{ll} u_r(R, z) = 0 & 0 \leqslant z \leqslant L \\ u(r, 0) = 0 & 0 \leqslant r \leqslant R \\ u(r, L) = f(r) & 0 \leqslant r \leqslant R \end{array}$$

We leave the solution to the reader as an exercise.

4-4. LEGENDRE POLYNOMIALS

Another important class of special functions occurs when the potential equation is separated in spherical coordinates. In

$$(r^2 u_r)_r + \frac{1}{\sin\theta}(\sin\theta\, u_\theta)_\theta + \frac{1}{\sin^2\theta} u_{\phi\phi} = 0 \tag{4-21}$$

we let

$$u(r, \theta, \phi) = v(r)\, w(\theta, \phi) \tag{4-22}$$

Then we get

$$\frac{(rv')'}{v} = \frac{1}{w}\left[-\frac{1}{\sin\theta}(\sin\theta\, w_\theta)_\theta - \frac{1}{\sin^2\theta} w_{\phi\phi}\right] = \lambda$$

whence

$$(r^2 v')' - \lambda v = 0 \tag{4-23}$$

and

$$\frac{1}{\sin\theta}(\sin\theta\, w_\theta)_\theta + \frac{1}{\sin^2\theta} w_{\phi\phi} + \lambda w = 0 \tag{4-24}$$

From (4-23)

$$r^2 v'' + 2rv' - \lambda v = 0 \tag{4-25}$$

This is an equation of the Euler-Cauchy type. Its solution is of the form

$$v = r^p \tag{4-26}$$

Substituting (4-26) into (4-25), we find

$$p(p-1) + 2p - \lambda = 0$$

or

$$p^2 + p - \lambda = 0$$

Rather than solve this for p in terms of λ, we will solve it for λ in terms of p. The reason will appear shortly.

$$\lambda = p(p+1) \tag{4-27}$$

If we separate variables in (4-24) by letting $w(\theta, \phi) = \alpha(\theta)\,\beta(\phi)$, we obtain

$$\frac{\sin^2\theta}{\alpha}\left[\left(\frac{1}{\sin\theta}\,\alpha'\right)' + \lambda\alpha\right] = -\frac{\beta''}{\beta} = n^2$$

where we have used the fact that the separation constant must be the square of an integer if we want solutions which are periodic in ϕ. The equation for α is

$$\sin\theta\,(\sin\theta\,\alpha')' + (\lambda\sin^2\theta - n^2)\alpha = 0$$

In it let $\cos\theta = z$. Then $-\sin\theta\,d\theta = dz$, and we get

$$\frac{d}{dz}\left[(1-z^2)\frac{d\alpha}{dz}\right] + \left(\lambda - \frac{n^2}{1-z^2}\right)\alpha = 0$$

This is the *generalized Legendre equation.* If u is independent of ϕ, then $n = 0$ and we get *Legendre's equation.* Its solutions are either singular at $+1$, at -1, or at both unless $\lambda = m(m+1)$, where m is an integer. In this case, one solution is a polynomial, called the *Legendre polynomial* of degree m. Thus, if we want regular solutions of the original problem, we have to take $p = m$ (an integer).

4-5. REMARKS ON THE METHOD OF SEPARATION OF VARIABLES

Separation of variables does not work for all equations. It works, in fact, for only a very few PDEs. If the PDE has variable coefficients, it will not usually be possible to assume a solution of the form $u(x, y) = \alpha(x)\,\beta(y)$ and be able to separate terms of the equation so that one side of the equation contains only functions of x and the other only functions of y. Even if the coefficients are constants, however, separation of variables is not always possible. For example, the biharmonic equation $u_{xxxx} + 2u_{xxyy} + u_{yyyy} = 0$

is not separable in rectangular coordinates. It is separable in polar coordinates, however.

The last remark illustrates another fact. Whether or not a PDE is separable depends, in general, on the coordinate system used. The few PDEs which are separable are separable only in a few special coordinate systems.

Even if a PDE is separable in a given, say $\xi\eta$, coordinate system, the fact will be of no avail unless the BCs for the problem at hand are given only on coordinate curves. For example, if a PDE is separable in the $\xi\eta$ coordinate system but one of the BCs is given on the line $\xi + \eta = 0$, then separability will not help unless there is a new coordinate system, say $\alpha\beta$, with $\alpha = \xi + \eta$, in which the PDE is separable and in which all other boundaries on which data are given have equations $\alpha =$ const or $\beta =$ const.

In addition, although it was not emphasized previously, an analysis of the method of separation of variables shows that the method will not usually work unless all but one of the BCs and ICs are zero. If there are several nonzero conditions, it is usually *necessary* (not merely convenient) to replace all but one of them by zero conditions, solve the several problems so obtained by taking all possible combinations of "all but one of them zero" conditions, and then use superposition to solve the original problem.

4-6. FUNDAMENTAL SOLUTION FOR THE REDUCED WAVE EQUATION

In the previous problems we have looked for solutions of

$$\nabla^2 u + \lambda u = 0 \qquad \text{in } \mathscr{R} \tag{4-28}$$

which vanished on the boundary of the domain, but which did not vanish identically inside. We found that such solutions existed only for certain special values of λ, called eigenvalues. Suppose λ is not one of these eigenvalues. Then the only solution of (4-28) vanishing on the boundary is $u = 0$. This means that the solution to (4-28) taking given boundary values is unique. In other words, it must be possible to solve the BVP

DE:

$$\Delta^2 u + \lambda u = 0 \qquad \text{in } \mathscr{R}$$

BC:

$$u = f \qquad \text{on } \partial\mathscr{R}$$

if λ is not an eigenvalue for $\mathscr{R}$.

It is natural to inquire if we can solve this first boundary-value problem for (4-28) by finding an appropriate Green's function. The answer is affirmative. Just as we did for the potential equation, we start by looking for spherically symmetric solutions of (4-28). If we assume $u = \psi(r)$ and substitute into (4-28), we find

$$\psi'' + \frac{n-1}{r}\psi' + \lambda\psi = 0 \tag{4-29}$$

If $n = 3$, (4-29) becomes

$$\psi'' + \frac{2}{r}\psi' + \lambda\psi = 0$$

or

$$r\psi'' + 2\psi' + \lambda r\psi = 0$$

This can be written in the form

$$(r\psi)'' + \lambda r\psi = 0$$

whence

$$\psi = \frac{e^{\pm ikr}}{r} \tag{4-30}$$

where, as usual, $\lambda = k^2$. It is necessary for later purposes to use complex exponentials instead of trigonometric functions.

Thus, just as for the potential equation, there are spherically symmetric solutions of the reduced wave equation that behave like $1/r$. Next we need an integral identity to use them in. The same Green's identity (3-4) as we used for the potential equation will do. If ϕ and ψ are both solutions of (4-28), then the left-hand side of (3-4) vanishes just as it did when ϕ and ψ were both harmonic. In Sec. 3-7 we applied (3-4) to a harmonic function and $1/r$ and thereby obtained (3-18), a formula which expresses a harmonic function at an interior point of a domain in terms of the boundary values of the function and of its normal derivative. If we carry out the same argument using a solution u of the reduced wave equation and one of the fundamental solutions (4-30), we will obtain the same formula as (3-17) with $1/r$ replaced by (4-30). In order to eliminate the unwanted $\partial u/\partial n$ from the boundary integral, we then replace (4-30) by the Green's function. In the present case, this must be a function $G(x, y, z, \xi, \eta, \zeta)$ such that $G = e^{\pm ikr}/r +$ regular solution of the reduced wave equation and such that $G = 0$ on ∂R. The rest of the analysis will be left to the reader.

If $n = 2$, then (4-29) is virtually Bessel's equation of index zero. Two linearly independent solutions of it are $J_0(r\sqrt{\lambda})$ and $N_0(r\sqrt{\lambda})$. It is known (see [48]) that $J_0(0) = 1$, but $N_0(z) = (2/\pi)\, J_0(z) \log z +$ regular function, near $z = 0$. Hence, N_0 is the desired fundamental solution. The Green's function in two dimensions must therefore satisfy $G = N_0 +$ regular function near $r = 0$.

Notice that the fundamental solution of (4-28) in two dimensions has a logarithmic singularity at $r = 0$. This means that when it is used in Green's identity, the fundamental solution of (4-28) will perform precisely the same picking out of the solution at one point that $\log r$ did for the potential equation.

For arbitrary n, (4-29) is reducible by means of a substitution to Bessel's equation with the result that

$$\psi(r) = r^{-(n-2)/2} Z_{(n-2)/2}(\sqrt{\lambda}\, r) \tag{4-31}$$

where $Z_{(n-2)/2}(\rho)$ is any solution of Bessel's equation of index $\frac{1}{2}(n - 2)$.

In order to have a fundamental solution of (4-28) we therefore have to take a singular Bessel function. It turns out that the singular Bessel function of order $\frac{1}{2}(n - 2)$ has a singularity of order $r^{-(n-2)/2}$ at the origin, and so the singularity of $\psi(r)$ is of order $r^{-(n-2)}$, which is the same as was the case for the potential equation.

4-7. OUTGOING AND INCOMING WAVES

As seen earlier, the reduced wave equation $\nabla^2 u + k^2 u = 0$ often arises from the time-dependent wave equation by splitting off a time-harmonic factor $e^{\pm ikct}$. If we agree to take the factor e^{-ikct} (the choice is purely arbitrary) then we have two spherically symmetric time-harmonic solutions to the three-dimensional time-dependent wave equation, namely,

$$\frac{e^{ik(r-ct)}}{r} \quad \text{and} \quad \frac{e^{-ik(r+ct)}}{r}$$

The first is an attenuated outgoing spherical wave, the second an incoming one. The situation would be reversed if we had chosen the time factor e^{+ikct}. Usually people choose the time factor e^{-ikct}, as we have. If it is clearly understood that this is the time factor, we can take the liberty of calling e^{ikr}/r an outgoing wave and e^{-ikr}/r an incoming wave. In fact, this is customary terminology. For example, in one dimension, e^{ikx} is often referred to as a wave going to the right even though this function obviously does not depend upon t at all. What is meant, of course, is that e^{ikx}, upon multiplication by the understood time factor e^{-ikct}, becomes $e^{ik(x-ct)}$, which is a wave going to the right.

In two dimensions even more liberty is taken. The time-dependent radially symmetric solutions are $H_0^{(1)}(kr)\, e^{-ikct}$ and $H_0^{(2)}(kr)\, e^{-ikct}$, where $H_0^{(1)} = J_0 + iN_0$ and $H_0^{(2)} = J_0 - iN_0$ are the Hankel functions of the first and second kind. Here J_0 and N_0 are the two Bessel functions mentioned previously. Neither of these solutions is a function of $r \pm ct$. However, for large r,

$$H_0^{(1)}(kr) \sim \sqrt{\frac{2}{\pi kr}}\, e^{i(kr-\pi/4)} \tag{4-32}$$

and

$$H_0^{(2)}(kr) \sim \sqrt{\frac{2}{\pi kr}}\, e^{-i(kr-\pi/4)} \tag{4-33}$$

Asymptotically, therefore, $H_0^{(1)}$ behaves like an outgoing wave and $H_0^{(2)}$ like an incoming wave in the sense explained above. Because of this, $H_0^{(1)}$ is usually called an outgoing wave and $H_0^{(2)}$ an incoming wave even though they are not waves in the precise sense of Chap. 2.

Even in three dimensions there are many solutions that do not have the form (4-30). Any solution that behaves at infinity like e^{ikr}/r is, however, called an outgoing wave, and any solution that behaves like e^{-ikr}/r there is called an incoming wave.

4-8. UNBOUNDED DOMAINS; RADIATION CONDITIONS

We saw that the Dirichlet problem for the potential equation did not have a unique solution in an unbounded domain unless some condition was imposed on the solution at infinity. In three dimensions we therefore required the solution to go to zero at infinity while in two dimensions mere boundedness there was sufficient. For the reduced wave equation different conditions are needed.

The reason for this can be seen from the fact that there are two linearly independent solutions, one of which behaves like e^{ikr}/r and the other of which behaves like e^{-ikr}/r at infinity but both of which obviously die out at infinity. It is therefore possible to have a solution of the reduced wave equation which vanishes on the boundary of an exterior domain and also tends to zero at infinity but which is nevertheless not identically zero.

What is needed is some sort of condition which will exclude one of the above two solutions. If such a condition is imposed, the remaining solution will be unable to satisfy the homogeneous boundary condition on the finite part of the boundary. Such a condition was found by Sommerfeld. Noting that in most physical problems the solution desired is the one which is outgoing at infinity, Sommerfeld devised a condition which excludes the incoming wave. He noticed that

$$\frac{\partial}{\partial r}\frac{e^{ikr}}{r} = ik\,\frac{e^{ikr}}{r} - \frac{e^{ikr}}{r^2}$$

and therefore

$$\frac{\partial}{\partial r}\frac{e^{ikr}}{r} - ik\,\frac{e^{ikr}}{r} = O\left(\frac{1}{r^2}\right)$$

Hence

$$\lim_{r\to\infty} r\left(\frac{\partial}{\partial r} - ik\right)\frac{e^{ikr}}{r} = 0$$

Accordingly, he formulated the *Sommerfeld radiation condition*

$$\lim_{r\to\infty} r\left(\frac{\partial u}{\partial r} - iku\right) = 0 \tag{4-34}$$

Notice that the incoming wave, which behaves like e^{-ikr}/r at infinity, does not satisfy (4-34). Notice, too, that if we wanted to keep the incoming wave and exclude the outgoing one, then instead of (4-34) we would impose the radiation condition

$$\lim_{r\to\infty} r\left(\frac{\partial u}{\partial r} + iku\right) = 0 \tag{4-35}$$

The radiation conditions (4-34) and (4-35) are the proper ones to use in three dimensions. In two dimensions, the right radiation conditions are

$$\lim_{r\to\infty} \sqrt{r}\left(\frac{\partial u}{\partial r} \pm iku\right) = 0 \tag{4-36}$$

By using the asymptotic formulas (4-32) and (4-39) it is easy to show that the Hankel functions satisfy (4-36). More precisely, $H_0^{(1)}$ satisfies (4-36) with the minus sign and $H_0^{(2)}$ satisfies (4-36) with the plus sign.

More generally, the radiation condition in n dimensions is

$$\lim_{r\to\infty} r^{(n-1)/2}\left(\frac{\partial u}{\partial r} \pm iku\right) = 0 \tag{4-37}$$

If the unbounded domain in question contains a full neighborhood of infinity, then any solution of the reduced wave equation which vanishes on the boundary and satisfies the radiation condition at infinity can be shown to be identically zero in the domain. If parts of the boundary of the domain extend to infinity, the situation is touchier. Stoker and Peters [269] have proved uniqueness if the part of the boundary that extends to infinity is a straight ray or the sides of a wedge. Other cases remain to be settled.

Magnus has shown that the condition (4-37) can be replaced by the somewhat weaker condition

$$\lim_{R\to\infty} \iint\limits_{r=R} \left|\frac{\partial u}{\partial r} \pm iku\right|^2 dS = 0 \tag{4-38}$$

For a particularly clear discussion of radiation conditions, the reader is referred to Wilcox [284]. For the more theoretically inclined reader, Levine [180] gives a very detailed treatment.

The formulation of a condition that guarantees the uniqueness of a solution to a problem does not necessarily represent the final word on the problem. In practice, for example, people are often interested in wave propagation problems in which plane waves are sent in from infinity. Such a plane wave has the form $\exp(ik\mathbf{n}\cdot\mathbf{r})$, where $\mathbf{n}$ is the normal to the wave front and $\mathbf{r} = x\mathbf{i} + y\mathbf{j} + z\mathbf{k}$ is the position vector from the origin to a point on the wavefront (remember that vectors are in boldface). Such a wave clearly does not die out at infinity, and so it does not satisfy the radiation condition. In such a case it is necessary to write the solution in the form $u = v + w$, where v is the given plane wave and w satisfies the radiation condition. Then uniqueness can be proved. Note that $\exp(ik\mathbf{n}\cdot\mathbf{r})$ is a solution of the reduced wave equation if $\mathbf{n} = \alpha\mathbf{i} + \beta\mathbf{j} + \gamma\mathbf{k}$ is a unit vector, because $\exp(ik\mathbf{n}\cdot\mathbf{r}) = \exp[ik(\alpha x + \beta y + \gamma z)]$ satisfies $\nabla^2 u + k^2 u = 0$ if $\alpha^2 + \beta^2 + \gamma^2 = 1$.

All of the above has referred to the equation $\nabla^2 u + k^2 u = 0$ because it is the equation that results from separating variables in the time-dependent wave equation. The equation

$$\nabla^2 u - k^2 u = 0 \tag{4-39}$$

is also of interest on occasion, though not in the same connection. Since the eigenvalues of ∇^2 are all negative, (4-39) always has a unique solution in a bounded domain. In an unbounded domain, nothing so sophisticated as a radiation condition is necessary. It turns out that a sufficient condition at infinity is that the solution merely go to zero there.

SUMMARY

Theorem S4-1 If u satisfies $\nabla^2 u + \lambda u = 0$ in a region $\mathscr{R}$ to which Green's identity is applicable, if $u = 0$ on $\partial\mathscr{R}$, and if λ (which may be a function of the independent variables in the PDE) satisfies $\lambda \leqslant 0$ in $\mathscr{R}$, then $u \equiv 0$ in $\mathscr{R}$.

Theorem S4-2 The solution to

DE:

$$\nabla^2 u - c^{-2} u_{tt} = 0 \qquad \begin{aligned} &0 < x < a \\ &0 < y < b \\ &0 < t < \infty \end{aligned}$$

BCs:

$$u(x, 0, t) = u(x, b, t) = u(0, y, t) = u(a, y, t) = 0$$

ICs:

$$u(x, y, 0) = f(x, y) \qquad u_t(x, y, 0) = g(x, y)$$

is

$$u(x, y, t) = \sum_{m,n} (a_{mn} \cos k_{mn}ct + b_{mn} \sin k_{mn}ct) \sin \frac{n\pi x}{a} \sin \frac{m\pi y}{b}$$

where

$$k_{mn} = \pi \sqrt{\frac{n^2}{a^2} + \frac{m^2}{b^2}} \qquad m, n = 1, 2, \ldots$$

and a_{mn} and $ck_{mn}b_{mn}$ are the Fourier coefficients of $f(x, y)$ and $g(x, y)$ respectively.

Theorem S4-3 The only spherically symmetric solutions to $\nabla^2 u + k^2 u = 0$ in three dimensions are $u = e^{\pm ikr}/r$ and linear combinations thereof. In two dimensions, the only radially symmetric solutions of this equation are $J_0(kr)$ and $N_0(kr)$ and linear combinations thereof, such as $H_0^{(1)}(kr)$, $H_0^{(2)}(kr)$, $K_0(kr)$.

Theorem S4-4 If u satisfies $\nabla^2 u + k^2 u = 0$ in a region $\mathscr{R}$ containing a full neighborhood of infinity, if $u = 0$ on $\partial\mathscr{R}$, and if

$$\lim_{r\to\infty} r^{(n-1)/2}(u_r \pm iku) = 0$$

then $u \equiv 0$ in $\mathscr{R}$.

PROBLEMS

1. Prove that the solution of the following problem, if it exists, is unique.

$$\nabla^2 u(x, y) - k^2 u = 0 \qquad \text{in a region } \mathscr{R}$$

$$u = f \qquad \text{on } \partial\mathscr{R}$$

2. Find $u(r, \theta, \phi)$ if

$$\nabla^2 u + k^2 u = 0 \qquad \begin{array}{l} \pi/k \leqslant r = \infty \\ 0 \leqslant \theta \leqslant \pi \\ 0 \leqslant \phi \leqslant 2\pi \end{array}$$

$$u\left(\frac{\pi}{k}, \theta, \varphi\right) = 15 \qquad \text{and} \qquad \lim_{r\to\infty} r(u_r - iku) = 0$$

3. The Schrödinger equation for the hydrogen atom (stationary solutions) requires $\psi(r, \theta, \phi)$ such that

$$-\frac{\hbar^2}{2m}\left[\frac{1}{r^2}\frac{\partial}{\partial r}\left(r^2\frac{\partial}{\partial r}\right)\right]\psi + \frac{L^2}{2mr^2}\psi - \frac{e^2}{r}\psi - E\psi = 0$$

where $\hbar$, m, e, and E are constants and

$$L^2 = -\hbar^2\left[\frac{1}{\sin^2\theta}\frac{\partial^2}{\partial\phi^2} + \frac{1}{\sin\theta}\frac{\partial}{\partial\theta}\left(\sin\theta\frac{\partial}{\partial\theta}\right)\right]$$

(*a*) Show that by assuming $\psi = R(r)\, Y(\theta, \phi)$ we are led to

$$L^2 Y = \lambda Y \qquad \lambda = \text{separation constant}$$

$$\lambda = \frac{1}{R} h^2 \left[\frac{\partial}{\partial r}\left(r^2 \frac{\partial R}{\partial r}\right)\right] + 2mre^2 + 2mr^2 E$$

(*b*) Prove that if we now assume $Y(\theta, \phi) = P(\theta)\, \Phi(\phi)$ and specify that Φ be periodic, we obtain

$$\Phi = Ae^{\pm in\varphi} \qquad n = \text{integer}$$

and P must satisfy

$$\frac{1}{P} \sin\theta \frac{\partial}{\partial\theta}\left(\sin\theta \frac{\partial P}{\partial\theta}\right) + \lambda \sin^2\theta - n^2 = 0$$

(*c*) Finally, if $\xi = \cos\theta$, show that we are led to

$$\frac{d}{d\xi}\left[(1 - \xi^2)\frac{dP}{d\xi}\right] - \frac{n^2}{1 - \xi^2} P + \lambda P = 0$$

which is the generalized Legendre equation. Since we want no singularities at θ and π, we must have $\lambda = l(l + 1)$ and then

$$Y_l^n(\theta, \phi) = BP_l^n(\cos\theta)\, e^{in\varphi}$$

where P_l^n is the associated Legendre function.

4. Let $\mathscr{R}$ be a cylinder of radius a and length L with its axis on the z axis. Solve:

DE:

$$\nabla^2 T(r, z) = 0 \qquad \begin{matrix} 0 \leqslant r \leqslant a \\ 0 \leqslant z \leqslant L \end{matrix}$$

BCs:

$$\frac{\partial T}{\partial r}(a, z) = 0 \qquad 0 \leqslant z \leqslant L$$

$$\left.\begin{matrix} T(r, 0) = 0 \\ T(r, L) = f(r) \end{matrix}\right. \qquad 0 \leqslant r \leqslant a$$

Hint:

$$J_0'(\lambda r) = J_1(\lambda r) \qquad \text{and} \qquad \int_0^a r J_0^2(\lambda_j r)\, dr = \tfrac{1}{2} a^2 J_1^2(\lambda_j a)$$

5. Show that the solution to the three-dimensional reduced wave equation $\nabla^2 u + k^2 u = 0$ satisfies the mean-value equation

$$u(P)\frac{\sin kR}{kR} = \frac{1}{4\pi R^2} \iint_{\mathscr{S}} u(S)\, dS$$

where $\mathscr{S}$ is a sphere of radius R centered at P.

6. Solve:

PDE:

$$u_{xx} + u_{yy} - c^{-2} u_{tt} = 0 \qquad \begin{matrix} 0 < x < 1 \\ 0 < y < 2 \\ 0 < t < \infty \end{matrix}$$

BC:

$$u = 0 \qquad \text{on boundary of above rectangle}$$

ICs:

$$u(x, y, 0) = 10xy(1 - x)$$
$$u_t(x, y, 0) = 0$$

7. Let r, θ, z be cylindrical coordinates. Solve:

DE:

$$\nabla^2 u = 0 \qquad \text{for } 0 \leqslant r < a \qquad \begin{array}{l} 0 \leqslant \theta \leqslant 2\pi \\ 0 < z < b \end{array}$$

BCs:

$$u(a, \theta, z) = 0$$
$$u(r, \theta, 0) = 0$$
$$u(r, \theta, b) = 0 = \text{const}$$

8. Solve:

DE:

$$u_{rr} + \frac{1}{r} u_r + \frac{1}{r^2} u_{\theta\theta} = c^{-2} u_{tt} \qquad \begin{array}{l} r_0 < r < R \\ 0 \leqslant \theta \leqslant 2\pi \\ 0 < t < \infty \end{array}$$

BCs:

$$u(r_0, \theta, t) = u(R, \theta, t) = 0$$

ICs:

$$u(r, \theta, 0) = (r - r_0)(r - R) \sin \theta \qquad u_t(r, \theta, 0) = 0$$

9. Solve:

PDE:

$$u_{xx} + u_{yy} = c^{-2} u_{tt} \qquad \begin{array}{l} |x| < a \\ |y| < a \\ t > 0 \end{array}$$

ICs:

$$u(x, y, 0) = x - y$$
$$u_t(x, y, 0) = x + y$$

BC:

$$u(\pm a, y, t) = u(x, \pm a, t) = 0$$

5
The Heat Equation

5-1. HEAT-DIFFUSION PROBLEMS

In Sec. 3-1 we considered the problem of the temperature distribution $u(x, y, z)$ in some region $\mathscr{R}$ when the boundary $\partial\mathscr{R}$ is held at some given temperature. It was assumed there that the boundary had been held at the given temperature for so long that the interior temperature had settled down to a steady state, i.e., was independent of time. Then, as was mentioned there, the temperature satisfies Laplace's equation, $\nabla^2 u = 0$ in $\mathscr{R}$, and the problem reduces to finding a solution of this partial differential equation (PDE) which satisfies the given boundary condition (BC). In the present chapter we are interested in finding out what happens if the BCs are not necessarily independent of time. In this case it turns out that $u(x, y, z, t)$ satisfies the PDE (cf. [48])

$$\nabla^2 u = u_t \qquad \text{in } \mathscr{R} \tag{5-1}$$

instead of the potential equation. Notice that (5-1) reduces to the potential equation if u is independent of t.

Again it is natural to assume that either u or $\partial u/\partial n$ or a linear combination of them is given on $\partial\mathscr{R}$, this time permitting these boundary values to depend upon t. However, on physical grounds, we realize that more information is needed. We also need to know what the initial temperature is throughout $\mathscr{R}$; that is, we need to know $u(x, y, z, 0)$ in $\mathscr{R}$. Thus, the proper problem for (5-1) seems to be one in which we give one BC and *one* initial condition (IC). This is in between the problems considered for the wave equation and the potential equation. For the wave equation we always gave two ICs, not just one. Of course, the wave equation contains a u_{tt} rather than just u_t. For the potential equation we gave no ICs, only a BC. Thus (5-1) appears to be of a type different from either the wave equation or the potential equation. This is in fact the case. As we will see in Chap. 6, the wave equation, the potential equation, and (5-1), which is called the *heat equation* or the *diffusion equation*, represent three essentially different types of second-order linear PDEs. In Chap. 6, we analyze these differences in detail. In the present chapter we merely make an introductory investigation into the nature of the heat equation (5-1).

As we did for the wave equation, we start with the one-dimensional case

$$u_{xx} = u_t \tag{5-2}$$

The first thing we notice about (5-2) is that it contains a first-derivative term as well as a second-derivative one. By itself, this fact is nothing new. The telegraph equation of Sec. 2-15 contained a u_t, too. In the telegraph equation we were able to get rid of the u_t term by a change of variable. A perusal of the method we used to get rid of it there, however, shows that we needed to have a u_{tt} term in the equation in order to get rid of the u_t term. The heat equation has no u_{tt} term in it. Therefore it is not possible to remove the u_t from it. This means that the differential operator in (5-2) has to remain one in which both first and second derivatives appear. If we try to factor this operator, we obtain a fractional derivative. Fractional derivatives can, in fact, be given a precise meaning (see Chap. 19), but at this stage we prefer more elementary methods. So we shall abandon the attempt to factor the operator in favor of separation of variables. Before doing this, however, we point out that if we were to conjecture that the second-derivative terms are the crucial ones in determining the nature of a second-order PDE, then the operator $\partial^2/\partial x^2$ would appear to be the appropriate one to consider for (5-2). This operator is remarkably easy to factor. However, the factors are equal and therefore yield only one family of characteristics, namely, $x = \text{const}$. Thus, as pointed out previously, the heat equation is indeed a sort of cross between the wave equation, which has two families of characteristics, and the potential equation, which has no (real) characteristics. Again we postpone the systematic investigation of these facts to Chap. 6.

5-2. HEAT CONDUCTION IN A FINITE BAR

Since x is the only space variable in (5-2), (5-2) can be thought of as describing the temperature distribution in a bar lying along the x axis. If the bar has finite length l, then an appropriate boundary-initial-value problem (BIVP) for (5-2) is

DE:

$$u_{xx} = u_t \qquad \begin{matrix} 0 < x < l \\ 0 < t < \infty \end{matrix}$$

BCs:

$$\begin{matrix} u(0, t) = h_1(t) \\ u(l, t) = h_2(t) \end{matrix} \qquad 0 < t < \infty$$

IC:

$$u(x, 0) = f(x) \qquad 0 < x < l$$

Other appropriate problems can be obtained from this one by replacing u by u_x or by a linear combination of u and u_x in either or both BCs.

The method of separation of variables will solve these problems, it being necessary, however, to replace some of the conditions by homogeneous ones and then superposing solutions to the various subproblems so obtained in order for the method to work. Since we have already met a similar situation in the case of the finite string, we leave the details to the reader as an exercise. Moreover, in Chap. 17, we will indicate how to solve problems for the heat equation by using the Laplace transform.

5-3. HEAT CONDUCTION IN AN INFINITE BAR

In order to get some idea as to the nature of the solutions to the heat equation, however, we will solve one special problem completely, namely, that of heat conduction in an infinite bar. The advantage here, just as for the infinite string, is the absence of any BCs. Only an IC need be prescribed. Thus, we consider the problem

PDE:

$$u_{xx} = u_t \qquad \begin{matrix} -\infty < x < \infty \\ 0 < t < \infty \end{matrix} \tag{5-3}$$

IC:

$$u(x, 0) = f(x) \qquad -\infty < x < \infty \tag{5-4}$$

In order to be able to carry out some of the techniques below, we will assume that $f(x) \to 0$ as $|x| \to \infty$ and that $|u(x, t)| < M$ as $|x| \to \infty$. These assumptions are also reasonable on physical grounds, since temperatures that grow ever greater are unrealistic.

As usual, we separate variables, writing

$$u(x, t) = \alpha(x)\,\beta(t) \tag{5-5}$$

Then we find

$$\frac{\alpha''}{\alpha} = \frac{\beta'}{\beta} = \lambda \tag{5-6}$$

where λ is the usual separation constant.

The equation for β implies $\beta = ce^{\lambda t}$. If $\lambda > 0$, then β, and hence u, grows exponentially in time. This is undesirable on physical grounds, and our assumption that u stays bounded enters at this stage to require that λ be less than 0. For clarity, we therefore put

$$\lambda = -\mu^2 \tag{5-7}$$

and then have

$$\beta(t) = ce^{-\mu^2 t} \tag{5-8}$$

Equations (5-6) and (5-7) imply that α satisfies the equation

$$\alpha'' + \mu^2\alpha = 0 \tag{5-9}$$

the (real-valued) solutions to which are sines and cosines. However, for purposes of technical convenience, it is preferable to use the complex-valued solutions $e^{\pm i\mu x}$ to (5-9).

In previous cases we always found that the separation constant μ had to be restricted to a discrete set of values, e.g., integers, or the zeros of a Bessel function, or the like. In the present case no such restriction is necessary. μ can take all real values from $-\infty$ to $+\infty$. It is therefore not necessary to use the solution $e^{-i\mu x}$. $e^{+i\mu x}$ is sufficient.

Previously, when there were only countably many μ's, say μ_n, $n = 1, 2, \ldots$, we multiplied each solution by a coefficient that depended on n and summed over n. This time we can multiply by a coefficient that depends on μ in any way we like. In other words, we can take an arbitrary function of μ, say $\omega(\mu)$, and thus form a solution of the form

$$u(x, t, \mu) = \omega(\mu)\, e^{i\mu x - \mu^2 t} \tag{5-10}$$

For every μ, (5-10) will be a solution of the heat equation which is bounded for all x and for all positive t. Since the parameter μ can take all values from $-\infty$ to $+\infty$, it is natural to superpose solutions by integrating over μ from $-\infty$ to $+\infty$ and then to try to determine the arbitrary function $\omega(\mu)$ so as to satisfy the initial condition (5-4). Thus we put

$$u(x, t) = \int_{-\infty}^{\infty} \omega(\mu)\, e^{i\mu x - \mu^2 t}\, d\mu \tag{5-11}$$

Then the IC requires that

$$f(x) = \int_{-\infty}^{\infty} \omega(\mu)\, e^{i\mu x}\, d\mu \tag{5-12}$$

and the problem now is how to solve the integral equation (5-12) for the unknown function $\omega(\mu)$ in terms of the given function $f(x)$. The solution to this problem is afforded by the Fourier inversion formula (see Chap. 17). This tells us that if $f(x)$ is expressed in terms of $\omega(\mu)$, as in (5-12), then $\omega(\mu)$ can be expressed in terms of $f(x)$ by means of the following formula:

$$\omega(\mu) = \frac{1}{2\pi} \int_{-\infty}^{\infty} f(x)\, e^{-i\mu x}\, dx \tag{5-13}$$

Substituting (5-13) into (5-11), we find

$$u(x, t) = \frac{1}{2\pi} \int_{-\infty}^{\infty} e^{i\mu x - \mu^2 t} \int_{-\infty}^{\infty} f(\xi)\, e^{-i\mu\xi}\, d\xi\, d\mu \tag{5-14}$$

or

$$u(x, t) = \frac{1}{2\pi} \iint\limits_{-\infty}^{\infty} f(\xi)\, e^{-t\mu^2 + i(x-\xi)\mu}\, d\xi\, d\mu \tag{5-15}$$

Since μ occurs only in the exponential, we can evaluate the integral over μ. To do this, we first complete the square in the exponent by writing

$$\begin{aligned} e^{-t\mu^2 + i(x-\xi)\mu} &= e^{-t\{\mu^2 - [i(x-\xi)/t]\mu\}} \\ &= e^{-t\{\mu - [i(x-\xi)/2t]\}^2}\, e^{-(x-\xi)^2/4t} \end{aligned}$$

Substituting this into (5-15), we have

$$u(x, t) = \frac{1}{2\pi} \int_{-\infty}^{\infty} e^{-(x-\xi)^2/4t} f(\xi)\, d\xi \int_{-\infty}^{\infty} e^{-t[\mu - i(x-\xi)/2t]^2}\, d\mu$$

Let

$$I = \int_{-\infty}^{\infty} e^{-t[\mu - i(x-\xi)/2t]^2}\, d\mu \tag{5-16}$$

and let

$$z = \mu - \frac{i(x-\xi)}{2t}$$

Then

$$I = \int_{-\infty+i\gamma}^{\infty+i\gamma} e^{-tz^2}\, dz$$

where

$$\gamma = -\frac{x-\xi}{2t}$$

Since the integrand is analytic and tends to zero as $|\,\mathrm{Re}(z)| \to \infty$, we can shift the path back to the real axis to get

$$I = \int_{-\infty}^{\infty} e^{-tz^2}\, dz$$

In this, let $z = y/\sqrt{t}$. Then $dz = dy/\sqrt{t}$ and

$$I = \frac{1}{\sqrt{t}} \int_{-\infty}^{\infty} e^{-y^2}\, dy = \sqrt{\frac{\pi}{t}}$$

Therefore

$$u(x, t) = \frac{1}{2\sqrt{\pi t}} \int_{-\infty}^{\infty} f(\xi)\, e^{-(x-\xi)^2/4t}\, d\xi \tag{5-17}$$

Just as in the case of Poisson's integral, the initial values are taken on only in the sense that

$$\lim_{t\to 0} u(x, t) = f(x) \tag{5-18}$$

The proof that (5-17) implies (5-18) can be found in any standard textbook. It makes use of the fact that the kernel

$$g(x, t, \xi, 0) = \frac{1}{2\sqrt{\pi t}}\, e^{-(x-\xi)^2/4t} \tag{5-19}$$

in (5-17) acts like a delta function. Indeed, for the benefit of those familiar with the delta function, we point out that $\lim_{t\to 0} g(x, t, \xi, 0) = \delta(x - \xi)$, and, if we replace t by $t - \tau$, then $\lim_{t\to\tau} g(x, t, \xi, \tau) = \delta(x - \xi)$. For those not familiar with the delta function, we insert the admonition that, contrary to terminology and some popular misconceptions, the delta function is not a function in the usual sense of the word. $\delta(x)$ does *not* assign any values to x. The symbol $\delta(x)$ all by itself is meaningless. What is defined is the *symbol* $\int_{-\infty}^{\infty} \delta(x) f(x)\, dx$. This complicated symbol is *defined* to be $f(0)$. No integration takes place here. The integral sign and the dx are only symbolic. Why write $f(0)$ in such a complicated way? The reason is that the symbolism can frequently be manipulated as if δ *were* a function and an integration really *were* taking place. For a clearer understanding of these concepts, see [113].

Because it has the character of a delta function, the kernel (5-19) is frequently called a *fundamental solution for the heat equation.*

SUMMARY

Theorem S5-1 The solution to

DE:

$$u_{xx} - u_t = 0 \qquad \begin{matrix} -\infty < x < \infty \\ 0 < t < \infty \end{matrix}$$

BC:

$$|u| < M \qquad \text{as } |x| \to \infty$$

IC:

$$\lim_{t \to 0} u(x, t) = f(x) \qquad \begin{matrix} -\infty < x < \infty \\ f(x) \to 0 \text{ as } |x| \to \infty \end{matrix}$$

is

$$u(x, t) = \frac{1}{2\sqrt{\pi t}} \int_{-\infty}^{\infty} f(\xi)\, e^{-(x-\xi)^2/4t}\, d\xi$$

PROBLEMS

1. Find the solution $u(x, t)$ of the following problem:

DE:

$$u_{xx} = u_t \qquad \begin{matrix} 0 < t < \infty \\ |x| < a \end{matrix}$$

IC:

$$u(x, 0) = x \qquad \lim_{t \to \infty} u(x, t) = 0$$

BC:

$$u(\pm a, t) = 0$$

2. Solve:

$$u_{xx} = u_t \qquad 0 \leqslant t < \infty$$

IC:

$$u(x, 0) = f(x)$$

BCs:

$$\begin{matrix} u(0, t) = u_1 \\ u(l, t) = u_2 \end{matrix} \qquad \text{both constants}$$

Hint: Put

$$u = u_1 + \frac{u_2 - u_1}{l} x + v$$

3. Find $u(x, t)$ such that

$$u_{xx} = u_t \qquad \begin{array}{l} 0 < x < 1 \\ 0 < t < \infty \end{array}$$

$u(x, 0) = cx$ where $c = \text{const}$
$u(0, t) = 0$
$u_x(1, t) = 0$

This corresponds to a flat slab of unit width such that:

(*a*) At $t = 0$, the temperature of the slab varies uniformly from zero at one end to c at the other end.

(*b*) One end is always at temperature zero.

(*c*) The other end is insulated for $t > 0$.

6
Classification of Equations

6-1. INTRODUCTION

We have so far discussed essentially three different second-order linear partial differential equations (PDEs), namely, the wave equation, the potential equation, and the heat equation. Each arises from a different physical situation than the other two, and for each a different mathematical problem was formulated. The differences in the formulations were twofold: (1) the type of domain in which the problem was formulated and (2) the type of boundary conditions (BCs) prescribed on the boundary of this domain. For the wave equation and for the heat equation the domain was always "open at the top," i.e., extended to infinity in the t direction. For the potential equation no such requirement was necessary. For the wave equation we gave one BC on lines parallel to the t axis but two BCs [called initial conditions (ICs) because t was interpreted as the time] on the x axis. For the heat equation we gave one BC all around the open boundary. For the potential equation we gave one BC on the entire boundary, closed or open.

Virtually all these problems were suggested by the physics of various situations which gave rise to these PDEs. On physical grounds it would be unrealistic to attempt to solve a Dirichlet or Neumann type of problem for the wave equation, just as it would be physically unrealistic to attempt to solve an initial-value problem (IVP) for the potential equation. However, there are a lot more than physical reasons for not attempting to interchange problems for some of the three PDEs under discussion. There are cogent mathematical reasons for this taboo. Some of these mathematical reasons are implicit in our results in preceding chapters. It is clear from Chap. 2 that just giving u (Dirichlet problem) or just giving u_t (Neumann problem) on the x axis would not be sufficient to enable us to solve the problem of the infinite string, i.e., the wave equation. Similarly, from Chap. 3 we know that since giving u or u_n alone on the boundary uniquely determines a potential function in a bounded domain, we cannot arbitrarily give *both* u and u_n on *any* part of the boundary and expect the resulting problem to be solvable for the potential equation.

We could easily cite many more examples to show that the type of problem appropriate for one of these three PDEs is not the same as the type appropriate for another one of them, and, indeed, as the opportunity arises in the future, we will. For the present, however, we merely want to convince the reader that the wave equation, the potential equation, and the heat equation are basically different mathematical animals even though all three are second-order linear PDEs in two independent variables. If the reader is not yet convinced of this fact, he should either continue the comparison we have started above or else patiently await the development of further evidence in the present and later chapters.

Thus, unlike ordinary differential equations (ODEs), PDEs cannot be classified by order alone. Even though two PDEs are both linear and second order and have the same number of independent variables, they may demand entirely different treatment. A type of problem that makes sense for one second-order linear PDE in two independent variables may be highly inappropriate for another second-order linear PDE in two independent variables.

Stating this another way, we can say that second-order linear PDEs in two independent variables come in various types. The natural question then is: How many different types of such equations are there, and how can they be classified?

6-2. CLASSIFICATION OF SECOND-ORDER LINEAR EQUATIONS IN TWO INDEPENDENT VARIABLES

The most general second-order linear PDE in two independent variables is

$$a(x, y)\, u_{xx} + 2b(x, y)\, u_{xy} + c(x, y)\, u_{yy} + d(x, y)\, u_x + e(x, y)\, u_y + f(x, y)\, u = F(x, y) \tag{6-1}$$

All the equations of the previous chapters fit into this form (in some cases with t renamed y). If $F \not\equiv 0$, then (6-1) is called *nonhomogeneous*. If $F \equiv 0$, then (6-1) is called *homogeneous*.

We have already seen in specific cases that the key problem is finding the solution to the homogeneous equation. Once we have a solution to the homogeneous equation, we can always cook up a solution to the nonhomogeneous equation out of it. Furthermore, it would be strange indeed if the differences between the nonhomogeneous wave, heat, and potential equations and others similar to them depended in any way on F. Thus, for the purpose of classifying PDEs of the type (6-1) we can restrict ourselves to the homogeneous case, i.e., the case when $F \equiv 0$.

We have already pointed out that the order of a PDE is not sufficient to determine its type. Nevertheless, we would expect that the terms involving the second derivatives, which differ from equation to equation, should be decisive in determining the type. We have already seen that in some special cases the first-derivative terms can be removed, but even if they cannot, as in the case of the heat equation, the feeling is that it is the way the second derivatives enter the equation that determines the type of second-order linear PDE. As just mentioned, the heat equation seems to violate this rule, but it really is the absence of u_{tt} rather than the presence of u_t in that equation which determines its type.

In any event we are going to define the type of second-order linear PDE in such a way that that type is entirely determined by the coefficients of the second-derivative terms. The plausibility argument just given for restricting the definition to just these terms might seem considerably short of a proof that this can be done. The final justification for the definition of type to be given below, however, is the fact that it can be proved rigorously that the same problem is appropriate for all equations of a given type. Thus, for example, since the Dirichlet problem is appropriate for the potential equation, it is also appropriate for all PDEs of the same type as the potential equation.

In order to be able to state things concisely, we give a name to the part of the operator consisting of the second-derivative terms.

DEFINITION 6-1

$$L[u] = a(x, y)\, u_{xx} + 2b(x, y)\, u_{xy} + c(x, y)\, u_{yy} \tag{6-2}$$

is called the *principal part* of the differential operator on the left side of (6-1).

The principal part is a second-order linear differential operator. Written in operational form, it reads

$$L = aD_x{}^2 + 2bD_xD_y + cD_y{}^2 \tag{6-3}$$

which is a quadratic form in the operators D_x and D_y. In order to simplify the notation, we let $D_x = \xi$ and $D_y = \eta$, so that (6-3) can be written as the quadratic form

$$Q(\xi, \eta) = a\xi^2 + 2b\xi\eta + c\eta^2 \tag{6-4}$$

Thus, associated with each PDE (6-1) there is a quadratic form (6-4). Now quadratic forms come in three different types: definite, semidefinite, or indefinite. Definite means that Q vanishes only if $\xi = \eta = 0$. It then must be either positive for all other values of ξ and η (positive definite) or else it must be negative for all other values (negative definite). Semidefinite means that Q is always of one sign but may vanish for values of ξ and η other than $\xi = \eta = 0$. Indefinite means that Q takes both positive and negative values and vanishes for a whole set of values of ξ and η.

Notice, however, that the coefficients of (6-4) depend upon x and y. Hence the type of quadratic form may vary from point to point in the xy plane; looked at in another way, (6-4) associates a different quadratic form with every point in the xy plane at which a, b, c are defined. Since each quadratic form is written as a form in the variables ξ, η, this means that at each point (x, y) we have a vector space with ξ and η as components of a generic vector in that vector space.

It is natural to classify the PDE (6-1) by means of the type of quadratic form (6-4) associated with it. Since, as has just been pointed out, the type of (6-4) depends upon x and y, this means that the classification of (6-4) will be a pointwise one. In other words, we shall not "classify (6-1)" but rather "classify (6-1) at a given point (x, y)."

Accordingly, we first give some definitions.

DEFINITION 6-2

The quadratic form (6-4) is called the *characteristic form* of the PDE (6-1).

DEFINITION 6-3

The PDE (6-1) is called *elliptic* at the point (x, y) if the characteristic form (6-4) is definite there. Equation (6-1) is called *parabolic* at (x, y) if (6-4) is semidefinite there and it is called *hyperbolic* at (x, y) if (6-4) is indefinite there.

Since $Q(\xi, \eta)$ contains only two independent variables, the criteria in Definition 6-3 can be simplified. If we factor η^2 out of Q and let

$$\zeta = -\frac{\xi}{\eta} \tag{6-5}$$

(the reason for introducing the minus sign into the definition of ζ will appear later), we can write

$$Q(\xi, \eta) = \eta^2 q(\zeta) \tag{6-6}$$

where

$$q(\zeta) = a\zeta^2 - 2b\zeta + c \tag{6-7}$$

From (6-6) it is evident that the type of Q is the same as the type of q. But q is just a quadratic in the single variable ζ. In the $q\zeta$ plane, q is a parabola. The quadratic $q(\zeta)$ is definite if this parabola stays either entirely above or entirely below the ζ axis, semidefinite if the parabola just touches the ζ axis at its vertex, and indefinite if the parabola crosses the ζ axis. This means that q is definite if the equation $q(\zeta) = 0$ has no real roots, semidefinite if that equation has one real root, and indefinite if the equation has two real roots. The following theorem follows from elementary algebra (cf. Definition 6-3).

Theorem 6-1 (Equivalent to Definition 6-3) (6-1) is elliptic at a point (x, y) if $b^2 - ac < 0$ there, parabolic at (x, y) if $b^2 - ac = 0$ there, and hyperbolic at (x, y) if $b^2 - ac > 0$ there.

Of course, if the coefficients a, b, c are constants, then the type of (6-1) is the same for all (x, y). Notice that this is true even if all the other coefficients in (6-1) are variable because our definition of type uses only the coefficients of the second derivatives.

Applying Theorem 6-1 to the wave equation, the potential equation, and the heat equation, respectively, we see that these are, respectively, hyperbolic, elliptic, and parabolic.

6-3. CHARACTERISTICS; REDUCTION OF A SECOND-ORDER LINEAR HYPERBOLIC EQUATION IN TWO INDEPENDENT VARIABLES TO CANONICAL FORM

Since, as we have just seen, the wave equation is hyperbolic, we look back to Chap. 2 for some ideas on how to analyze hyperbolic equations in general. The very first thing we did for the wave equation was to write it in operational form and then factor the operator. We have already written the principal part, (6-2), of (6-1) in operational form in (6-3). From the analysis of Sec. 6-2 we know that $q(\zeta)$ has two real distinct roots, say ζ_+ and ζ_- in the hyperbolic case. This means that the operator L given by (6-3) can be factored into the product of two first-order linear operators each of which represents a directional derivative, one in the direction whose slope is ζ_+ and the other in the direction whose slope is ζ_-. We have already done this in the special case of the one-dimensional wave equation (Sec. 2-2). The difference in the present more complicated case is that the roots $\zeta_\pm$ now depend upon x and y. Therefore, the directions of the directional derivatives vary from point to point. This is not a new situation either, however, for in Chap. 1 we encountered just such a case. There we introduced the curves whose directions were the same as those of the directional derivative. So we do the same here. We define two families of curves in the xy plane by means of the two ODEs

$$\frac{dy}{dx} = \zeta_\pm(x, y) \tag{6-8}$$

and call them *characteristic curves* of the PDE (6-1). The $\zeta_\pm$ are called the *characteristic directions* of (6-1). Note that (6-8) defines two distinct families of curves. The two ODEs in (6-8) are completely independent of each other. Each ODE defines a different family of characteristics.

Again let us emphasize that the characteristics defined by (6-8) exist only in those regions of the xy plane in which (6-1) is hyperbolic, i.e., where

$$b^2 - ac > 0 \tag{6-9}$$

Suppose that (6-9) does hold in some region of the xy plane and that we can solve the characteristic equations (6-8) there. What can we do with these solutions? Well, what did we do with the characteristics in the case of the wave equation? We used them as new coordinates there and they simplified the PDE considerably. Suppose we do the same thing here. We write the solutions to (6-8) in implicit form, letting

$$\phi(x, y) = \alpha \tag{6-10}$$

be the one-parameter (α) family of solutions of $dy/dx = \zeta_+$ and

$$\psi(x, y) = \beta \tag{6-11}$$

be the one-parameter (β) family of solutions of $dy/dx = \zeta_-$. Then we use the chain rule to introduce α and β as new independent variables in (6-2). Omitting the messy details of the calculation, we state the result:

$$L = \hat{a}(\alpha, \beta)\, u_{\alpha\alpha} + 2\hat{b}(\alpha, \beta)\, u_{\alpha\beta} + \hat{c}(\alpha, \beta)\, u_{\beta\beta} \tag{6-12}$$

where

$$\hat{a} = a\phi_x^2 + 2b\phi_x\phi_y + c\phi_y^2 = Q(\phi_x, \phi_y) \tag{6-13}$$

$$\hat{b} = a\phi_x\psi_x + b(\phi_x\psi_y + \phi_y\psi_x) + c\phi_y\psi_y \tag{6-14}$$

$$\hat{c} = a\psi_x^2 + 2b\psi_x\psi_y + c\psi_y^2 = Q(\psi_x, \psi_y) \tag{6-15}$$

Now from (6-10) we have

$$\phi_x + \frac{dy}{dx}\phi_y = 0$$

or

$$\frac{dy}{dx} = -\frac{\phi_x}{\phi_y} \tag{6-16}$$

Comparing this with (6-8), we see that $\zeta_+ = -\phi_x/\phi_y$. Then using (6-4) to (6-7) we see that

$$Q(\phi_x, \phi_y) = 0$$

Similarly,

$$Q(\psi_x, \psi_y) = 0$$

Hence $\hat{a} = \hat{c} = 0$ and the PDE assumes the canonical form

$$u_{\alpha\beta} + \text{lower-order terms} = 0 \tag{6-17}$$

Thus, a hyperbolic equation has two one-parameter families of characteristics. If these characteristics are chosen as new curvilinear coordinates, then the PDE assumes the form (6-17).

We have already seen in Chap. 2 that the expression $u_{xx} - u_{yy}$ can be transformed into u_{xy} and vice versa by a simple change of variables. Hence the canonical form for a hyperbolic equation can, if desired, be taken to be

$$u_{xx} - u_{yy} + \text{lower-order terms} = 0 \tag{6-18}$$

instead of (6-17).

6-4. PARABOLIC EQUATIONS

Next suppose that the characteristic equation has only one real root and there is only one family of characteristics. We choose their equation as $\phi = \alpha$. ψ can then be chosen as anything just so long as $\psi = \beta$ gives a different family of curves. In this case, the new PDE will have $\hat{a} = 0$ but $\hat{c} \neq 0$. It will, however, have $\hat{b} = 0$, too. This follows from the fact, verifiable by direct computation, that

$$\hat{b}^2 - \hat{a}\hat{c} = (b^2 - ac)(\phi_x\psi_y - \phi_y\psi_x) \tag{6-19}$$

Since one real root means $b^2 - ac = 0$, (6-19) implies $\hat{b}^2 - \hat{a}\hat{c} = 0$. Hence $\hat{a} = 0$ entails $\hat{b} = 0$, and we get the canonical form

$$u_{\beta\beta} + \text{lower-order terms} = 0 \tag{6-20}$$

6-5. ELLIPTIC EQUATIONS

Finally, we consider the case in which the roots of the characteristic equation are complex, i.e., the quadratic form is definite. In this case there are no characteristics at all, and it would at first seem that no simplification of the equation is possible. If we use complex numbers temporarily, however, we shall be able to see the correct canonical form in this case.

The two roots ζ_+ and ζ_- are complex conjugates in the present case. If we determine ϕ and ψ from them, we will have $\psi = \bar{\phi}$ and the canonical form (6-17). However, the coefficients in (6-17) will now be complex. If we put $x = \frac{1}{2}(\phi + \bar{\phi})$ and $y = (1/2i)(\phi - \bar{\phi})$, we then obtain a real transformation that takes the PDE into

$$u_{\tilde{x}\tilde{x}} + u_{\tilde{y}\tilde{y}} + \text{lower-order terms} = 0 \tag{6-21}$$

Having seen the result, we can, if we wish, eliminate the reliance on complex numbers by merely trying to determine ϕ and ψ so that $\hat{a} = \hat{c}$ and $\hat{b} = 0$. It can be proved that this is always possible if the characteristic form is definite and the coefficients are sufficiently "nice."

From a practical point of view, however, it is preferable to retain the complex notation. If we let $\zeta_+ = p + iq$, then $\zeta_- = p - iq = \bar{\zeta}_+$. Hence the ODEs (6-8) can be written in the form $dy/dx = p \pm iq$. Suppose we consider only the first of these, namely, $dy/dx = p + iq$. It will have a complex solution $y(x)$ given implicitly by an equation of the form $\phi(x, y) = \gamma$, where ϕ and γ are complex. Let $\phi(x, y) = \Phi(x, y) + i\Psi(x, y)$, and let $\gamma = \alpha + i\beta$, where α and β are real constants. Then $\eta(x, y) = \bar{\phi}(x, y) = \bar{\gamma}$ is obviously the solution to the other ODE, namely, $dy/dx = p - iq$. Moreover, $\Phi = \frac{1}{2}(\phi + \bar{\phi}) = \text{Re}\,(\phi)$ and $\Psi = (1/2i)(\phi - \bar{\phi}) = \text{Im}\,(\phi)$.

This shows how to find a real transformation that takes an elliptic equation into canonical form. The method is as follows: find one of the two complex roots of the characteristic equation (6-8). Call it ζ; put $dy/dx = \zeta(x, y)$ and solve this ODE for y as a function of x and a constant γ, writing the solution in implicit form as $\phi(x, y) = \gamma$. Since ζ is complex, ϕ and γ will be too. Let $\Phi(x, y) = \text{Re}\,[\phi(x, y)]$, $\Psi(x, y) = \text{Im}\,[\phi(x, y)]$, $\alpha = \text{Re}\,(\gamma)$ and $\beta = \text{Im}\,(\gamma)$. Then

$$\begin{aligned} \alpha &= \Phi(x, y) \\ \beta &= \Psi(x, y) \end{aligned}$$

is a real transformation which takes the given elliptic PDE into the canonical form

$$u_{\alpha\alpha} + u_{\beta\beta} + \text{lower-order terms} = 0$$

As emphasized previously, the type of equation may vary from point to point. If the coefficients of the PDE are continuous, however, an equation which is elliptic at a point will be so in the neighborhood of the point. The same can be said for hyperbolicity, but not for parabolicity. It is quite possible for an equation to be of different types in different regions. For example, Tricomi's equation, $yu_{xx} + u_{yy} = 0$, is elliptic in the upper half-plane, hyperbolic in the lower half-plane, and parabolic on the x axis.

From the canonical forms we see that the wave equation is the prototype of hyperbolic equations, the potential equation the prototype of elliptic equations, and the heat equation the prototype of parabolic equations.

It should also be pointed out that the transformation which carries a given PDE into canonical form is by no means unique. There are many such transformations. For example, if the equation is elliptic, then the canonical form is $\nabla^2 u +$ lower-order terms $= 0$. If we choose any transformation which carries the given elliptic equation into canonical form and follow that transformation with a conformal mapping, then ∇^2 will be multiplied by the Jacobian of the conformal mapping. If we divide the new equation by this Jacobian, then we will again have an equation which is in canonical form. Thus the transformation of an elliptic equation into canonical form is determined only to within an arbitrary conformal transformation. Similarly, in the hyperbolic and parabolic cases, any mapping which transforms a family of characteristics into itself will leave the equation in canonical form.

6-6. SECOND-ORDER LINEAR EQUATIONS IN n INDEPENDENT VARIABLES

All of the preceding has been done for a PDE with only two independent variables, x and y. If there are more than two independent variables, the situation becomes considerably more complex. Suppose that the PDE is still second order linear but that there are n independent variables, $x_1, \ldots, x_n$. Then the solution $u = u(x_1, \ldots, x_n)$ can be written in the abbreviated form $u = u(\mathbf{x})$, where $\mathbf{x}$ is an n-dimensional vector and the most general linear second-order PDE can be written in the form

$$\sum_{j,k=1}^{n} a_{jk}(\mathbf{x})\, u_{x_j x_k} + \sum_{j=1}^{n} b_j(\mathbf{x})\, u_{x_j} + c(\mathbf{x})\, u = F(\mathbf{x}) \tag{6-22}$$

The *principal part* is

$$L = \sum_{j,k=1}^{n} a_{jk} u_{x_j x_k} \tag{6-23}$$

The *characteristic form* is obtained by writing

$$u_{x_j x_k} = \frac{\partial}{\partial x_j}\frac{\partial}{\partial x_k} u$$

dropping u, and replacing $\partial/\partial x_j$ by ξ_j. Thus,

$$Q(\boldsymbol{\xi}) = \sum_{j,k=1}^{n} a_{jk}\xi_j\xi_k \tag{6-24}$$

is *defined* to be the characteristic form for (6-23).

In two dimensions we were able to find a change of independent variables,

$$y_j = \phi_j(\mathbf{x}) \qquad j = 1, \ldots, n \tag{6-25}$$

which transformed the principal part, (6-23), of the PDE into the canonical form

$$\hat{L}u = \sum_{j=1}^{n} \lambda_j u_{y_j y_j} \qquad \lambda_j = 0, \pm 1 \tag{6-26}$$

i.e., into a form that contains no mixed derivatives. In order to achieve the canonical form (6-26) it is necessary to remove from (6-23) all the coefficients a_{jk}, with $j \neq k$. There are $\frac{1}{2}n(n-1)$ such coefficients but only n functions ϕ_j with which to achieve this. It follows, therefore, that if $n > 3$, then the canonical form (6-26) cannot, in general, be achieved by means of a change of independent variables, (6-25). If $n = 3$, then the canonical form (6-26) can be achieved, but there are no free variables left to satisfy the additional conditions that all λ's be either 0 or ± 1. Hence in the case $n = 3$ (6-23) can be reduced to the canonical form (6-26) but with λ's that will, in general, be functions of the new independent variables $\mathbf{y}$.

Having failed to achieve canonical form for (6-23) by means of a change of the independent variables, (6-25), we remain faced with the problem of classifying PDEs such as (6-22) by other means. The "other means" are obtained by noticing what happens to the characteristic form Q under a change of variables such as (6-25). Obviously Q goes over into a new quadratic form with new coefficients. If the transformation (6-25) is successful in achieving the canonical form (6-26) (disregarding the conditions on the λ's), then the characteristic form for the new PDE will be in diagonal form. If the transformation (6-25) fails to achieve the canonical form (6-26), then the new characteristic form will not be diagonal. The latter is the case we have to circumvent. We do it by giving up the attempt to classify the PDE (6-22) by means of a transformation in the space of independent variables $\mathbf{x}$ of the PDE. Instead, given the PDE (6-22), we consider its corresponding characteristic form, (6-24), and we fix a definite point $\mathbf{x}$ at which we wish to classify the PDE. We do not attempt to change $\mathbf{x}$. $\mathbf{x}$ is fixed for the rest of the discussion. Hence the coefficients a_{jk} in Q are fixed. Now consider the problem of transforming the quadratic form (6-24), with the a_{jk}'s evaluated at the fixed point $\mathbf{x}$, into diagonal form. This is a standard problem in linear algebra. From that subject we know that there always exists an orthogonal linear transformation

$$\eta_j = \sum_{j,k=1}^{n} \gamma_{jk} \xi_k \tag{6-27}$$

that transforms (6-25) into the diagonal form

$$Q(\boldsymbol{\eta}) = \sum_{j=1}^{n} \lambda_j \eta_j^2 \tag{6-28}$$

Generalizing the definitions from the case $n = 2$, we lay down a definition.

DEFINITION 6-4

If all the λ's in (6-28) are different from zero and of the same sign, i.e., if Q is definite, then the PDE is called *elliptic*. If some of the λ's are zero, the PDE is called *parabolic*. If none of the λ's are zero and all but one are of the same sign, the equation is called *hyperbolic*. If none of the λ's are zero but at least two are positive and at least two negative, the equation is called *ultrahyperbolic*.

Notice that the classification of (6-22) has been achieved by using transformations in the vector space of the quadratic form and not in the space of the independent variables of the PDE. It has already been pointed out that transformations in the latter space will not usually suffice to achieve the desired canonical form. However, if the coefficients of the principal part of the PDE are constants, then the transformation (6-27) can be used to construct a transformation of the form (6-25) which does work in n dimensions. This is the subject of the next section.

Notice, too, that thus far no reference to characteristics has been made in the present section. They will be discussed in Chap. 11. Their discussion at this point would lead us too far afield from the subject at hand, the mere classification of PDEs.

Finally, we point out that the classification of (6-22) depends, according to Definition 6-4, only on the λ's in (6-28). Since the λ's are the eigenvalues of the real symmetric matrix (a_{jk}), the type of PDE (6-22) can be determined at a point by merely determining the eigenvalues of the matrix of the coefficients of the principal part of (6-22) at that point. Finding the coefficients γ_{jk} in (6-27) is unnecessary if all you want to do is to determine the type of PDE at the point in question.

6-7. SECOND-ORDER LINEAR EQUATIONS WITH CONSTANT COEFFICIENTS

Suppose that the coefficients of the principal part, (6-23), of (6-22) are all constants and that (6-27) is the transformation that reduces (6-24) to the canonical form (6-28). Put

$$y_j = \sum_{k=1}^{n} \gamma_{jk} x_k \tag{6-29}$$

From the chain rule for partial differentiation, it follows immediately that the change of variables (6-29) transforms (6-22) into the canonical form

$$\sum_{j=1}^{n} \lambda_j u_{y_j y_j} + \sum_{j=1}^{n} \hat{b}_j(\mathbf{y})\, u_{y_j} + c(\mathbf{y})\, u = F(\mathbf{y}) \tag{6-30}$$

If the $\hat{b}_j$'s are all constants, (6-30) can be simplified even further. Generalizing what we did for the telegraph equation, we put

$$u(\mathbf{y}) = v(\mathbf{y}) \exp\left(\sum_{k=1}^{n} \alpha_k y_k\right) = v(\mathbf{y}) \exp(\boldsymbol{\alpha} \cdot \mathbf{y}) \tag{6-31}$$

where $v(\mathbf{y})$ is the new unknown function and the α_k's are to be determined so as to simplify the resulting PDE.

Inserting (6-31) into (6-30), we find

$$\sum_{j=1}^{n} \lambda_j v_{y_j y_j} + \sum_{j=1}^{n} (2\lambda_j \alpha_j + \hat{b}_j)\, v_{y_j} + \tilde{c}(\mathbf{y})\, v = \tilde{F}(\mathbf{y}) \tag{6-32}$$

where $\tilde{c}$ is a new coefficient and $\tilde{F}$ a new nonhomogeneous term.

Equation (6-32) shows that the first derivative term corresponding to any $\lambda_j \neq 0$ can be removed by choosing

$$\alpha_j = -\frac{\hat{b}_j}{2\lambda_j} \tag{6-33}$$

in (6-31).

In particular, if all λ's are different from zero, the transformation (6-31) with α_j given by (6-33) takes (6-30) into the form

$$\sum_{j=1}^{n} \lambda_j v_{y_j y_j} + cv = F \tag{6-34}$$

If a particular λ, say λ_k, is zero, then, naturally, no $v_{y_k y_k}$ will occur in (6-32). Then, however, a term v_{y_k} will, as a rule, be present.

Equation (6-34) can be simplified still further. If $\lambda_j > 0$, then we can put $y_j = z_j \sqrt{\lambda_j}$ and thereby absorb the λ_j into a new independent variable. If $\lambda_j < 0$, we write $\lambda_j = -\mu_j$, where $\mu_j > 0$, and put $y_j = z_j \sqrt{\mu_j}$. The result is that (6-34) can be written in the form

$$\sum_{j=1}^{n} \epsilon_j v_{z_j z_j} + cv = F \tag{6-35}$$

where $\epsilon_j = \pm 1$.

In particular, for an elliptic equation either all the λ's are positive or all are negative. Then either $\epsilon_j = +1$ for all j or $\epsilon_j = -1$ for all j. In the latter case, we can multiply (6-35) by -1 and thus achieve a form in which all coefficients of $v_{z_j z_j}$ are $+1$. In any event, we see that a second-order linear elliptic equation in n independent variables *with constant coefficients* can always be transformed into the form

$$\nabla^2 u + cu = F \tag{6-36}$$

i.e., the reduced wave equation.

In the hyperbolic case, either all the λ's will be positive except one which will be negative, or, vice versa, all will be negative except one which will be positive. Since one of these cases goes into the other if we multiply the PDE by -1, we lose no generality in assuming that all λ's are positive except one which is negative. Then, assuming for notational convenience that there are $n+1$ independent variables, labeled $x_1, \ldots, x_n, t$, and that t is the variable corresponding to the negative λ, we see that the most general second-order linear hyperbolic PDE in $n+1$ independent variables can always be transformed into the form

$$\nabla^2 u - u_{tt} + cu = F \tag{6-37}$$

i.e., the telegraph equation.

6-8. *m*TH-ORDER LINEAR EQUATIONS

Still more complicated is the case of linear equations of order higher than the second. The most general equation of order m in n independent variables is

$$\sum_{j_1,\ldots,j_m=1}^{n} a^{(1)}_{j_1\cdots j_m}(\mathbf{x})\, u_{x_{j_1}\cdots x_{j_m}} + \sum_{j_1,\ldots,j_{m-1}=1}^{n} a^{(2)}_{j_1\cdots j_{m-1}} u_{x_{j_1}\cdots x_{j_{m-1}}} + \cdots = 0 \tag{6-38}$$

The principal part is

$$L = \sum_{j_1,\ldots,j_m=1}^{n} a^{(1)}_{j_1\cdots j_m} u_{x_{j_1}\cdots x_{j_m}} \tag{6-39}$$

and the characteristic form, obtained as before for (6-23), is

$$Q(\xi) = \sum_{j_1,\ldots,j_m=1}^{n} a^{(1)}_{j_1\cdots j_m} \xi_{j_1} \cdots \xi_{j_m} \tag{6-40}$$

Expression (6-40) is a form, i.e., a homogeneous polynomial, of degree m in the n variables $\xi_1, \ldots, \xi_n$. If $m > 2$, then it is no longer possible to diagonalize the n^mth array of coefficients $(a^{(1)}_{j_1\ldots j_m})$, and the method of classification developed in Sec. 6-4 fails. In order to classify equations such as (6-38) when $m > 2$, it is necessary to return once again to the notion of characteristic directions. This will be done in Chap. 11. Until then, we shall content ourselves with the only case in which characteristics play no role.

DEFINITION 6-5

If (6-40) is definite, then (6-38) is called *elliptic*. Note that (6-40) can be definite only if m is even.

6-9. NONLINEAR EQUATIONS

Since the definitions of principal part, characteristic form, and type of PDE all depend only on the terms involving the highest derivatives, the analysis of the preceding sections all goes through even if the lower derivatives occur nonlinearly, so long as they do not occur in the coefficients of the highest derivatives. Thus, the second-order PDE

$$a(x, y)\, u_{xx} + 2b(x, y)\, u_{xy} + c(x, y)\, u_{yy} + F(x, y, u, u_x, u_y) = 0 \qquad (6\text{-}41)$$

can be classified exactly as (6-1) was and brought into the appropriate canonical form just as (6-1) was. Similar remarks apply to equations of higher order. Because the characteristics of (6-41) are independent of the solution u considered, (6-41) is called *semilinear.*

If, however, the coefficients a, b, c in (6-41) are permitted to contain u and/or u_x and u_y, then the situation is far more complicated. As an example, consider the *quasi-linear equation* (so called because it is linear in the highest derivatives occurring)

$$a(x, y, u, u_x, u_y)\, u_{xx} + 2b(x, y, u, u_x, u_y)\, u_{xy} + c(x, y, u, u_x, u_y)\, u_{yy} + F(x, y, u, u_x, u_y) = 0 \qquad (6\text{-}42)$$

The type of (6-42) at any point (x, y) depends upon the sign of

$$\Delta(x, y, u, u_x, u_y) = b^2 - ac \qquad (6\text{-}43)$$

As indicated by the notation, however, Δ depends on u as well as upon x and y, with the result that the type of PDE (6-42), in general, depends upon the particular solution of (6-42) considered. For some solutions (6-42) may be elliptic, for some it may be parabolic, and for some hyperbolic. Since in a given problem the solution to (6-42) is not known in advance, the type is not either, except in special cases. Such a special case could be a PDE whose discriminant is of the same sign regardless of the solution considered. For example,

$$u^2 u_{xx} + 3u_x u_y u_{xy} - u^2 u_{yy} = 0 \qquad (6\text{-}44)$$

is clearly hyperbolic for every solution u. On the other hand, the equation

$$(1 - u_x^2)\, u_{xx} - 2u_x u_y u_{xy} + (1 - u_y^2)\, u_{yy} = 0 \qquad (6\text{-}45)$$

for the velocity potential in certain types of compressible flow is hyperbolic for those solutions u such that $|\nabla u| = \sqrt{u_x^2 + u_y^2} > 1$ and elliptic for those for which $|\nabla u| < 1$.

6-10. EIGENFUNCTION EXPANSIONS

While we are dealing with rather general equations we can place some of our previous work in a more general framework. To this end, let $L[u]$ be a second-order linear elliptic operator in n independent variables. Then the PDE

$$L[u] = c^{-2}u_{tt} \qquad \begin{matrix} \mathbf{x} \text{ in } \mathscr{R} \\ t \geqslant 0 \end{matrix} \tag{6-46}$$

is clearly hyperbolic and represents a generalized wave equation. It is therefore natural to try to impose the

BC:

$$u = 0 \qquad \begin{matrix} \text{for } \mathbf{x} \text{ on } \partial\mathscr{R} \\ t \geqslant 0 \end{matrix}$$

and

ICs:

$$\begin{matrix} u(\mathbf{x}, 0) = f(\mathbf{x}) \\ u_t(\mathbf{x}, 0) = g(\mathbf{x}) \end{matrix} \qquad \text{for } \mathbf{x} \text{ in } \mathscr{R} \tag{6-47}$$

A special case of this problem was considered in Sec. 4-1. Note that the coefficients of L are not permitted to depend upon t and that L does not contain any derivatives with respect to t.

We separate variables:

$$u(\mathbf{x}, t) = v(\mathbf{x})\, w(t) \tag{6-48}$$

It is again possible to prove that all the eigenvalues are negative. Hence

$$\frac{L[v]}{v} = \frac{w''}{c^2 w} = -k^2$$

Therefore

$$L[v] + k^2 v = 0 \qquad \text{in } \mathscr{R} \tag{6-49}$$

and

$$v = 0 \qquad \text{on } \partial\mathscr{R}$$

In general there will be a sequence of eigenvalues $-k_n^2$ and corresponding eigenfunctions v_n. Then

$$u(\mathbf{x}, t) = \sum_{n=0}^{\infty} v_n(\mathbf{x})(a_n \cos k_n ct + b_n \sin k_n ct) \tag{6-50}$$

and to satisfy the ICs (6-47) it is necessary to be able to expand $f(\mathbf{x})$ into a series

$$f(\mathbf{x}) = \sum_{n=0}^{\infty} a_n v_n(\mathbf{x}) \tag{6-51}$$

and $g(\mathbf{x})$ into a similar series.

If L is self-adjoint, then the v_n's are orthogonal, i.e.,

$$\int v_n v_m \, d\mathbf{x} = 0 \qquad \text{if } n \neq m \tag{6-52}$$

If L is not self-adjoint, then replace v_m by v_m^*, the eigenfunction of L^*. To show this take $\phi = v_n$ and $\psi = v_m$ in (3-4) and use (6-49) and the fact that the eigenfunctions vanish on $\partial\mathscr{R}$; (6-52) follows immediately.

If we put

$$N_n = \int_{\mathscr{R}} v_n^2(\mathbf{x}) \, d\mathbf{x} \tag{6-53}$$

then

$$a_n = \frac{1}{N_n} \int_{\mathscr{R}} f(\mathbf{x}) \, v_n(\mathbf{x}) \, d\mathbf{x} \tag{6-54}$$

If we let

$$\phi_n = N_n^{-1/2} v_n$$

then

$$\int_{\mathscr{R}} \phi_n^2 \, d\mathbf{x} = 1$$

and the coefficients α_n in the series

$$f(\mathbf{x}) = \sum_n \alpha_n \phi_n(\mathbf{x})$$

are given by

$$\alpha_n = \int_{\mathscr{R}} f(\mathbf{x}) \, \phi_n(\mathbf{x}) \, d\mathbf{x}$$

Thus far the eigenfunctions have been used to solve homogeneous equations. They can also be used to obtain a particular solution to a nonhomogeneous equation. In so using them we shall discover an interesting connection between the eigenfunctions and the Green's function. We have already seen (Sec. 3-11) that the Green's function can be used to obtain a particular solution of a nonhomogeneous equation. Now we shall show how to use the eigenfunctions to obtain a particular solution. To do this let $\phi_1, \ldots, \phi_n, \ldots$ be the normalized eigenfunctions of L corresponding to the eigenvalues $-k_1^2, \ldots, -k_n^2, \ldots$ and consider the nonhomogeneous equation

$$L[v] + k^2 v = F$$

Write

$$F = \sum_{n=0}^{\infty} A_n \phi_n$$

and

$$v = \sum_{n=0}^{\infty} c_n \phi_n$$

Since F is given, the A_n's are known. Since v is sought, the c_n's are unknown. Inserting these expansions into the DE (6-49), we get

$$L[\Sigma c_n \phi_n] + k^2 \Sigma c_n \phi_n = \Sigma A_n \phi_n$$

or

$$\Sigma c_n L[\phi_n] + \Sigma k^2 c_n \phi_n = \Sigma A_n \phi_n$$

From the DE we have $L[\phi_n] = -k_n{}^2 \phi_n$. So we get

$$-\Sigma c_n k_n{}^2 \phi_n + \Sigma k^2 c_n \phi_n = \Sigma A_n \phi_n$$

whence

$$(k^2 - k_n{}^2)\, c_n = A_n$$

or

$$c_n = \frac{A_n}{k^2 - k_n{}^2}$$

Hence

$$v(\mathbf{x}) = \sum_{n=0}^{\infty} \frac{A_n \phi_n(\mathbf{x})}{k^2 - k_n{}^2} \tag{6-55}$$

If $k = k_n$, there is no solution unless $A_n = 0$.
Since

$$A_n = \int_{\mathscr{R}} F(\xi)\, \phi_n(\xi)\, d\xi$$

we have from (6-55)

$$v(\mathbf{x}) = \sum_{n=0}^{\infty} \int_{\mathscr{R}} F(\xi) \frac{\phi_n(\mathbf{x})\, \phi_n(\xi)}{k^2 - k_n{}^2}\, d\xi$$

or, if the order of summation and integration can be interchanged,

$$v(\mathbf{x}) = \int_{\mathscr{R}} \left[\sum_{n=0}^{\infty} \frac{\phi_n(\mathbf{x})\, \phi_n(\boldsymbol{\xi})}{k^2 - k_n^{\,2}} \right] F(\boldsymbol{\xi})\, d\boldsymbol{\xi} \tag{6-56}$$

Comparing (3-36) and (6-56), we see that

$$g(\mathbf{x}, \boldsymbol{\xi}) = \sum_{n=0}^{\infty} \frac{\phi_n(\mathbf{x})\, \phi_n(\boldsymbol{\xi})}{k^2 - k_n^{\,2}} \tag{6-57}$$

Formula (6-57) establishes an interesting connection between the eigenfunctions ϕ_n and the Green's function g. It shows that if we have all the eigenfunctions, we can find the Green's function from them. The converse is also true in the sense that knowledge of the Green's function $g(\mathbf{x}, \boldsymbol{\xi}, \lambda)$ *for all complex values of the parameter* $\lambda = k^2$ leads to the eigenfunction expansion. That expansion is obtained by integrating g around a large contour in the complex λ plane and letting the contour expand to infinity. The eigenvalues turn out to be poles of g, and the eigenfunctions are obtained by finding the residues at those poles. Since a complete discussion of the details would take us too far afield, we refer the interested reader to Friedman [113].

6-11. REMARKS ON VARIOUS TYPES OF BOUNDARY-VALUE PROBLEMS

We have seen in the present chapter that PDEs can be classified into several different types. We have also seen in many of the previous chapters that the type of problem posed for a PDE can vary widely from equation to equation. For example, in Chap. 2 we solved IVPs and BIVPs, but in Chap. 4 we solved BVPs only. It is natural to ask how one knows what constitutes a sensible problem for a PDE. Unfortunately, the complete answer to this question is not yet known. For example, if the PDE is ultrahyperbolic, very few results exist showing just how many and what kind of boundary conditions are needed to determine a unique solution. Similarly for many nonlinear equations the current status of results still leaves much to be desired. Nevertheless, some statements can be made which give at least a measure of insight into the question of how to determine what conditions to impose on the solution of a PDE and where to impose them. We shall indicate here some of the guiding principles involved. Most of these principles, however, will be laid down for rather specific cases. Nevertheless, they should be of some use in practice.

Consider first the potential equation $u_{xx} + u_{yy} = 0$. We have seen in Chap. 3 that knowing u on the boundary of a bounded domain uniquely determines u inside that domain. Hence we cannot, as we did in some cases

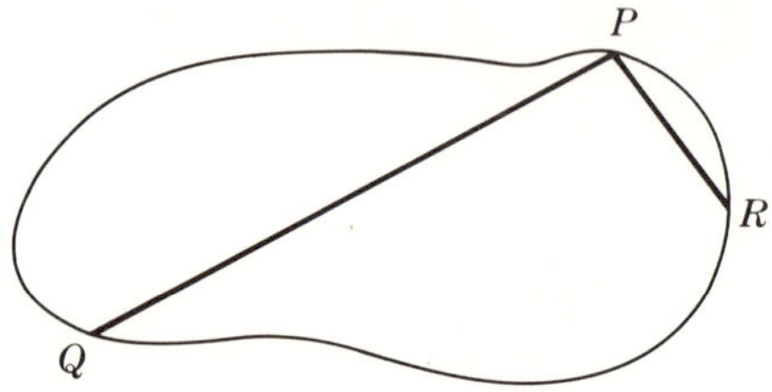

Figure 6-1

for the wave equation, give both u and an outgoing derivative of u on the boundary. We can give either u or an outward derivative but not both. In an unbounded domain, we need a condition at infinity, too. More generally, for any second-order elliptic equation one boundary condition on the boundary of a bounded domain is all that can be given. For higher-order elliptic equations (we can assume the order is even or else the equation could not be elliptic in the first place) it turns out that n boundary conditions should be prescribed if the elliptic PDE is of order $2n$.

For hyperbolic equations in two dimensions the situation is different. If u is the solution to a hyperbolic equation, then u cannot be prescribed arbitrarily on the entire boundary of a bounded domain because the characteristics of the equation will carry the prescribed boundary values from one point of the boundary to points on the other side. For example, if u is a solution to $u_{xx} - u_{yy} = 0$ in the domain in Fig. 6-1, then the value of u at P is completely determined by its values on the segment of the boundary between Q and R and hence cannot be prescribed at all at P. It is also clear that if two data are given for u on one part of the boundary of a domain, e.g., on the bottom of the semi-infinite strip in Fig. 6-2, then only one datum can be given on any other part of the boundary which is hit by a characteristic emanating from the part on which the two data are given. For further remarks on this subject, see Sec. 9-6.

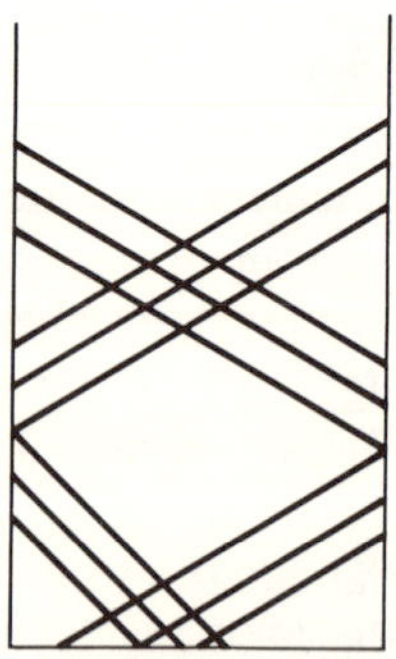

Figure 6-2

SUMMARY

The most general second-order linear PDE in two independent variables (x, y) is

$$a(x, y)\, u_{xx} + b(x, y)\, u_{xy} + c(x, y)\, u_{yy} + d(x, y)\, u_x + e(x, y)\, u_y + f(x, y)\, u = F(x, y)$$

If $F \equiv 0$, this PDE is called *homogeneous*. Otherwise it is called *nonhomogeneous*.

DEFINITION S6-1

The PDE

$$au_{xx} + bu_{xy} + cu_{yy} = F(x, y, u, u_x, u_y) \tag{S6-1}$$

is called *semilinear* if a, b, c are functions of the independent variables (x, y) only. If a, b, c are functions of (x, y, u, u_x, u_y), then (S6-1) is called *quasi-linear*.

DEFINITION S6-2

The PDE (S6-1) is called *hyperbolic* if $b^2 - 4ac > 0$, *parabolic* if $b^2 - 4ac = 0$, and *elliptic* if $b^2 - 4ac < 0$.

If (S6-1) is semilinear, the type of (S6-1) as defined here is a function of the point (x, y). If (S6-1) is quasi-linear, then the type also depends upon the solution u for which the type is being computed.

DEFINITION S6-3

Let $F(x, y, u, p, q, r, s, t)$ possess continuous partial derivatives with respect to all 8 independent variables. A function $u(x, y)$ that satisfies

$$F[x, y, u(x, y), u_x(x, y), u_y(x, y), u_{xx}(x, y), u_{xy}(x, y), u_{yy}(x, y)] = 0$$

identically in (x, y) is called a *solution* of the PDE

$$F(x, y, u, u_x, u_y, u_{xx}, u_{xy}, u_{yy}) = 0$$

This PDE is called *hyperbolic* at those points and for those solutions for which $F_s^2 - 4F_rF_t > 0$, *parabolic* for those for which this expression is zero, and *elliptic* for those for which it is negative.

DEFINITION S6-4

The *characteristic directions* of the PDE (S6-1) at a point (x, y) are the roots $\zeta_\pm(x, y)$ of the quadratic equation $a\zeta^2 + b\zeta + c = 0$. If (S6-1) is quasi-linear, these roots may depend on the solution $u(x, y)$ chosen as well as on the point (x, y).

DEFINITION S6-5

The *characteristic curves* of (S6-1) are the solutions of the ODEs $dy/dx = \zeta_{\pm}(x, y)$.

If (S6-1) is semilinear and hyperbolic, there are two independent families of characteristics for it. If $\phi(x, y) = \alpha$ and $\psi(x, y) = \beta$ are equations for these families, then the change of variables

$$\begin{aligned} \alpha &= \phi(x, y) \\ \beta &= \psi(x, y) \end{aligned} \tag{S6-2}$$

converts (S6-1) into $\hat{u}_{\alpha\beta} = \hat{F}(\alpha, \beta, \hat{u}, \hat{u}_\alpha, \hat{u}_\beta)$.

The further change of variables

$$\begin{aligned} \xi &= \alpha + \beta \\ \eta &= \alpha - \beta \end{aligned}$$

converts this into $\tilde{u}_{\xi\xi} - \tilde{u}_{\eta\eta} = \tilde{F}(\xi, \eta, \tilde{u}, \tilde{u}_\xi, \tilde{u}_\eta)$.

If (S6-1) is parabolic, then there exists a single family of characteristics. If $\phi(x, y) = \alpha$ is a representation for this family and $\psi(x, y) = \beta$ is any other independent family of curves (which are therefore noncharacteristic), then the change of variables (S6-2) converts (S6-1) into $\hat{u}_{\beta\beta} = \hat{F}(\alpha, \beta, \hat{u}, \hat{u}_\alpha, \hat{u}_\beta)$.

If (S6-1) is elliptic, then the characteristic curves will be complex. Let $\Phi(x, y) = \phi(x, y) + i\psi(x, y)$; let $\gamma = \alpha + i\beta$, where $\phi, \psi, \alpha, \beta$ are real, and let $\Phi(x, y) = \gamma$ be an equation for one family of these complex characteristics. The representation for the other family could be obtained by taking complex conjugates, but we do not need it because the transformation (S6-2), where now ϕ and ψ are defined as just described, converts (S6-1) into $\hat{u}_{\alpha\alpha} + \hat{u}_{\beta\beta} = \hat{F}(\alpha, \beta, \hat{u}, \hat{u}_\alpha, \hat{u}_\beta)$.

The most general second-order linear PDE in n independent variables, $\mathbf{x} = (x_1, \ldots, x_n)$, is

$$\sum_{j,k=1}^{n} a_{jk}(\mathbf{x})\, u_{x_j x_k} + \sum_{j=1}^{n} b_j(\mathbf{x})\, u_{x_j} + c(\mathbf{x})\, u = F(\mathbf{x}) \tag{S6-3}$$

DEFINITION S6-6

Let $\lambda_j(\mathbf{x})$, $j = 1, \ldots, n$ be the eigenvalues of the matrix $(a_{jk}(\mathbf{x}))$. If at least one $\lambda_j = 0$, then (S6-3) is called *parabolic*. If all $\lambda_j \neq 0, j = 1, \ldots, n$, then (S6-3) is called *elliptic* if all λ_j's have the same sign, *hyperbolic* if all but one have the same sign, and *ultrahyperbolic* if at least two λ's are positive and at least two are negative.

DEFINITION S6-7

The *principal part* of (S6-3) is the linear homogeneous differential operator

$$L[u] = \sum_{j,k=1}^{n} a_{jk}(\mathbf{x})\, u_{x_j x_k}$$

DEFINITION S6-8

The *characteristic form* of (S6-3) (obtained by writing

$$u_{x_j x_k} = \frac{\partial}{\partial x_j}\frac{\partial}{\partial x_k} u$$

dropping u, and replacing $\partial/\partial x_j$ by ξ_j) is the quadratic form

$$Q(\boldsymbol{\xi}) = \sum_{j,k=1}^{n} a_{jk}(\mathbf{x})\, \xi_j \xi_k$$

If the matrix (a_{jk}) in (S6-3) is constant, if λ_j, $j = 1, \ldots, n$, are its eigenvalues, and if (γ_{jk}) is the matrix that diagonalizes (a_{jk}), then the change of variables

$$y_j = \sum_{k=1}^{n} \gamma_{jk} x_k$$

converts (S6-3) into the form

$$\sum_{j=1}^{n} \lambda_j \hat{u}_{y_j y_j} + \sum_{j=1}^{n} \hat{b}_j(\mathbf{y})\, \hat{u}_{y_j} + \hat{c}(\mathbf{y})\, \hat{u} = \hat{F}(\mathbf{y}) \tag{S6-4}$$

If, in addition, the $\hat{b}_j$'s are all constants and the λ_j's are all different from zero, then the further change of variables

$$\hat{u}(\mathbf{y}) = v(\mathbf{y}) \exp\left(-\frac{1}{2}\sum_{j=1}^{n} \lambda_j^{-1}\hat{b}_j\right) \tag{S6-5}$$

reduces this to

$$\sum_{j=1}^{n} \lambda_j v_{y_j y_j} + c(\mathbf{y})\, v = F(\mathbf{y})$$

If a particular λ, say λ_J, is zero, then the Jth term has to be omitted from the sum in (S6-5) and the resulting PDE will lack a term in $u_{y_J y_J}$ but will contain one in u_{y_J}.

As corollaries of the above we have two theorems.

Theorem S6-1 The most general second-order linear elliptic PDE with constant coefficients can always be reduced by a change of variables to the form $\nabla^2 u + \lambda u = 0$, where ∇^2 is the n-dimensional Laplacian.

Theorem S6-2 The most general second-order linear hyperbolic PDE with constant coefficients can, by an appropriate change of variables, always be reduced to the form $\nabla^2 u - c^{-2}u_{tt} + \lambda u = 0$, where ∇^2 is the $(n - 1)$-dimensional Laplacian.

Theorems S6-1 and S6-2 are *false* if the coefficients are variable. Even a reduction to the form (S6-4) is generally impossible if the number of independent variables exceeds three.

PROBLEMS

1. Reduce to canonical form:

(*a*) $u_{rr} + \frac{1}{r} u_r + \frac{1}{r^2} u_{\theta\theta} = 0$

(*b*) $x^2u_{xx} - 2xyu_{xy} + y^2u_{yy} = 17{,}986{,}111$

2. Give an example of:

(*a*) A quasi-linear second-order PDE which is elliptic everywhere.
(*b*) A second-order PDE which is parabolic everywhere.
(*c*) A third-order PDE which is elliptic.

3. Classify the following equations as to type in the xy plane:

(*a*) $x^2yu_{xx} + xyu_{xy} - y^2u_{yy} = 0$
(*b*) $(x \log y)\, u_{xx} + 4yu_{yy} = 0$
(*c*) $\sqrt{y^2 + x^2}\, u_{xx} + 2(x - y)\, u_{xy} + \sqrt{y^2 + x^2}\, u_{yy} = 0$
(*d*) $yu_{xx} + x^2y^2u_{yy} = 0$
(*e*) $(\sin x)\, u_{xx} + 2(\cos x)\, u_{xy} + (\sin x)\, u_{yy} = 0$
(*f*) $u_{xy} - xyu_{yy} = 0$
(*g*) $x^2yu_{xx} - xy^2u_{yy} = 0$

4. Reduce to canonical form:

(*a*) $u_{xx} + 2u_{xy} + u_{yy} + u_x + u_y = 0$
(*b*) $u_{xx} + 2u_{xy} + 5u_{yy} + 3u_x + u = 0$
(*c*) $3u_{xx} + 10u_{xy} + 3u_{yy} = 0$
(*d*) $u_{xx} + 2u_{yy} + u_{zz} = 2u_{xy} + 2u_{yz}$
(*e*) $u_{xx} + u_{yy} + u_{zz} + 2u_{yz} = 0$

5. Find the equations of the characteristics of
$x^2u_{xx} - 2xyu_{xy} + y^2u_{yy} = 2$

6. Put Tricomi's equation, $yu_{xx} + u_{yy} = 0$, into canonical form for (*a*) $y < 0$ and (*b*) $y > 0$.

7. For what values of k is $u_{xx} - kxu_{xy} + 4x^2u_{yy} = 0$ (*a*) elliptic, (*b*) parabolic, or (*c*) hyperbolic? For $k = 0$, reduce to canonical form.

8. Given the PDE $u_{xx} + 2au_{xy} + 4u_{yy} + bu_{zz} = 0$:

(*a*) For what values of a and b is the PDE parabolic?
(*b*) If $b = 4$, for what value(s) of a is the PDE elliptic?
(*c*) If $b = 4$, for what value(s) of a is the PDE hyperbolic?

9. Classify the PDE $u_{xx} + u_{yy} + u_{zz} + u_{xy} + u_{xz} + u_{yz} = 0$.

10. Reduce $x^2u_{xx} + 4xyu_{xy} + y^2u_{yy} = 98$ to canonical form.

11. Find the characteristic curves and reduce the following PDE to canonical form: $x^2yu_{xx} + xy^2u_{yy} = 0$

Indicate in which region of the xy plane the equation (and its corresponding canonical form) is (*a*) elliptic, (*b*) hyperbolic, (*c*) parabolic.

12. Classify the following equations as to type:

(*a*) $xyu_{xx} + xtu_{xt} + z^2u_{tt} + u_{yy} + tzu_{zz} = 0$, where $u = u(x, y, z, t)$
(*b*) $u_{xx} + u_{yy} + u_{zz} + u_{tt} = 0$, where $u = u(x, y, z, t)$
(*c*) $u_{xx} + u_{yy} - u_{zz} + u_{tt} = 0$, where $u = u(x, y, z, t)$

13. Find the general solution of $u_{xx} - 5u_{xy} + 6u_{yy} = 0$. *Hint*: Reduce to canonical form.

7
The Wave Equation in Higher Dimensions

7-1. SOLUTION OF THE INITIAL-VALUE PROBLEM IN THREE SPACE DIMENSIONS

In Chap. 2 we were able to solve many different problems for the one-dimensional wave equation by using the representation (2-8) for the general solution. In higher dimensions no such general representation can be obtained. It is therefore necessary to consider specific problems and to develop methods suited to their solution. First we consider an initial value problem (IVP) for the three-dimensional wave equation in all of space, i.e.,

DE:

$$\nabla^2 u - c^{-2} u_{tt} = 0 \qquad \begin{array}{l} \text{all } (x, y, z) \\ \text{all } t > 0 \end{array} \tag{7-1}$$

ICs:

$$\left.\begin{array}{l} u(x, y, z, 0) = f(x, y, z) \\ u_t(x, y, z, 0) = g(x, y, z) \end{array}\right. \qquad \text{all } (x, y, z) \tag{7-2}$$

Just as we have done in previous cases where we were unable to find the general solution to a PDE, we look for special solutions. In particular, we look for solutions possessing spherical symmetry, that is, $u = v(r, t)$. Then we find

$$v_{rr} + \frac{2}{r} v_r - c^{-2} v_{tt} = 0 \tag{7-3}$$

This can be written in the form

$$\frac{1}{r} [(rv)_{rr} - c^{-2}(rv)_{tt}] = 0 \tag{7-4}$$

which is the one-dimensional wave equation for rv. Hence

$$v = \frac{\phi(r - ct)}{r} + \frac{\psi(r + ct)}{r} \tag{7-5}$$

So we see that the spherically symmetric solutions are outgoing and incoming waves with decreasing and increasing amplitudes, respectively. The distance r can be measured from an arbitrary point (ξ, η, ζ):

$$r^2 = (x - \xi)^2 + (y - \eta)^2 + (z - \zeta)^2 \tag{7-6}$$

Then, since (ξ, η, ζ) are parameters in the special solutions (7-5) of the PDE (7-1), we can, as we did when we separated variables for the heat equation in Chap. 5, multiply the solutions in (7-5) by arbitrary functions of (ξ, η, ζ) and integrate over (ξ, η, ζ). Thus, for any functions $h_1(\xi, \eta, \zeta)$ and $h_2(\xi, \eta, \zeta)$ for which the integrals below exist and can be differentiated twice with respect to (x, y, z), we have

$$\begin{aligned} u(x, y, z, t) = &\iiint h_1(\xi, \eta, \zeta) \frac{\phi(r - ct)}{r} \, d\xi \, d\eta \, d\zeta \\ &+ \iiint h_2(\xi, \eta, \zeta) \frac{\psi(r + ct)}{r} \, d\xi \, d\eta \, d\zeta \end{aligned} \tag{7-7}$$

as a large family of solutions of (7-1). We do not claim that (7-7) is the most general solution of (7-1), however. Nevertheless, it should be possible to determine the arbitrary functions h_1, h_2, ϕ, and ψ so that the ICs (7-2) are satisfied. Since this appears to be a formidable problem, we attempt an inverse approach. Instead of trying to determine the arbitrary functions in (7-7) so as to satisfy the given ICs (7-2), we try specializing the arbitrary functions in (7-7) and seeing just what ICs the solution obtained by our specialization does satisfy.

Our specialization is heuristic. We imagine that we have only an outgoing wave, i.e., we assume that $\psi \equiv 0$, and that the outgoing disturbance

ϕ is spread out over a very thin shell. In fact, we assume that ϕ is a delta function. Renaming the arbitrary function h_1 by calling it g, we then have

$$u(x, y, z, t) = \iiint_{-\infty}^{\infty} g(\xi, \eta, \zeta) \frac{\delta(r - ct)}{r} \, d\xi \, d\eta \, d\zeta \tag{7-8}$$

as a special solution of (7-1).

In order to exploit the "picking-out" property of the delta function, we have to use r as one of the integration variables. According to (7-6), r represents the distance from the fixed point (x, y, z) to the variable point of integration (ξ, η, ζ). It is therefore natural to introduce spherical polar coordinates with the pole at (x, y, z) and r as the distance from the pole. In order to preserve the symmetry present in the problem, we use angles different from the usual ones. Since r is to be one of the new coordinates, the others must vary on the surfaces $r =$ const. The surfaces $r =$ const are spheres centered at (x, y, z). If α, β, γ are the angles between a radius vector from (x, y, z) to (ξ, η, ζ), then $(r \cos \alpha, r \cos \beta, r \cos \gamma)$ is a point on a sphere of radius r about (x, y, z). Hence α, β, γ will serve as coordinates on the sphere. They are not all independent but are connected by the relation

$$\cos^2 \alpha + \cos^2 \beta + \cos^2 \gamma = 1 \tag{7-9}$$

To summarize, we introduce new integration variables, r, α, β, γ, by putting

$$\begin{aligned} \xi &= x + r \cos \alpha \\ \eta &= y + r \cos \beta \\ \zeta &= z + r \cos \gamma \end{aligned}$$

where α, β, γ satisfy (7-9). Thus, $(\cos \alpha, \cos \beta, \cos \gamma)$ is a point on the unit sphere. If $d\Omega$ is the element of surface area of the unit sphere, then

$$d\xi \, d\eta \, d\zeta = r^2 \, d\Omega \, dr$$

and (7-8) becomes

$$u(x, y, z, t) = \iint_{\Omega}^{\infty} \int_0^{\infty} r \, g(x + r \cos \alpha, y + r \cos \beta, z + r \cos \gamma) \, \delta(r - ct) \, dr \, d\Omega$$

or, by the fundamental property of the delta function,

$$u(x, y, z, t) = ct \iint_{\Omega} g(x + ct \cos \alpha, y + ct \cos \beta, z + ct \cos \gamma) \, d\Omega$$

Obviously $u(x, y, z, 0) = 0$ and $u_t(x, y, z, 0) = 4\pi c g(x, y, z)$. This leads us to conjecture the following theorem.

Theorem 7-1

$$u(x, y, z, t) = \frac{t}{4\pi} \iint_{\Omega} g(x + ct \cos \alpha, y + ct \cos \beta, z + ct \cos \gamma)\, d\Omega \tag{7-10}$$

is the solution to the following IVP:

DE:

$$\nabla^2 u - c^{-2} u_{tt} = 0 \qquad \begin{matrix} \text{all space} \\ t \geqslant 0 \end{matrix} \tag{7-11}$$

ICs:

$$\begin{aligned} u(x, y, z, 0) &= 0 \\ u_t(x, y, x, 0) &= g(x, y, z) \end{aligned} \qquad \text{all } (x, y, z) \tag{7-12}$$

Rather than attempt to justify the nonrigorous method by which this theorem was derived, we prove it by direct verification. That (7-10) satisfies the ICs (7-12) is, as mentioned above, obvious. That it satisfies the PDE (7-11) is far less obvious, and the direct proof of this fact is lengthy and untransparent. We therefore omit it.

It remains to solve the IVP (7-2) with $f \neq 0$. This turns out to be easy to do once we have Theorem 7-1. In order to facilitate matters, let

$$\mu[g] = \frac{1}{4\pi} \iint_{\Omega} g(x + ct \cos \alpha, y + ct \cos \beta, z + ct \cos \gamma)\, d\Omega \tag{7-13}$$

μ, which is a function of (x, y, z, t) and a functional of g, is a sort of mean value of g. (To see this multiply by c^2t^2 inside and divide by it outside the integral.)

The solution, (7-10), of (7-12) can be written in the abbreviated form

$$u = t\mu[g] \tag{7-14}$$

Now suppose we wish to find a solution of (7-11) satisfying the ICs

$$\begin{aligned} u(x, y, z, 0) &= f(x, y, z) \\ u_t(x, y, z, 0) &= 0 \end{aligned} \tag{7-15}$$

instead of (7-12). Let us put

$$v = t\mu[f] \tag{7-16}$$

Then

$$\begin{aligned} v(x, y, z, 0) &= 0 \\ v_t(x, y, z, 0) &= f(x, y, z) \end{aligned} \tag{7-17}$$

The second condition suggests that we try

$$u = v_t \tag{7-18}$$

for then u satisfies the first condition of (7-15). As mentioned above, any function of the form (7-16) satisfies the wave equation. Since the wave equation is linear and has constant coefficients, the derivative of any solution is again a solution. Hence u is a solution of the wave equation because v and therefore v_t is. All that remains is to see what $u_t(x, y, z, 0)$ is. We shall show that it is zero. From (7-18), we have

$$u_t = v_{tt} = c^2 \nabla^2 v \tag{7-19}$$

because v is a solution of the wave equation.

From the first equation of (7-17) it follows that $\nabla^2 v(x, y, z, 0) = 0$. Then the second condition of (7-15) follows from (7-19).

To summarize, we have proved another theorem.

Theorem 7-2 Let $v(x, y, z, t)$ satisfy the

PDE:

$$\nabla^2 v - c^{-2} v_{tt} = 0 \qquad \text{all space}$$

and the

ICs:

$$\begin{aligned} v(x, y, z, 0) &= 0 \\ v_t(x, y, z, 0) &= f(x, y, z) \end{aligned} \tag{7-17}$$

Put

$$u = v_t \tag{7-18}$$

Then u satisfies the

PDE:

$$\nabla^2 u - c^{-2} u_{tt} = 0 \qquad \text{all space}$$

and the

ICs:

$$\begin{aligned} u(x, y, z, 0) &= f(x, y, z) \\ u_t(x, y, z, 0) &= 0 \end{aligned} \tag{7-15}$$

From (7-15) and (7-17) we have explicitly

$$u = \frac{\partial}{\partial t} t\mu[f]$$

as the solution to the IVP (7-15).

Next we combine Theorems 7-1 and 7-2.

Theorem 7-3 The solution to the IVP (7-2) is

$$u(x, y, z) = \frac{\partial}{\partial t} t\mu[f] + t\mu[g] \tag{7-20}$$

where $\mu[\ ]$ is given by (7-13).

We recall that in one dimension, if g was identically zero and if f was different from zero only in a bounded interval, then the waves arrived at any given point at a finite time and left at a definite later time. In other words, sharp signals could be sent.

Formula (7-20) shows that in three dimensions this phenomenon, called *Huygens' principle*, obtains even if g is not identically zero. To see this, note from (7-13) that only the values of g on the surface of a sphere of radius ct about the point (x, y, z) enter into the computation of the solution at the point (x, y, z) at the time t. Therefore, if f and g are different from zero only in a bounded region $\mathscr{R}$, the solution will be zero at any point (x, y, z) at any time t such that the surface of the sphere centered at (x, y, z) and with radius ct does not intersect the region $\mathscr{R}$ in which f and g are different from zero. If (x, y, z) is not in the region $\mathscr{R}$ where f and g are different from zero, then the solution will be zero at (x, y, z) until t is big enough so that the sphere of radius ct about (x, y, z) intersects $\mathscr{R}$ and zero again once t is so big that the sphere of radius ct about (x, y, z) includes the whole region $\mathscr{R}$ in its interior (see Fig. 7-1).

Another way of putting it is this. Suppose we station ourselves at a given point (x, y, z) outside the region $\mathscr{R}$ where the initial values are different from zero. The solution at (x, y, z) at any time t is given by certain integrals of f and g over a sphere of radius ct centered at (x, y, z). At time $t = 0$ this sphere has zero radius; hence only the values of f and g at (x, y, z) are involved then. By hypothesis f and g are zero at (x, y, z). Hence the solution is zero at (x, y, z) at $t = 0$. Suppose we now let t increase. Then the sphere of radius ct about (x, y, z) expands. The values of f and g will remain zero on the surface of this expanding sphere until that surface first comes into contact with the region $\mathscr{R}$ in which f and g are different from zero. As t increases still further, the surface of the sphere will penetrate this region and the solution will then cease to be zero at (x, y, z). When t has increased sufficiently, however, the entire region $\mathscr{R}$ where f and g are different from zero will be inside the sphere and the surface of the sphere will again contain only points where f and g are zero. The solution at (x, y, z) will therefore once again be zero and will obviously remain so for all future time.

The situation in which the solution at a point depends only on the values of the initial data on the *surface* of a sphere about that point and not on the values throughout the sphere is referred to as *Huygens' principle*.

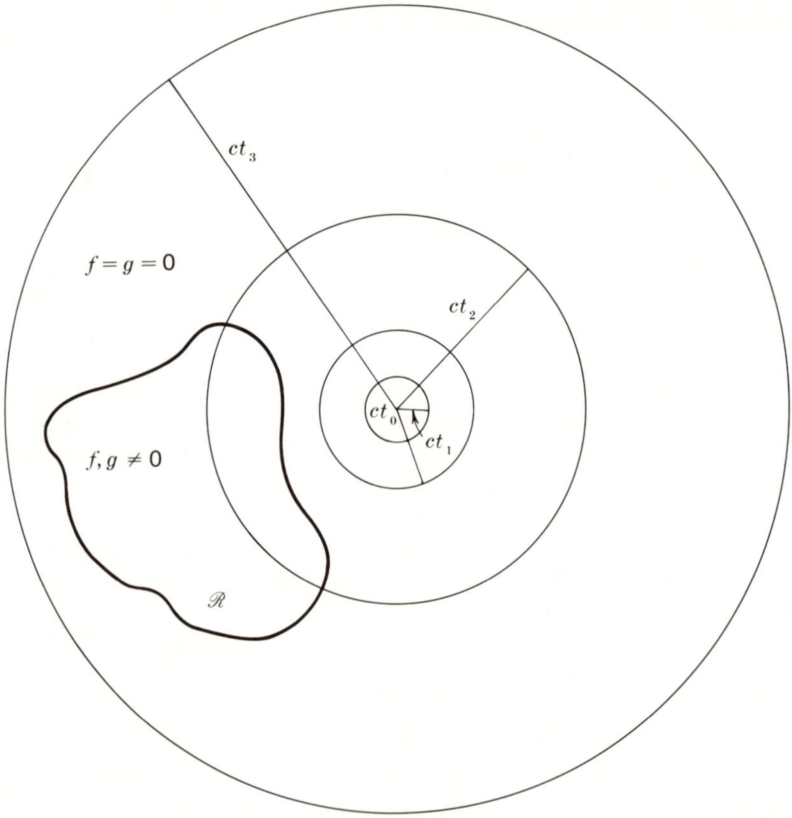

Figure 7-1

7-2. HADAMARD'S METHOD OF DESCENT

In Chap. 2 we solved an IVP for the wave equation in one dimension, and in the present chapter we have solved it in three dimensions. What about two dimensions? Clearly the formula (7-20) for the solution in three dimensions must yield the solution in two dimensions if we let it be independent of one variable. Accordingly, we assume that the initial values $f(x, y, z)$ and $g(x, y, z)$ in (7-2) are independent of z. Then the IVP (7-1) and (7-2) becomes

DE:

$$u_{xx} + u_{yy} - c^{-2}u_{tt} = 0 \qquad \begin{array}{l}\text{all } (x, y)\\ \text{all } t > 0\end{array}$$

ICs:

$$\begin{array}{l} u(x, y, 0) = f(x, y) \\ u_t(x, y, 0) = g(x, y) \end{array} \qquad \text{all } (x, y) \tag{7-21}$$

Formula (7-13) now reads

$$\mu[g] = \frac{1}{4\pi} \iint_{\Omega} g(x + ct \cos \alpha, y + ct \cos \beta)\, d\Omega \tag{7-22}$$

The integral in (7-22) is still over the unit sphere in three-dimensional space. Since the IVP it solves is stated in the xy plane, we wish to reduce the surface integral in (7-22) to a double integral in the xy plane. This is easy because the projection of the unit sphere on the xy plane is just the unit circle covered twice. The relation between $d\Omega$, the element of area on the sphere, and $d\omega$, the area of its projection on the xy plane, is $\cos \gamma \, d\Omega = d\omega$. Since

$$\cos \gamma = \sqrt{1 - \cos^2 \alpha - \cos^2 \beta}$$

$$d\Omega = \frac{d\omega}{\sqrt{1 - \cos^2 \alpha - \cos^2 \beta}}$$

Therefore (7-21) reduces to (recall the double covering mentioned above)

$$\mu[g] = \frac{1}{2\pi} \iint_{\cos^2\alpha + \cos^2\beta \leqslant 1} \frac{g(x + ct \cos \alpha, y + ct \cos \beta)}{\sqrt{1 - \cos^2 \alpha - \cos^2 \beta}}\, d\omega$$

If we let $\xi = \cos \alpha$, $\eta = \cos \beta$, this can be written as

$$\mu[g] = \frac{1}{2\pi} \iint_{\xi^2 + \eta^2 \leqslant 1} \frac{g(x + ct\xi, y + ct\eta)}{\sqrt{1 - \xi^2 - \eta^2}}\, d\xi\, d\eta \tag{7-23}$$

A similar expression obtains for $\mu[f]$, and the solution u is then given by (7-20) again. There is, however, one very important difference between the solution in three dimensions and that in two. In three dimensions the solution depended only on the initial values on the boundary of a sphere. In two dimensions the solution depends upon the initial values throughout the entire circle of radius ct, not just on the initial values on the rim of that circle. The result of this situation is the loss of part of the sharp signal property. It will still be true that the wave arrives at a definite time. It will not leave at a definite time, however. Once the circle of radius ct about the point (x, y) includes the entire region in which the initial values are different from zero, the solution will retain a constant and in general nonzero value, namely, the value determined by the integrals over the nonzero region of the initial values. We have already seen this phenomenon in one dimension when $g(x) \neq 0$ (Sec. 2-3).

The method described above in which the solution to the IVP for the two-dimensional wave equation is obtained by starting out with the solution in three dimensions and "descending" to two dimensions is called *Hadamard's method of descent*. Hadamard's method of descent can be used in higher dimensions, too, but the solution to the IVP for the wave equation in higher dimensions is quite complicated and will be omitted here. The interested reader should consult Courant and Hilbert [65, vol. 2].

EXERCISE Descend one more dimension to one space dimension and thereby rederive the formulas of Chap. 2.

7-3. THE NONHOMOGENEOUS EQUATION; DUHAMEL'S METHOD; RETARDED POTENTIAL

To solve the nonhomogeneous equation

$$u_{tt} - c^2 \nabla^2 u = F(x, y, z, t) \qquad 0 < t < \infty \tag{7-24}$$

we first proceed heuristically by looking for a Green's function $g(x, y, z, t, \tau)$ which has a singularity only in t. More precisely, we look for $g(x, y, z, t, \tau)$ satisfying:

DE:

$$g_{tt} - c^2 \nabla^2 g = \delta(t - \tau) \qquad 0 < t < \infty \tag{7-25}$$

ICs:

$$g(x, y, z, 0, \tau) = g_t(x, y, z, 0, \tau) = 0 \tag{7-26}$$

This means that we look at (7-25) as an ODE in t. For a second-order ODE the Green's function must be continuous in t, and its t derivative must have a jump of magnitude 1 at $t = \tau$ (see Friedman [113] for details). Because of the ICs (7-26), it is obvious that $g = 0$ for $t < \tau$. By the assumed continuity of g, it follows that $g(x, y, z, \tau, \tau) = 0$. Thus, if we think of the process as starting at $t = \tau$ instead of at $t = 0$, we see that (7-25) and (7-26) are equivalent to the following IVP:

DE:

$$g_{tt} - c^2 \nabla^2 g = 0 \qquad \tau < t < \infty$$

ICs:

$$\begin{aligned} g(x, y, z, \tau, \tau) &= 0 \\ g_t(x, y, z, \tau, \tau) &= 1 \end{aligned} \tag{7-27}$$

Having g, we obtain the solution u to (7-24) by

$$u(x, y, z, t) = \int_0^t g(x, y, z, t, \tau)\, F(x, y, z, \tau)\, d\tau \tag{7-28}$$

The upper limit on the integral in (7-28) is t rather than ∞ because $g = 0$ if $t < \tau$.

The above results are usually stated slightly differently by including the $F(x, y, z, t)$ with the delta function.

Theorem 7-4 Duhamel's principle If v is a solution to the following problem:

DE:

$$v_{tt} - c^2 \nabla^2 v = 0$$

ICs:

$$\begin{aligned} v(x, y, z, \tau, \tau) &= 0 \\ v_t(x, y, z, \tau, \tau) &= F(x, y, z, \tau) \end{aligned} \tag{7-29}$$

and if u is given by

$$u(x, y, z, t) = \int_0^t v(x, y, z, t, \tau)\, d\tau \tag{7-30}$$

then u satisfies (7-24).

Proof From (7-30) we have

$$u_t = v(x, y, z, t, t) + \int_0^t v_t(x, y, z, t, \tau)\, d\tau \tag{7-31}$$

But, by (7-28),

$$v(x, y, z, t, t) = 0$$

Then

$$u_{tt} = v_t(x, y, z, t, t) + \int_0^t v_{tt}(x, y, z, t, \tau)\, d\tau$$

Hence

$$u_{tt} - c^2 \nabla^2 u = v_t(x, y, z, t, t)$$

Therefore

$$v_t(x, y, z, t, t) = F(x, y, z, t) \qquad \text{Q.E.D.}$$

The function $v(x, y, z, t)$ in Theorem 7-4 is frequently called the *pulse function.*

The problem (7-29) to which we have reduced (7-24) has already been solved in Sec. 7-1. According to (7-10), the solution is

$$v(x, y, z, t, \tau) = \frac{t - \tau}{4\pi} \times \iint_\Omega F[x + c(t - \tau) \cos \alpha, y + c(t - \tau) \cos \beta, z + c(t - \tau) \cos \gamma, \tau]\, d\Omega \tag{7-32}$$

Hence, by (7-30),

$$u(x, y, z, t) = \frac{1}{4\pi} \int_0^t (t - \tau) \iint_\Omega F[x + c(t - \tau)$$

$$\times \cos\alpha, y + c(t - \tau)\cos\beta, z + c(t - \tau)\cos\gamma, \tau]\, d\Omega\, dt \tag{7-33}$$

If we let

$$r = c(t - \tau) \tag{7-34}$$

that is,

$$\tau = t - \frac{r}{c} \tag{7-35}$$

then we see that

$$u(x, y, z, t) = \frac{1}{4\pi c}$$

$$\times \int_0^{ct} \iint_\Omega rF\left(x + r\cos\alpha, y + r\cos\beta, z + r\cos\gamma, t - \frac{r}{c}\right) d\Omega\, dr \tag{7-36}$$

is just an integral over $r \leqslant ct$ in three-space polar coordinates. In terms of rectangular coordinates,

$$\xi = x + r\cos\alpha$$
$$\eta = y + r\cos\beta$$
$$\zeta = z + r\cos\gamma$$

$$d\Omega\, dr = \frac{d\xi\, d\eta\, d\zeta}{r^2}$$

and (7-36) becomes

$$u(x, y, z) = \frac{1}{4\pi c} \iiint_{\xi^2+\eta^2+\zeta^2 \leqslant c^2t^2} \frac{F(\xi, \eta, \zeta, t - r/c)}{r}\, d\xi\, d\eta\, d\zeta \tag{7-37}$$

The integral occurring in (7-37) is called a *retarded potential.* It gives the solution to (7-24) and is analogous to (3-32).

7-4. BOUNDARY-INITIAL-VALUE PROBLEMS FOR THE WAVE EQUATION

In all of the preceding analysis in this chapter we have assumed that the space variables could range over the full space. Suppose we wished to solve a boundary-initial-value problem (BIVP) for the wave equation?

More precisely, suppose we choose some simple, i.e., non-self-intersecting, closed curve σ in the xy plane and consider the BIVP:

PDE:

$$u_{tt} - u_{xx} - u_{yy} = 0 \qquad \begin{array}{l} \text{for } (x, y) \text{ inside } \sigma \\ 0 < t < \infty \end{array}$$

ICs:

$$\begin{aligned} u(x, y, 0) &= f(x, y) \\ u_t(x, y, 0) &= g(x, y) \end{aligned} \qquad (x, y) \text{ inside } \sigma$$

BC:

$$u \text{ given on } \sigma \tag{7-38}$$

Is this a well-set problem? In other words, is the solution to this problem unique? The answer is yes under proper assumptions on the functions involved. In a very lucidly written paper, Zauderer [302] proves uniqueness under the assumption that σ is a strictly convex smooth curve. He is also able to show uniqueness when the BC (7-38) is replaced by $\partial u/\partial n$ given on σ or by $\partial u/\partial n + \rho(x, y)u$ given on σ, provided $\rho(x, y) \geqslant 0$ on σ. The proof consists in showing that the only solution satisfying zero ICs and BCs is identically zero. In carrying out the proof it is necessary to use a little differential geometry and to analyze how the characteristic cones for (7-3) intersect the cylinder $\sum$ erected in xyt space over σ. Since Zauderer's exposition is very clear, the interested reader is advised to read [302] for the details. In a later section of [302] Zauderer extends his results to the case of n space dimensions.

SUMMARY

DEFINITION S7-1

The *spherical mean* $\mu[f]$ of a function $f(x, y, z)$ is defined to be

$$\mu[f] = \frac{1}{4\pi} \iint_{\Omega} f(x + ct \cos \alpha, y + ct \cos \beta, z + ct \cos \gamma)\, d\Omega \tag{S7-1}$$

where Ω is the unit sphere and $(\cos \alpha, \cos \beta, \cos \gamma)$ is a point on that sphere. Note that (S7-1) is a function of t.

Theorem S7-1 The solution $u(x, y, z, t)$ to

DE:

$$u_{xx} + u_{yy} + u_{zz} - c^{-2}u_{tt} = 0 \qquad \begin{array}{l} t > 0 \\ \text{all } (x, y, z) \end{array}$$

ICs:

$$\begin{aligned} u(x, y, z, 0) &= f(x, y, z) \\ u_t(x, y, z, 0) &= g(x, y, z) \end{aligned} \qquad \text{all } (x, y, z)$$

is

$$u(x, y, z, t) = \frac{\partial}{\partial t} t\mu[f] + t\mu[g] \tag{S7-2}$$

where $\mu[\ ]$ is defined by (S7-1).

Theorem S7-2 The solution $u(x, y, t)$ to

DE:

$$u_{xx} + u_{yy} - c^{-2}u_{tt} = 0 \qquad \begin{matrix} t > 0 \\ \text{all } (x, y, z) \end{matrix}$$

ICs:

$$\begin{aligned} u(x, y, 0) &= f(x, y) \\ u_t(x, y, 0) &= g(x, u) \end{aligned} \qquad \text{all } (x, y)$$

is given by (S7-2), where now

$$\mu[f] = \frac{1}{2\pi} \iint_{\xi^2+\eta^2 \leqslant 1} \frac{f(x + ct\xi, y + ct\eta)}{\sqrt{1 - \xi^2 - \eta^2}}\, d\xi\, d\eta$$

Theorem S7-3 A particular solution $u(x, y, z, t)$ to

$$u_{xx} + u_{yy} + u_{zz} - c^{-2}u_{tt} = F(x, y, z, t)$$

is given by

$$u(x, y, z, t) = \frac{1}{4\pi c} \iiint_{\xi^2+\eta^2+\zeta^2 \leqslant c^2t^2} \frac{F(\xi, \eta, \zeta, t - r/c)}{r}\, d\xi\, d\eta\, d\zeta$$

where

$$r^2 = (\xi - x)^2 + (\eta - y)^2 + (\zeta - z)^2$$

PROBLEMS

1. A solution of the three-dimensional wave equation of the form $h(r)\,f(r - vt)$, where $r^2 = x^2 + y^2$, is known as a *cylindrical wave*. Show that relatively undistorted cylindrical waves are in general impossible.

2. In spherical coordinates, if u is a spherical wave, that is, $u = u(r, t)$, then the wave equation becomes

$$\frac{1}{r^2}\frac{\partial}{\partial r}\left(r^2 \frac{\partial u}{\partial r}\right) = \frac{1}{c^2}\frac{\partial^2 u}{\partial t^2}$$

Find solutions of this equation by the method of separation of variables.

3. Find the form taken by the three-dimensional wave equation in cylindrical coordinates if $u = u(r, t)$, $r^2 = x^2 + y^2$, and find solutions of it by the method of separation of variables.

4. Solve:

DE:

$$u_{xx} + u_{yy} - c^{-2}u_{tt} = 0 \qquad \begin{matrix} \text{all } (x, y) \\ t > 0 \end{matrix}$$

ICs:

$$\left.\begin{aligned} u(x, y, 0) &= f(x) \\ u_t(x, y, 0) &= g(x) \end{aligned}\right\} \quad \text{all } (x, y)$$

where

$$f(x) = \begin{cases} x, & x > 0 \\ 0, & x \leqslant 0 \end{cases} \qquad g(x) = 1$$

5. Find the solution to the following IVP:

DE:

$$\nabla^2 u(x, y, z, t) - c^{-2}u_{tt}(x, y, z, t) = 0 \qquad \begin{matrix} \text{all space} \\ t > 0 \end{matrix}$$

ICs:

$$\left.\begin{aligned} u(x, y, z, 0) &= xyz \\ u_t(x, y, z, 0) &= 1 \end{aligned}\right\} \quad \text{all } (x, y, z)$$

6. Find the solution to the following IVP:

DE:

$$\nabla^2 u(x, y, t) - c^{-2}u_{tt}(x, y, t) = 0 \qquad \begin{matrix} \text{all space} \\ t > 0 \end{matrix}$$

ICs:

$$\left.\begin{aligned} u(x, y, 0) &= xy \\ u_t(x, y, 0) &= 1 \end{aligned}\right\} \quad \text{all } (x, y)$$

7. Show that $u(x, y, z, t) = s^{-1}v(s, z)$, where $s^2 = c^2t^2 - x^2 - y^2$ and $v_{ss} - v_{zz} = 0$, is a solution of $\nabla^2 u - c^{-2}u_{tt} = 0$.

8. Solve the BVP:

DE:

$$u_{xx} + u_{yy} - c^{-2}u_{tt} = -c^{-2}f(x, y)\sin\omega t \qquad \begin{matrix} 0 < x < a \\ 0 < y < b \\ 0 < t < \infty \end{matrix}$$

ICs:

$$u(x, y, 0) = u_t(x, y, 0) = 0 \qquad \begin{matrix} 0 < x < a \\ 0 < y < b \end{matrix}$$

BCs:

$$u(0, y, t) = u(a, y, t) = u(x, 0, t) = u(x, b, t) = 0 \qquad 0 < t < \infty$$

8
Single First-order Equations in General

8-1. INTRODUCTION

In Chap. 1 we started, naturally enough, with a special type of first-order partial differential equation (PDE). After that, however, most of our work has been on second-order ones. The reasons are twofold: (1) second-order equations arise more frequently in applications, and (2) the analysis of Chaps. 2 to 6 is simpler and easier to understand than that of the current chapter. This second reason itself has two basic reasons: (1) the equations of Chaps. 2 to 6 were virtually all linear whereas those of the present chapter are highly nonlinear, and (2) the heuristic reasoning that guided the analysis of Chaps. 2 to 5 was mainly physical whereas the heuristics underlying the present chapter are mainly geometrical. In itself this would not be any drawback, but some of the geometry involved is not simple and may cause some readers some difficulties. Of course, the geometry is intended to serve only as a guide. For that reason anyone having difficulties with it should merely concentrate on the analytical statements that result from it.

Having been warned of impending difficulties, the reader may wonder why he should undertake the study of single first-order PDEs in the first place. There are many reasons for studying them. First of all, they do arise in practice (see [253] for a first-order PDE that arises in a birth and death process), although the number of physical problems giving rise to single first-order PDEs is not so great as those engendering second-order ones.

Second, they are of considerable importance in theoretical physics, e.g., in Hamiltonian mechanics. Third, just as ordinary differential equations (ODEs) arise in the analysis of first-order PDEs, so too on some occasions do first-order PDEs arise in the analysis of certain second-order PDEs (see Chap. 11). In the fourth place, *systems* of first-order PDEs, as distinguished from single equations, are extremely important nowadays in applications. The analysis of single first-order PDEs constitutes a first step in the direction of the analysis of systems of first-order PDEs. Finally, the results of the present chapter give some further insight into the nature of PDEs and, in particular, into some of the intricacies that full nonlinearity entails.

Of course, if first-order PDEs are really this important, the reader may well wonder why we didn't continue their analysis right after Chap. 1 instead of waiting until now to take them up again. Indeed, most books on PDEs do start with a full analysis of first-order PDEs. We have not done so primarily for the two reasons mentioned at the outset of this chapter. We have preferred to start with the easiest and most useful PDEs and gradually work our way up to more difficult ones as the course progresses. Logically, we should have completed the treatment of first-order PDEs first. Pedagogically, experience has shown that applied scientists do better with the order followed here.

8-2. QUASI-LINEAR EQUATIONS IN TWO INDEPENDENT VARIABLES

In Chap. 1 we solved the most general first-order linear homogeneous PDE in two independent variables, namely,

$$a(x, y)\, u_x + b(x, y)\, u_y = c(x, y)\, u \tag{8-1}$$

and we indicated that the same methods could be carried over to the *semilinear* PDE

$$a(x, y)\, u_x + b(x, y)\, u_y = c(x, y, u) \tag{8-1a}$$

Notice that (8-1*a*) includes the nonhomogeneous linear equation as a special case. [Put $c(x, y, u) = c(x, y)u + f(x, y)$.] Recall that (8-1*a*) is called *semilinear* because its characteristics are independent of u.

Now we are interested in the case where the characteristics are not independent of u. We start with the simplest such case, namely, that in which u_x and u_y still enter the PDE linearly but x, y, and u can enter in any way at all. Such an equation must be of the form

$$a(x, y, u)\, u_x + b(x, y, u)\, u_y = c(x, y, u) \tag{8-1b}$$

Since (8-1*b*) is linear in the highest derivatives of u occurring, namely, u_x and u_y, it is called *quasi-linear*. As we shall see, quasi-linear equations are far simpler to treat than fully nonlinear ones.

At first sight, (8-1*b*) looks a lot more difficult to deal with than (8-1*a*). Our method of solving (8-1*a*) relied upon solving the characteristic equations

$$\frac{dx}{dt} = a \qquad \frac{dy}{dt} = b \tag{8-2}$$

for the characteristic curves. For (8-1*a*), Eqs. (8-2) represent a system of two ODEs for x and y as functions of t. For (8-1*b*), however, they do not. For (8-1*b*), a and b contain u as well as x and y, and u is an *unknown* function of x and y. Hence (8-2) cannot be solved for x and y if a and b depend upon u as well as upon x and y. For higher-order equations in Chap. 6 we met a situation in which the characteristic curves depended upon the solution u, but since there we were not trying to solve anything, we did not discuss it further.

The dilemma in the present case is not as bad as it seems, however. Equations (8-2) represent a system of two equations for three unknowns. We need another ODE. Since we have equations for dx/dt and dy/dt, it is natural to look for one for du/dt. But we can get an ODE for du/dt in the same way as we got one for it in Chap. 1. We merely insert (8-2) into (8-1) and use the notion of directional derivative again. Then, just as we did in Chap. 1, we obtain the ODE

$$\frac{du}{dt} = c \tag{8-2a}$$

The only difference between the present situation and that of Chap. 1 is that Eqs. (8-2) were independent of (8-2*a*) in Chap. 1 so that we could solve (8-2) first and then solve (8-2*a*), whereas now the three equations (8-2) and (8-2*a*) have to be solved together as a system.

Thus, to solve (8-1*b*) we should still write down the system of ODEs

$$\begin{aligned}\frac{dx}{dt} &= a(x, y, u)\\ \frac{dy}{dt} &= b(x, y, u)\\ \frac{du}{dt} &= c(x, y, u)\end{aligned} \tag{8-3}$$

only this time we have to solve all three simultaneously.

It is natural to call (8-3) the *characteristic equations* for the PDE (8-1*b*) and their solutions the *characteristic curves* for (8-1*b*), but notice that this constitutes a *redefinition* of the characteristic equations and characteristic curves from Chap. 1. Now the characteristic equations include the one for du/dt, and now the characteristic curves are curves in xyu space rather than in the xy plane. This means that the characteristic curves of Chap. 1 are the projections onto the xy plane of the characteristic curves as presently defined. For this reason the characteristics defined in Chap. 1 are frequently called the *ground characteristics* or *base characteristics*.

8-3. CAUCHY PROBLEM FOR QUASI-LINEAR EQUATIONS

The formulation of the Cauchy problem also requires a slight alteration. Previously we gave a curve Γ: $x = \phi(\tau)$, $y = \psi(\tau)$ in the xy plane, and on it we gave the values of u: $u = \omega(\tau)$. In the present case we merely give a space curve

$$\Gamma: \begin{cases} x = \phi(\tau) \\ y = \psi(\tau) \\ u = \omega(\tau) \end{cases} \tag{8-4}$$

and require that the integral surface (the solution to a first-order PDE is frequently called an *integral surface*) $u = F(x, y)$ of the PDE contain Γ, that is, that $\omega(\tau) = F[\phi(\tau), \psi(\tau)]$. As in the previous case, we shall find a condition has to be imposed on the initial curve Γ to ensure that the method of solution works.

Geometrically it is obvious how to solve the Cauchy problem (8-4) for (8-1*b*). We merely have to lay the characteristic curves through the initial curve (8-4). The resulting one-parameter family of characteristic curves will then generate the desired integral surface. Analytically this means we have to find solutions $x(t, \tau)$, $y(t, \tau)$, $u(t, \tau)$ of (8-3) satisfying the initial conditions (ICs)

$$\begin{aligned} x(0, \tau) &= \phi(\tau) \\ y(0, \tau) &= \psi(\tau) \\ u(0, \tau) &= \omega(\tau) \end{aligned} \tag{8-5}$$

Suppose we have found such solutions. In practice, finding them is usually far from a trivial problem, but suppose, nevertheless, that we have three functions

$$\begin{aligned} x &= f(t, \tau) \\ y &= g(t, \tau) \\ u &= h(t, \tau) \end{aligned} \tag{8-6}$$

which satisfy (8-3) and (8-5). Then we can obtain the solution to (8-1*b*) from them by solving the first two equations of (8-6) for t and τ in terms of x and y and substituting the resulting expressions into the third equation of (8-6). This will yield the solution u of (8-1b) as a function of x and y. In practice, carrying out this method may be quite difficult. In theory, all that is required is the knowledge that $\partial(x, y)/\partial(t, \tau) \neq 0$ on Γ. By continuity this Jacobian will then be different from zero in a neighborhood of Γ. Just as in Chap. 1, we have

$$\frac{\partial(x, y)}{\partial(t, \tau)} = x_t y_\tau - x_\tau y_t = a\psi' - b\phi' \tag{8-7}$$

The nonvanishing of this Jacobian means that the direction of the projection of the initial curve on the xy plane is nowhere parallel to the direction of the projection of the tangent to a characteristic.

In perusing the above method of solving the Cauchy problem the reader may be a little puzzled by one point. The system of ODEs (8-3) has a two-parameter family of solution curves, i.e., characteristics. In (8-6), however, we have only a one-parameter (τ) family of solutions to (8-3). How were we able to solve the PDE with such a restricted family of characteristics? The answer is simple. We did not obtain a general solution of the PDE. Instead we solved an initial-value problem (IVP) for it. Instead of looking for all integral surfaces of the PDE, we looked for those integral surfaces which contained the given curve Γ. We chose only those characteristics which went through Γ. If we pick a definite point (x_0, y_0, u_0) on Γ, then the theory of ODEs tells us that through this point there goes exactly one solution of (8-3), i.e., one characteristic. Since the points of Γ, like the points of any curve, form a one-parameter set, it follows that there must be a one-parameter family of characteristics through Γ. The surface generated by this family must be an integral surface for the following reason: the characteristic curves have direction numbers (a, b, c). The PDE (8-1b) says that these curves are perpendicular to the lines with direction numbers $(u_x, u_y, -1)$; that is, they are perpendicular to the normals to the integral surface and hence parallel to the integral surface. Since one point on each of them lies on the surface and since, as just shown, they are parallel to the integral surface, the characteristics must lie entirely on the surface. The integral surface is therefore composed of a one-parameter family of characteristics laid through the initial curve.

This answers the question of why a one-parameter family of solutions to (8-3) sufficed for solving the Cauchy problem. It also indicates, however, the significance of the two-parameter family of characteristics which are the solutions of (8-3). We have just seen that the integral surfaces of (8-1b) are composed of one-parameter families of characteristics. Since there exist two-parameter families of characteristics, there must therefore be a one-parameter family of integral surfaces of (8-1b).

8-4. A METHOD FOR FINDING THE GENERAL SOLUTION

If we are merely interested in generating integral surfaces, we can do so in several ways. One way is to vary the initial curve Γ. For every Γ we get an integral surface. Another way is to take a two-parameter family of characteristics and from them choose a one-parameter family. That one-parameter family will then generate an integral surface. In practice the latter method is usually the better one. Suppose, for example, we write the characteristic equations (8-3) in ratio form (to do this, solve each for dt and equate the results):

$$\frac{dx}{a} = \frac{dy}{b} = \frac{du}{c} \tag{8-8}$$

The usual method of solving such a system of equations is to find two independent integrals

$$\begin{aligned} \Phi(x, y, u) &= \alpha \\ \Psi(x, y, u) &= \beta \end{aligned} \tag{8-9}$$

each of which represents a one-parameter family of surfaces. Together they give the equations of a two-parameter family of curves, namely, their curves of intersection. Thus, (8-9) are the equations of the two-parameter family of characteristics of (8-1*b*). If we let one parameter be an arbitrary function of the other or, more symmetrically, let

$$F(\alpha, \beta) = 0$$

where F is arbitrary, we obtain a one-parameter family of curves each of which lies on the surface

$$F[\Phi(x, y, u), \Psi(x, y, u)] = 0 \tag{8-10}$$

Equation (8-10) therefore gives an implicit representation of an integral of (8-1*b*). Since it contains an arbitrary function F, it is called the *general solution* of (8-1*b*). Just as in the case of ODEs, it is not claimed that all solutions of the PDE are included in the general solution. Most are, but a few stray ones may, on occasion, get left out.

The two integrals (8-9) represent surfaces in xyu space on which the characteristic curves lie. Taken together, the two equations (8-9) define a two-parameter family of curves. It is obvious, however, that a space curve can be defined by the intersection of many different surfaces. Accordingly, there will, in general, exist many different pairs of independent integrals of (8-8). Use of any two independent ones in (8-10) will yield a solution of the PDE. In solving problems, therefore, the reader should bear in mind that two different people may obtain solutions which look quite different from each other if they choose different independent integrals. Since there is an arbitrary function $F(\alpha, \beta)$ in the solution, seemingly different solutions can be reconciled by appropriate choices of F. This situation will be illustrated in the example below.

EXAMPLE

$$u(xu_x - yu_y) = y^2 - x^2 \tag{8-11}$$

The characteristic equations are

$$\frac{dx}{ux} = \frac{dy}{-uy} = \frac{du}{y^2 - x^2} \tag{8-12}$$

From the first two of these we have

$$\frac{dx}{x} = \frac{dy}{-y}$$

whence

$$\log x + \log y = \log \alpha$$

or

$$xy = \alpha \tag{8-13}$$

To obtain a second integral, multiply numerator and denominator of the first equation in (8-12) by x and of the second by y and add. Since by elementary algebra this does not change the value of the ratio, we can equate the result to the third equation and get

$$\frac{x\,dx + y\,dy}{u(x^2 - y^2)} = \frac{-du}{x^2 - y^2}$$

Hence

$$x\,dx + y\,dy = -u\,du$$

and so

$$x^2 + y^2 + u^2 = \beta \tag{8-14}$$

From (8-13) and (8-14) we have, by the method described above,

$$F(xy, x^2 + y^2 + u^2) = 0$$

as an implicit representation of the general solution u. In the present case, if we choose the relation between the parameters more judiciously, we can obtain an explicit representation for u. Merely take $\beta = G(\alpha)$. Then

$$x^2 + y^2 + u^2 = G(xy)$$

Hence

$$u = \sqrt{G(xy) - x^2 - y^2} \tag{8-15}$$

In general it will not be possible to solve for u explicitly as we have done in the above example. Only the fact that α was independent of u in the example permitted us to obtain an explicit instead of an implicit expression for u in terms of x, y, and an arbitrary function of x and y.

As remarked above, the two integrals (8-13) and (8-14) are by no means unique. For instance, if we add the first two equations of (8-12) and equate the result to the third, we obtain

$$\frac{dx + dy}{u(x - y)} = \frac{du}{y^2 - x^2}$$

Hence

$$(x + y)^2 + u^2 = \gamma \tag{8-16}$$

is another integral. Using it with (8-13), we get

$$(x + y)^2 + u^2 = H(xy) \tag{8-17}$$

where H is an arbitrary function of a single variable, say $H = H(\alpha)$.

From (8-17) we have

$$u = \sqrt{H(xy) - (x + y)^2} \tag{8-18}$$

The solutions (8-15) and (8-18) become the same if we take $H(\alpha) = G(\alpha) + 2\alpha$.

8-5. THE COMPLETE INTEGRAL

The result of the preceding section is that the general solution to a first-order PDE depends upon an arbitrary function. This is in contrast to the situation for ODEs, where the solution to a first-order ODE depends on an arbitrary constant. In fact, for ODEs there is a certain equivalence between a first-order ODE and a one-parameter family of curves in the following sense: given a first-order ODE $F(x, y, y') = 0$, it is well known that the general solution is a one-parameter family of curves which can be written implicitly in the form $\phi(x, y, c) = 0$, where c is the parameter, and, vice versa, given a one-parameter family of curves $\phi(x, y, c) = 0$, it is in general possible, by eliminating the parameter, to find a differential equation $F(x, y, y') = 0$ of which that family is the general solution. It is natural to attempt to establish the same type of equivalence for first-order PDEs. We have just shown that given a first-order PDE, we can find a general solution of it that depends upon an arbitrary function. Now we shall show that given an arbitrary function $F(\alpha, \beta)$ of two fixed functions $\Phi(x, y, u)$ and $\Psi(x, y, u)$, $F[\Phi(x, y, u), \Psi(x, y, u)]$, we can find a PDE for which $F(\Phi, \Psi) = 0$ yields the general solution. In doing this, it is important to bear in mind that F is an arbitrary function (it plays the role that the parameter c does for ODEs) while Φ and Ψ are two given fixed definite functions. The PDE that results will therefore depend upon the functions Φ and Ψ chosen but will be independent of the arbitrary function F. Note, also, that whereas in the preceding section we started out with the PDE and tried to find its general solution, now we are starting out with the general solution and trying to find the PDE of which it is the general solution.

Accordingly, suppose we are given an arbitrary function $F(\alpha, \beta)$ and two definite functions $\Phi(x, y, u)$ and $\Psi(x, y, u)$. Can we then eliminate F from (8-10) and obtain a PDE for u as a function of x and y? Obviously we should differentiate (8-10) with respect to x and y, looking at u as a function of x and y. This gives

$$F_\alpha(\Phi_x + \Phi_u u_x) + F_\beta(\Psi_x + \Psi_u u_x) = 0$$
$$F_\alpha(\Phi_y + \Phi_u u_y) + F_\beta(\Psi_y + \Psi_u u_y) = 0$$

Since $F_\alpha^2 + F_\beta^2 \neq 0$, it follows that

$$\begin{vmatrix} \Phi_x + \Phi_u u_x & \Psi_x + \Psi_u u_x \\ \Phi_y + \Phi_u u_y & \Psi_y + \Psi_u u_y \end{vmatrix} = 0 \tag{8-19}$$

If (8-19) is multiplied out, the term involving the product $u_x u_y$ cancels out and the result is a quasi-linear PDE for u. Thus we have a sort of equivalence between quasi-linear PDEs and families of functions depending on an arbitrary function.

It is possible to obtain PDEs in another seemingly simpler way. Suppose we have a two-parameter family of functions $u = f(x, y, \alpha, \beta)$ defined implicitly by

$$F(x, y, u, \alpha, \beta) = 0 \tag{8-20}$$

where F is a given (not arbitrary) function. By differentiating with respect to x and y we obtain

$$F_x + F_u u_x = 0$$
$$F_y + F_u u_y = 0$$

From these two equations and the original equation, (8-20), we can (in theory) eliminate α and β. The result will be a PDE (*not necessarily quasi-linear*) for u as a function of x and y.

Thus, just as (8-10), when F is arbitrary and Φ and Ψ fixed, gives rise to a PDE, so does (8-20) where F is fixed and only the two constants α and β are arbitrary. Since (8-10) contains an arbitrary *function* whereas (8-20) contains only two arbitrary *constants*, (8-10) appears to be much more general a solution of its PDE than (8-20) is of its PDE. Nevertheless, it turns out that a lot more solutions can be obtained from (8-20) than at first appears. In fact, so many solutions can be obtained from (8-20) that it merits special treatment.

DEFINITION 8-1

A function of two parameters which is an integral of a first-order PDE in two independent variables is called a *complete integral* of the PDE.

The amazing connection between a complete integral and the general solution is given by the following.

Theorem 8-1 The general solution of a PDE can, in theory, be obtained from a complete integral of the PDE. In the present theorem the words "in theory" are not avoidable. An explicit formula for the general solution in terms of the given complete integral is, in general, impossible to write down.

Proof Let (8-20) be a complete integral. Geometrically, (8-20) represents a two-parameter family of surfaces. Somehow or other, we have to get an arbitrary function into (8-20). We do it by letting one parameter, say β, be an arbitrary function, say $\beta = G(\alpha)$, of the other. The u defined implicitly by

$$F[x, y, u, \alpha, G(\alpha)] = 0 \tag{8-21}$$

thus depends upon an arbitrary function G. The only trouble is that G is an arbitrary function of the parameter α whereas we want u to contain an arbitrary function of x and y. We achieve such a switch as follows. Equation (8-21) defines a one-parameter family of surfaces depending upon the parameter α. Because the PDE contains only first derivatives of u, every envelope of (8-21) will also provide a function u which satisfies the PDE. Recall that the envelope of (8-21) is obtained by first differentiating (8-21) with respect to α and then eliminating α between (8-21) and the resulting equation.

From (8-21) we have

$$F_\alpha + G'F_\beta = 0 \tag{8-22}$$

In (8-22) the variables u, y, u, and α occur. In theory (8-22) can be solved for α as a function of x, y, u: $\alpha = \alpha(x, y, u)$. If this expression is substituted back into (8-21), the result is

$$F\{x, y, u, \alpha(x, y, u), G[\alpha(x, y, u)]\} = 0$$

which, since G is an arbitrary function, is the general solution of the PDE. Note, however, that (8-22) cannot in practice be solved for α unless we have been given a specific function G. Hence we get a solution depending on an arbitrary function, but we have to choose our arbitrary function early in the calculation. Note, too, that Theorem 8-1 is not restricted to quasi-linear PDEs. It holds for any first-order PDE.

EXAMPLE A special case of Clairaut's equation

$$u = xu_x + yu_y + u_xu_y \tag{8-23}$$

(This is not quasi-linear.) An obvious complete integral is

$$u = \alpha x + \beta y + \alpha\beta \tag{8-24}$$

In order to obtain other solutions it is necessary to assume a specific relation between α and β. As an example, we take $\beta = 1/\alpha$. Then

$$u = \alpha x + \frac{y}{\alpha} + 1$$

The partial of this with respect to α is

$$0 = x - \frac{y}{\alpha^2}$$

whence

$$\alpha = \sqrt{\frac{y}{x}}$$

Inserting this back into the complete integral, we have

$$u = 2\sqrt{xy} + 1$$

as a solution of the PDE.

8-6. THE GENERAL NONLINEAR FIRST-ORDER PARTIAL DIFFERENTIAL EQUATION IN TWO INDEPENDENT VARIABLES

We have seen that the treatment of a quasi-linear first-order PDE is not greatly different from that of a linear one. The same is not true for the fully nonlinear equation. As we shall soon see, the treatment of the general first-order PDE involves complications not encountered in the quasi-linear case.

In order to avoid confusion it is frequently necessary to use abbreviations for the partial derivatives of the unknown function u. The standard abbreviations are

$$u_x = p \qquad \text{and} \qquad u_y = q \tag{8-25}$$

The most general first-order PDE in two independent variables can then be written as

$$F(x, y, u, p, q) = 0 \tag{8-26}$$

In the quasi-linear case, F is linear in p and q. In the present case we make no such assumption.

It should be pointed out that there is usually no analog of (8-26) in ODEs. From the theory of ODEs, you will recall that all the existence and uniqueness theorems there are proved for an equation which has been solved for the highest derivative. Such an equation is clearly linear in the highest derivative, hence quasi-linear. In the case of the PDE (8-26) there are two derivatives, and it is quite obviously impossible to solve the single equation (8-26) for both of them. Thus, it is not possible to reduce a general PDE to a quasi-linear one as it is in the case of an ODE.

Having convinced ourselves of the futility of attempting to reduce (8-26) to (8-1*b*), we proceed to analyze the situation further by comparing the current case with the quasi-linear one. Suppose we have a quasi-linear equation $au_x + bu_y = c$, and suppose we pick a point P_0: (x_0, y_0, u_0) and evaluate the coefficients a, b, c there. Let $a_0 = a(x_0, y_0, u_0)$, $b_0 = b(x_0, \ldots)$, etc. Then the PDE tells us that the normal $(p_0, q_0, -1)$ to every integral surface through P_0 is perpendicular to the direction (a_0, b_0, c_0). In other words, every integral surface through P_0 is tangent to the direction (a_0, b_0, c_0). Thus, at every point we have a one-parameter family of integral surfaces all going through the "axis" (a_0, b_0, c_0) at that point.

In the fully nonlinear case the situation is more complicated. The normals to the integral surfaces through a given point no longer are all perpendicular to a fixed direction at that point. Rather they form a one-parameter family of lines all going through the fixed point, namely, the family of directions $(p_0, q_0, -1)$, where p_0 and q_0 are connected by the equation $F(x_0, y_0, u_0, p_0, q_0) = 0$.

Since all the normals begin at the same point (x_0, y_0, u_0), their end points form a one-parameter family and therefore fill out a curve. The normals themselves therefore sweep out a conical surface. The tangent planes to the integral surfaces at (x_0, y_0, u_0), being a one-parameter family of planes all going through the same point, also envelop a conical surface, called the *Monge cone*. In the quasi-linear case this cone degenerates into a single line, as described above. Note that the Monge cone is the envelope of the *tangent planes*, *not* of the normals to the integral surface.

In the quasi-linear case there is at each point in space only one tangential direction, namely, the characteristic direction (a, b, c). Through each point there goes a whole one-parameter family of integral surfaces, but all are tangent to the same direction (a, b, c) at the point. In the fully nonlinear case there is a whole Monge cone of characteristic directions at each point. Each integral surface through that point is tangent to the Monge cone at that point. The generator of the cone along which the integral surface is tangent is a characteristic. In order to find the equations of the characteristics we therefore have to find the equations of the generators of the Monge cone at each point. Since the Monge cone is the envelope of a one-parameter family of tangent planes, we begin by applying the standard procedure for finding the envelope of a one-parameter family of surfaces. We recall again that if $\phi(x, y, u, \alpha) = 0$ is a one-parameter family of surfaces, then their envelope, if it exists, is obtained by eliminating the parameter α between the original equation $\phi = 0$ and the equation $\phi_\alpha = 0$. In our case the equation of the tangent plane to the integral surface u at the point (x, y, u) is given by

$$p(\xi - x) + q(\eta - y) = \zeta - u \tag{8-27}$$

where (ξ, η, ζ) are the running coordinates on the tangent plane and where p and q are parameters connected by (8-26). We assume that (8-26) can be solved for q as a function of p: $q = q(p)$ and that this function has been inserted into (8-27). Then (8-27) contains only the parameter p, and we can take the partial of (8-27) with respect to p to obtain

$$(\xi - x) + q'(\eta - y) = 0 \tag{8-28}$$

The equation of the Monge cone would now be obtained by eliminating p between Eqs. (8-27) and (8-28). This cannot be done explicitly, but our objective is to obtain the equations of the characteristics, i.e., the generators of the Monge cone. These are the curves along which the Monge cone is tangent to the tangent planes of the integral surfaces. They are given implicitly by (8-27) and (8-28). In order to obtain their equations explicitly, we first eliminate q' from (8-28). For fixed (x, y, u), q is defined as an implicit function of p by the PDE (8-26). If we differentiate (8-26) with respect to p, we get

$$F_p + q'F_q = 0$$

From this and (8-28) it follows that (solve both for q' and equate results)

$$\frac{\xi - x}{F_p} = \frac{\eta - y}{F_q} \tag{8-29}$$

From (8-27) we have

$$\frac{p(\xi - x)}{\eta - y} + q = \frac{\zeta - u}{\eta - y}$$

Using (8-29) in this, we get

$$\frac{pF_p}{F_q} + q = \frac{\zeta - u}{\eta - y}$$

or

$$\frac{\eta - y}{F_q} = \frac{\zeta - u}{pF_p + qF_q}$$

Altogether we have

$$\frac{\xi - x}{F_p} = \frac{\eta - y}{F_q} = \frac{\zeta - u}{pF_p + qF_q} \tag{8-30}$$

Now both (ξ, η, ζ) and (x, y, z) lie in the tangent plane. Hence $\xi - x$, $\eta - y$, and $\zeta - z$ are direction numbers of a line in the tangent plane. Because of the way the Monge cone is constructed as an envelope of tangent planes, these three numbers are also direction numbers of a generator of the Monge cone. Since all the concepts involved are really concepts "in the small," we should use dx, dy, and du instead of $\xi - x$, $\eta - y$, and $\zeta - u$. If we do this in (8-30), we obtain

$$\frac{dx}{F_p} = \frac{dy}{F_q} = \frac{du}{pF_p + qF_q} \tag{8-31}$$

as the equations for the generators of the Monge cone. By calling these ratios dt we can also write (8-31) in the parametric form

$$\begin{aligned} \frac{dx}{dt} &= F_p \\ \frac{dy}{dt} &= F_q \\ \frac{du}{dt} &= pF_p + qF_q \end{aligned} \tag{8-32}$$

Equations (8-32) are the equations for the characteristic curves of (8-26). As in the quasi-linear case, they depend on u, the unknown solution, but, unlike the quasi-linear case, they also depend upon p and q, the unknown direction of the unknown solution. This was to be expected since our geometrical analysis indicated that through each point there was a whole cone of characteristics, not just one as in the quasi-linear case. Thus, Eqs. (8-32) constitute three equations for five unknowns, x, y, u, p, and q.

In the quasi-linear case, $F = ap + bq - c$. Hence in that case, $F_p = a$, $F_q = b$, and $pF_p + qF_q = ap + bq = c$ [the last step following from the fact that $F = 0$ is the PDE (8-1*b*)]. Therefore p and q are not present in (8-32) when F is quasi-linear.

If F is not quasi-linear, we need two more equations. It is natural to look for equations for dp/dt and dq/dt to complete the system (8-32). Since $p = p(x, y)$ and $q = q(x, y)$, we have, by the chain rule and (8-32),

$$\begin{aligned} \frac{dp}{dt} &= p_x \frac{dx}{dt} + p_y \frac{dy}{dt} = p_x F_p + p_y F_q \\ \frac{dq}{dt} &= q_x \frac{dx}{dt} + q_y \frac{dy}{dt} = q_x F_p + q_y F_q \end{aligned} \tag{8-33}$$

We differentiate the PDE (8-26) to get

$$\begin{aligned} F_x + pF_u + p_xF_p + q_xF_q &= 0 \\ F_y + qF_u + p_yF_p + q_yF_q &= 0 \end{aligned} \tag{8-34}$$

In (8-34) we use (8-33) and the fact that $p_y = q_x$ [cf. (8-25)]. Then (8-34) reduces to

$$F_x + pF_u + \frac{dp}{dt} = 0$$

$$F_y + qF_u + \frac{dq}{dt} = 0$$

Thus we have found our additional two equations

$$\frac{dp}{dt} = -F_x - pF_u$$

$$\frac{dq}{dt} = -F_y - qF_u$$

Altogether we now have five ODEs for five unknown functions

$$\begin{aligned} \frac{dx}{dt} &= F_p \\ \frac{dy}{dt} &= F_q \\ \frac{du}{dt} &= pF_p + qF_q \\ \frac{dp}{dt} &= -F_x - pF_u \\ \frac{dq}{dt} &= -F_y - qF_u \end{aligned} \tag{8-35}$$

or, written in ratio form,

$$\frac{dx}{F_p} = \frac{dy}{F_q} = \frac{du}{pF_p + qF_q} = \frac{dp}{-F_x - pF_u} = \frac{dq}{-F_y - qF_u} \tag{8-36}$$

8-7. CAUCHY PROBLEM FOR $F(x, y, u, p, q) = 0$

The system of ODEs (8-35) or (8-36) is called the set of *characteristic equations* of the PDE (8-26). They differ from the characteristic equations for the quasi-linear equation (8-1*b*) in the inclusion of the functions p and q. In order to obtain a sensible problem for (8-35) we need to give not only the initial values of x, y and u, as we did in the quasi-linear case; we need to give the initial values of p and q, also. This means that the Cauchy problem for the nonlinear PDE (8-26) should not be posed for an initial curve but rather for what is called an *initial strip*. A *strip* is a set of five functions $x(\tau)$, $y(\tau)$, $u(\tau)$, $p(\tau)$, $q(\tau)$ which are not all independent but which, however, satisfy

$$u' = px' + qy' \tag{8-37}$$

where the prime stands for $d/d\tau$, called the *strip condition*. Thus, a strip is a curve plus a pair of direction numbers p and q. Since we want p to be the x derivative of u along the curve and q to be u_y there, the reason for condition (8-37) on the functions is evident. Intuitively, a strip can be looked at as a curve with small pieces of tangent planes attached.

The reason why additional conditions are needed has already been mentioned above, where it was pointed out that in the general nonlinear case, there is not just one characteristic through each point of xyu space but rather a whole (Monge) cone of characteristics at each point. In the quasi-linear case, the requirement that an integral surface contain a given free initial curve determined the integral surface uniquely, because there was, at each point of the initial curve, only one characteristic direction in which the characteristic curves building up the integral surface could leave the initial curve. In the general nonlinear case, there is a whole Monge cone of characteristic directions along which an integral-surface-building characteristic curve could leave the initial curve. Prescribing p and q as additional data at a point on the initial curve means specifying which of the characteristic generators of the Monge cone the departing characteristic curve is to be tangent to.

It is obvious that the characteristic directions, i.e., the values of p and q, thus prescribed at each point on the initial curve must be so prescribed that they fit together in some coherent fashion. We cannot arbitrarily select a generator of each Monge cone at each point and expect the characteristics tangent to the selected generators to fit together smoothly to form an integral surface unless they satisfy some additional condition. This additional condition is the strip condition, [see (8-37)].

For those who are uneasy about the seemingly heavy reliance on geometry, especially geometry in the small, it is important to point out that, actually, the geometry serves only as a guide. It does not prove anything rigorously. The rigorous proof proceeds analytically by proving that the solution of the IVP for the system of ODEs (8-35) yields the solution to the Cauchy problem for the PDE. From a logical point of view, it is not necessary to *derive* the characteristic equations (8-35). One could just as well start out by saying "given the PDE (8-26), consider the system of ODEs (8-35)" and proceed to demonstrate analytically that solving (8-26) is equivalent to solving (8-35). From a pedagogical point of view, however, it seems preferable to give some intuitive reasons why (8-35) should be considered in the first place. The preceding discussion constitutes an attempt in that direction.

To summarize, we have arrived at the following formulation of the Cauchy problem for (8-26): prescribe the initial strip, i.e., the five functions

$$\begin{aligned} x &= \phi(\tau) \\ y &= \psi(\tau) \\ u &= \omega(\tau) \\ p &= \rho(\tau) \\ q &= \sigma(\tau) \end{aligned} \tag{8-38}$$

subject to the strip condition (8-37). Find a solution of the PDE (8-26) which contains the strip (8-38).

The solution is accomplished in the same way as for the quasi-linear case. We write down the characteristic equations (8-35) and solve them subject to the initial conditions

$$\begin{aligned} x(0, \tau) &= \phi(\tau) \\ y(0, \tau) &= \psi(\tau) \\ \cdots&\cdots \end{aligned} \tag{8-39}$$

We obtain, at least in theory, the solutions

$$\begin{aligned} x &= f(t, \tau) \\ y &= g(t, \tau) \\ \cdots&\cdots \end{aligned} \tag{8-40}$$

This time the solvability condition is $\phi' F_q - \psi' F_p \neq 0$ [cf. Eq. (8-7) and the remarks above it].

We solve the first two of Eqs. (8-40) for t and τ in terms of x and y and substitute into the third. Then we get u as a function of x and y. We can also get p and q as functions of x and y this way. However, it is not a priori clear that once we have functions $u(x, y)$, $p(x, y)$, and $q(x, y)$ obtained from (8-40) by eliminating t and τ, the functions $p(x, y)$ and $q(x, y)$ will be identical with $u_x(x, y)$ and $u_y(x, y)$. This latter fact needs a proof.

To prove that $p(x, y)$ and $q(x, y)$ obtained from (8-40) really are the x and y derivatives of u, respectively, it suffices to prove that, in the variables t and τ, $u_\tau(t, \tau) = p(t, \tau)\, x_\tau(t, \tau) + q(t, \tau)\, y_\tau(t, \tau)$, because then, by the chain rule, the coefficients of x_τ and y_τ must be u_x and u_y.

Initially, i.e., when $t = 0$, the strip condition tells us that

$$u_\tau(0, \tau) = p(0, \tau)\, x_\tau(0, \tau) + q(0, \tau)\, y_\tau(0, \tau) \tag{8-41}$$

We wish to prove that this holds for all t, not just for $t = 0$. Put

$$U(t, \tau) = u_\tau(t, \tau) - p(t, \tau)\, x_\tau(t, \tau) - q(t, \tau)\, y_\tau(t, \tau) \tag{8-42}$$

Then (8-41) tells us that $U(0, \tau) = 0$. We wish to prove that $U(t, \tau) \equiv 0$, that is, that U does not vary with t. To do this, we compute U_t.

From (8-42) we have

$$U_t = u_{\tau t} - p_t x_\tau - p x_{\tau t} - q_t y_\tau - q y_{\tau t}$$

If we use the characteristic equations (8-35), we can write this

$$U_t = (pF_p + qF_q)_\tau + (F_x + pF_u)\, x_\tau - p(F_p)_\tau + (F_y + qF_u)\, y_\tau - q(F_q)_\tau$$

or

$$U_t = p_\tau F_p + q_\tau F_q + F_x x_\tau + F_y y_\tau + F_u(p x_\tau + q y_\tau)$$

On the other hand, from the PDE (8-26) we have

$$F_x x_\tau + F_y y_\tau + F_u u_\tau + F_p p_\tau + F_q q_\tau = 0$$

Hence

$$U_t = F_u(p x_\tau + q y_\tau - u_\tau)$$

that is,

$$U_t = -F_u U$$

Since F does not contain U explicitly, this can be looked at as an ODE for U as a function of t with τ entering as a parameter. As an ODE it is first order and linear. Since U vanishes when $t = 0$, it follows that $U \equiv 0$; Q.E.D.

Having shown that the solutions of (8-40) are related to each other as expected, it remains to show that they actually give us a solution of the PDE (8-26). We first show that they almost do regardless of the ICs. From (8-26) we have

$$F_t = F_x x_t + F_y y_t + F_u u_t + F_p p_t + F_q q_t$$

Inserting the characteristic equations (8-35) into this we find

$$F_t = 0 \qquad \text{along a characteristic}$$

Thus, F is constant along a characteristic. If we choose a characteristic that goes through the initial strip, then F will be zero at the point on that characteristic where it intersects the curve that "carries" the initial strip and hence zero on the entire characteristic.

In Sec. 8-4 we indicated a practical method of solving a quasi-linear PDE by finding certain integrals of the characteristic equations. From those integrals we were then able to derive the general solution to the PDE. For the nonlinear PDE (8-26) things are more complicated. There does exist a method, however, due to Charpit, that yields a complete integral of (8-26). According to Theorem 8-1, which holds for any first-order PDE, not just for quasi-linear ones, it is sufficient to find a complete integral, for most solutions can then be obtained from it.

We present Charpit's method for obtaining a complete integral of (8-26). Suppose that we can find an integral of the characteristic equations (8-36):

$$\Phi(x, y, u, p, q) = \alpha \tag{8-43}$$

Suppose, furthermore, that we can solve (8-43) and the given PDE (8-26) for p and q in terms of x, y, u, and α: $p = P(x, y, u, \alpha)$, $q = Q(x, y, u, \alpha)$. Then a computation shows that $du = P\,dx + Q\,dy$ is exact. The omitted computation is complicated by the fact that P and Q depend upon u as well as on x and y.

Integration of the expression for du will yield a second constant β in addition to the constant α in (8-43). The result is a solution $u(x, y, \alpha, \beta)$ that depends upon two parameters and hence is a complete integral.

EXAMPLE

$$p^2 + qy - u = 0 \tag{8-44}$$

The characteristic equations are

$$\frac{dx}{2p} = \frac{dy}{y} = \frac{du}{2p^2 + qy} = \frac{dp}{p} = \frac{dq}{0}$$

The last of these yields the immediate integral

$$q = \alpha$$

Then, solving the PDE for p, we have

$$p = \sqrt{u - \alpha y}$$

Hence

$$du = \sqrt{u - \alpha y}\, dx + \alpha\, dy$$

or

$$\frac{du - \alpha\, dy}{\sqrt{u - \alpha y}} = dx$$

and so

$$2\sqrt{u - \alpha y} = x + \beta$$

or

$$u = \alpha y + \tfrac{1}{4}(x + \beta)^2$$

Note that the fact that the expression for du is exact is not as trivial as it might at first seem because the coefficients of dx and dy may contain u. As a result, the integration of the expression may require a certain amount of manipulation before it can be accomplished in practice.

8-8. FIRST-ORDER QUASI-LINEAR EQUATIONS IN n INDEPENDENT VARIABLES

If, as usual, we let $x_1, \ldots, x_n$ be the independent variables and write $f(x_1, \ldots, x_n) = f(\mathbf{x})$ then the most general first-order quasi-linear PDE can be written in the form

$$\sum_{j=1}^{n} a_j(\mathbf{x}, u)\, u_{x_j} = a(\mathbf{x}, u) \tag{8-45}$$

The corresponding characteristic equations are

$$\begin{aligned} \frac{dx_j}{dt} &= a_j(\mathbf{x}, u) \qquad j = 1, \ldots, n \\ \frac{du}{dt} &= a(\mathbf{x}, u) \end{aligned} \tag{8-46}$$

Their solutions are an n-parameter family of curves. The fact that t is missing from the right-hand sides makes the family of solutions an n- instead of an $(n + 1)$-parameter family.

Since an integral surface of (8-45) is an n-dimensional surface lying in an $(n + 1)$-dimensional space, an integral surface of (8-45) must be generated by an $(n - 1)$-parameter family of characteristic curves, i.e., solutions of (8-46). Thus, in order to obtain an integral surface of (8-45) we have to devise a method for selecting an $(n - 1)$-parameter family of solutions of (8-46). This is easy, however, because the solutions to a system of ODEs such as (8-46) are uniquely determined by their ICs. Hence the number of parameters in any particular set of solutions to (8-46) will be determined by the number of parameters we put into the ICs for (8-46). Since we want an $(n - 1)$-parameter family of solutions of (8-46), we prescribe the

ICs at $t = 0$:

$$\begin{aligned} x_j &= \Phi_j(\tau_1, \ldots, \tau_{n-1}) \qquad j = 1, \ldots, n \\ u &= \Phi(\tau_1, \ldots, \tau_{n-1}) \end{aligned} \tag{8-47}$$

Then (8-46) will have solutions

$$x_j = \phi_j(t, \tau_1, \ldots, \tau_{n-1}) \qquad j = 1, \ldots, n \tag{8-48}$$

$$u = \phi(t, \tau_1, \ldots, \tau_{n-1}) \tag{8-49}$$

If

$$\frac{\partial(\phi_1, \ldots, \phi_n)}{\partial(t, \tau_1, \ldots, \tau_{n-1})} \neq 0 \tag{8-50}$$

then (8-48) can be inverted to give t and the τ's in terms of the x_j's.

Substituting the resulting expressions into (8-49), we get the desired integral surface $u(\mathbf{x})$.

Prescribing $n + 1$ functions of $n - 1$ variables, as in (8-47), means prescribing an $(n - 1)$-dimensional surface in the $(n + 1)$-dimensional space. Solving the IVP for (8-46) under these conditions means finding an integral surface containing this $(n - 1)$-dimensional manifold. Such an IVP clearly constitutes a generalization of the Cauchy problem of Sec. 8-3, where the special case $n = 2$ was considered.

If (8-50) is satisfied on the initial manifold, then that initial manifold is called *free* or *noncharacteristic*. Just as was the case in lower dimensions, a manifold is free if it is not tangent to the characteristic direction $(a_1, \ldots, a_n, a)$.

The alternate method of obtaining integrals of the characteristic equations also carries over. If we write the characteristic equations in the ratio form

$$\frac{dx_1}{a_1} = \frac{dx_2}{a_2} = \cdots = \frac{dx_n}{a_n} = \frac{du}{a} \tag{8-51}$$

then any integral $\Phi(\mathbf{x}, u) = \alpha$ of them represents a surface on which the characteristics lie. If we can find n independent integrals

$$\Phi_j(\mathbf{x}, u) = \alpha_j, \qquad j = 1, \ldots, n \tag{8-52}$$

of (8-51), then, taken together, Eqs. (8-52) represent an n-parameter family of characteristic curves, namely, the curves of intersection of the surfaces (8-52). If, from this n-parameter family of characteristic curves, we select an $(n - 1)$-parameter family by choosing an arbitrary function $F(\boldsymbol{\alpha})$ of n independent variables, $\boldsymbol{\alpha} = (\alpha_1, \ldots, \alpha_n)$, and then putting

$$F(\boldsymbol{\alpha}) = 0 \tag{8-53}$$

we obtain an implicit representation of an integral surface of (8-45) as an arbitrary function of n definite functions

$$F[\Phi_1(\mathbf{x}, u), \ldots, \Phi_n(\mathbf{x}, u)] = 0 \tag{8-54}$$

Equation (8-54) gives an implicit representation of the so-called general solution of (8-45). As in the lower-dimensional case, the Φ_j's are far from unique.

The notion of complete integral also carries over from the case of two independent variables. A complete integral of (8-45) is any solution containing n independent arbitrary constants. We leave it to the reader as an exercise to show that the so-called general solution can, in principle, be obtained from a complete integral.

8-9. LINEAR HOMOGENEOUS EQUATIONS

Just as was the case for two independent variables, the ideas are easiest to conceive when we consider the special case of a PDE which is linear and homogeneous in the first derivatives of u, in other words, the generalization of (1-6) to the case of n independent variables. This is the equation

$$\sum_{j=1}^{n} a_j(\mathbf{x})\, u_{x_j} = 0 \tag{8-55}$$

If we introduce the characteristic curves as the solutions to the system of n ODEs

$$\frac{dx_j}{dt} = a_j(\mathbf{x}) \tag{8-56}$$

then the characteristics form an $(n-1)$-parameter family of curves in the n-dimensional space.

The solutions to (8-56) can be written in the form

$$x_j = f_j(t, \tau_1, \ldots, \tau_{n-1}) \qquad j = 1, \ldots, n \tag{8-57}$$

where the τ's are the $n - 1$ parameters. Since the PDE (8-55) says $u_t = 0$ on the characteristics, we have

$$u(t, \tau_1, \ldots, \tau_{n-1}) = u(0, \tau_1, \ldots, \tau_{n-1}) = F(\tau_1, \ldots, \tau_{n-1}) \tag{8-58}$$

where $F(\tau_1, \ldots, \tau_{n-1})$, which represents the initial values of u on any noncharacteristic $(n-1)$-dimensional initial manifold, is an arbitrary function of its $n - 1$ variables.

If we solve (8-57) for the τ's in terms of the x's, we obtain

$$\tau_j = g_j(\mathbf{x}), \qquad j = 1, \ldots, n-1 \tag{8-59}$$

Inserting (8-59) into (8-58), we find that

$$u(\mathbf{x}) = F[g_1(\mathbf{x}), \ldots, g_{n-1}(\mathbf{x})] \tag{8-60}$$

that is, u is an arbitrary function of $n - 1$ determined functions. We get only $n - 1$ determined functions instead of n because we have written the solution explicitly in terms of the x's instead of defining it implicitly as in (8-54). There is, however, still exactly one arbitrary function in the solution.

Although (8-55) appears to be much more special than (8-45), it is interesting to note that, by means of a device to be introduced in Sec. 8-10, any quasi-linear PDE in n independent variables can be reduced to a linear homogeneous PDE in $n + 1$ independent variables. Thus, if we are willing to pay the price of adding one more independent variable, it is sufficient to deal with equations of form (8-55) instead of (8-45). Fully nonlinear equations, however, cannot be reduced to the form (8-55); only quasi-linear ones can.

8-10. GENERAL NONLINEAR FIRST-ORDER EQUATION IN n INDEPENDENT VARIABLES

The general first-order equation in n variables is

$$F(\mathbf{x}, u, \mathbf{p}) = 0 \tag{8-61}$$

where

$$\mathbf{p} = (p_1, \ldots, p_n)$$

and

$$p_j = u_x \tag{8-62}$$

The characteristic equations are

$$\begin{aligned} \frac{dx_j}{dt} &= F_{p_j} \qquad j = 1, \ldots, n \\ \frac{du}{dt} &= \sum_{j=1}^{n} p_j F_{p_j} \\ \frac{dp_j}{dt} &= -F_{x_j} - p_j F_u \qquad j = 1, \ldots, n \end{aligned} \tag{8-63}$$

There are thus $2n + 1$ of them.

It is possible to write (8-63) in a more symmetric form. If the PDE (8-61) does not contain the dependent variable u explicitly, then (8-63) looks much neater because the last n equations there reduce to $dp_j/dt = -F_{x_j}$. We therefore attempt to get rid of the dependent variable u in the PDE (8-61). We do it by an artifice which introduces an additional independent variable. Since we have n independent variables already, one more can hardly hurt.

The artifice consists in considering the dependent variable u to be a new independent variable x_{n+1}. In order to do this, we first think of u as being defined implicitly as a function of $\mathbf{x}$ instead of explicitly as such a function; i.e., we imagine the solution u of (8-61), defined implicitly by the equation

$$U(\mathbf{x}, u) = 0 \tag{8-64}$$

Then finding u is equivalent to finding U. We shall therefore convert the

PDE (8-61) into a PDE for the new dependent variable U as a function of the $n + 1$ independent variables $(x_1, \ldots, x_n, u)$. To do this, we have to express the u_{x_j}'s in terms of the U_{x_j}'s and U_u. Since we are thinking of the u in (8-64) as $u(\mathbf{x})$, we have, by differentiating (8-64) with respect to x_j,

$$U_{x_j} + U_u u_{x_j} = 0 \qquad j = 1, \ldots, n$$

or

$$u_{x_j} = -\frac{U_{x_j}}{U_u} \tag{8-65}$$

If we insert (8-65) into the PDE (8-61), we obtain a new PDE of the form $G(\mathbf{x}, u, P_1, \ldots, P_n, P) = 0$, where $P_j = U_{x_j}$ and $P = U_u$. To simplify the notation, we let

$$u = x_{n+1} \tag{8-66}$$

$$U_u = U_{x_{n+1}} = P_{n+1}$$

$$\mathbf{X} = (\mathbf{x}, x_{n+1}) \tag{8-67}$$

and

$$\mathbf{P} = (P_1, \ldots, P_n, P_{n+1}) \tag{8-68}$$

Then we obtain a PDE for $U(\mathbf{X})$ which does not contain U explicitly:

$$G(\mathbf{X}, \mathbf{P}) = 0 \tag{8-69}$$

For (8-69) the characteristic equations (8-63) take the more symmetric form

$$\frac{dX_j}{dt} = G_{P_j} \qquad j = 1, \ldots, n + 1$$

$$\frac{dU}{dt} = \sum_{j=1}^{n+1} P_j G_{P_j} \tag{8-70}$$

$$\frac{dP_j}{dt} = -G_{X_j} \qquad j = 1, \ldots, n + 1$$

In addition to symmetry, the characteristic equations (8-70) have another feature. Since $G = G(\mathbf{X}, \mathbf{P})$ does not contain U, the first $n + 1$ and the last $n + 1$ equations of (8-70) form a complete system of $2n + 2$ equations in themselves for **X** and **P**. Once **X** and **P** have been found from them, the middle equation of (8-70) determines U. The characteristic curves can therefore be looked at as curves in the $(n+1)$-dimensional **X** space instead of in the $(n+2)$-dimensional **X**U space. Of course, this is not surprising since the original problem was stated in the $(n+1)$-dimensional **x**u space and u is now x_{n+1}.

In any event, we see that we would lose no generality in confining ourselves to first-order PDEs in which the dependent variable itself does not appear explicitly, i.e., PDEs of the form (8-69). Nevertheless, sometimes retaining the form (8-61), in which u enters the PDE explicitly, yields results that would be missed by restricting ourselves to the form (8-69). Thus although we can always replace a first-order PDE of the form (8-61) by one of the form (8-69), there will be occasions in the future on which we will retain the form (8-61). On the other hand, there also will be occasions when we will find it convenient to use the form (8-69).

Sometimes we will find it convenient to write (8-69) in a slightly different form. In applications, principally in analytical mechanics, PDEs of the form (8-69) frequently arise directly, i.e., in a form in which the dependent variable is missing. Then no transformation of the form (8-64) is necessary. We merely have a PDE of the form (8-69) in which all the independent variables are independent variables right from the start. Furthermore, one of these independent variables is frequently interpreted as the time. For the sake of definiteness, suppose we have a PDE for a function $u(\mathbf{x}, t)$, where, as usual, $\mathbf{x} = (x_1, \ldots, x_n)$. Also as usual, we let $p_j = u_{x_j}$, $j = 1, \ldots, n$ and $\mathbf{p} = (p_1, \ldots, p_n)$. Since t is a distinguished variable, we let $p = u_t$. Then we can write the PDE in the form

$$F(\mathbf{x}, t, \mathbf{p}, p) = 0 \tag{8-71}$$

If we assume that (8-71) can be solved (algebraically) for p, then we can write (8-71) in the special form

$$p + H(\mathbf{x}, t, \mathbf{p}) = 0 \tag{8-72}$$

The PDE (8-72) is called the *Hamilton-Jacobi equation*. It plays a central role in the development of analytical mechanics, and we shall have a lot more to say about it later on. The analysis of the present section shows that every first-order PDE can be reduced to a Hamilton-Jacobi equation. Thus, in spite of its somewhat special appearance, the Hamilton-Jacobi equation includes all first-order PDEs in the sense described above.

Usually we have denoted the parameter on the characteristics by t, but in the present case t is one of the independent variables in the PDE. Therefore, in order to avoid confusion, we shall call the parameter on the characteristics τ here. Then the characteristic equations (8-70) applied to the present situation read

$$\begin{aligned} \frac{dx_j}{d\tau} &= H_{p_j} \\ \frac{dp_j}{d\tau} &= -H_{x_j} \end{aligned} \qquad j = 1, \ldots, n$$

$$\frac{du}{d\tau} = \sum_{j=1}^{n} p_j H_{p_j} + p \tag{8-73}$$

$$\frac{dp}{d\tau} = -H_t$$

$$\frac{dt}{d\tau} = 1$$

The last equation of (8-73) shows that by choosing $\tau = 0$ when $t = 0$ we can take $\tau = t$; that is, the independent variable t can be chosen as the running parameter on the characteristics. Then the last equation of (8-73) is no longer needed, and the system (8-73) reduces to

$$\begin{aligned} \frac{dx_j}{dt} &= H_{p_j} \\ \frac{dp_j}{dt} &= -H_{x_j} \end{aligned} \qquad j = 1, \ldots, n$$

$$\frac{du}{dt} = \sum_{j=1}^{n} p_j H_{p_j} + p \tag{8-73a}$$

$$\frac{dp}{dt} = -H_t$$

Since $H = H(\mathbf{x}, \mathbf{p})$, the first $2n$ of Eqs. (8-73a) form a complete system in themselves. Suppose they have been solved for $\mathbf{x}(t)$ and $\mathbf{p}(t)$. From the PDE (8-72) we have $p = -H_t$. Making this replacement in the equation for du/dt in (8-73a), we have

$$\frac{du}{dt} = \sum_{j=1}^{n} p_j H_{p_j} - H \tag{8-73b}$$

If we insert the values of $\mathbf{x}(t)$ and $\mathbf{p}(t)$ found by solving the first $2n$ equations of (8-73a) into H_{p_j} and into H in (8-73b), then the right side of (8-73b) becomes a known function of t, and u can then be found from (8-73b) merely by integrating. Similarly, p can be found by integrating the equation $dp/dt = -H_t$, but, as we will see later, this last equation is of importance for other reasons. It plays an important role in the Hamilton-Jacobi theory to be developed later.

To summarize, we can say that the characteristics for the Hamilton-Jacobi equation (8-72) can be found by solving the system of $2n$ ODEs

$$\begin{aligned} \frac{dx_j}{dt} &= H_{p_j} \\ \frac{dp_j}{dt} &= -H_{x_j} \end{aligned} \qquad j = 1, \ldots, n \tag{8-74}$$

for the $2n$ functions $\mathbf{x}(t)$ and $\mathbf{p}(t)$.

The ODEs (8-74), which have arisen here as the characteristic equations for the PDE (8-72), also arise directly in dynamics, where they are known as *Hamilton's equations* or *Hamilton's canonical equations of motion*. Their solutions represent the generalized coordinates and generalized momenta of a dynamical system whose Hamiltonian is $H(\mathbf{x}, t, \mathbf{p})$. So we can say that the solutions of Hamilton's equations are the characteristics of the Hamilton-Jacobi equation.

8-11. HAMILTON-JACOBI THEORY: CONSTRUCTION OF CHARACTERISTICS FROM A COMPLETE INTEGRAL

Since this is a book on PDEs, all our efforts so far have been devoted to attempts to solve PDEs. In treating first-order PDEs we have shown that the problem of solving them can be reduced to the problem of solving a system of first-order ODEs. However, we have also mentioned that when written in the form (8-74) this system of ODEs is identical with one that arises in dynamics. Indeed, in dynamics, the ODEs arise first, and the problem is to solve them. Since we now know that these ODEs are the characteristic equations of a PDE, this raises an intriguing question: Can we turn our previous procedures around; i.e., instead of using the ODEs (8-74) to solve the PDE (8-72), can we use the PDE (8-72) to solve the ODEs (8-74)? This idea is due to Jacobi, and the resulting theory is therefore called the Hamilton-Jacobi theory.

Although its principal application is to dynamics, the Hamilton-Jacobi theory can be developed in a geometrical way. The solutions of the Hamilton-Jacobi equation are integral surfaces, and those of Hamilton's equations are their characteristics. In previous sections we have used the characteristics, i.e., the solutions of Hamilton's equations, to construct the integral surfaces. Now we want to do the reverse. We want to use the integral surfaces to construct the characteristics. In order to obtain all the characteristics, we shall need a fairly broad class of integral surfaces. Clearly we cannot use the characteristics to construct the general solution when the characteristics are what we are looking for. So we look for another way of obtaining a general solution. We do not have to look very far. From Sec. 8-5 we recall that the general solution of a first-order PDE can be obtained from a complete integral. It is obtained from a complete integral by forming envelopes of one-parameter families of integral surfaces formed from that complete integral. An envelope of such a family will be tangent to each member of the family along a curve. Since this curve lies on two different integral surfaces, it must be a characteristic. This fact will be seen more clearly below. Granting its plausibility, we see that we can generate characteristics by forming envelopes of one-parameter families of complete integrals and finding the curves along which the envelopes are tangent to the members of a family.

For simplicity we return to the case of two independent variables, and for completeness we take the most general first-order PDE in two independent variables, permitting the unknown function u to enter explicitly. We will see later that there are some essential differences between the case in which u enters the equation explicitly and that in which it does not.

In any event, we begin by considering a PDE of the form

$$F(x, y, u, p, q) = 0 \tag{8-75}$$

The characteristic equations for (8-75) are (8-35). Since t is missing from the right-hand sides of (8-35), the solutions of (8-35) form a four-parameter family of curves. However, if we insert the solutions of (8-35) into (8-75), we find a relationship among these parameters so that only three of them are independent unless (8-75) is an identity in t. It nearly is, but not quite. So there is only a three-parameter family of characteristics. When we say that (8-75) nearly holds identically in t, we mean that although F does not vanish identically in t on a characteristic curve, it is constant on one. This fact follows immediately from differentiation of (8-75) along a characteristic, i.e., differentiation of (8-75) with respect to t with the solutions of (8-35) inserted; for then we have

$$\begin{aligned}\frac{dF}{dt} &= F_x\frac{dx}{dt} + F_y\frac{dy}{dt} + F_u\frac{du}{dt} + F_p\frac{dp}{dt} + F_q\frac{dq}{dt}\\ &= F_xF_p + F_yF_q + F_u(pF_p + qF_q) - F_p(F_x + pF_u) - F_q(F_y + qF_u) = 0\end{aligned}$$

Thus for the case of (8-75) we have to construct a three-parameter family of characteristics. As explained above, we will attempt to do this by using a complete integral of (8-75). According to Sec. 8-5, a complete integral of (8-75) can be represented (implicitly) in the form

$$\Phi(x, y, u, \alpha, \beta) = 0 \tag{8-76}$$

If we put $\beta = \phi(\alpha)$, then we obtain a one-parameter family of integral surfaces

$$\Phi[x, y, u, \alpha, \phi(\alpha)] = 0 \tag{8-77}$$

The envelope of the family (8-77) is the surface $u = u(x, y)$ that results when the parameter α is eliminated between (8-77) and

$$\Phi_\alpha + \phi' \Phi_\beta = 0 \tag{8-78}$$

Now α indexes the family; i.e., for each value of α we get a different integral surface (8-77). If we choose a definite value of α, we are on a definite integral surface of the family (8-77) and this integral surface will be tangent to the envelope along a curve whose equations are (8-77) and (8-78). (Recall that a curve in three dimensions is described by *two* equations.) Thus, for fixed α, (8-77) and (8-78) represent a curve. If we let α vary, we get a one-parameter family of curves. However, the function $\beta = \phi(\alpha)$ occurring in (8-77) and (8-78) is arbitrary. We can therefore assign an arbitrary value of β to each curve. If we let this arbitrary value of β vary too, we get a two-parameter family of curves. Finally, we observe that if $\phi(\alpha)$ is arbitrary, then so is $\phi'(\alpha)$. Therefore we can assign an arbitrary value γ to it. Then we let γ vary. All in all we get a three-parameter family of curves whose equations are

$$\begin{aligned} &\Phi(x, y, u, \alpha, \beta) = 0 \\ &\Phi_\alpha(x, y, u, \alpha, \beta) + \gamma \Phi_\beta(x, y, u, \alpha, \beta) = 0 \end{aligned} \tag{8-79}$$

We claim that (8-79) represent all characteristic curves of the given PDE. In order to avoid duplication of effort, we first generalize to the case of n independent variables and then give a proof in that more general case. We leave it to the reader as an instructive exercise to prove directly that Eqs. (8-79) give the characteristics in the case of two independent variables.

In the case of the general first-order PDE, suppose that

$$u = \Phi(\mathbf{x}, \boldsymbol{\alpha}) \tag{8-80}$$

is a complete integral of

$$F(\mathbf{x}, u, \mathbf{p}) = 0 \tag{8-81}$$

i.e., a solution of (8-81) depending on n arbitrary parameters, $\boldsymbol{\alpha} = (\alpha_1, \ldots, \alpha_n)$. We would like to obtain the characteristic curves of (8-81) by forming envelopes of (8-80) and finding the curves of tangency of these envelopes with the members of (8-80). In the $(n + 1)$-dimensional $\mathbf{x}u$ space, a curve can be described by the intersection of n surfaces, i.e., by n equations. In forming the envelope of (8-80) we therefore have to generate $n - 1$ additional equations. This requires that we use an $(n - 1)$-parameter subfamily of the family (8-80). Thus, we put

$$\alpha_j = \phi_j(\beta_1, \ldots, \beta_{n-1}) \qquad j = 1, \ldots, n \tag{8-82}$$

where the β's are new parameters. Then

$$u = \Psi(\mathbf{x}, \beta_1, \ldots, \beta_{n-1}) = \Phi[\mathbf{x}, \boldsymbol{\phi}(\beta_1, \ldots, \beta_{n-1})] \tag{8-83}$$

and the envelope is obtained by eliminating the β's from (8-83) and

$$\sum_{j=1}^{n} \Phi_{\alpha_j} \frac{\partial \phi_j}{\partial \beta_k} = 0 \qquad k = 1, \ldots, n-1 \tag{8-84}$$

If we fix the β's, then (8-83) and (8-84) represent the n equations of a curve which lies both on the envelope and on (8-83). Since the ϕ_j's are arbitrary functions of the β's, however, we can choose their values $\boldsymbol{\alpha}$ arbitrarily at any point. Hence the curves so obtained are functions of $\boldsymbol{\alpha}$.

Furthermore, just as in the two-dimensional case, we can assign arbitrary values to the $\partial\phi_j/\partial\beta_k$'s. Since there are only $n - 1$ equations in (8-84), they can always be solved for the n Φ_{α_j}'s. This means that the Φ_{α_j}'s can also be given arbitrary values, say γ_j. Moreover, because (8-84) are homogeneous, we can write the general solution of (8-84) in the form

$$\Phi_{\alpha_j} = \tau\gamma_j \qquad j = 1, \ldots, n \tag{8-85}$$

where τ is an additional parameter that will be put to use shortly as the running parameter on the curves.

Written more fully, Eqs. (8-85) read

$$\Phi_{\alpha_j}(\mathbf{x}, \boldsymbol{\alpha}) = \tau\gamma_j \qquad j = 1, \ldots, n \tag{8-86}$$

If $\det(\Phi_{\alpha_j x_k}) \neq 0$, which we shall assume to be the case, then Eqs. (8-86) define implicitly n functions

$$\mathbf{x} = \boldsymbol{\psi}(\tau, \boldsymbol{\alpha}, \boldsymbol{\gamma}) \tag{8-87}$$

of τ and the $2n$ parameters $\boldsymbol{\alpha}$ and $\boldsymbol{\gamma}$. As such, they represent a $2n$-parameter family of curves with τ as the curve parameter. We claim that the curves so obtained are the characteristic curves of the PDE (8-81). To prove this assertion, we have to produce a set of n functions

$$\mathbf{p} = \mathbf{p}(\tau, \boldsymbol{\alpha}, \boldsymbol{\gamma}) \tag{8-88}$$

and a function

$$u = \omega(\tau, \boldsymbol{\alpha}, \boldsymbol{\gamma}) \tag{8-89}$$

which, together with the functions (8-87), satisfy the characteristic equations (8-74) with an appropriate parameter $t = \mu(\tau)$. This is easy, however, for we have merely to insert the functions (8-87) into (8-80) to obtain (8-89), and into

$$p_j = \Phi_{x_j}(\mathbf{x}, \boldsymbol{\alpha}) \tag{8-90}$$

which is the usual definition of p, to obtain (8-88).

Having defined the functions (8-87) to (8-89), we now have to show that they satisfy the characteristic equations (8-74). To do this, we first insert

(8-80) into (8-81). For arbitrary $\mathbf{x}$ and $\boldsymbol{\alpha}$ [not just for the x's given by (8-87)] we then have the identity

$$F(\mathbf{x}, \Phi, \Phi_{x_1}, \ldots, \Phi_{x_n}) = 0 \tag{8-91}$$

which just expresses the fact that (8-86) is a solution of the PDE and has nothing to do with any of our foregoing envelope and characteristic-curve construction. In (8-91) $\mathbf{x}$ and $\boldsymbol{\alpha}$ are independent variables, with the α's entering (8-91) only through Φ and its partial derivatives. Consequently, if we differentiate (8-86) with respect to α_j, $j = 1, \ldots, n$, we obtain

$$F_u \Phi_{\alpha_j} + \sum_{k=1}^{n} F_{p_k} \Phi_{x_k \alpha_j} = 0 \qquad j = 1, \ldots, n \tag{8-92}$$

Having obtained (8-92) in this way, we now insert the functions (8-87) to (8-89) into the arguments of F and Φ. Then we make the same insertions into (8-85) and differentiate (8-85) with respect to τ. This gives

$$\sum_{k=1}^{n} \Phi_{\alpha_j x_k} \frac{dx_k}{d\tau} = \gamma_j \qquad j = 1, \ldots, n \tag{8-93}$$

According to (8-85), $\Phi_{\alpha_j} = \tau\gamma_j$. Using this for Φ_{α_j} in (8-92) and transposing a term, we can write (8-92) in the form

$$\sum_{k=1}^{n} \Phi_{\alpha_j x_k} F_{p_k} = -\tau F_u \gamma_j \qquad j = 1, \ldots, n \tag{8-94}$$

Now we have assumed that det $(\Phi_{\alpha_j x_k}) \neq 0$. Therefore any system of linear equations with the matrix $(\Phi_{\alpha_j x_k})$ must have a unique solution. If we multiply (8-93) by $-\tau F_u$, then the right-hand side of the resulting system is identical with that of (8-94). Hence

$$-\tau F_u \frac{dx_k}{d\tau} = F_{p_k} \qquad k = 1, \ldots, n \tag{8-95}$$

The arguments in F are (8-87) to (8-89) and are thus functions of τ. If we introduce a new parameter t by putting

$$d\tau = -\tau F_u \, dt \tag{8-96}$$

then Eqs. (8-95) read

$$\frac{dx_k}{dt} = F_{p_k} \qquad k = 1, \ldots, n \tag{8-97}$$

the desired form for the first set of characteristic equations (8-74). To get the equations for the dp_k/dt's, we differentiate the complete integral (8-90) and the PDE (8-91) with respect to t and compare results. From (8-90) we have

$$\frac{dp_j}{dt} = \sum_{k=1}^{n} \Phi_{x_j x_k} \frac{dx_k}{dt} \qquad j = 1, \ldots, n \tag{8-98}$$

and from (8-91)

$$F_{x_j} + F_u \Phi_{x_j} + \sum_{k=1}^{n} F_{p_k} \Phi_{x_j x_k} = 0 \qquad j = 1, \ldots, n \tag{8-99}$$

We have just shown [(8-97)] that $dx_k/dt = F_{p_k}$. Hence, comparing (8-98) and (8-99) and remembering that $\Phi_{x_j} = p_j$, we see that

$$\frac{dp_j}{dt} = -F_{x_j} - p_j F_u \tag{8-100}$$

the desired equations. The remaining characteristic equation for du/dt follows just as easily.

Thus, we have shown how to obtain a $2n$-parameter family of characteristic curves from a complete integral. Since the solutions of (8-74) depend upon exactly $2n$ arbitrary constants, we have thus found all characteristic curves of the PDE.

One thing must be noted, however. If the PDE does not contain u explicitly, then $F_u = 0$ and the foregoing method fails. So a separate derivation has to be given for the very case of paramount interest, namely the Hamilton-Jacobi equation.

If $F_u = 0$, then one of the arbitrary constants in the complete integral can be taken as an additive one and we therefore need merely find a solution of the PDE that depends on only $n - 1$ arbitrary constants. If

$$u = \Phi(\mathbf{x}, \alpha_1, \ldots, \alpha_{n-1}) \tag{8-101}$$

is such a solution of

$$F(\mathbf{x}, \mathbf{p}) = 0 \tag{8-102}$$

then obviously

$$u = \Phi(\mathbf{x}, \alpha_1, \ldots, \alpha_{n-1}) + \alpha_n \tag{8-103}$$

is a complete integral of (8-102).

Again, we want to form envelopes of $(n - 1)$-parameter families of solutions derived from the complete integral. Since Φ, in the case of (8-103), already has exactly $n - 1$ parameters in it, it is natural to take $\alpha_1, \ldots, \alpha_{n-1}$ as these parameters and to form the envelope by setting

$$\alpha_n = \phi(\alpha_1, \ldots, \alpha_{n-1}) \tag{8-104}$$

where ϕ is an arbitrary function of its $n - 1$ arguments. Thus we form the $(n - 1)$-parameter family of integral surfaces

$$u = \Phi(\mathbf{x}, \alpha_1, \ldots, \alpha_{n-1}) + \phi(\alpha_1, \ldots, \alpha_{n-1}) \tag{8-105}$$

The envelopes of (8-105) satisfy (8-105) and

$$\Phi_{\alpha_j} + \phi_{\alpha_j} = 0 \qquad j = 1, \ldots, n - 1 \tag{8-106}$$

To find the characteristic curves, we again use the fact that the values of $\phi_{\alpha_j}, j = 1, \ldots, n - 1$, are arbitrary. This implies that

$$\Phi_{\alpha_j} = \beta_j \qquad j = 1, \ldots, n - 1 \tag{8-107}$$

where the β's are arbitrary constants.

Written more fully, Eqs. (8-107) read

$$\Phi_{\alpha_j}(\mathbf{x}, \alpha_1, \ldots, \alpha_{n-1}) = \beta_j \qquad j = 1, \ldots, n - 1 \tag{8-108}$$

As such, they constitute $n - 1$ equations for the n x's. Their solution will therefore be a function of the $2n - 2$ α's and β's and of one additional parameter, say τ:

$$\mathbf{x} = \boldsymbol{\psi}(\tau, \alpha_1, \ldots, \alpha_{n-1}, \beta_1, \ldots, \beta_{n-1}) \tag{8-109}$$

This is just the right number of parameters for the general solution of the characteristic equations of (8-102). We leave it to the reader to finish the argument by showing that (8-109) are indeed the characteristic curves of (8-102).

We can summarize our results in the following theorems.

Theorem 8-2 Let $u = \Phi(\mathbf{x}, \boldsymbol{\alpha})$ be a complete integral of the PDE $F(\mathbf{x}, u, \mathbf{p}) = 0$. Assume that $F_u \neq 0$ and $\det(\Phi_{x_j\alpha_k}) \neq 0$. Put $\Phi_{\alpha_j} = \beta_j$, $j = 1, \ldots, n$, where the β_j's are arbitrary constants. The solutions $\mathbf{x}$ of the equations $\Phi_{\alpha_j} = \beta_j$ then constitute the complete set of characteristic curves of the PDE, with the corresponding p_j's being obtained from the equations $p_j = \Phi_{x_j}, j = 1, \ldots, n$.

Theorem 8-3 Let $u = \Phi(\mathbf{x}, \alpha_1, \ldots, \alpha_{n-1})$ be a solution of the PDE $F(\mathbf{x}, \mathbf{p}) = 0$ depending on $n - 1$ parameters $\alpha_1, \ldots, \alpha_{n-1}$ and such that the matrix $(\Phi_{x_j\alpha_k})$ has rank $n - 1$. Set $\Phi_{\alpha_j} = \beta_j, j = 1, \ldots, n - 1$, and $\Phi_{x_j} = p_j, j = 1, \ldots, n$. If these sets of equations are solved for $\mathbf{x}$ and $\mathbf{p}$, the resulting functions yield all characteristics of the PDE.

8-12. EXAMPLE: THE TWO-BODY PROBLEM

Consider the motion of two particles attracting each other according to the inverse-square law. It is not difficult to show that the origin can be chosen at one of the particles and that the motion is planar. The system is then described by the two coordinates (x, y) of the particle not at the origin and its corresponding momenta (p, q). The total kinetic energy is $\frac{1}{2}(p^2 + q^2)$, and the potential energy is $-k^2/\sqrt{x^2 + y^2}$, where k is a constant. In the present case the Hamiltonian is the total energy. Hence

$$H(x, y, p, q) = \frac{1}{2}(p^2 + q^2) - \frac{k^2}{\sqrt{x^2 + y^2}}$$

and the equations of motion are

$$\begin{aligned} \frac{dx}{dt} &= p \\ \frac{dy}{dt} &= q \\ \frac{dp}{dt} &= \frac{-k^2 x}{(x^2 + y^2)^{3/2}} \\ \frac{dq}{dt} &= \frac{-k^2 y}{(x^2 + y^2)^{3/2}} \end{aligned} \tag{8-110}$$

The corresponding PDE is

$$u_t + \frac{1}{2}(u_x{}^2 + u_y{}^2) = \frac{k^2}{\sqrt{x^2 + y^2}} \tag{8-111}$$

The idea is to solve the system (8-110) by solving the PDE (8-111). By "solve (8-111)" we mean find a solution of it that depends upon two arbitrary constants. We find such a solution by first transforming (8-111) to polar coordinates, a step suggested by the radial symmetry of the problem, and then separating variables. In polar coordinates (8-111) reads

$$u_t + \frac{1}{2}\left(u_r{}^2 + \frac{1}{r^2}u_\theta{}^2\right) = \frac{k^2}{r} \tag{8-112}$$

We look for solutions of (8-112) which have the form

$$u(r, \theta, t) = \rho(r) + \psi(\theta) + \omega(t)$$

Then we find

$$\omega' + \frac{1}{2}\left[(\rho')^2 + \frac{1}{r^2}(\psi')^2\right] = \frac{k^2}{r}$$

or

$$\frac{1}{2}\left[(\rho')^2 + \frac{1}{r^2}(\psi')^2\right] - \frac{k^2}{r} = -\omega' = \alpha \qquad \alpha = \text{const}$$

and so

$$\omega = -\alpha t$$

and

$$r^2(\rho')^2 - 2k^2 r - 2\alpha r^2 = -(\psi')^2 = -\beta^2$$

whence

$$\rho' = \sqrt{2\alpha + \frac{2k^2}{r} - \frac{\beta^2}{r^2}}$$

For later convenience we take

$$\psi' = -\beta$$

and therefore

$$\psi = -\beta\theta$$

We also take the negative square root in the expression for ρ'. Then we have

$$u = -\alpha t - \beta\theta - \int_{r_0}^{r} \sqrt{2\alpha + \frac{2k^2}{z} - \frac{\beta^2}{z^2}}\, dz$$

To apply Theorem 8-3 we need

$$u_\alpha = -t - \int_{r_0}^{r} \frac{dz}{\sqrt{2\alpha + 2k^2/z - \beta^2/z^2}}$$

and

$$u_\beta = -\theta + \beta \int_{r_0}^{r} \frac{dz}{\sqrt{2\alpha + 2k^2/z - \beta^2/z^2}}$$

We then choose arbitrary constants t_0 and θ_0 and put

$$u_\alpha = -t_0 \qquad u_\beta = -\theta_0$$

We get

$$t - t_0 = -\int_{r_0}^{r} \frac{dz}{\sqrt{2\alpha + 2k^2/z - \beta^2/z^2}} \tag{8-113}$$

$$\theta - \theta_0 = \beta \int_{r_0}^{r} \frac{dz}{\sqrt{2\alpha + 2k^2/z - \beta^2/z^2}} \tag{8-114}$$

Equation (8-113) expresses r implicitly as a function of t and the parameters α, β, and t_0. Equation (8-114) then gives θ as a function of t, α, β, t_0 and θ_0. Together, (8-113) and (8-114) therefore give (as the theory predicts) the general solution to (8-110) in polar coordinates. Particular solutions are obtained by specializing the parameters so as to satisfy given conditions. In the present case (8-114) is a relation between r and θ; hence it is the equation of the path of the moving particle.

8-13. CANONICAL TRANSFORMATIONS

In the previous section, the connection between Hamilton's equations of motion (8-74) and the Hamilton-Jacobi equation (8-72) was developed through the theory of first-order PDEs. It is interesting that there are several other ways of developing this connection. One of these other ways will be discussed in Chap. 16. In the present section we will discuss a method based on so-called *canonical transformations*. In doing so, we will introduce some expressions that are particularly useful in the transition from classical to quantum mechanics, namely, Lagrange brackets and Poisson brackets. The latter will occur again in a different context in Chap. 15.

Our aim in this section, as in the previous one, is to develop a method for solving for $\mathbf{x}(t)$ and $\mathbf{p}(t)$ the system of $2n$ ODEs

$$\begin{aligned} x_j' &= H_{p_j} \\ p_j' &= -H_{x_j} \end{aligned} \qquad j = 1, \ldots, n \tag{8-115}$$

where $H(\mathbf{x}, t, \mathbf{p})$ is a given function of its $2n + 1$ arguments and the prime stands for d/dt. In the previous section we did this by observing that (8-115) can be looked at as the characteristic equations for a certain first-order PDE and then deriving the solutions of (8-115) from a complete integral of that PDE. In the present section, we do not want to use this knowledge but rather to see if we can derive in a different way a PDE associated with (8-115).

Although (8-115) is the usual way that Hamilton's equations occur in mechanics, it is more convenient for the purposes of the theory about to be developed to return to the form (8-70) of the characteristic equations in which G is independent of the curve parameter. This means that if we are given a Hamiltonian $H(\mathbf{x}, t, \mathbf{p})$, we merely let $t = x_{n+1}$ and put

$$G(\mathbf{x}, x_{n+1}, \mathbf{p}, p_{n+1}) = p_{n+1} + H(\mathbf{x}, t, \mathbf{p}) \tag{8-116}$$

This places t on an equal footing with the other independent variables and gives us a system of $2n + 2$ ODEs of the form

$$\begin{aligned} x_j' &= G_{p_j} \\ p_j' &= -G_{x_j} \end{aligned} \qquad j = 1, \ldots, n + 1 \tag{8-117}$$

in which $G = G(\mathbf{x}, \mathbf{p})$, the boldface type now indicating vectors with $n + 1$ components instead of n.

Given G, (8-117) is a system of $2n + 2$ ODEs for the $2n + 2$ functions $\mathbf{x}(t)$, $\mathbf{p}(t)$. Since the parameter t does not occur explicitly in G, the solutions of (8-117) form a $(2n + 1)$-parameter family. Notice that (8-117) is a special system of ODEs in that all the right-hand sides are determined from a single function $G(\mathbf{x}, \mathbf{p})$.

It is natural to inquire into the possibility of simplifying (8-117) by a change of variables

$$\begin{aligned} x_j &= x_j(\tilde{\mathbf{x}}, \tilde{\mathbf{p}}) \\ p_j &= p_j(\tilde{\mathbf{x}}, \tilde{\mathbf{p}}) \end{aligned} \qquad j = 1, \ldots, n + 1 \tag{8-118}$$

We use the same symbol, x_j, for both variable and function sign because it actually clarifies rather than beclouds the ensuing computations.

Under the change of variables (8-118) the system of ODEs (8-117) will be transformed into a new system of ODEs for $\tilde{\mathbf{x}}$ and $\tilde{\mathbf{p}}$. It will not in general be true, however, that when solved for $\tilde{\mathbf{x}}_j'$ and $\tilde{\mathbf{p}}_j'$, the new system will have right-hand sides that can be expressed as the partial derivatives of a given function of $\tilde{\mathbf{x}}$ and $\tilde{\mathbf{p}}$. In other words, the nice symmetrical form of Hamilton's equations will, in general, be ruined by a transformation such as (8-118). We would like to know under what circumstances the symmetrical form of Hamilton's equations is preserved, and we shall call transformations of the form (8-118) which do preserve that form *canonical*.

DEFINITION 8-2

A transformation of the form (8-118) is called *canonical* if there exists a function $\tilde{G}(\tilde{\mathbf{x}}, \tilde{\mathbf{p}})$ such that equations (8-117) transform into

$$\begin{aligned} \tilde{x}_j' &= \tilde{G}_{\tilde{p}_j} \\ \tilde{p}_j' &= -\tilde{G}_{\tilde{x}_j} \end{aligned} \qquad j = 1, \ldots, n + 1 \tag{8-119}$$

We are interested in finding conditions on the functions (8-118) that guarantee that the transformation (8-118) is canonical. To find such conditions, we merely transform both sides of Hamilton's equations (8-117) and compare results. This gives (all sums run from 1 to $n + 1$, as do all indices, whether summed or not)

$$\sum_k \left(\frac{\partial x_j}{\partial \tilde{x}_k} \tilde{x}_k' + \frac{\partial x_j}{\partial \tilde{p}_k} \tilde{p}_k' \right) = \sum_k \left(\tilde{G}_{\tilde{x}_k} \frac{\partial \tilde{x}_k}{\partial p_j} + \tilde{G}_{\tilde{p}_k} \frac{\partial \tilde{p}_k}{\partial p_j} \right) \tag{8-120}$$

and

$$\sum_k \left(\frac{\partial p_j}{\partial \tilde{x}_k} \tilde{x}_k' + \frac{\partial p_j}{\partial \tilde{p}_k} \tilde{p}_k' \right) = - \sum_k \left(\tilde{G}_{\tilde{x}_k} \frac{\partial \tilde{x}_k}{\partial x_j} + \tilde{G}_{\tilde{p}_k} \frac{\partial \tilde{p}_k}{\partial x_j} \right) \tag{8-121}$$

If the transformation is canonical, then Eqs. (8-119) hold and can be used to replace the $\tilde{G}$'s in (8-120) and (8-121). Doing this and transposing terms, we get

$$\sum_k \left(\frac{\partial x_j}{\partial \tilde{x}_k} - \frac{\partial \tilde{p}_k}{\partial p_j} \right) \tilde{x}'_k + \sum_k \left(\frac{\partial x_j}{\partial \tilde{p}_k} + \frac{\partial \tilde{x}_k}{\partial p_j} \right) \tilde{p}'_k = 0 \tag{8-122}$$

from (8-120), and a similar expression from (8-121). Since (8-122) and the unwritten "similar expression" must be identities in $\tilde{\mathbf{x}}$ and $\tilde{\mathbf{p}}$, we conclude that

$$\frac{\partial x_j}{\partial \tilde{x}_k} = \frac{\partial \tilde{p}_k}{\partial p_j} \tag{8-123}$$

$$\frac{\partial x_j}{\partial \tilde{p}_k} = - \frac{\partial \tilde{x}_k}{\partial p_j} \tag{8-124}$$

$$\frac{\partial p_j}{\partial \tilde{x}_k} = - \frac{\partial \tilde{p}_k}{\partial x_j} \tag{8-125}$$

and

$$\frac{\partial p_j}{\partial \tilde{p}_k} = \frac{\partial \tilde{x}_k}{\partial x_j} \tag{8-126}$$

with (8-123) and (8-124) following from (8-122) and (8-125) and (8-126) from the "similar expression."

This shows that if the transformation (8-118) is to be canonical, then the functions (8-118) must satisfy Eqs. (8-123) to (8-126), and, vice versa, it is obvious from (8-120), (8-121), and (8-123) to (8-126) that if Eqs. (8-123) to (8-126) are satisfied, then the transformation (8-118) is canonical. Thus, we have proved the following theorem.

Theorem 8-4 Conditions (8-123) to (8-126) are necessary and sufficient for (8-118) to be canonical.

8-14. LAGRANGE BRACKETS

From Eqs. (8-123) to (8-126) we shall derive another set of relations which are also necessary and sufficient for (8-118) to be canonical. The fact that derivatives of the inverse functions to (8-118) enter (8-123) to (8-126) suggests that we attempt to compute these derivatives by implicit differentiation of (8-118). Accordingly, we differentiate (8-118) with respect to x_k. This gives

$$\delta_{jk} = \sum_l \left(\frac{\partial x_j}{\partial \tilde{x}_l} \frac{\partial \tilde{x}_l}{\partial x_k} + \frac{\partial x_j}{\partial \tilde{p}_l} \frac{\partial \tilde{p}_l}{\partial x_k} \right) \tag{8-127}$$

and

$$0 = \sum_l \left(\frac{\partial p_j}{\partial \tilde{x}_l} \frac{\partial \tilde{x}_l}{\partial x_k} + \frac{\partial p_j}{\partial \tilde{p}_l} \frac{\partial \tilde{p}_l}{\partial x_k} \right) \tag{8-128}$$

If we use (8-123) to replace $\partial x_j/\partial \tilde{x}_l$ in the first term under the sum in (8-127) and (8-124) to replace $\partial x_j/\partial \tilde{p}_l$ in the second, then (8-127) becomes

$$\sum_l \left(\frac{\partial \tilde{x}_l}{\partial x_k}\frac{\partial \tilde{p}_l}{\partial p_j} - \frac{\partial \tilde{x}_l}{\partial p_j}\frac{\partial \tilde{p}_l}{\partial x_k}\right) = \delta_{jk} \tag{8-129}$$

The quantity on the left in (8-129) is called a *Lagrange bracket* and is denoted by $[x_k, p_j]$.

DEFINITION 8-3

The *Lagrange bracket* $[x_j, p_k]$ is defined to be

$$[x_j, p_k] = \sum_l \left(\frac{\partial \tilde{x}_l}{\partial x_j}\frac{\partial \tilde{p}_l}{\partial p_k} - \frac{\partial \tilde{x}_l}{\partial p_j}\frac{\partial \tilde{p}_l}{\partial x_k}\right) \tag{8-130}$$

More precisely, let F_j, $j = 1, \ldots, n$, and G_j, $j = 1, \ldots, n$, be a given set of functions depending on a number of independent variables and let u and v be two of these variables. Then, by definition, the Lagrange bracket $[u, v]$ is

$$[u, v] = \sum_j \left(\frac{\partial F_j}{\partial u}\frac{\partial G_j}{\partial v} - \frac{\partial F_j}{\partial v}\frac{\partial G_j}{\partial u}\right) \tag{8-131}$$

Using Lagrange brackets, we can write (8-129) in the form

$$[x_j, p_k] = \delta_{jk} \tag{8-132}$$

By using Eqs. (8-123) to (8-126) appropriately in (8-128) and in the equations resulting from (8-118) by differentiating with respect to p_k, it is not difficult to show that

$$[x_j, x_k] = 0 \tag{8-133}$$

and

$$[p_j, p_k] = 0 \tag{8-134}$$

When the reader has done the following exercise, the proof of the theorem that follows will be complete.

EXERCISE Show that Eqs. (8-132) to (8-134) imply Eqs. (8-123) to (8-126).

Theorem 8-5 (8-132) to (8-134) constitute a set of necessary and sufficient conditions that the transformation (8-118) be canonical.

8-15. POISSON BRACKETS

In using Eqs. (8-123) to (8-126) to reduce (8-127), etc., to Lagrange brackets, we made replacements in (8-127) and other equations in such a way as to obtain expressions in which the tilded variables were differentiated with respect to the untilded ones. If we make replacements in the opposite way, then the resulting expressions will have untilded variables differentiated with respect to the tilded ones and the sums will be over the independent rather than the dependent variables. For example, by using (8-126) to replace $\partial\tilde{x}_l/\partial x_k$ by $\partial p_k/\partial\tilde{p}_l$ and (8-123) to replace $\partial\tilde{p}_l/\partial x_k$ by $-\partial p_k/\partial\tilde{x}_l$, we can change (8-128) into the form

$$\sum_l \left(\frac{\partial p_j}{\partial\tilde{x}_l}\frac{\partial p_k}{\partial\tilde{p}_l} - \frac{\partial p_j}{\partial\tilde{p}_l}\frac{\partial p_k}{\partial\tilde{x}_l}\right) = 0 \tag{8-135}$$

As mentioned above, the sum in (8-135) is over the independent variables. The expression on the left side of (8-135) is called a *Poisson bracket* and denoted by (p_j, p_k). More generally, we have the following definition.

DEFINITION 8-4

Let $F = F(x_1, \ldots, x_{n+1}, p_1, \ldots, p_{n+1})$ and

$G = G(x_1, \ldots, x_{n+1}, p_1, \ldots, p_{n+1})$

Then the *Poisson bracket* (F, G) is defined to be

$$(F, G) = \sum_l \left(\frac{\partial F}{\partial x_l}\frac{\partial G}{\partial p_l} - \frac{\partial F}{\partial p_l}\frac{\partial G}{\partial x_l}\right) \tag{8-136}$$

EXERCISE Prove the following Theorem.

Theorem 8-6 A necessary and sufficient condition that (8-118) be canonical is that

$$(x_j, p_k) = \delta_{jk} \tag{8-137}$$

$$(x_j, x_k) = 0 \tag{8-138}$$

$$(p_j, p_k) = 0 \tag{8-139}$$

Thus, the conditions that the transformation be canonical can be expressed either in terms of Lagrange brackets or in terms of Poisson brackets. Notice that in a Lagrange bracket the sum is over all the *dependent* variables. Partial derivatives with respect to only two of the *independent* variables occur in a given Lagrange bracket, and it is these two independent variables that are used in the notation for that bracket.

In a Poisson bracket the sum is over all the *in*dependent variables. Only two *de*pendent variables enter a given Poisson bracket, and it is these two dependent variables which appear in the notation for that bracket.

8-16. GENERATING FUNCTIONS FOR CANONICAL TRANSFORMATIONS

Although Theorems 8-4 to 8-6 give us necessary and sufficient conditions for a transformation to be canonical, they do not give us any means of constructing canonical transformations. In the present section we will derive such a means. To do this, we use the Lagrange bracket formulation. We first write out the condition (8-133)

$$[x_j, x_k] = \sum_l \left(\frac{\partial \tilde{x}_l}{\partial x_j} \frac{\partial \tilde{p}_l}{\partial x_k} - \frac{\partial \tilde{x}_l}{\partial x_k} \frac{\partial \tilde{p}_l}{\partial x_j} \right) = 0 \tag{8-140}$$

Then we rewrite (8-140) in the form

$$\frac{\partial}{\partial x_k} \left(\sum_l \tilde{p}_l \frac{\partial \tilde{x}_l}{\partial x_j} \right) - \frac{\partial}{\partial x_j} \left(\sum_l \tilde{p}_l \frac{\partial \tilde{x}_l}{\partial x_k} \right) = 0 \tag{8-141}$$

because (8-141) is just the condition that the differential expression $\sum_j a_j(\mathbf{x}, \mathbf{p})\, dx_j$, where

$$a_j(\mathbf{x}, \mathbf{p}) = \sum_l \tilde{p}_l \frac{\partial \tilde{x}_l}{\partial x_j} \tag{8-142}$$

be exact if the x's are looked at as the variables and the p's as parameters. Thus, there exists a function $\phi(\mathbf{x}, \mathbf{p})$ such that $\phi_{x_j} = a_j$, but we shall not make explicit use of this fact. All we will need is the exactness condition (8-141), which, when written in terms of the abbreviation (8-142), reads

$$\frac{\partial a_j}{\partial x_k} - \frac{\partial a_k}{\partial x_j} = 0 \tag{8-143}$$

Similarly, we deduce from $[p_j, p_k] = 0$ that

$$\frac{\partial b_j}{\partial p_k} - \frac{\partial b_k}{\partial p_j} = 0 \tag{8-144}$$

where

$$b_j(\mathbf{x}, \mathbf{p}) = \sum_l \tilde{p}_l \frac{\partial \tilde{x}_l}{\partial p_j} \tag{8-145}$$

Finally, we write $[x_j, p_k] = \delta_{jk}$ in the form

$$\frac{\partial a_j}{\partial p_k} - \frac{\partial b_k}{\partial x_j} = \delta_{jk} \tag{8-146}$$

Because the right-hand side is not zero, (8-146) is not an integrability condition. This shortcoming can be easily remedied, however, by writing

$$\delta_{jk} = \frac{\partial p_j}{\partial p_k}$$

Then the right-hand side can be transposed and combined with the terms on the left side to give

$$\frac{\partial(a_j - p_j)}{\partial p_k} - \frac{\partial(b_k - p_k)}{\partial x_j} = 0 \tag{8-147}$$

Since conditions (8-143) and (8-144) can also be written in the form

$$\frac{\partial(a_j - p_j)}{\partial x_k} - \frac{\partial(a_k - p_k)}{\partial x_j} = 0 \tag{8-148}$$

and

$$\frac{\partial(b_j - p_j)}{\partial p_k} - \frac{\partial(b_k - p_k)}{\partial p_j} = 0 \tag{8-149}$$

we see that in Eqs. (8-147) to (8-149) we have the complete set of necessary and sufficient conditions for the exactness of the differential expression

$$\sum_j [(a_j - p_j)\,dx_j + (b_j - p_j)\,dp_j]$$

in all $2n + 2$ variables. Thus, there exists a function $\hat{\Phi}(\mathbf{x}, \mathbf{p})$ such that

$$d\hat{\Phi} = \sum_j [(a_j - p_j)\,dx_j + (b_j - p_j)\,dp_j] \tag{8-150}$$

Since $\sum_j p_j\,dp_j$ is always exact, its presence in (8-150) will not add any new information. So we transpose it to the left side, absorb it into $\hat{\Phi}$, and name the new left-hand side Φ. Then we have

$$d\Phi = \sum_j [(a_j\,dx_j + b_j\,dp_j) - p_j\,dx_j] \tag{8-151}$$

From Eqs. (8-142) and (8-145) we see that

$$\sum_j (a_j\,dx_j + b_j\,dp_j) = \sum_l \tilde{p}_l \sum_j \left(\frac{\partial \tilde{x}_l}{\partial x_j}\,dx_j + \frac{\partial \tilde{x}_l}{\partial p_j}\,dp_j\right) = \sum_l \tilde{p}_l\,d\tilde{x} \tag{8-152}$$

Hence (8-151) can be written in the form

$$d\Phi = \sum_j (\tilde{p}_j\,d\tilde{x}_j - p_j\,dx_j) \tag{8-153}$$

DEFINITION 8-5

The function Φ given by (8-153) is called the *generating function* of the canonical transformation (8-118).

Thus, we have shown that if (8-118) is canonical, then a generating function Φ exists. Conversely, it is easy to show that choosing any function $\Phi(\mathbf{x}, \mathbf{p})$ and determining $\mathbf{x}$ and $\mathbf{p}$ from (8-153) yields a canonical transformation. This makes it look as if we could produce canonical transformations with great ease by merely choosing any function $\Phi(\mathbf{x}, \mathbf{p})$ and calculating its partial derivatives. Unfortunately, things are not quite that simple. If we choose a function $\Phi(\mathbf{x}, \mathbf{p})$, we can easily compute its differential $d\Phi = \sum_j (\Phi_{x_j}\, dx_j + \Phi_{p_j}\, dp_j)$, but (8-153) contains dx_j and $d\tilde{x}_j$, not dx_j and dp_j. Switching differentials in this instance leads to complications. Rather than attempt this, we use another line of attack. Equations (8-118) represent $2n + 2$ equations for the $4n + 4$ quantities $\mathbf{x}$, $\mathbf{p}$, $\tilde{\mathbf{x}}$, and $\tilde{\mathbf{p}}$. Up to now we have looked at (8-118) as expressing $\mathbf{x}$ and $\mathbf{p}$ in terms of $\tilde{\mathbf{x}}$ and $\tilde{\mathbf{p}}$. As a rule, however, we could choose any $2n + 2$ of the $4n + 4$ variables in (8-118) as the independent variables and the other $2n + 2$ as the dependent ones. Cases where there are some restrictions on which variables can be chosen as independent will have to be investigated too. First, though, let us assume that it is possible to choose $\mathbf{x}$ and $\tilde{\mathbf{x}}$ as the independent variables. Then we denote the generating function by $\Phi^{(1)}(\mathbf{x}, \tilde{\mathbf{x}})$, and we have

$$d\Phi^{(1)} = \sum_j (\Phi^{(1)}_{x_j}\, dx_j + \Phi^{(1)}_{\tilde{x}_j}\, d\tilde{x}_j) \tag{8-154}$$

Comparison of (8-153) with (8-154) immediately gives

$$\begin{aligned} \tilde{p}_j &= \Phi^{(1)}_{\tilde{x}_j} \\ p_j &= -\Phi^{(1)}_{x_j} \end{aligned} \tag{8-155}$$

Equation (8-155) still does not give us an explicit representation of our canonical transformation, however, because $\Phi^{(1)}$ contains $\mathbf{x}$ in it. Equations (8-154) therefore represent a set of $2n + 2$ algebraic equations which have to be solved for $\mathbf{x}$ and $\mathbf{p}$ in terms of $\tilde{\mathbf{x}}$ and $\tilde{\mathbf{p}}$.

As mentioned above, the derivation of Eqs. (8-155) relies on the assumption that $\mathbf{x}$ and $\tilde{\mathbf{x}}$ can be chosen as independent variables in (8-118). This assumption is not always valid. For example, the pure point transformation $\mathbf{x} = \mathbf{x}(\tilde{\mathbf{x}})$, $\mathbf{p} = \tilde{\mathbf{p}}$, obviously does not have $\mathbf{x}$ and $\tilde{\mathbf{x}}$ as independent variables, hence cannot be generated by a generating function of the form $\Phi^{(1)}(\mathbf{x}, \tilde{\mathbf{x}})$. It can, however, be generated by a function of the form $\Phi^{(2)}(\mathbf{x}, \tilde{\mathbf{p}})$ or by one of the form $\Phi^{(3)}(\tilde{\mathbf{x}}, \mathbf{p})$, because $\mathbf{x}$ and $\tilde{\mathbf{p}}$ can be chosen as independent variables in this case, as could $\tilde{\mathbf{x}}$ and $\mathbf{p}$. Thus, we need to know how to obtain (8-118) from a function of the form $\Phi^{(2)}(\mathbf{x}, \tilde{\mathbf{p}})$ and also from one of the form $\Phi^{(3)}(\tilde{\mathbf{x}}, \mathbf{p})$. We shall show how to do this for $\Phi^{(2)}(\mathbf{x}, \tilde{\mathbf{p}})$. The method for $\Phi^{(3)}(\tilde{\mathbf{x}}, \mathbf{p})$ is entirely analogous.

Since

$$d\Phi^{(2)}(\mathbf{x}, \tilde{\mathbf{p}}) = \sum_j (\Phi^{(2)}_{x_j}\, dx_j + \Phi^{(2)}_{\tilde{p}_j}\, d\tilde{p}_j) \tag{8-156}$$

we write (8-153) in the form

$$d\Phi(\mathbf{x}, \mathbf{p}) = \sum_j [d(\tilde{p}_j \tilde{x}_j) - p_j\, dx_j - \tilde{x}_j\, d\tilde{p}_j] \tag{8-157}$$

or

$$d\left[\sum_j \tilde{x}_j \tilde{p}_j - \Phi(\mathbf{x}, \mathbf{p})\right] = \sum_j (p_j\, dx_j + \tilde{x}_j\, d\tilde{p}_j) \tag{8-158}$$

because then with

$$\Phi^{(2)}(\mathbf{x}, \tilde{\mathbf{p}}) = \sum_j \tilde{x}_j \tilde{p}_j - \Phi(\mathbf{x}, \mathbf{p}) \tag{8-159}$$

we see that

$$\begin{aligned} p_j &= \Phi^{(2)}_{x_j}(\mathbf{x}, \tilde{\mathbf{p}}) \\ \tilde{x}_j &= \Phi^{(2)}_{\tilde{p}_j}(\mathbf{x}, \tilde{\mathbf{p}}) \end{aligned} \tag{8-160}$$

Equations (8-159) and (8-160) constitute a Legendre transformation (see Sec. 8-22).

Again the equations determining the canonical transformation from the generating function come out in a form in which they are solved for neither set of variables in terms of the others. Some algebra is necessary to obtain (8-118) from (8-160).

The cases in which the generating function is $\Phi^{(3)}(\tilde{\mathbf{x}}, \mathbf{p})$, as well as that in which it is of the form $\Phi^{(4)}(\mathbf{p}, \tilde{\mathbf{p}})$, present no new difficulties. For that reason we leave their treatment to the reader.

EXAMPLES If

$$\Phi^{(1)}(\mathbf{x}, \tilde{\mathbf{x}}) = \sum_j x_j \tilde{x}_j \tag{8-161}$$

then (8-155) become

$$\begin{aligned} \tilde{p}_j &= x_j \\ p_j &= -\tilde{x}_j \end{aligned} \tag{8-162}$$

that is,

$$\begin{aligned} x_j &= \tilde{p}_j \\ p_j &= -\tilde{x}_j \end{aligned} \tag{8-163}$$

so that the x's and p's are essentially interchanged.

If

$$\Phi^{(2)}(\mathbf{x}, \tilde{\mathbf{p}}) = \sum_j x_j \tilde{p}_j \tag{8-164}$$

then Eqs. (8-160) read

$$\begin{aligned} p_j &= \tilde{p}_j \\ \tilde{x}_j &= x_j \end{aligned} \tag{8-165}$$

i.e., (8-164) gives rise to the identity transformation.

8-17. CONTACT TRANSFORMATIONS

The analysis in the preceding sections was facilitated by the symmetric representation obtained by treating the time t on an equal footing with the x's. In order to apply the results to mechanics, it is necessary to restore t to its original status as a special variable and to return to the unsymmetric Hamiltonian $H(\mathbf{x}, t, \mathbf{p})$. Rccall that

$$x_{n+1} = t \tag{8-166}$$

From this, (8-116), and the last equation in (8-117), we have

$$p'_{n+1} = -H_t \tag{8-167}$$

We cannot integrate (8-167) immediately because H_t is the partial rather than the total derivative of H with respect to t. Since $\mathbf{x}$ and $\mathbf{p}$ are functions of t, we have

$$H' = \sum_{j=1}^{n} (H_{x_j}x'_j + H_{p_j}p'_j) + H_t \tag{8-168}$$

But Hamilton's equations (8-115) imply that the expression under the sum in (8-168) vanishes, and we therefore have

$$H' = H_t \tag{8-169}$$

after all, it being understood that (8-169) holds under the assumption that the arguments $\mathbf{x}$ and $\mathbf{p}$ in H have been replaced by the solutions of (8-115).

Using (8-169), we can integrate (8-167) immediately to get

$$p_{n+1} = -H \tag{8-170}$$

Thus, the Hamiltonian plays the role of canonical coordinate conjugate to t, and, in unsymmetric notation, Hamilton's equations can be written, as before, in the form (8-73):

$$\begin{aligned} x'_j &= H_{p_j} \\ p'_j &= -H_{x_j} \\ t' &= 1 \\ p' &= -H_t \end{aligned} \tag{8-171}$$

In applying canonical transformations to (8-171) we want to keep t unchanged. We therefore restrict our canonical transformations to those of the form

$$\begin{aligned} x_j &= x_j(\tilde{\mathbf{x}}, \tilde{\mathbf{p}}) \\ p_j &= p_j(\tilde{\mathbf{x}}, \tilde{\mathbf{p}}) \end{aligned} \qquad j = 1, \ldots, n \tag{8-172}$$

where now $\mathbf{x}$ and $\mathbf{p}$ are n-dimensional vectors again. Transformations like (8-172), which preserve the form of Hamilton's equations, are called *contact transformations*. The terminology on canonical and contact transformations is not universal, but it seems preferable to have some nomenclature that distinguishes those canonical transformations in which t is transformed from those in which it is not. The terminology adopted here seems to be the prevalent one.

All the results of the previous sections are applicable to (8-172) if we merely use the facts that $x_{n+1} = t$, that $p_{n+1} = -H$, and that (8-172) is independent of t and H, that is, of x_{n+1} and p_{n+1}. In particular, the relation (8-153) for the generating function reads

$$d\Phi = \sum_{j=1}^{n} (\tilde{p}_j \, d\tilde{x}_j - p_j \, dx_j) - (\tilde{H} - H) \, dt \tag{8-173}$$

8-18. THE HAMILTON-JACOBI EQUATION AGAIN

It remains to put the transformation theory we have developed to practical use in solving Hamilton's equations. We do this by attempting to choose a contact transformation that substantially simplifies Hamilton's equations. The simpler the new Hamiltonian H is, the easier the resulting Hamiltonian system should be to solve. In fact, the Hamiltonian which would produce the simplest form for Hamilton's equations is a constant, because then the equations would read

$$\begin{aligned} \tilde{x}_j' &= 0 \\ \tilde{p}_j' &= 0 \end{aligned} \qquad j = 1, \ldots, n \tag{8-174}$$

and there is little doubt about our ability to solve (8-174).

In order to obtain (8-174) it is sufficient to find a contact transformation such that the new Hamiltonian $\tilde{H}(\tilde{\mathbf{x}}, t, \tilde{\mathbf{p}})$ is a constant. We can do even better. We will show how to find a generating function such that the contact transformation to which it gives rise has $\tilde{H} \equiv 0$. It turns out to be best to take the generating function of the form $\Phi^{(2)}(\mathbf{x}, t, \tilde{\mathbf{p}})$. Then the corresponding contact transformation is determined [cf. (8-160) and (8-173)] from

$$p_j = \Phi^{(2)}_{x_j}(\mathbf{x}, t, \tilde{\mathbf{p}}, \tilde{H}) \qquad j = 1, \ldots, n \tag{8-175}$$

$$\tilde{x}_j = \Phi^{(2)}_{\tilde{p}_j}(\mathbf{x}, t, \tilde{\mathbf{p}}, \tilde{H}) \qquad j = 1, \ldots, n \tag{8-176}$$

$$-\tilde{H} = \Phi^{(2)}_t + H(\mathbf{x}, t, \mathbf{p}, \tilde{H}) \tag{8-177}$$

$$-t = \Phi^{(2)}_{\tilde{H}}(\mathbf{x}, t, \mathbf{p}, \tilde{H}) \tag{8-178}$$

Therefore, if we want $\tilde{H} \equiv 0$, then $\Psi(\mathbf{x}, t, \tilde{\mathbf{p}}) = \Phi^{(2)}(\mathbf{x}, t, \tilde{\mathbf{p}}, 0)$ must satisfy

$$\Psi_t + H(\mathbf{x}, t, \mathbf{p}) = 0 \tag{8-179}$$

Using (8-176) in (8-179), we see that this means that Ψ must be chosen so as to satisfy

$$\Psi_t + H(\mathbf{x}, t, \Psi_{x_1}, \ldots, \Psi_{x_n}) = 0 \tag{8-180}$$

which is a first-order PDE for $\Psi(\mathbf{x}, t, \tilde{\mathbf{p}})$ as a function of the $n + 1$ independent variables $(\mathbf{x}, t)$ with the n $\tilde{\mathbf{p}}_j$'s as parameters. Ψ is thus an integral of (8-180) when (8-180) is looked upon as a PDE, with the p's having their usual meaning, that is $p_j = \Psi_{x_j}$.

Thus, if we have a complete integral $\Psi(\mathbf{x}, t, \tilde{\mathbf{p}})$ of (8-180), then, by putting

$$\Psi_{\tilde{p}_j} = \tilde{x}_j \tag{8-181}$$

where the $\tilde{x}_j$'s are new parameters, we get in (8-175) and (8-181) an implicit solution of Hamilton's equations (8-171). But the solutions of Hamilton's equations (8-171) are the characteristics of the PDE (8-180), and the equations (8-175) and (8-181) for determining them are, if we call $\tilde{\mathbf{p}} = \boldsymbol{\alpha}$ and $\tilde{\mathbf{x}} = \boldsymbol{\beta}$, precisely the same equations as we found in Sec. 8-6 through a geometrical approach to the subject. Thus, we have now arrived at Theorem 8-3 in an entirely different fashion. In Chap. 16 we discuss still another method.

8-19. GENERAL SYSTEMS OF FIRST-ORDER ORDINARY DIFFERENTIAL EQUATIONS

It is interesting to note that the foregoing reduction of a problem in ODEs to one in PDEs can be carried out for a general first-order system of ODEs by writing that system as part of a Hamiltonian system. To do this, we assume, without loss of generality, that the independent variable is not present in the right-hand sides. Then a general first-order system has the form

$$x_j' = f_j(\mathbf{x}) \qquad j = 1, \ldots, n \tag{8-182}$$

To convert (8-182) to the form (8-117), we introduce n additional dependent variables $\mathbf{p}(t)$ and define $G(\mathbf{x}, \mathbf{p})$ by

$$G(\mathbf{x}, \mathbf{p}) = \sum_j f_j(\mathbf{x})\, p_j \tag{8-183}$$

Then (8-182) are the first n equations of the Hamiltonian system

$$\begin{aligned} x_j' &= G_{p_j} \\ p_j' &= -G_{x_j} \end{aligned} \qquad j = 1, \ldots, n \tag{8-184}$$

8-20. THE LEGENDRE TRANSFORMATION

In some applications a change of variables known as the *Legendre transformation* is useful. In theoretical mechanics and in thermodynamics the Legendre transformation is frequently of greater use in the development of the theory than it is in actually solving any PDEs. It is chiefly of interest in changing one PDE into another one from which facts of interest may be read off more readily.

The geometrical idea behind the Legendre transformation is that of changing from point coordinates to line coordinates or, in three dimensions, to plane coordinates. Instead of looking at a surface as composed of points, we look upon it as the envelope of its tangent planes. If (x, y, u) is a typical point on the surface $u = u(x, y)$, then the tangent plane at that point has the equation

$$\alpha\xi + \beta\eta - \zeta = \gamma \tag{8-185}$$

where (ξ, η, ζ) are the running coordinates of a point on the tangent plane, where

$$\alpha = u_x(x, y) \tag{8-186}$$

$$\beta = u_y(x, y) \tag{8-187}$$

$$\gamma = \alpha x + \beta y - u = xu_x + yu_y - u \tag{8-188}$$

(x, y, u) being given, Eqs. (8-186) to (8-188) determine α, β, γ uniquely. Conversely, if

$$\frac{\partial(\alpha, \beta)}{\partial(x, y)} = u_{xx}u_{yy} - u_{xy}^2 \neq 0 \tag{8-189}$$

which means that the surface $u(x, y)$ is not a developable, then (8-186) and (8-187) can be inverted to give x and y as functions of α and β. When these functions are inserted into (8-188), they yield u as a function of α, β, and γ. Thus, the transformation is one to one if (8-189) is satisfied.

The transformation given by Eqs. (8-186) to (8-188) is called a *Legendre transformation.* Since (8-186) is symmetric in (x, y, u) and (α, β, γ), it leads us to suspect that the inverse to a Legendre transformation is another Legendre transformation. We can verify this immediately, for from (8-188) we have

$$\begin{aligned} \gamma_\alpha &= \alpha x_\alpha + x + \beta y_\alpha - u_\alpha \\ \gamma_\beta &= \alpha x_\beta + y + \beta y_\beta - u_\beta \end{aligned} \tag{8-190}$$

But

$$u_\alpha = u_x x_\alpha + u_y y_\alpha = \alpha x_\alpha + \beta y_\alpha$$

and

$$u_\beta = u_x x_\beta + u_y y_\beta = \alpha x_\beta + \beta y_\beta$$

Hence Eqs. (8-190) reduce to

$$\begin{aligned} \gamma_\alpha &= x \\ \gamma_\beta &= y \end{aligned}$$

and so the inverse transformation to (8-186) to (8-188) is, as expected, another Legendre transformation

$$\begin{aligned} x &= \gamma_\alpha \\ y &= \gamma_\beta \\ u &= \alpha x + \beta y - \gamma \end{aligned} \tag{8-191}$$

From Eqs. (8-186) to (8-188) and (8-190) it is clear what the Legendre transformation does to the PDE $F(x, y, u, u_x, u_y) = 0$: it interchanges the roles of the partial derivatives and the independent variables. Thus, if u_x and u_y enter the PDE in a very complicated way but x and y enter in a very simple way, then the Legendre transformation will simplify the PDE. For example, the PDE $xu_x{}^2 + yu_y{}^2 = 1$ becomes $\alpha^2\gamma_\alpha + \beta^2\gamma_\beta = 1$, which is linear.

The extension of the Legendre transformation to higher dimensions is rather obvious. It is

$$\begin{aligned} \alpha_j &= u_{x_j} \qquad j = 1, \ldots, n \\ \gamma &= \sum_{j=1}^{n} \alpha_j x_j - u \end{aligned} \tag{8-192}$$

with the obvious inverse.

To apply the Legendre transformation to equations of higher order requires the computation of u_{xx}, etc. We refer the reader to Courant and Hilbert [65, vol. II, p. 34] for details.

SUMMARY

Cauchy problem for quasi-linear first-order partial differential equation in two independent variables Let $\phi(\tau)$, $\psi(\tau)$, and $\omega(\tau)$ belong to $C^{(1)}$ in some interval $\tau_1 \leqslant \tau \leqslant \tau_2$, and let $a(x, y, u)$, $b(x, y, u)$, and $c(x, y, u)$ be continuous in some region of (x, y, u) space containing the curve $\Gamma: x = \phi(\tau)$, $y = \psi(\tau)$, $u = \omega(\tau)$. If $a[\phi(\tau), \psi(\tau), \omega(\tau)]\, \psi'(\tau) - b[\phi(\tau), \psi(\tau), \omega(\tau)]\, \phi'(\tau) \neq 0$ in $\tau_1 \leqslant \tau \leqslant \tau_2$, then there exists a unique solution $u = h(x, y)$ of the PDE $au_x + bu_y = c$ in the neighborhood of Γ satisfying $h[\phi(\tau), \psi(\tau)] = \omega(\tau)$.

This solution can be obtained as follows: solve the *characteristic equations*, i.e., the system of ODEs

$$\frac{dx}{dt} = a(x, y, u)$$
$$\frac{dy}{dt} = b(x, y, u) \tag{S8-1}$$
$$\frac{du}{dt} = c(x, y, u)$$

subject to the ICs: $x = \phi(\tau)$, $y = \psi(\tau)$, $u = \omega(\tau)$, when $t = 0$. Let the solution be

$$x = f(t, \tau)$$
$$y = g(t, \tau)$$
$$u = h(t, \tau)$$

Solve the first two of these equations for t and τ as functions of x and y and insert the result into the third. The resulting function of x and y is the desired solution to the Cauchy problem for the PDE.

Method for obtaining a general solution of $au_x + bu_y = c$ Write the characteristic equations (S8-1) in ratio form

$$\frac{dx}{a} = \frac{dy}{b} = \frac{du}{c} \tag{S8-2}$$

Find two independent integrals of (S8-2), i.e., two functions $\Phi(x, y, u) = \alpha$ and $\Psi(x, y, u) = \beta$ such that no functional relation exists between them. Choose an arbitrary function $F(\alpha, \beta)$. Then the general solution of the PDE is the function $u(x, y)$ defined implicitly by $F[\Phi(x, y, u), \Psi(x, y, u)] = 0$.

DEFINITION S8-1

A solution of a first-order PDE in n independent variables is called a *complete integral* of that PDE if that solution contains, in addition to the independent variables, n independent arbitrary constants. This definition of a complete integral is applicable to *any* first-order PDE, *not* just to quasi-linear ones.

Cauchy problem for the general first-order partial differential equation in two independent variables Let $\phi(\tau)$, $\psi(\tau)$, $\omega(\tau)$, $\rho(\tau)$, $\sigma(\tau)$ all belong to $C^{(1)}$ for $\tau_1 \leqslant \tau \leqslant \tau_2$, and let $F(x, y, u, p, q)$ possess continuous first partial derivatives with respect to all of its five independent variables in some region in $xyupq$ space containing the curve (or, looked at in xyu space, the strip) Γ: $x = \phi(\tau)$, $y = \psi(\tau)$, $u = \omega(\tau)$, $p = \rho(\tau)$, $q = \sigma(\tau)$, $\tau_1 \leqslant \tau \leqslant \tau_2$. Assume that

$$F_p[\phi(\tau), \psi(\tau), \omega(\tau), \rho(\tau), \sigma(\tau)]\, \psi'(\tau) - F_q[\phi(\tau), \psi(\tau), \omega(\tau), \rho(\tau), \sigma(\tau)]\, \phi'(\tau) \neq 0$$

and that (*strip condition*) $\omega'(\tau) = \rho(\tau)\, \phi'(\tau) + \sigma(\tau)\, \psi'(\tau)$, for $\tau_1 \leqslant \tau \leqslant \tau_2$. Then there exists, in the neighborhood of Γ, a function $u = h(x, y)$ satisfying the following:

$$F[x, y, h(x, y), h_x(x, y), h_y(x, y)] \equiv 0$$

$$h[\phi(\tau), \psi(\tau)] = \omega(\tau),\ h_x[\phi(\tau), \psi(\tau)] = \rho(\tau), \text{ and } h_y[\phi(\tau), \psi(\tau)] = \sigma(\tau).$$

Charpit's method to obtain complete integral for $F(x, y, u, u_x, u_y) = 0$ Find an integral

$$\Phi(x, y, u, p, q) = \alpha \tag{S8-3}$$

of the *characteristic equations*

$$\frac{dx}{F_p} = \frac{dy}{F_q} = \frac{du}{pF_p + qF_q} = \frac{dp}{-F_x - pF_u} = \frac{dq}{-F_y - qF_u}$$

Solve (S8-3) and

$$F(x, y, u, p, q) = 0 \tag{S8-4}$$

for p and q as functions of x, y, u, and α. The differential $du = p\,dx + q\,dy$ will be exact, and its integration will yield a solution u depending on a second arbitrary parameter β which will therefore constitute a complete integral of (S8-4).

First-order quasi-linear partial differential equation in n independent variables This is the equation

$$\sum_{j=1}^{n} a_j(\mathbf{x}, u)\, u_{x_j} = a(\mathbf{x}, u) \tag{S8-5}$$

where $\mathbf{x} = (x_1, \ldots, x_n)$. Its *characteristic equations* are

$$\left.\begin{aligned} \frac{dx_j}{dt} &= a_j(\mathbf{x}, u) \\ \frac{du}{dt} &= a(\mathbf{x}, u) \end{aligned}\right. \qquad j = 1, \ldots, n \tag{S8-6}$$

or, in ratio form,

$$\frac{dx_1}{a_1} = \cdots = \frac{dx_n}{a_n} = \frac{du}{a} \tag{S8-7}$$

The *Cauchy problem* for (S8-5) consists in choosing a *free*, i.e., noncharacteristic, $(n - 1)$-dimensional manifold in the n-dimensional $\mathbf{x}$ space, prescribing the values of u on this manifold, and then asking for a solution $u(\mathbf{x})$ of (S8-5) that takes the given initial values. Equivalently, we could look at this as prescribing an $(n - 1)$-dimensional manifold in $(n + 1)$-dimensional $\mathbf{x}u$ space and asking for a surface $u(\mathbf{x})$ containing it.

Analytically, prescribing an $(n - 1)$-dimensional manifold in $\mathbf{x}$ space means giving n independent functions

$$x_j = \phi_j(\tau_1, \ldots, \tau_{n-1}) \qquad j = 1, \ldots, n \tag{S8-8}$$

of $n - 1$ independent parameters, $\tau_1, \ldots, \tau_{n-1}$. Prescribing u on this manifold means giving u as a function,

$$u = \psi(\tau_1, \ldots, \tau_{n-1}) \tag{S8-9}$$

As mentioned above, we could look at (S8-8) and (S8-9) together as defining an $(n - 1)$-dimensional manifold in $\mathbf{x}u$ space.

The condition that the initial manifold be free is that it be nowhere tangent to the *characteristic direction* $\mathbf{a} = (a_1, \ldots, a_n)$. Analytically, this means that $\mathbf{a}$ is not a linear combination of the tangent vectors to the manifold defined by (S8-8) and thus that the determinant of the matrix formed by adjoining $\mathbf{a}$ to the matrix $(\phi_{j\tau_k})$ as an nth column is different from zero.

To solve the Cauchy problem for (S8-5), we solve the *characteristic equations* (S8-6) subject to the ICs (S8-8) and (S8-9) when $t = 0$. The result will be functions

$$x_j = f_j(t, \tau_1, \ldots, \tau_{n-1}) \qquad j = 1, \ldots, n \tag{S8-10}$$

and

$$u = g(t, \tau_1, \ldots, \tau_{n-1}) \tag{S8-11}$$

Because of the requirement that the initial manifold be free, (S8-10) can be solved for t and the τ's in terms of the x's. When these solutions are inserted into (S8-11), the resulting function of $\mathbf{x}$ is the desired solution to the Cauchy problem.

DEFINITION S8-2

A *general solution* of a first-order PDE (*any* first-order PDE in *any* number of independent variables) is one containing an arbitrary function. A general solution does not necessarily contain all solutions of the PDE.

Method for obtaining a general solution of (S8-5) Find any n independent integrals, $\Phi(\mathbf{x}, u) = \alpha_j$, $j = 1, \ldots, n$, of the characteristic equations (S8-7); choose an arbitrary function $F(\boldsymbol{\alpha})$ of n independent variables, $\boldsymbol{\alpha} = (\alpha_1, \ldots, \alpha_n)$; and then set $F(\Phi_1, \ldots, \Phi_n) = 0$ to obtain an implicit representation for the general solution u as a function of $\mathbf{x}$ and the arbitrary function F.

As mentioned in the text, a complete integral, although it seems much less general than a general solution, can nevertheless be used to generate all solutions of the PDE. This is done by putting one of the α's in the complete integral equal to an arbitrary function of the others and forming envelopes of the $(n-1)$-parameter family of integral surfaces thus obtained, i.e., eliminating the α's between the complete integral and the equations $\Phi_{\alpha_j} = 0$.

The most general first-order PDE is

$$F(\mathbf{x}, u, \mathbf{p}) = 0 \qquad \text{(S8-12)}$$

where, as usual, $p_j = u_{x_j}$. Its characteristic equations are

$$\frac{dx_j}{dt} = F_{p_j} \qquad j = 1, \ldots, n$$

$$\frac{du}{dt} = \sum_{j=1}^{n} p_j F_{p_j}$$

$$\frac{dp_j}{dt} = -F_{x_j} - p_j F_u \qquad j = 1, \ldots, n$$

The Cauchy problem for (S8-12) consists in choosing an $(n-1)$-dimensional initial manifold $x_j = \phi_j(\tau_1, \ldots, \tau_{n-1})$ in $\mathbf{x}u$ space, prescribing $u = \omega(\tau_1, \ldots, \tau_{n-1})$ and $u_{x_j} = p_j = \psi_j(\tau_1, \ldots, \tau_{n-1})$ on it, subject to the *strip conditions*

$$\omega_{\tau_j} = \sum_{k=1}^{n} \psi_k \frac{\partial \phi_k}{\partial \tau_j} \qquad j = 1, \ldots, n-1$$

and the condition that the *initial strip* so formed is free, i.e., that the matrix formed by adjoining the column F_{p_j} to $(\phi_{j\tau_k})$ be nonsingular when the initial data are inserted into the arguments of F.

In dealing with (S8-12) it is always permissible to assume that the dependent variable u itself is not present in F. If u does occur explicitly in F, then (S8-12) can be replaced by the PDE $G(\mathbf{X}, \mathbf{P}) = 0$ for a new unknown function $U(\mathbf{X})$, which does not occur explicitly in the new PDE. Here $\mathbf{X} = (x_1, \ldots, x_{n+1})$ stands for $n+1$ independent variables, and $\mathbf{P}$ has the obvious meaning. G is obtained from F by leaving the first n x's in F alone, calling $u = x_{n+1}$, and setting $p_j = -P_j/P_{n+1}$, $j = 1, \ldots, n$. Once we have found $U(\mathbf{X})$, we get the solution $u(\mathbf{x})$ to the original PDE by solving the algebraic equation $U(\mathbf{X}, u) = 0$ for u.

In the special case in which the dependent variable u is missing from the general equation and in which one independent variable has been distinguished, the PDE can be written in the form

$$p + H(\mathbf{x}, t, \mathbf{p}) = 0 \tag{S8-13}$$

where $p = u_t$ and $\mathbf{x}$ and $\mathbf{p}$ have their usual meanings. The PDE (S8-13) is called the *Hamilton-Jacobi equation*. The significant ones of its characteristic equations, namely,

$$\begin{aligned} \frac{dx_j}{dt} &= H_{p_j} \\ \frac{dp_j}{dt} &= -H_{x_j} \end{aligned} \qquad j = 1, \ldots, n \tag{S8-14}$$

are called *Hamilton's equations*. The *Hamilton-Jacobi theory* consists in finding the solutions of Hamilton's equations from a complete integral of the Hamilton-Jacobi equation, in other words, of finding the characteristics of a PDE from a complete integral of that PDE. The method for doing this can be developed in several ways. The result is the following prescription.

In order to find the general solution $\mathbf{x} = \mathbf{x}(t, \boldsymbol{\alpha}, \boldsymbol{\beta})$, $\mathbf{p} = \mathbf{p}(t, \boldsymbol{\alpha}, \boldsymbol{\beta})$ of Hamilton's equations (S8-14), where $\boldsymbol{\alpha}$ and $\boldsymbol{\beta}$ together represent the $2n$ parameters that a general solution of a system of ODE's like (S8-14) must have, first find a solution, $\Phi(\mathbf{x}, t, \boldsymbol{\alpha})$, of the Hamilton-Jacobi equation (S8-13) that depends, as indicated, on n arbitrary constants $\boldsymbol{\alpha}$. Set

$$\Phi_{\alpha_j} = \beta_j \qquad j = 1, \ldots, n \tag{S8-15}$$

where the β's are another n arbitrary constants. The solutions $\mathbf{x} = \mathbf{x}(t, \boldsymbol{\alpha}, \boldsymbol{\beta})$ of (S8-15) constitute the "first half" of the set of solutions to (S8-14). The "other half," $\mathbf{p} = \mathbf{p}(t, \boldsymbol{\alpha}, \boldsymbol{\beta})$, is then obtained immediately from the equations

$$p_j = \Phi_{x_j}$$

without the necessity of solving anything further.

DEFINITION S8-3

Let F_j, $j = 1, \ldots, n$ and G_j, $j = 1, \ldots, n$ be a given set of functions depending on a number of independent variables, and let u and v be two of these variables. Then, by definition, the *Lagrange bracket* $[u, v]$ is

$$[u, v] = \sum_{j=1}^{n} \left(\frac{\partial F_j}{\partial u} \frac{\partial G_j}{\partial v} - \frac{\partial F_j}{\partial v} \frac{\partial G_j}{\partial u} \right)$$

DEFINITION S8-4

Let $F = F(x_1, \ldots, x_n, p_1, \ldots, p_n)$, and $G = G(x_1, \ldots, x_n, p_1, \ldots, p_n)$. Then the *Poisson bracket* (F, G) is defined to be

$$(F, G) = \sum_{j=1}^{n} \left(\frac{\partial F}{\partial x_j} \frac{\partial G}{\partial p_j} - \frac{\partial F}{\partial p_j} \frac{\partial G}{\partial x_j} \right)$$

DEFINITION S8-5

A transformation $\mathbf{x} = \mathbf{x}(\tilde{\mathbf{x}}, \tilde{\mathbf{p}})$, $\mathbf{p} = \mathbf{p}(\tilde{\mathbf{x}}, \tilde{\mathbf{p}})$ is called *canonical* if there exists a function $\tilde{G}(\tilde{\mathbf{x}}, \tilde{\mathbf{p}})$, such that the ODEs $x_j' = G_{p_j}$, $p_j' = -G_{x_j}$ transform into $\tilde{x}_j' = \tilde{G}_{\tilde{p}_j}$, $\tilde{p}_j' = -\tilde{G}_{\tilde{x}_j}$.

DEFINITION S8-6

A function Φ of any $2n$ independent variables of the $4n$ variables related by a canonical transformation is called a *generating function* if

$$d\Phi = \sum_{j=1}^{n} (\tilde{p}_j \, d\tilde{x}_j - p_j \, dx_j)$$

Theorem S8-1 A transformation is canonical if and only if

$$\frac{\partial x_j}{\partial \tilde{x}_k} = \frac{\partial \tilde{p}_k}{\partial p_j} \qquad \frac{\partial x_j}{\partial \tilde{p}_k} = -\frac{\partial \tilde{x}_k}{\partial p_j} \qquad \frac{\partial p_j}{\partial \tilde{x}_k} = -\frac{\partial \tilde{p}_k}{\partial x_j} \qquad \frac{\partial p_j}{\partial \tilde{p}_k} = \frac{\partial \tilde{x}_k}{\partial x_j}$$

Theorem S8-2 A transformation is canonical if and only if

$$[x_j, x_k] = [p_j, p_k] = 0 \qquad \text{and} \qquad [x_j, p_k] = \delta_{jk} \qquad j, k = 1, \ldots, n$$

Theorem S8-3 A transformation is canonical if and only if

$$(x_j, x_k) = (p_j, p_k) = 0 \qquad \text{and} \qquad (x_j, p_k) = \delta_{jk} \qquad j, k = 1, \ldots, n$$

Theorem S8-4 A transformation is canonical if and only if a generating function exists.

Theorem S8-5 If $\Phi(\mathbf{x}, \boldsymbol{\alpha})$ is a solution of the Hamilton-Jacobi equation, then $\Phi(\mathbf{x}, \mathbf{p})$ is a generating function for a canonical transformation which converts the new Hamiltonian into identically zero.

DEFINITION S8-7

Given a function $u(\mathbf{x})$, the change of variables $\xi_j = u_{x_j}$, $v = \sum_{j=1}^{n} \xi_j x_j - u$ from $u(\mathbf{x})$ to $v(\boldsymbol{\xi})$ is called a *Legendre transformation*.

PROBLEMS

1. Find the general solution:
   (*a*) $px(u - 2y^2) = (u - qy)(u - y^2 - 2x^3)$
   (*b*) $p + xq = xy\sqrt{u}$
   (*c*) $yu_x - xu_y + yzu_z = x$
   (*d*) $(x^2 - y^2 - z^2)\, z_x + 2xyz_y = 2xz$
   (*e*) $(\cos x \cos y)\, u_x - (\sin x \sin y)\, u_y + (\sin x \cos y)\, u_z = 0$
   (*f*) $yzu_x + zxu_y + xyu_z + xyz = 0$
   (*g*) $u_x + u^2 u_y = 1$

2. Find complete integrals:

(*a*) $p(q^4 + 4) + a(1 - 2u) = 0$
(*b*) $p^2 + q^2 - px - qy + \frac{1}{2}xy = 0$
(c) $\sqrt{p} + \sqrt{q} = 2x$
(*d*) $p^2u + q^2 = 4$
(*e*) $u - x + \log pq = 0$

3. Find the general solution:

(*a*) $(x^2 + y^2)\, u_x + 2xyu_y = (x + y)u$
(*b*) $(y + 1)\, u_x + (x + 1)\, u_y = u$
(c) $yuu_x + xuu_y = x + y$
(*d*) $(x^2 + y^2 + yu)\, u_x + (x^2 + y^2 - xu)\, u_y - (x + y)u = 0$

9
Second-order Hyperbolic Equations in Two Independent Variables

9-1. INTRODUCTION: CAUCHY PROBLEM

In Chap. 6 we considered the equation

$$\begin{aligned} a(x, y)\, u_{xx} + 2b(x, y)\, u_{xy} + c(x, y)\, u_{yy} + d(x, y)\, u_x \\ + e(x, y)\, u_y + f(x, y)\, u = g(x, y) \end{aligned} \tag{9-1}$$

and called it hyperbolic at any point (x, y) where

$$b^2 - ac > 0 \tag{9-2}$$

Condition (9-2) was arrived at in the process of transforming (9-1) into a canonical form. In that process the concept of characteristic curves arose naturally. In fact the characteristic curves turned out to be the natural ones to choose as new coordinate curves in order to simplify the appearance of (9-1). In Chaps. 1 and 8, however, for first-order partial differential equations (PDEs), the origin and use of characteristics were different from their introduction in Chap. 6. We wish now to introduce the concept of characteristics for (9-1) in a way more closely in harmony with that of Chap. 8. In doing this we can, just as we did in Chap. 1, deal with a quasi-linear instead of linear equation. Accordingly, we consider the most general second-order quasi-linear PDE in two independent variables. Using the abbreviations $u_x = p$ and $u_y = q$ introduced in the previous chapter, we can write this PDE in the form

$$a(x, y, u, p, q)\, u_{xx} + 2b(x, y, u, p, q)\, u_{xy} + c(x, y, u, p, q)\, u_{yy} - d(x, y, u, p, q) = 0 \tag{9-3}$$

As pointed out in Sec. 6-9, this equation is not only nonlinear in u, it is in general nonlinear in the first derivatives of u also. It is linear only in the highest derivatives of u occurring in the equation, i.e., the second derivatives in the present case.

The hyperbolicity condition (9-2), as mentioned in Sec. 6-9, now depends upon the solution and could be violated for some solutions. In general, an equation such as (9-3) will not be hyperbolic for all solutions. It will be hyperbolic for some, elliptic for some, and parabolic for some. In addition, the type of equation, of course, also depends upon the region in the xy plane.

To deal with (9-3) it is sometimes convenient to add to the abbreviations (8-25) by introducing

$$u_{xx} = r \qquad u_{xy} = s \qquad u_{yy} = t \tag{9-4}$$

Then (9-3) can be written

$$ar + 2bs + ct = d \tag{9-5}$$

Our first objective is to formulate a Cauchy problem for (9-5) which will be the analog of the one already treated in Chap. 8 for first-order equations. To do this we need an initial curve Γ: $x = \phi(\tau)$, $y = \psi(\tau)$ in the xy plane. On it we prescribe the value of u: $u = \omega(\tau)$. Since (9-5) is second order, we expect to have to prescribe more data on Γ. To get an idea of what additional data should be prescribed on Γ, we look at the special case of the wave equation. In the initial-value problem (IVP) for it, we prescribed the function and its t derivative. Since the t derivative is a normal derivative on the x axis, we are led to conjecture that the correct Cauchy problem for (9-5) consists in prescribing u and u_ν on Γ, where u_ν means the normal derivative of u on Γ. Either normal can be taken as positive on Γ as long as we are consistent. By analogy with the first-order case, we expect that the Cauchy problem as just formulated will have a unique solution if Γ is not characteristic.

In the first-order case, the PDE could be looked at as defining a unique directional derivative at each point, and a curve was called characteristic at a point if it was tangent to that direction at that point. Thus, a characteristic curve of a first-order quasi-linear PDE has the property that the differential operator is tangential to the curve at every point.

For second-order hyperbolic operators, the situation is not quite so simple. At every point in its region of hyperbolicity, the PDE defines *two* characteristic directions; hence it cannot be looked at as a directional derivative at a point. Nevertheless, it is possible to generalize the definition of a tangential operator to operators of higher order than the first, but the generalization will require the surrender of some of the properties possessed by first-order tangential operators.

9-2. INNER AND OUTER DERIVATIVES; TANGENTIAL OPERATORS

In order to arrive at the generalization proposed above, we start with some considerations which at first sight seem to have relatively little to do with PDEs. Suppose we are given a curve Γ in the xy plane and on it the values of some function $u(x, y)$. Let the parametric equations of Γ be $x = \phi(\tau)$, $y = \psi(\tau)$. Then the given values of u on Γ can be represented by a function $\omega(\tau)$ given by $\omega(\tau) = u[\phi(\tau), \psi(\gamma)]$.

Since u is a given function of τ on Γ, we can compute as many τ derivatives of u as we like on Γ. For example, $u_\tau = \omega'$, $u_{\tau\tau} = \omega''$, etc., on Γ. It is clear that u_τ is a tangential derivative of u on Γ. It is also clear that no normal derivative u_ν, $u_{\nu\nu}$, etc., could be found solely from the values of u on Γ. In fact, it is obvious that no derivative of u in any direction not tangential to Γ could be found using solely the values of u on Γ. Calling a derivative of u in a direction not tangential to Γ an *outer derivative*, we can say that the values of u on Γ are insufficient to enable us to determine any outer derivative of u on Γ. All that can be determined from them are the tangential derivatives of u.

Now suppose that in addition to the value of u on Γ, we are given the value of u_ν on Γ. From the values of u on Γ we can, as already pointed out, compute u_τ on Γ. Thus we know both u_τ and u_ν on Γ. Since both u_τ and u_ν are linear combinations of u_x and u_y, we can solve the two equations $u_\tau = u_x \cos + \cdots$ and $u_\nu = u_x \cos + \cdots$ for u_x and u_y and then, using u_x and u_y, we can compute any outer first derivative of u we please on Γ. In other words, knowing u and u_ν on Γ enables us to find all outer first derivatives of u on Γ. Similarly, knowing u and any outer derivative of u on Γ would enable us to find all other outer first derivatives of u on Γ because any directional derivative is a linear combination of any two linearly independent directional derivatives.

To summarize, we have shown that given u and u_ν on Γ, we can find all first derivatives of u on Γ. What about second derivatives? By the remark above, we can find $u_{\tau\tau}$ and $u_{\nu\tau}$. However, $u_{\nu\nu}$ cannot be found from the given data. In order to find $u_{\nu\nu}$ we need additional information. Suppose that additional information consists in knowing that u satisfies the PDE (9-5). Will this additional information enable us to determine $u_{\nu\nu}$? In order to answer this question, we change notation so that the equation of Γ is written in the form $\phi(x, y) = 0$. We embed Γ into a one-parameter family of curves $\phi(x, y) = \nu$ and let $\psi(x, y) = \tau$ be their orthogonal trajectories. Then we introduce, as in Chap. 6, τ and ν as new independent variables in the PDE.

The PDE becomes

$$Au_{\nu\nu} + Bu_{\nu\tau} + Cu_{\tau\tau} = D \tag{9-6}$$

where

$$A = a\phi_x^2 + 2b\phi_x\phi_y + c\phi_y^2 \qquad B = \cdots \tag{9-7}$$

[cf. Eqs. (6-10) to (6-15)].

Since $u_{\nu\tau}$ and $u_{\tau\tau}$ are already known, (9-6) clearly determines $u_{\nu\nu}$ if and only if $A \neq 0$. According to Chap. 6, $A = 0$ means Γ is characteristic. Thus we see that the PDE together with the initial data will determine the second normal derivative if and only if the initial curve is noncharacteristic. If the initial curve is characteristic, then not only does the PDE not determine the missing normal derivative, it gives an additional relation among the prescribed data. The initial data therefore cannot be prescribed arbitrarily for a solution of the PDE in that case.

A noncharacteristic curve is called *free*. Thus, the Cauchy problem has a solution for arbitrary initial data only if the initial curve is free.

To summarize, we have discovered the following: if u and u_ν are given on a curve Γ, then these initial data together with the PDE (9-5) uniquely determine all second derivatives of u on Γ if and only if Γ is free, i.e., noncharacteristic. In fact, they determine more. We shall show that the *Cauchy data* (as u and u_ν are called) together with the PDE determine all derivatives of all orders on Γ if Γ is free.

To show this, it is sufficient to show that they determine all third derivatives. The reader will then see how to show the rest by induction. By a previous remark, all τ derivatives of any order can be determined because they are tangential. Therefore all we have to do is to show how to find $u_{\nu\nu\nu}$. We do it by differentiating (9-6) with respect to ν and observing that $u_{\nu\nu\nu}$ appears only in the term $Au_{\nu\nu\nu}$. All other terms contain only known quantities (u, u_τ, u_ν, $u_{\tau\tau}$, $u_{\nu\tau}$, $u_{\nu\nu}$, $u_{\tau\tau\tau}$, $u_{\tau\tau\nu}$, $u_{\nu\nu\tau}$). Since Γ is free, $A \neq 0$ and the equation $Au_{\nu\nu\nu} + \cdots = 0$ therefore determines $u_{\nu\nu\nu}$.

Thus, all third derivatives of u can be found on a free curve. It is clear that the process can be continued and all higher derivatives of u can be determined step by step. Thus, if a curve is free, all derivatives of a solution of (9-5) are uniquely determined on it if the solution and its first derivatives are known there. This suggests that the solution to an IVP with analytic initial data is analytic in the neighborhood of an analytic free curve. To prove this conjecture requires considerable additional work, since the mere existence of all derivatives of the solution does not yet prove its analyticity. As usual, we omit the proof. We do wish to point out, however, that this special case of the *Cauchy-Kowalewski theorem* does not settle the question of analyticity of the solutions to (9-5). The reason for this is that other types of problems can be posed for (9-5) and the solutions to these other problems are not necessarily analytic. All that can be proved is that the solution to a very special type of IVP for (9-5) is analytic in a very small neighborhood of a free analytic initial curve. In practice quite different problems arise, and we shall say more on the subject later on.

The foregoing enables us to give the promised generalization of tangential derivative. Since it will contain normal first derivatives, the word "tangential" is a little misleading; so we call it an *inner derivative* instead.

DEFINITION 9-1

A second-order differential operator in two independent variables is called an *inner operator with respect to a curve* Γ if the data u, u_ν on Γ together with the operator fail to determine any outer second derivatives of u on Γ.

Notice that whether an operator is inner or outer is always relative to a given curve. (The definition of *outer operator* is obvious in view of Definition 9-1.)

9-3. RIEMANN'S METHOD FOR HYPERBOLIC EQUATIONS

Suppose we have a PDE which is hyperbolic in some region, and suppose we are given a free initial curve Γ. The discussion above indicates that we ought to be able to prescribe the values of u and u_ν on Γ and then find a solution of (9-3) which takes these initial values. Back in Chap. 2 we solved a number of such IVPs for a very simple hyperbolic equation, namely, the one-dimensional wave equation $u_{xx} - c^{-2}u_{tt} = 0$. We are now interested in extending our results to more general equations.

However, (9-3) is too general. We assume instead that we have a *linear* hyperbolic equation in two independent variables. According to Chap. 6, such an equation can always be reduced to the canonical form

$$u_{xy} + a(x, y)\, u_x + b(x, y)\, u_y + c(x, y)\, u = F(x, y) \tag{9-8}$$

We assume that such a reduction has been carried out. This means, in practice, that the reader who wishes to apply the results of the present section to hyperbolic equations which are not in the canonical form (9-8) will first have to reduce them to that form. It also means that the results will not be directly applicable to equations in more independent variables or to equations of higher order.

In Chap. 2, we solved the IVP for the wave equation by drawing the characteristics backward to the initial curve and then determining the solution at the point in question from the initial values on the segment of the initial curve cut out by those characteristics. The characteristics of (9-8) are the straight lines $x = \text{const}$ and $y = \text{const}$. We assume the initial curve Γ to be free at each of its points. This means that Γ can never be parallel to the x or y axis and hence that no characteristic can cross Γ more than once. If, to find the solution at the point $Q(\xi, \eta)$, we draw the characteristics from Q to Γ, then we expect that $u(\xi, \eta)$ will depend upon the initial values on the segment of Γ between P_1 and P_2 (see Fig. 9-1).

Next we do something more analogous to what we did for the potential equation. We derive a sort of Green's formula for (9-8).

When written in the form

$$\iint_{\mathscr{R}} \phi \, \nabla^2 \psi \, dA = \iint_{\mathscr{R}} \psi \, \nabla^2 \phi \, dA + \int_{\partial\mathscr{R}} \left(\phi \frac{\partial \psi}{\partial n} - \psi \frac{\partial \phi}{\partial n} \right) ds$$

the two-dimensional analog of (3-4) looks like a sort of generalization of integration by parts in which the differential operator ∇^2 got shifted from ψ to ϕ. Given any linear differential operator L, we can always form $\iint \phi \, L\psi \, dA$ and then integrate by parts, i.e., use an appropriate Green's theorem, to shift the derivatives onto ϕ. In so doing we obtain a formula that looks like

$$\iint_{\mathscr{R}} \phi L \psi \, dA = \iint_{\mathscr{R}} \psi L^* \phi \, dA + \int_{\partial\mathscr{R}} B(\phi, \psi) \, ds \tag{9-9}$$

where L^* is a new linear differential operator, called the (formal) *adjoint* of L, and $B(\phi, \psi)$ is an expression containing ϕ, ψ, and their derivatives up to an order one lower than that of L.

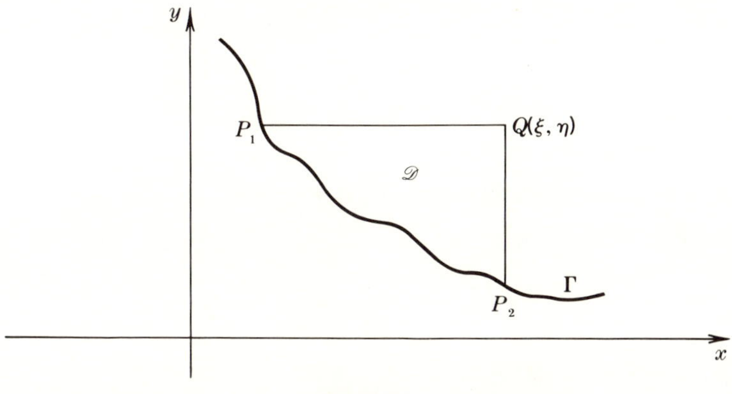

Figure 9-1

Our aim is to derive an identity such as (9-9) for the operator on the left of (9-8). To do this we first recall two of Green's formulas in the plane

$$\iint_{\mathscr{R}} f_y \, dA = -\int_{\partial\mathscr{R}} f \, dx \tag{9-10}$$

and

$$\iint_{\mathscr{R}} g_x \, dA = \int_{\partial\mathscr{R}} g \, dy \tag{9-11}$$

In these put $f = \phi\psi$ and $g = \phi\psi$. The we get the two-dimensional analogs of integration by parts

$$\iint_{\mathscr{R}} \phi\psi_y \, dA = -\iint_{\mathscr{R}} \psi\phi_y \, dA - \int_{\partial\mathscr{R}} \phi\psi \, dx \tag{9-12}$$

and

$$\iint_{\mathscr{R}} \phi\psi_x \, dA = -\iint_{\mathscr{R}} \psi\phi_x \, dA + \int_{\partial\mathscr{R}} \phi\psi \, dy \tag{9-13}$$

In (9-12) take $\phi = v$ and $\psi = u_x$. Then

$$\iint_{\mathscr{R}} vu_{xy} \, dA = -\iint_{\mathscr{R}} u_x v_y \, dA - \int_{\partial\mathscr{R}} vu_x \, dx \tag{9-14}$$

In (9-13) take $\phi = v_y$ and $\psi = u$. Then

$$\iint_{\mathscr{R}} u_x v_y \, dA = -\iint_{\mathscr{R}} uv_{xy} \, dA + \int_{\partial\mathscr{R}} uv_y \, dy$$

Inserting this into (9-14), we obtain

$$\iint_{\mathscr{R}} vu_{xy} \, dA = \iint_{\mathscr{R}} uv_{xy} \, dA - \int_{\partial\mathscr{R}} uv_y \, dy - \int_{\partial\mathscr{R}} vu_x \, dx$$

that is,

$$\iint_{\mathscr{R}} vu_{xy} \, dA = \iint_{\mathscr{R}} uv_{xy} \, dA - \int_{\partial\mathscr{R}} (vu_x \, dx + uv_y \, dy) \tag{9-15}$$

Similarly, by obvious choices of ϕ and ψ we obtain

$$\iint_{\mathscr{R}} avu_x \, dA = -\iint_{\mathscr{R}} u(av)_x \, dA + \int_{\partial\mathscr{R}} auv \, dy \tag{9-16}$$

and

$$\iint_{\mathscr{R}} bvu_y \, dA = -\iint_{\mathscr{R}} u(bv)_y \, dA - \int_{\partial\mathscr{R}} buv \, dx \tag{9-17}$$

If we multiply (9-8) by an arbitrary function $v(x, y)$ and integrate over any region $\mathscr{R}$, we have

$$\iint_{\mathscr{R}} v(u_{xy} + au_x + bu_y + cu)\, dA = \iint_{\mathscr{R}} vF\, dA \tag{9-18}$$

Using (9-15) to (9-17), we can write this as

$$\iint_{\mathscr{R}} u[v_{xy} - (av)_x - (bv)_y + cv]\, dA$$
$$= \int_{\partial\mathscr{R}} \Big[(vu_x\, dx + uv_y\, dy) + \int_{\partial\mathscr{R}} uv(b\, dx - a\, dy)\Big] + \iint_{\mathscr{R}} vF\, dA$$

or

$$\iint_{\mathscr{R}} u[v_{xy} - (av)_x - (bv)_y + cv]\, dA$$
$$= \int_{\partial\mathscr{R}} [v(u_x + bu)\, dx + u(v_y - av)\, dy)] + \iint_{\mathscr{R}} vF\, dA \tag{9-19}$$

If we let

$$L[u] = u_{xy} + au_x + bu_y + cu \tag{9-20}$$

and

$$L^*[v] = v_{xy} - (av)_x - (bv)_y + cv \tag{9-21}$$

then (9-18) and (9-19) show that

$$\iint_{\mathscr{R}} (vL[u] - uL^*[v])\, dA = -\int_{\partial\mathscr{R}} v(u_x + bu)\, dx + u(v_y - av)\, dy \tag{9-22}$$

which represents a generalization of the Green's identity (3-4).

The operator L^*, which, as seen above, results from integrating the left side of (9-18) by parts, i.e., applying Green's identity, is, as mentioned above, called the (formal) adjoint of L. If $L^* = L$, then L is called *formally self-adjoint*. The operator ∇^2 is an example of a formally self-adjoint operator.

Having obtained a Green's identity for (9-8), we next try to pick the function v so that it will play a roll analogous to that of the Green's function. From Chap. 3 we recall that the Green's function is a solution of the DE which has a singularity of just the strength necessary to pick out the integrand from part of the boundary integral occurring in Green's identity. Since (9-8) is not self-adjoint, we choose v to be a solution of the adjoint equation

$L^*[v] = 0$. With v so chosen, and with u a solution of (9-8), (9-22) reduces to

$$0 = \int_{\partial\mathscr{R}} [v(u_x + bu)\,dx + u(v_y - av)\,dy] - \iint_{\mathscr{R}} vF\,dA \tag{9-23}$$

If, following our intention stated above, we take $\mathscr{R}$ to be the region indicated in Fig. 9-1, then, since $dx = 0$ on P_2Q and $dy = 0$ on QP_1, we see that (9-23) takes the form

$$0 = \int_{\Gamma} [v(u_x + bu)\,dx + u(v_y - av)\,dy] + \int_{P_2}^{Q} u(v_y - av)\,dy + \int_{Q}^{P_1} v(u_x + bu)\,dx + \iint_{\mathscr{R}} vF\,dA \tag{9-24}$$

Next we have to pick v so that $u(Q)$ appears on the right-hand side of (9-24) and so that all unknown data in (9-24) disappear from the resulting formula. Since the integral over Γ involves only known data on the initial curve, it is clear that the second and third integrals in (9-24) are the two integrals to be dealt with. Moreover, it is clear that no singularity in v is necessary. Indeed, since the characteristics $x = \xi$ and $y = \eta$ depend upon the variable point (ξ, η), requiring v to be singular on either of these two characteristics would be tantamount to requiring v to be singular everywhere. Thus v *must* be regular everywhere. In fact, from (9-24) we see that integrating the first term in the second or third integral in (9-24) will yield a term involving $u(Q)$. Notice, however, that we should integrate only one of these two integrals by parts. If we integrate both by parts, the terms involving $u(Q)$ cancel and we get nowhere. So we integrate the first half of the third integral in (9-24) by parts. This gives

$$\int_{Q}^{P_1} v(u_x + bu)\,dx = v(P_1)\,u(P_1) - v(Q)\,u(Q) - \int_{Q}^{P_1} u(v_x - bv)\,dx \tag{9-25}$$

Inserting this into (9-24) and transposing a term, we get

$$v(Q)\,u(Q) = v(P_1)\,u(P_1) + \int_{\Gamma} [v(u_x + bu)\,dx + u(v_y - av)\,dy] + \int_{P_2}^{Q} u(v_y - av)\,dy - \int_{Q}^{P_1} u(v_x - bv)\,dx + \iint_{\mathscr{R}} vF\,dA \tag{9-26}$$

Now suppose the values of u and its normal derivative are known on Γ. This means that u, u_x, and u_y are all known on Γ (see sec. 9-2). Formula (9-26) expresses the value of u at Q in terms of its value at the point P_1 on Γ and integrals involving the values of u, u_x, and u_y on Γ and on the two characteristics. Also involved are the values of v at various points. v, however, will be a known function which we are now in the process of picking. The only unknown quantities in (9-26) are therefore the values of u at Q and on the characteristics. If we could eliminate the need for knowing the values of u on the characteristics, then (9-26) would provide us with a formula expressing the value of u at the arbitrary point Q in terms of known values. In other words, it would provide the solution to the IVP.

Thus, all we have to do is to pick v so that the values of u on the characteristics are not needed. From (9-26) it is evident that choosing v so that

$$v_y - av = 0 \qquad \text{on } P_2Q\text{, that is, when } x = \xi \tag{9-27}$$

and

$$v_x - bv = 0 \qquad \text{on } P_1Q\text{, that is, when } y = \eta \tag{9-28}$$

will do this. Again we note the analogy with the use of Green's functions for the potential equation.

Let us summarize the conditions on v. First of all note that conditions (9-27) and (9-28) depend on the location of the point Q. For different Q's the characteristics will be different, and therefore the lines on which (9-27) and (9-28) are to hold will be different. Since Q depends upon ξ and η, so therefore will v. v also depends upon x and y, of course. Like the Green's function, it is thus a function of four independent variables, x, y, ξ, η. We proceed to state precisely the conditions on $v(x, y, \xi, \eta)$.

1. As a function of (x, y), v satisfies $L^*[v] = 0$, where L^*, given by (9-21), is the adjoint operator to (9-20).
2. $v_y(\xi, y, \xi, \eta) = a(\xi, y)\, v(\xi, y, \xi, \eta)$ (9-27*a*)
3. $v_x(x, \eta, \xi, \eta) = b(x, \eta)\, v(x, \eta, \xi, \eta)$ (9-28*a*)

Conditions 2 and 3 are just (9-27) and (9-28) restated. To simplify (9-26) a little more, we add a mild additional condition, namely,

4. $v(\xi, \eta, \xi, \eta) = 1$ (9-29)

This reduces the left-hand side of (9-26) to $u(\xi, \eta)$ without any complicated coefficient.

The function satisfying conditions 1 to 4, whose existence we assume, is called the *Riemann function* of the PDE (9-8). Sometimes it is called the *Riemann-Green function* and occasionally just the *Green's function.* Note that the Riemann function depends only on the equation and not on the curve Γ or the initial values prescribed on Γ. It can therefore be used to solve an arbitrary IVP for the PDE on an arbitrary free initial curve.

If v is the Riemann function, then (9-26) reduces to

$$u(\xi, \eta) = v(P_1)\, u(P_1) + \int_\Gamma [v(u_x + bu)\, dx + u(v_y - av)\, dy] + \iint_{\mathscr{D}} vF\, dA \tag{9-30}$$

This is not quite the analog of (2-17) since it involves the value of u at only one of the end points of the characteristic triangle. In order to obtain a formula that is symmetric in P_1 and P_2 we let

$$\begin{aligned} I &= \int_\Gamma v(u_x + bu)\, dx + u(v_y - av)\, dy \\ &= \int_\Gamma vu_x\, dx + uv_y\, dy + \int_\Gamma uv(b\, dx - a\, dy) \end{aligned}$$

and observe that

$$\begin{aligned} \int_\Gamma vu_x\, dx + uv_y\, dy &= \int_\Gamma d(uv) - uv_x\, dx - vu_y\, dy \\ &= u(P_2)\, v(P_2) - u(P_1)\, v(P_1) - \int_\Gamma uv_x\, dx + vu_y\, dy \end{aligned}$$

Hence

$$I = u(P_2)\, v(P_2) - u(P_1)\, v(P_1) - \int_\Gamma [u(v_x - bv)\, dx + v(u_y + au)\, dy]$$

Inserting this into (9-30), we find

$$u(\xi, \eta) = u(P_2)\, v(P_2) - \int_\Gamma [u(v_x - bv)\, dx + v(u_y + au)\, dy] + \iint_{\mathscr{D}} vF\, dA \tag{9-31}$$

Finally, adding (9-30) and (9-31), we obtain the desired symmetric formula for u

$$\begin{aligned} u(\xi, \eta) &= \frac{1}{2}[v(P_1)\, u(P_1) + v(P_2)\, u(P_2)] \\ &\quad + \frac{1}{2}\int_\Gamma [(vu_x - uv_x + 2buv)\, dx + (uv_y - vu_y - 2auv)\, dy] + \iint vF\, dA \end{aligned} \tag{9-32}$$

Let us check that (9-32) reduces to (2-17) in the special case of the one-dimensional wave equation. To carry out the reduction we need the Riemann function for the homogeneous wave equation in characteristic coordinates, i.e., for the equation $u_{xy} = 0$. Obviously it has the form

$$v(x, y, \xi, \eta) = \phi(x, \xi, \eta) + \psi(y, \xi, \eta) \tag{9-33}$$

From (9-27a) and (9-28a) we see that $\phi_x(x, \xi, \eta) = 0$ and $\psi_y(y, \xi, \eta) = 0$. Hence $\phi(x, \xi, \eta) = \alpha(\xi, \eta)$ and $\psi(y, \xi, \eta) = \beta(\xi, \eta)$, where α and β are arbitrary functions of ξ and η. If we put $\gamma(\xi, \eta) = \alpha(\xi, \eta) + \beta(\xi, \eta)$, then $v(x, y, \xi, \eta) = \gamma(\xi, \eta)$. But according to (9-32), $v(\xi, \eta, \xi, \eta) = 1$. Hence $\gamma(\xi, \eta) = 1$, and therefore $v(x, y, \xi, \eta) = 1$ is the Riemann function for the one-dimensional wave equation. Using it in (9-32), we have

$$u(\xi, \eta) = \frac{1}{2}[u(P_1) + u(P_2)] + \frac{1}{2}\int_\Gamma u_x\,dx - u_y\,dy \tag{9-34}$$

It is easy to check that transforming back to the usual noncharacteristic coordinates yields (2-17). For simplicity, suppose $c = 1$. Then

$$u_x\,dx - u_y\,dy = u_t\,dt$$

and the initial curve $x = 0$ in the xt plane corresponds to $x + y = 0$ in the xy plane. From these facts the reader should be able to complete the reduction.

9-4. RIEMANN FUNCTION FOR THE TELEGRAPH EQUATION

In characteristic coordinates the telegraph equation (see Sec. 2-15) can be written in the form

$$u_{xy} + cu = 0 \tag{9-35}$$

We assume that c is a constant.

Since (9-35) is self-adjoint, the Riemann function $v(x, y, \xi, \eta)$ must be a solution of it. In addition [see (9-27a) and (9-28a)] v must satisfy

$$\begin{aligned} v_y(\xi, y, \xi, \eta) &= 0 \\ v_x(x, \eta, \xi, \eta) &= 0 \end{aligned} \tag{9-36}$$

and (cf. (9-29)]

$$v(\xi, \eta, \xi, \eta) = 1 \tag{9-37}$$

From (9-36) it follows that

$$v_x(\xi, \eta, \xi, \eta) = v_y(\xi, \eta, \xi, \eta) = 0 \tag{9-38}$$

Also from (9-36)

$$\begin{aligned} v_{xx}(x, \eta, \xi, \eta) &= 0 \\ v_{yy}(\xi, y, \xi, \eta) &= 0 \end{aligned} \tag{9-39}$$

Hence

$$v_{xx}(\xi, \eta, \xi, \eta) = v_{yy}(\xi, \eta, \xi, \eta) = 0 \tag{9-40}$$

Since v satisfies (9-35), we have

$$v_{xy}(\xi, \eta, \xi, \eta) = -cv(\xi, \eta, \xi, \eta) = -c \qquad \text{by (9-37)} \tag{9-41}$$

From the PDE we have

$$v_{xyy} = -cv_y \qquad v_{xyx} = -cv_x \qquad \text{etc.}$$

From (9-36), (9-37), and (9-39),

$$v_{xxy}(\xi, \eta, \xi, \eta) = v_{xyy}(\xi, \eta, \xi, \eta) = 0 \qquad \text{etc.}$$

v_{xxyy} turns out to be different from zero, while all other fourth derivatives vanish at (ξ, η, ξ, η).

Let us look at $v(x, y, \xi, \eta)$ as a function of x and y depending upon two parameters ξ and η. Let us expand v as a function of x and y into a Taylor series centered at $(x, y) = (\xi, \eta)$, namely,

$$\begin{aligned} v(x, y, \xi, \eta) = v(\xi, \eta, \xi, \eta) &+ (x - \xi)\, v_x(\xi, \eta, \xi, \eta) \\ &+ (y - \eta)\, v_y(\xi, \eta, \xi, \eta) + \tfrac{1}{2}(x - \xi)^2\, v_{xx}(\xi, \eta, \xi, \eta) \\ &+ (x - \xi)(y - \eta)\, v_{xy}(\xi, \eta, \xi, \eta) + \cdots \end{aligned}$$

The above analysis seems to indicate that all terms except those involving $(x - \xi)^n (y - \eta)^n$ (same exponent on both) vanish. Rather than prove this conjecture, we merely assume that the Riemann function has this property. After all, our aim is to find the Riemann function. Any method that leads us to it is all right as long as we can verify that the function we find has all the properties of the Riemann function.

In any event, we assume that the Riemann function for (9-35) is a function of $(x - \xi)(y - \eta)$, namely,

$$v(x, y, \xi, \eta) = w[(x - \xi)(y - \eta)] = w(s) \tag{9-42}$$

where

$$s = (x - \xi)(y - \eta) \tag{9-43}$$

In order to insert this into (9-35) we need

$$v_x = (y - \eta)\, w'(s) \tag{9-44}$$

and then

$$v_{xy} = (x - \xi)(y - \eta)\, w''(s) + w'(s) = sw''(s) + w'(s) \tag{9-45}$$

Then (9-35) becomes

$$sw'' + w' + cw = 0 \tag{9-46}$$

The fact that we have been able to express all x's and y's in terms of s augurs well for our assumption (9-42).

Equation (9-46) can be reduced to Bessel's equation of index zero [Eq. (4-13) with $n = 0$] by setting

$$\sigma = \sqrt{4cs}$$

Then

$$w = J_0(\sigma) = J_0(\sqrt{4cs}) = J_0[\sqrt{4c(x - \xi)(y - \eta)}]$$

We leave it to the reader to check that this w has all the properties necessary to make it the Riemann function for (9-35).

In introducing the Riemann function we were guided in part by analogs of Green's identity and of Green's functions. If we use complex variables, we can see an interesting analogy between the Green's function and the Riemann function.

Consider the reduced wave equation

$$u_{xx} + u_{yy} + k^2 u = 0 \tag{9-47}$$

and let $z = x + iy$ and $\bar{z} = x - iy$. Changing variables formally, we find

$$u_{z\bar{z}} + \tfrac{1}{4}k^2 u = 0 \tag{9-48}$$

which is the telegraph equation. The Riemann function for (9-48) is

$$J_0[\sqrt{k^2(z - \zeta)(\bar{z} - \bar{\zeta})}] = J_0(k \mid z - \zeta \mid)$$

On the other hand, the fundamental solution of (9-47) is

$$H_0^{(1)}(kr) = H_0^{(1)}(k \mid z - \zeta \mid)$$

(cf. Sec. 4-7). Hence the Riemann function is the real part of the fundamental solution. Naturally the analogy is not complete. It hardly could be. The Green's function depends upon the domain, whereas the Riemann function is independent of the initial curve. Therefore the Riemann function is more closely related to the fundamental solution than to the Green's function since both depend upon the equation and not the region.

Vekua has used the above connection to generalize to (9-48) the representation of potential functions as real parts of analytic functions. He found that the solution to (9-48) can be represented in the form

$$u = \operatorname{Re} \left\{ \phi(z) - \int_{z_0}^{z} \phi(t) \frac{\partial}{\partial t} J_0[k \sqrt{\bar{z}(z - t)}]\, dt \right\} \tag{9-49}$$

where ϕ is an arbitrary analytic function of z.

9-5. BOUNDARY-INITIAL-VALUE PROBLEMS; UNIQUENESS

Not all the problems considered in Chap. 2 satisfy the conditions of the present chapter. For example, the problem for the finite string involved an initial curve which was free but which did not have the property that every characteristic cut it only once. As a result we were not able to prescribe both u and u_ν on the boundary. Recall that the xt region (see Fig. 9-2) for the finite string was the semi-infinite strip $0 \leqslant x \leqslant l, 0 \leqslant t < \infty$. On the bottom boundary, $t = 0$, we prescribed both u and u_t. On the vertical sides, however, we prescribed either u or u_x or a linear combination of the two but never more than one condition there. The reason for this has already been explained in Sec. 2-4.

Back in Chap. 2 we solved problems for the finite string by separation of variables. The idea of that method is to put together enough special solutions of the PDE to be able to satisfy all the BCs and ICs in the problem. Since there is no compulsion to choose any particular set of solutions, the method fails to show whether or not the solution so obtained is unique. We shall therefore prove directly from the PDE and the BCs and ICs that the solutions to the various problems posed for the finite string are unique. Although the proof is very easy, it illustrates a method that can be applied to uniqueness proofs for much more complicated hyperbolic equations.

It suffices to prove that the problem with homogeneous initial and boundary values has only the solution $u = 0$. Accordingly, let u satisfy

DE:

$$u_{xx} - c^{-2}u_{tt} = 0 \qquad \begin{array}{l} 0 < x < l \\ 0 < t < \infty \end{array} \tag{9-50}$$

ICs:

$$u(x, 0) = u_t(x, 0) = 0 \qquad 0 < x < l \tag{9-51}$$

BCs:

$$u(0, t) \text{ or } u_x(0, t) = 0 \quad \text{and} \quad u(l, t) \text{ or } u_x(l, t) = 0 \quad 0 < t < \infty \tag{9-52}$$

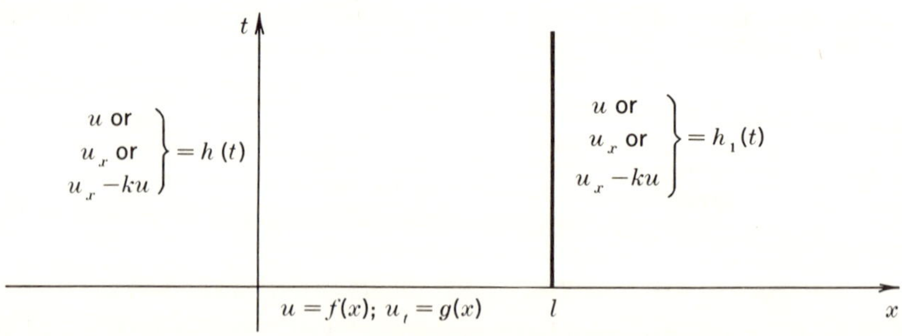

Figure 9-2

Any of the four possible combinations of the BCs is permissible. A BC of the third type can also be handled but requires a slight additional alteration in the proof. We will point out this alteration below.

The proof that the only solution of the problem (9-50) to (9-52) is $u = 0$ derives from a consideration of the energy of the system. The expression

$$E(t) = \frac{1}{2}\int_0^l (u_x^2 + c^{-2}u_t^2)\,dx \tag{9-53}$$

can be interpreted as the sum of the potential and kinetic energies of the string. If a third BC, i.e., one of the type $u_x - ku = 0$, were imposed instead of (9-52) at one or both boundaries, then an additional term would have to be added to (9-53) to take account of the fact that energy is being gained or lost through the boundaries. This is due to the fact that the third BC corresponds to attaching the end points to springs which may therefore feed energy into the system or absorb it from the system, depending upon the sign of the constant in the BC. In order to bring out the main ideas, we restrict ourselves to BCs of the type (9-52).

Since (9-53) represents the total energy of the system, and since we expect the solution u to be zero, we expect that the energy (9-53) will not change in time. In order to verify this conjecture, we compute the derivative of $E(t)$ with respect to t and show that it is zero.

From (9-53) we have

$$E'(t) = \int_0^l (u_x u_{xt} + c^{-2}u_t u_{tt})\,dx$$

From the DE (9-50), however, $c^{-2}u_{tt} = u_{xx}$. Hence

$$E'(t) = \int_0^l (u_x u_{xt} + u_t u_{xx})\,dx = \int_0^l (u_x u_t)_x\,dx = (u_x u_t)\Big|_0^l$$

Thus

$$E'(t) = u_x(l, 0)\,u_t(l, t) - u_x(0, t)\,u_t(0, t)$$

If the BCs are $u_x(l, t) = u_x(0, t) = 0$, then obviously $E'(t) = 0$. $E'(t)$ also equals zero if $u(l, t) = u(0, t) = 0$, however, because then $u_t(l, t) = u_t(0, t) = 0$. The result follows similarly in all other cases. Hence

$$E'(t) = 0$$

Moreover, the ICs (9-54) imply that u_x and u_t are zero at $t = 0$. Hence $E(0) = 0$. Therefore $E(t) \equiv 0$. This implies that $u_x = u_t = 0$. Hence u is constant. Since $u(x, 0) = 0$, it follows that $u \equiv 0$; Q.E.D.

The above proof illustrates a method that works in proving uniqueness for many different BIVPs for hyperbolic equations. The idea is to set up an energy integral. The integrand of this energy integral will be nonnegative by nature. We differentiate the energy integral with respect to time and substitute from the differential equation for some of the quantities in the new integrand. The result, if written properly, is usually a divergence expression, i.e., a perfect differential in one variable or a divergence in more variables. The integration can then be carried out and the BCs and ICs used to show that the derivative of the energy integral is zero. The integral is therefore a constant. Initially it vanishes, however. Hence the constant is zero. Since the integrand is nonnegative, the integrand is therefore zero. From this one can usually conclude that the solution u is identically zero.

9-6. SPACELIKE AND TIMELIKE INITIAL CURVES

The reader is forewarned that the remarks in this section apply strictly to single second-order linear hyperbolic PDEs in two independent variables (for single hyperbolic equations in more than two independent variables, see Chap. 11, and for hyperbolic systems, see Chap. 12).

The dilemma of how many ICs can be prescribed on a given piece of initial curve can be somewhat mitigated by the introduction of the concepts of spacelike and timelike arcs. Their method of introduction also serves to clarify another apparent anomaly in BIVPs for hyperbolic equations.

For the semi-infinite string, the domain in which the wave equation held was the first quadrant of the xt plane. We gave two conditions on the x axis and only one condition on the t axis. The homogeneous wave equation is symmetric in x and t, however. If we were to interchange their roles, the wave equation would not even know the difference. Why, therefore, could we not impose only one condition on the x axis and then give two conditions on the t axis? Mathematically, the latter appears to give as reasonable a problem as the former, the solution to which should exist and be unique. Mathematically, we were forced to give only one condition on the t axis because the C^- characteristics carried part of the initial data on the x axis up to the t axis. However, if we were to think of data as being prescribed first on the t axis, we could prescribe two data there and then the C^- characteristics would carry part of that data down to the x axis and force us to prescribe only one datum on the x axis. Mathematically, therefore, there is no difference between the x and t axes as far as the one-dimensional wave equation is concerned. In higher dimensions, this statement is *false* (see Sec. 11-3). The only reason we cannot prescribe two data on both axes is that the C^- characteristics connect them. It is this connection of the two different parts of the initial curve by characteristics and not the fact that one is the x axis and the other the t axis that restricts the way we can prescribe data on the initial curve. Physically, of course, prescribing two conditions on the t axis and only one on the x axis is unreasonable. At first sight it therefore appears that there is a dichotomy between the physics and the mathematics of the problem. This dichotomy is illusory, however. It arises solely because we have considered a hyperbolic PDE in two independent variables x and t and elected to interpret one of them, t, as the time. If we had chosen to interpret x instead of t as time, then the "physically unreasonable" problem would have become the "physically reasonable" one and vice versa. The way out of this dilemma is simple. Given a hyperbolic PDE in two independent variables, we first select one of the independent variables and call it the time. Which once we elect to call "time" is inessential mathematically (if and only if there are *two* independent variables), but it is important for the physical interpretation. Once we have chosen, both mathematically and physically, which variable is to be called "time," and once we have given the definitions below with reference to this "time" variable, there is no longer any possibility of ambiguity, either in the mathematics or in the physics, and the previous dichotomy between mathematics and physics will evanesce.

Having chosen one variable as the time, we define the positive direction on each characteristic to be that direction in which the time increases. For example, if x and t are the independent variables in the PDE, if t is the time, and the t axis is drawn vertically upward, then the positive direction on each characteristic is the upward one. Now choose an arc of a curve Γ in the xt plane and a point P on Γ. If Γ is not characteristic at P, then two characteristics, one from each family, must cross Γ at P (see Fig. 9-3). If both these characteristics when traversed in the positive direction on them leave Γ from the same side (of Γ), then Γ is called *spacelike* at P. If one characteristic when traced in the positive direction on it leaves Γ from one side but the other characteristic when traced in its positive direction leaves Γ from the other side, then Γ is called *timelike* at P. According to this definition, the x axis is spacelike for the wave equation, while the t axis is timelike for it.

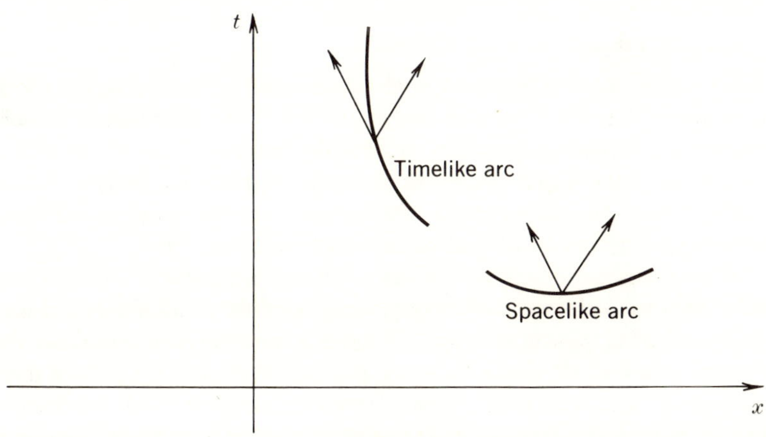

Figure 9-3

It follows from a look at the geometry of the situation that two initial data cannot be prescribed everywhere on an initial curve which contains both spacelike and timelike portions because there are always characteristics connecting the space- and timelike parts. If two data are prescribed on a spacelike part of an initial curve, only one datum can be prescribed on a timelike part of that curve, and vice versa, if two data are prescribed on a timelike part of the curve, only one datum can be prescribed on the spacelike part. In order to obtain agreement with physics it is necessary to say that by convention two data will always be given on any spacelike portion of any initial curve. It then follows that only one datum can be given on a timelike part because the characteristics will carry data from the spacelike to the timelike part. A look at the geometry of a curve with both spacelike and timelike parts will verify this. Again we emphasize that it is not really the space- or timelike character of an initial curve that determines the number of data that can be prescribed on this curve. It is rather the fact that the space- and timelike parts of the curve must, because of their nature, be connected by characteristics that restrict the number of data that can be prescribed. As for which part merits the prescription of two data and which can take only one, that is entirely arbitrary mathematically. It is solely for the purpose of physics that we agree to prescribe two data on the spacelike and not on the timelike parts.

Again we emphasize that the definitions of spacelike and timelike as well as the statements about the arbitrary choice of data *must be modified in more dimensions*.

Finally, we note that it is clear from the nature of the characteristics that only one datum can be prescribed on any part of an initial curve that is characteristic.

9-7. DISCONTINUOUS SOLUTIONS

We have seen in Sec. 9-2 that the Cauchy data u and u_ν on a free curve, together with a linear hyperbolic second-order PDE in two independent variables, determine the solution u uniquely in a neighborhood of that curve provided the curve, the data, and the coefficients in the PDE are all analytic. Suppose we were to drop the requirement of analyticity for the solution and inquire about the possibility of discontinuities in the solution.

If we assume that u and/or u_ν jump across some curve C, then, using as Cauchy data the values to which they jump, we could find a solution on the "other side" of C. Since the Cauchy data can be prescribed arbitrarily on C, the jumps of u and u_ν across C can be prescribed arbitrarily. Hence there is no way of determining them. This shows that conditions on the jumps of quantities involving u across a curve can be derived only under the assumption that u and u_ν are continuous across that curve. Accordingly, we suppose that

$u(x, y)$ is a solution of a second-order linear hyperbolic PDE with continuous coefficients and that there exists some curve C across which u and its first partial derivatives are continuous but across which some second partial derivative of u suffers a jump. More specifically, we assume that an outer derivative of u suffers a jump in crossing C whereas all tangential derivatives of u are continuous across C. There is no loss of generality in assuming that the discontinuous outer derivative is the normal one. Thus, in our usual notation, we assume that u, u_τ, u_ν, $u_{\tau\tau}$, and $u_{\nu\tau}$ are all continuous across C while $u_{\nu\nu}$ suffers a jump in crossing C. If C were free, i.e., noncharacteristic, then the results of Sec. 9-2 show that $u_{\nu\nu}$ is uniquely determined in a neighborhood of C and necessarily continuous across C. This proves the following theorem.

Theorem 9-1 If u satisfies a second-order linear hyperbolic PDE in two independent variables, and if u, u_τ, u_ν, $u_{\tau\tau}$, $u_{\nu\tau}$ are all continuous across a curve C but $u_{\nu\nu}$ suffers a jump upon crossing C, then C is necessarily a characteristic of the PDE.

If, as the notation indicates, τ is the parameter on C, then the jump in $u_{\nu\nu}$ across C is obviously a function of τ. Call it $j(\tau)$. By introducing, as in Sec. 9-2, ν and τ as new independent variables, with $\phi(x, y) = \nu$ for $\nu = 0$ being the equation of the characteristic C along which $j(\tau)$ is defined, we can write the PDE in the form [cf. Eqs. (6-10) to (6-15) with $\alpha = \nu$ and $\beta = \tau$]

$$\hat{a}u_{\nu\nu} + \hat{b}u_{\nu\tau} + \hat{c}u_{\tau\tau} + \hat{d}u_\nu + \cdots = 0 \tag{9-54}$$

Equation (9-54) holds in a neighborhood of C. In that neighborhood, staying off C, we can differentiate (9-54) with respect to τ to obtain

$$\hat{a}u_{\nu\nu\tau} + \hat{a}_\tau u_{\nu\nu} + \hat{b}u_{\nu\tau\tau} + \hat{b}_\tau u_{\nu\tau} + \cdots + \hat{d}u_{\nu\nu} + \cdots = 0 \tag{9-55}$$

All terms except the first two and $\hat{d}u_{\nu\nu}$ in (9-55) are continuous across C. Hence, on C, $j(\tau)$ satisfies the linear ODE

$$\hat{a}j' + (\hat{a}_\tau + \hat{d})\,j = 0 \tag{9-56}$$

where j' means $dj/d\tau$. Equation (9-56) is linear because j does not occur in any of the coefficients.

More will be said on the subject of discontinuities in solutions in later chapters. It will turn out that nonlinear equations can have discontinuities across curves other than characteristics and that in certain types of problems for them discontinuities will have to be permitted in order to obtain any solutions at all.

SUMMARY

DEFINITION S9-1

The *Riemann function* for the PDE

$$u_{xy} + a(x, y)\, u_x + b(x, y)\, u_y + c(x, y)\, u = F(x, y) \tag{S9-1}$$

is the function $R(x, y, \xi, \eta)$ satisfying

$$\begin{aligned} &R_{xy} - (aR)_x - (bR)_y + cR = F \\ &R_y(\xi, y, \xi, \eta) = a(\xi, y)\, R(\xi, y, \xi, \eta) \\ &R_x(x, \eta, \xi, \eta) = b(x, \eta)\, R(x, \eta, \xi, \eta) \end{aligned} \tag{S9-2}$$

and

$$R(\xi, \eta, \xi, \eta) = 1$$

If R is known, the solution to (S9-1) satisfying the ICs $u(x, y) = f(x, y)$, $u_x(x, y) = p(x, y)$, $u_y(x, y) = q(x, y)$ on a free initial curve Γ, it being assumed that the initial data are compatible with each other, is given by

$$u(\xi, \eta) = \tfrac{1}{2}[R(x_1, y_1, \xi, \eta)\, f(x_1, y_1) + R(x_2, y_2, \xi, \eta)\, f(x_2, y_2)]$$
$$+ \frac{1}{2} \int_\Gamma [(Rp - R_x f + 2bRf)\, dx + (R_y f - Rq - 2aRf)\, dy] + \iint_{\mathscr{D}} RF\, dx\, dy$$

Here (x_1, y_1) is the point in which the characteristic $y = \eta$ intersects Γ, and (x_2, y_2) is the point in which the characteristic $x = \xi$ intersects Γ.

The Riemann function for the PDE $u_{xy} = 0$ is $R(x, y, \xi, \eta) \equiv 1$.

The Riemann function for the PDE $u_{xy} + \lambda u = 0$ ($\lambda = \text{const}$) is $J_0[\sqrt{4\lambda(x - \xi)(y - \eta)}]$, where J_0 is the usual Bessel function.

PROBLEM

1. Solve the IVP:

DE:

$$u_{xx} + 4u_{xy} + 3u_{yy} = 0$$

ICs:

$$u = 1 \qquad u_x = x \qquad u_y = y \qquad \text{on the curve } x = 2y$$

10
Linear Elliptic Equations

10-1. INTRODUCTION

As pointed out in Chap. 6, the most general linear elliptic second-order partial differential equation (PDE) in n independent variables is

$$\sum_{j,k=1}^{n} a_{jk}(\mathbf{x})\, u_{x_j x_k} + \sum_{j=1}^{n} b_j(\mathbf{x})\, u_{x_j} + c(\mathbf{x})\, u = F(\mathbf{x}) \tag{10-1}$$

where the matrix $(a_{jk}(\mathbf{x}))$ is positive definite and symmetric, and boldface, as before, is used for vectors. In the first few sections of the present chapter we discuss (10-1), starting with the simplest cases and gradually going to more complicated cases. In later sections, we will say a little about linear elliptic equations of higher order.

10-2. THE POTENTIAL EQUATION AND POISSON'S EQUATION

The special case of (10-1) in which $a_{jk}(\mathbf{x}) = \delta_{jk}$, $b_j \equiv 0$, $c \equiv 0$, and $F \equiv 0$ is, of course, the potential equation in n dimensions. Since the integral identities used in Chap. 3 can all be extended in the obvious ways to higher dimensions, all the results of Chap. 3 can be extended to n dimensions. The Dirichlet and Neumann problems still have unique solutions in bounded domains and unique ones in unbounded domains if proper conditions are imposed at infinity. The mean-value property still obtains for harmonic functions, and if r^{-1} is replaced by r^{2-n}, then Green's functions exist and have the usual picking-out property.

Similarly, if $F \not\equiv 0$, then we have Poisson's equation, and a particular solution of it is given by

$$u(\mathbf{x}) = \frac{1}{\omega_n} \int_{\mathscr{R}} F(\boldsymbol{\xi})\, r^{2-n}(\mathbf{x}, \boldsymbol{\xi})\, d\boldsymbol{\xi} \tag{10-2}$$

where ω_n is the surface area of the sphere $\sum_{j=1}^{n} x_j^2 = 1$ and

$$r^2(\mathbf{x}, \boldsymbol{\xi}) = \sum_{j=1}^{n} (x_j - \xi_j)^2 \tag{10-3}$$

10-3. CONSTANT COEFFICIENTS

In Chap. 6 we saw that the most general second-order linear homogeneous elliptic PDE in n independent variables could be reduced to the form

$$\nabla^2 u + \lambda u = 0 \tag{10-4}$$

where, as usual,

$$\nabla^2 = \sum_{j=1}^{n} \frac{\partial^2}{\partial x_j^2} \tag{10-5}$$

As pointed out in Chap. 6, (10-4) is the n-dimensional reduced wave equation. So all theorems about it, including the fact that its eigenvalues are negative, carry over from the three-dimensional case. The fundamental solution, as pointed out in Sec. 4-6, is

$$\rho(r) = r^{-(n-2)/2} Z_{(n-2)/2}(\sqrt{\lambda}\, r) \tag{10-6}$$

where r is again given by (10-3) and where Z_ν is a singular solution of Bessel's equation of index ν. We note that $\rho(r) \sim r^{2-n}$ for r close to 0.

10-4. SECOND-ORDER EQUATIONS WITH VARIABLE COEFFICIENTS

If the coefficients in (10-1) are variable, then (cf. Chap. 6) (10-1) can be reduced to a form in which the principal part is the Laplacian only in the case of two dimensions. In higher dimensions, we can make a change of variables

$$\tilde{x}_j = \phi_j(\mathbf{x}) \qquad j = 1, \ldots, n \tag{10-7}$$

in which case the PDE (10-1) will become

$$\sum_{j,k=1}^{n} \tilde{a}_{jk}(\tilde{\mathbf{x}})\, u_{\tilde{x}_j\tilde{x}_k} + \sum_{j=1}^{n} \tilde{b}_j(\tilde{\mathbf{x}})\, u_{\tilde{x}_j} + c(\tilde{\mathbf{x}})\, u = F(\tilde{\mathbf{x}}) \tag{10-8}$$

where, as a computation shows,

$$\tilde{a}_{jk}(\tilde{\mathbf{x}}) = \sum_{l,m=1}^{n} a_{lm}(\mathbf{x}) \frac{\partial \tilde{x}_j}{\partial x_l} \frac{\partial \tilde{x}_k}{\partial x_m} \tag{10-9}$$

but, in general, we will not be able to choose the ϕ_j's so that $\tilde{a}_{jk} = \delta_{jk}$.

Equation (10-9) does suggest something, however. If we change notation appropriately by using superscripts instead of subscripts everywhere in (10-9), it will then read

$$\tilde{a}^{jk}(\tilde{\mathbf{x}}) = \sum_{l,m=1}^{n} a^{lm}(\mathbf{x}) \frac{\partial \tilde{x}^j}{\partial x^l} \frac{\partial \tilde{x}^k}{\partial x^m} \tag{10-10}$$

which shows the reader acquainted with tensor analysis that the coefficients $a^{jk}(\mathbf{x})$ are the components of a second-order twice contravariant tensor. Since the matrix (a^{jk}) is symmetric and positive definite due to the assumed ellipticity of the PDE, we can use the inverse matrix (a_{jk}) to define a Riemannian metric by

$$ds^2 = \sum_{j,k=1}^{n} a_{jk}(\mathbf{x})\, dx^j\, dx^k \tag{10-11}$$

From tensor analysis, we recall that the Laplacian, which is the divergence of a gradient, can be written in the invariant form

$$\nabla^2 u = \frac{1}{\sqrt{a}} \sum_{j,k=1}^{n} \frac{\partial}{\partial x^j} \left(\sqrt{a}\, a^{jk} \frac{\partial u}{\partial x^k} \right) \tag{10-12}$$

or

$$\nabla^2 u = \sum_{j,k=1}^{n} a^{jk} \frac{\partial^2 u}{\partial x^j\, \partial x^k} + \frac{1}{\sqrt{a}} \sum_{j,k=1}^{n} \frac{\partial}{\partial x^j} (\sqrt{a}\, a^{jk}) \frac{\partial u}{\partial x^k}$$

where $a = \det(a_{jk})$. Hence, calling the new coefficients of the first-order derivatives in the PDE b^j, we see that we can write (10-1) in the form

$$\nabla^2 u + \sum_{j=1}^{n} b^j \frac{\partial u}{\partial x^j} + cu = F \tag{10-13}$$

where now ∇^2 is defined by (10-12).

For reasons that will become apparent later (Chap. 17), (10-13) is called *self-adjoint* if all the b^j's vanish. This means that the most general second-order self-adjoint linear elliptic PDE in n independent variables can always be written in the form

$$\nabla^2 u + cu = F \tag{10-14}$$

i.e., a reduced wave equation with variable coefficients in a Riemannian metric determined by the coefficients of the principal part of the original PDE.

Having got a ∇^2 into the PDE, we next look around for a Green's formula to use on it. Since Green's identities involve only Laplacians, divergences, and gradients, Green's identities are easy to write down. All we need are the expressions for these differential operators in terms of the metric coefficients. In fact, we have already combined the expressions for gradient and divergence in writing down (10-12). Now we need the separate expressions for them. In particular, we need

$$(\nabla\phi)^2 = \sum_{j,k=1}^{n} a^{jk} \frac{\partial \phi}{\partial x^j}\frac{\partial \phi}{\partial x^k} \tag{10-15}$$

and

$$\nabla\phi \cdot \nabla\psi = \sum_{j,k=1}^{n} a^{jk} \frac{\partial \phi}{\partial x^j}\frac{\partial \psi}{\partial x^k} \tag{10-16}$$

We also need $\partial\phi/\partial n$. Defining it requires a little more effort. Suppose that the equation of the surface $\partial\mathscr{R}$ on which we need $\partial\phi/\partial n$ is $\Phi(\mathbf{x}) = 0$. Then $\nabla\Phi$ is a vector in the direction of the normal to $\partial\mathscr{R}$. Hence $\nabla\Phi/|\nabla\Phi|$ is a (covariant) unit normal to $\partial\mathscr{R}$, where $|\nabla\Phi|$ is the square root of the right side of (10-15) with ϕ replaced by Φ. $\partial\phi/\partial n$ is then defined as the dot product of this unit normal with $\nabla\phi$, that is,

$$\frac{\partial \phi}{\partial n} = \frac{\sum_{j,k=1}^{n} a^{jk} \frac{\partial \phi}{\partial x^j}\frac{\partial \Phi}{\partial x^k}}{\sum_{j,k=1}^{n} a^{jk} \frac{\partial \Phi}{\partial x^j}\frac{\partial \Phi}{\partial x^k}} \tag{10-17}$$

With these definitions, the integral identities of Chap. 3 carry over virtually verbatim, and their use as in Chaps. 3 and 4 yields the expected theorems on the solutions of (10-14). In using them, however, the customary care must be taken to make sure that the quantities involved have the right variance before the particular operator needed is applied. As usual, the metric tensor and its determinant must be used when needed to alter the variance or weight of a given quantity.

The only thing that does not carry over immediately is the definition of a Green's function, because it is not clear offhand what constitutes a fundamental solution of (10-14) when ∇^2 is defined with respect to a Riemannian metric. Certainly r^{2-n} is no longer a particular solution. The question is what plays the role of r^{n-2} in the n-dimensional Riemannian space? Obviously we need to use the distance between the points $\mathbf{x}$ and $\boldsymbol{\xi}$ but measured now in the Riemannian metric. Moreover, it is natural to use the "shortest" distance between the two points. In other words, we should replace $r(\mathbf{x}, \boldsymbol{\xi})$, as defined by (10-3), by the function $s(\mathbf{x}, \boldsymbol{\xi})$ representing the length of the geodesic joining $\mathbf{x}$ and $\boldsymbol{\xi}$. It would be natural to call the function $s(\mathbf{x}, \boldsymbol{\xi})$ just defined the fundamental solution of (11-14) if it were not for one unfortunate shortcoming. $s(\mathbf{x}, \boldsymbol{\xi})$ is not, in general, a solution of the homogeneous form of (10-14). Nevertheless, we expect the fundamental solution of (10-14) to have a singularity of the same order as $s(\mathbf{x}, \boldsymbol{\xi})$. We need such a singularity in order to have the requisite picking-out property in the Green's function to be built out of it. An appropriate definition turns out to be the following.

DEFINITION 10-1

A fundamental solution of (10-3) is a solution $\rho(\mathbf{x}, \boldsymbol{\xi})$ of the homogeneous version of (10-3) of the form

$$\rho(\mathbf{x}, \boldsymbol{\xi}) = \alpha(\mathbf{x}, \boldsymbol{\xi})\, s(\mathbf{x}, \boldsymbol{\xi}) + \beta(\mathbf{x}, \boldsymbol{\xi}) \tag{10-18}$$

where $s(\mathbf{x}, \boldsymbol{\xi})$ is the geodetic distance defined above, α is a regular function of its arguments in a neighborhood of $\mathbf{x} = \boldsymbol{\xi}$, and β is either regular in that neighborhood or else behaves like $\log s(\mathbf{x}, \boldsymbol{\xi})$ there.

For the reasons for this, see Garabedian [115]. The potential equation in euclidean space is the special case in which $\alpha \equiv 1$ and β is harmonic.

DEFINITION 10-2

A *Green's function* for a boundary-value problem (BVP) for (10-14) is a fundamental solution which satisfies the corresponding homogeneous boundary condition (BC).

We have pointed out above that the geodetic distance $s(\mathbf{x}, \boldsymbol{\xi})$ as a function of $\mathbf{x}$ is not generally a solution of the homogeneous form of (10-14). Nevertheless, since it provides the correct singularity for a fundamental solution, it deserves a name of its own. It is customarily called a *parametrix*. A parametrix, then, is not a solution of a given PDE but rather a function whose singularity is of the right order for use in a Green's identity.

The parametrix $s(\mathbf{x}, \boldsymbol{\xi})$, while not satisfying the given PDE, does nevertheless satisfy a PDE of its own. Because s represents arc length on a geodesic, ∇s must be a unit tangent vector to a geodesic; that is, $(\nabla s)^2 = 1$. In other words, s satisfies the PDE

$$\sum_{j,k=1}^{n} a^{jk} \frac{\partial s}{\partial x^j} \frac{\partial s}{\partial x^k} = 1 \tag{10-19}$$

The reader should not be deceived by all the attractive notation of the present section into thinking that the results permit the immediate solution of (10-14). Although all derivatives in (10-14) are contained in the ∇^2, ∇^2 is not a simple operator. It is given by (10-12). Similarly, $s(\mathbf{x}, \boldsymbol{\xi})$ would not generally be a simple function to find in practice. The main advantage of the methods of the present section is the insight they give into the nature of the solutions of the PDE (10-1).

10-5. THE BIHARMONIC EQUATION

As mentioned in Chap. 6, elliptic equations must always have even order. For an elliptic equation of order $2m$, the correct number of BCs is m. In order to prove uniqueness, complicated generalizations of Green's identity are needed. As an illustration, we consider the *biharmonic equation*

$$\nabla^2 \nabla^2 u = \nabla^4 u = 0 \tag{10-20}$$

In two dimensions, (10-20) describes the motion of a plate, i.e., a membrane that resists shearing (cf. Sec. 4-1). One appropriate Green's identity for ∇^4 turns out to be

$$\iiint_{\mathscr{R}} (\phi \, \nabla^4 \psi - \psi \, \nabla^4 \phi) \, dV = \iint_{\mathscr{R}} \left(\phi \frac{\partial \nabla^2 \psi}{\partial n} - \nabla^2 \psi \frac{\partial \phi}{\partial n} + \nabla^2 \phi \frac{\partial \psi}{\partial n} - \psi \frac{\partial \nabla^2 \phi}{\partial n} \right) dS \tag{10-21}$$

There are several others. We mention one other below.

A natural BVP for (10-20) is

DE:

$$\nabla^4 u = 0 \qquad \text{in } \mathscr{R} \tag{10-22}$$

BCs:

$$u = f(S) \qquad \partial u/\partial n = g(S) \qquad \text{on } \partial\mathscr{R} \tag{10-23}$$

For the circle, (10-22) can be solved with the BCs (10-23) by separation of variables in polar coordinates. For the square it has yet to be solved in closed form due to the fact that (10-22) is not separable in rectangular coordinates.

Although it is of less physical interest than (10-23), mathematically easy BCs for (10-22) are

$$u = f(S) \qquad \nabla^2 u = g(S) \qquad \text{on } \partial\mathscr{R} \tag{10-24}$$

To solve (10-22) subject to the BCs (10-24) we merely put

$$v = \nabla^2 u \tag{10-25}$$

Then v satisfies

DE:

$$\nabla^2 v = 0 \qquad \text{in } \mathscr{R} \tag{10-26}$$

BC:

$$v = g \qquad \text{on } \partial\mathscr{R} \tag{10-27}$$

Having solved this potential problem for v, we find u by solving the Poisson's equation (10-25)(with v now known) subject to the BCs (10-24), that is, $u = f$ on $\partial\mathscr{R}$.

A plate satisfying the BCs (10-24) is called *simply supported.*

10-6. NONHOMOGENEOUS EQUATIONS

Just as in previous cases, the superposition principle enables us to solve a nonhomogeneous equation by finding a particular solution of it and then reducing the BVP for it to one for a homogeneous equation, and, just as in the simpler cases, a particular solution can be obtained by formula (3-33), where g is the appropriate Green's function for the BVP at hand.

10-7. MAXIMUM PRINCIPLE

For the linear elliptic equation

$$\sum_{j,k=1}^{n} a_{jk}(\mathbf{x})\, u_{x_j x_k} + \sum_{j=1}^{n} b_j(\mathbf{x})\, u_{x_j} + c(\mathbf{x})\, u = 0 \tag{10-28}$$

the following theorem holds.

Theorem 10-1 If $c \leqslant 0$, in $\mathscr{R}$, then no solution of (10-28) can have a positive maximum or a negative minimum in $\mathscr{R}$.

PROBLEMS

1. Solve $\nabla^4 u(r, \theta) = 0$, $0 \leqslant \theta \leqslant 2\pi$, $0 \leqslant r < 1$, subject to the BCs $u(1, \theta) = f(\theta)$, $u_r(1, \theta) = 0$, $0 < \theta < 2\pi$. $f(\theta)$ is a given function of θ.

2. Find all radially symmetric solutions of $\nabla^4 u = 0$ in two dimensions. *Hint*: Write

$$\nabla^2 = \frac{1}{r}\frac{d}{dr}\left(r\frac{d}{dr}\right)$$

3. Show that ω_n, the surface area of the sphere $\sum_{j=1}^n x_j^2 = 1$, is given by

$$\omega_n = \frac{2\pi^{n/2}}{\Gamma(n/2)}$$

where $\Gamma(x)$ is the *gamma function*, defined by $\Gamma(x) = \int_0^\infty e^{-t} t^{x-1}\, dt$.

11
More General Linear Hyperbolic Equations

11-1. CHARACTERISTICS FOR SECOND-ORDER LINEAR EQUATIONS

In Chap. 10 we considered the general second-order linear partial differential equation (PDE)

$$\sum_{j,k=1}^{n} a_{jk}(\mathbf{x})\, u_{x_j x_k} + \sum_{j=1}^{n} b_j(\mathbf{x})\, u_{x_j} + c(\mathbf{x})\, u = F(\mathbf{x}) \tag{11-1}$$

in the case when the matrix (a_{jk}) was positive definite at a point $\mathbf{x}$, that is, when (11-1) was elliptic there. (Boldface symbols stand for vectors.) In that case the *characteristic form* for (11-1)

$$Q(\boldsymbol{\xi}) = \sum_{j,k=1}^{n} a_{jk}(\mathbf{x})\, \xi_j \xi_k \tag{11-2}$$

was positive definite at $\mathbf{x}$, and the *characteristic equation*

$$Q(\boldsymbol{\xi}) = 0 \tag{11-3}$$

had no real solutions $\boldsymbol{\xi}$. Now we are interested in the opposite case, namely, that in which (11-3) has n real linearly independent solutions.

In the special case $n = 2$ we have already seen in Sec. 6-2 that this property is equivalent to hyperbolicity for (11-1); i.e., if $n = 2$, then (11-1) is hyperbolic if and only if (11-3) has two real linearly independent solutions. This leads us to conjecture that the analogous result holds in any number of dimensions, i.e., that (11-1) is hyperbolic if and only if (11-3) has n real linearly independent roots. Unfortunately, this conjecture is false. It is true, as we shall see shortly, that if (11-1) is hyperbolic, then (11-3) has n real linearly independent solutions, but the converse is false. For example, the ultrahyperbolic equation whose characteristic form is $\xi_1^2 + \xi_2^2 - \xi_3^2 - \xi_4^2$ has the four linearly independent solutions (1, 0, 0, 1), (0, 1, 1, 0), (1, 0, 1, 0), and (1, 0, 0, -1), and the parabolic equation with characteristic form $\xi_1^2 - \xi_2^2 + 0 \cdot \xi_3^2$ has (1, 1, 0), (1, -1, 0) and (0, 0, 1) as three linearly independent solutions. Since we do not wish to deal with these equations, we make the explicit assumption in this chapter that (11-1) is hyperbolic at $\mathbf{x}$. According to Sec. 6-6, this means that there is an orthogonal linear transformation (in $\boldsymbol{\xi}$ space) that transforms (11-3) into the canonical form

$$Q(\boldsymbol{\eta}) = \sum_{j=0}^{n-1} \lambda_j \eta_j^2 - \lambda_n \eta_n^2 \tag{11-4}$$

where all λ's are different from zero and all have the same sign, say $+$. There is no question about (11-3)'s having n real linearly independent solutions when Q is given by (11-4) because we can write them down explicitly. For the first $n - 1$ of them we take $\boldsymbol{\eta}^{(1)} = (1, 0, \ldots, 0, 1)$, $\boldsymbol{\eta}^{(2)} = (0, 1, 0, \ldots, 0, 1)$, etc., the jth $\boldsymbol{\eta}$ having 1s in the jth and nth places and zeros everywhere else. This gives $n - 1$ $\boldsymbol{\eta}$'s. For the nth we take $\boldsymbol{\eta}^{(n)} = (1, 0, \ldots\ 0, -1)$.

This shows that in a coordinate system in which Q is diagonal (11-3) has n real linearly independent solutions. However, the linear transformation that "undiagonalizes" Q is orthogonal, hence preserves linear independence. Therefore (11-3) must have n real linearly independent solutions for any hyperbolic PDE of the form (11-1).

It is natural to call these n linearly independent solutions the characteristic directions of (11-1) at that point. It is natural, but unfortunately it is wrong. As we will see below, the solutions $\boldsymbol{\xi}$ to (11-3) are *normal* rather than tangential to what will emerge as the characteristics of (11-1). The situation here is analogous to that for first-order equations in Sec. 8-6, where the directions that arose first were those normal to the characteristic directions.

In order to define characteristics for (11-1) we return to the notion of inner operator introduced in Chap. 9. There we called a second-order linear partial differential operator L in two independent variables an inner operator with respect to a curve of the Cauchy data; that is, u and u_ν on that curve, together with the PDE, failed to determine the solution u off the curve. A curve on which the operator was an inner operator was called a characteristic, and this definition was shown to agree with the others we had given.

In order to determine whether a given curve was a characteristic or not, we wrote the equation of that curve in the form $\phi(x, y) = 0$, embedded this curve in the family of curves $\phi(x, y) = \nu$, and introduced ν as a new (curvilinear) coordinate. If we wish to determine characteristics in the analogous way in n dimensions, we first have to write the equation of a potential, i.e., possible, characteristic in the form $\phi(\mathbf{x}) = 0$. This means that a characteristic in n dimensions will be an $(n - 1)$-dimensional *surface* lying in an n-dimensional space. To be precise, suppose that we have a surface $\mathscr{S}$ given by $\phi(\mathbf{x}) = 0$. Then $\mathscr{S}$ can always be represented parametrically in the form $\mathbf{x} = \mathbf{x}(\tau_1, \ldots, \tau_{n-1})$, where the τ's are independent parameters. If we introduce the additional parameter $\nu = \phi(\mathbf{x})$, then $(\tau_1, \ldots, \tau_{n-1}, \nu)$ form a curvilinear coordinate system in the n-dimensional $\mathbf{x}$ space, with $\mathscr{S}$ being the coordinate surface $\nu = 0$. Prescribing a function $u(\mathbf{x})$ on $\mathscr{S}$ is therefore equivalent to giving u as a function of the τ's: $u = u(\tau_1, \ldots, \tau_{n-1})$. If in addition to giving $u(\tau_1, \ldots, \tau_{n-1})$, we also give the normal derivative $u_\nu(\tau_1, \ldots, \tau_{n-1})$ on $\mathscr{S}$, then we say that we have prescribed *Cauchy data* for (11-1) on $\mathscr{S}$. We then call $\mathscr{S}$ *free* with respect to (11-1) if the Cauchy data together with (11-1) uniquely determine u on, and in a neighborhood of, $\mathscr{S}$. If (11-1), when supplemented by the Cauchy data on $\mathscr{S}$, fails to determine u on, and in a neighborhood of, $\mathscr{S}$, then we call $\mathscr{S}$ *characteristic* for (11-1). Just as in Chap. 9, whether $\mathscr{S}$ is free or characteristic for (11-1) depends upon whether the term $u_{\nu\nu}$ is present or absent in the equation that results when $\mathbf{x}$ has been replaced by $(\tau_1, \ldots, \tau_{n-1}, \nu)$ in (11-1). If the coefficient of $u_{\nu\nu}$ is different from zero, then $\mathscr{S}$ is free. If the coefficient of $u_{\nu\nu}$ is zero, then $\mathscr{S}$ is characteristic for (11-1).

Just as in Chap. 9, a computation shows that the coefficient of $u_{\nu\nu}$ in (11-1) is

$$Q(\nabla\phi) = \sum_{j,k=1}^{n} a_{jk}(\mathbf{x})\, \phi_{x_j}\phi_{x_k} \tag{11-5}$$

Hence $\phi(\mathbf{x}) = 0$ is a characteristic surface of (11-1) if and only if

$$Q(\nabla\phi) = 0 \tag{11-6}$$

Comparing (11-3) and (11-6) [cf. (11-2) and (11-5)], we see that $\boldsymbol{\xi} = \nabla\phi$, so that, as mentioned before, the roots $\boldsymbol{\xi}$ of the characteristic equation are the components of the normals to the characteristic surfaces. The assumption that (11-1) is hyperbolic at a point $\mathbf{x}$ thus implies that it has n characteristic surfaces through that point.

In two dimensions we were able to put $\zeta = -\xi_2/\xi_1$, or, what is the same thing, $dy/dx = -\phi_y/\phi_x$, and thereby reduce the characteristic equation, i.e., the equation for the characteristics, to an ordinary differential equation (ODE). Since this ODE was quadratic, it split into two ODEs, each of which determined one of the two families of characteristics. When $n > 2$, we can no longer perform such a reduction of (11-6) to an ODE. Thus for $n > 2$ the characteristic equation seems to be a PDE. Of course, (11-6) is only a first-order PDE, and it could be solved by the methods of Chap. 8. Before we do this, however, we have to point something out: Eq. (11-6) is not quite a PDE because the variables $x_1, \ldots, x_n$ in it are not all independent. They are connected by the relation $\phi(\mathbf{x}) = 0$ because (11-6) was derived under the assumption that the Cauchy data were prescribed on the surface $\phi - 0$.

In order to transform (11-6) into an honest-to-goodness PDE we have to eliminate one of the x's. So we suppose that the equation $\phi(\mathbf{x}) = 0$ has been solved for one of the x's, say x_n, in terms of the others:

$$x_n = \psi(x_1, \ldots, x_{n-1}) \tag{11-7}$$

More precisely, there are many different equations $\phi(\mathbf{x}) = 0$ that describe the same surface $\mathscr{S}$. We choose the particular representation

$$\hat{\phi}(\mathbf{x}) = \psi(x_1, \ldots, x_{n-1}) - x_n \tag{11-8}$$

so that the equation $\hat{\phi}(\mathbf{x}) = 0$ is equivalent to (11-7).

Substituting (11-8) into (11-6), we find

$$\sum_{j,k=1}^{n-1} a_{jk}(\mathbf{x})\, \psi_{x_j}\psi_{x_k} - 2\sum_{j=1}^{n-1} a_{jn}(\mathbf{x})\, \psi_{x_j} + a_{nn} = 0 \tag{11-9}$$

as the PDE for the characteristic surfaces when their equations are written in the form (11-7).

From the form of (11-6) and (11-8) we notice that if ψ is a solution of (11-9), then $\phi(\mathbf{x}, k) = \psi(x_1, \ldots, x_{n-1}) - x_n - k$ satisfies (11-6) for any constant k. Thus, from any solution of (11-9) we immediately obtain a whole one-parameter family of characteristic surfaces given by the equation

$$\hat{\phi}(\mathbf{x}) = \psi(x_1, \ldots, x_{n-1}) - x_n = k \tag{11-10}$$

Equation (11-10) shows that when $\phi(\mathbf{x})$ is written in the special form (11-8), i.e., in a form in which one variable enters linearly, not only is $\phi = 0$ the equation of a characteristic surface, but so, too, is $\phi = k$ for any constant k. This is not true for ϕ's which are not linear in one of the variables. If ϕ is not linear in one of its variables, then $\phi = 0$ may still be a characteristic surface, i.e., may still satisfy (11-6) when the x's are connected by the equation $\phi = 0$, but $\phi = k$ may fail to satisfy (11-6) when the x's are connected by the equation $\phi = k$ if $k \neq 0$.

Nevertheless, if we have a ϕ such that $\phi = 0$ is a solution to (11-6) when the x's are connected by the equation $\phi(\mathbf{x}) = 0$, we can, by solving the equation $\phi(\mathbf{x}) = 0$ for x_n in terms of the other x's, always replace this ϕ by a ϕ of the form (11-8). Since x_n enters this new ϕ linearly, we can get a one-parameter family of characteristic surfaces by putting this ϕ equal to k. Thus, given any single characteristic surface, we can always embed it in a one-parameter family of characteristic surfaces.

To illustrate the foregoing ideas, consider the wave equation in three space dimensions

$$u_{xx} + u_{yy} + u_{zz} - c^{-2}u_{tt} = 0 \tag{11-11}$$

For (11-11), (11-6) reads

$$Q(\nabla\phi) = \phi_x^2 + \phi_y^2 + \phi_z^2 - c^{-2}\phi_t^2 = 0 \tag{11-12}$$

From our work on the wave equation, it is natural to conjecture that the characteristic surfaces for (11-11) will be cones with equations

$$x^2 + y^2 + z^2 - c^{-2}t^2 = k \tag{11-13}$$

Accordingly, we investigate whether

$$\tilde{\phi}(x, y, z, t) = x^2 + y^2 + z^2 - c^{-2}t^2 \tag{11-14}$$

is a solution of (11-12) when (11-13) holds, i.e., when $\tilde{\phi} = k$. From (11-14) we have

$$\tilde{\phi}_x^2 + \tilde{\phi}_y^2 + \tilde{\phi}_z^2 - c^{-2}\tilde{\phi}_t^2 = 4(x^2 + y^2 + z^2 - c^{-2}t^2) = 4\tilde{\phi}$$

Thus (11-14) satisfies

$$Q(\nabla\tilde{\phi}) = 4\tilde{\phi}$$

and hence satisfies

$$Q(\nabla\tilde{\phi}) = 0$$

only when $\tilde{\phi} = 0$, that is, when (11-13) holds with $k = 0$. Thus, the family $\tilde{\phi} = k$ has only one member that is a characteristic surface, namely, the one for which $k = 0$. This was to be expected in view of our previous discussion, because (11-14) does not contain any variable linearly. However, if, as indicated in our previous discussion, we solve the equation $\tilde{\phi} = 0$ for t to get

$$t = c\sqrt{x^2 + y^2 + z^2} \tag{11-15}$$

and replace $\tilde{\phi}$ by

$$\hat{\phi} = t - c\sqrt{x^2 + y^2 + z^2} \tag{11-16}$$

then all surfaces of the family $\hat{\phi} = k$ are characteristics.

Returning to our general discussion, we see that although (11-6) is not a PDE, we can nevertheless obtain the characteristic surfaces of (11-1) by solving (11-6) as if it were a PDE for $\phi(\mathbf{x})$ as a function of the n independent variables $\mathbf{x}$, then setting $\phi(\mathbf{x}) = 0$, and solving this algebraic equation for x_n in terms of the other x's. If (11-7) is the solution to $\phi(\mathbf{x}) = 0$, then, by defining ϕ as in (11-8), we obtain the one-parameter family of characteristic surfaces $\phi = k$.

Thus we can confine our future analysis to (11-6), looked at as an honest-to-goodness PDE for $\phi(\mathbf{x})$. Once we have the solutions to it, we can proceed as described above to get families of characteristic surfaces from them.

11-2. BICHARACTERISTICS

As a PDE, (11-6) is first order but nonlinear. Using the notation of Chap. 8, we can write it in the form

$$Q(\mathbf{x}, \mathbf{p}) = 0 \tag{11-17}$$

where, as in Chap. 8, $\mathbf{p} = \nabla\phi$.

Since (11-17) has ϕ missing, its characteristic equations (cf. Sec. 8-10) have the symmetrical form (Hamilton's equations)

$$\begin{aligned} \frac{dx_j}{dt} &= Q_{p_j} \\ \frac{dp_j}{dt} &= -Q_{x_j} \end{aligned} \qquad j = 1, \ldots, n \tag{11-18}$$

The solutions to (11-18) are the characteristics of (11-17). Since the solutions of (11-17) are themselves characteristics of (11-1), the solutions of (11-18) are the characteristics of the characteristics of (11-1) and are therefore called the *bicharacteristics* of (11-1). Notice that the solutions of (11-18) are the characteristics of (11-17) and are the bicharacteristics of (11-1). Notice, too, that the characteristics of (11-1) are *surfaces* in $\mathbf{x}$ space. The characteristics of (11-18), i.e., the bicharacteristics of (11-1), are *curves* in $\mathbf{x}$ space.

Using the terminology just introduced, we can say that the characteristic surfaces of the hyperbolic PDE (11-1) are generated by its bicharacteristics, i.e., by the solutions of the system of $2n$ ODEs (11-18).

Because Q has the form (11-4), i.e.,

$$Q(\mathbf{x}, \mathbf{p}) = \sum_{j,k=1}^{n} a_{jk}(\mathbf{x})\, p_j p_k \tag{11-19}$$

we can carry the analysis a little further. From (11-19) we have

$$Q_{p_l} = \sum_{j,k=1}^{n} a_{jk} p_j \frac{\partial p_k}{\partial p_l} + \sum_{j,k=1}^{n} a_{jk} p_k \frac{\partial p_j}{\partial p_l}$$

or

$$Q_{p_l} = \sum_{j,k=1}^{n} a_{jk} p_j \delta_{kl} + \sum_{j,k=1}^{n} a_{jk} p_k \delta_{jl} = \sum_{j=1}^{n} a_{jl} p_j + \sum_{k=1}^{n} a_{lk} p_k$$

But $a_{jk} = a_{kj}$ [the matrix (a_{jk}) is symmetric]. Hence

$$Q_{p_l} = 2 \sum_{k=1}^{n} a_{lk} p_k$$

Thus, the first set of equations in (11-18) reduces to

$$\frac{dx_j}{dt} = 2 \sum_{k=1}^{n} a_{jk}(\mathbf{x})\, p_k \qquad j = 1, \ldots, n \tag{11-20}$$

In the special case when the a_{jk}'s are constants, the bicharacteristics of (11-1) can be found explicitly. If $a_{jk} = \text{const}$, then $Q_{x_j} = 0, j = 1, \ldots, n$. Hence (11-18) tells us that $dp_j/dt = 0$, $j = 1, \ldots, n$. So the p's are constants. But then (11-20) tells us that $dx_j/dt = \text{const}, j = 1, \ldots, n$. Put

$$p^{(j)} = \sum_{k=1}^{n} a_{jk} p_k \tag{11-21}$$

Then $dx_j/dt = p^{(j)}$, and therefore $x_j = p^{(j)}t + x_0^{(j)}$, where the $x_0^{(j)}$'s are constants. Thus, when the coefficients of the second derivatives are constants, the bicharacteristics are straight lines. For example, for the wave equation (7-1), if we choose the parameter t on the bicharacteristic curves to be the independent variable t of the PDE, then the equations of the bicharacteristics for (7-1) can be written in the form

$$\begin{aligned} x &= \alpha(t - t_0) + x_0 \\ y &= \beta(t - t_0) + y_0 \\ z &= \gamma(t - t_0) + z_0 \\ t &= (t - t_0) + t_0 \end{aligned} \tag{11-22}$$

where α, β, γ, x_0, y_0, z_0, t_0 are all arbitrary parameters.

If from this family of bicharacteristics we choose the special subfamily for which $\alpha^2 + \beta^2 + \gamma^2 = 1$, then from (11-22) we see that this subfamily satisfies

$$(x - x_0)^2 + (y - y_0)^2 + (z - z_0)^2 = c^{-2}(t - t_0)^2$$

in other words, they generate the characteristic cone through (x_0, y_0, z_0, t_0).

Even if the a_{jk}'s are not constants, we can still define $p^{(j)}$ by (11-21), only now $p^{(j)}$ will be a function of **x**. We still have

$$\frac{dx_j}{dt} = p^{(j)} \qquad j = 1, \ldots, n \tag{11-23}$$

only these equations can no longer be integrated explicitly. Nevertheless, if we solve (11-21) for p_j to get

$$p_j = \sum_{k=1}^{n} b_{jk} p^{(k)} \qquad j = 1, \ldots, n \tag{11-24}$$

where (b_{jk}) is the inverse matrix to (a_{jk}), substitute (11-24) into (11-19), and simplify, we find

$$Q(\mathbf{x}, \mathbf{q}) = \sum_{j,k=1}^{n} b_{jk} p^{(j)} p^{(k)} \tag{11-25}$$

where, in order to distinguish the contravariant vector with components $p^{(j)}$ from the corresponding covariant one with components p_j, we have used **q** to denote the contravariant one, keeping **p** for the covariant one.

Using (11-23) in (11-25), we have

$$Q\left(\mathbf{x}, \frac{d\mathbf{x}}{dt}\right) = \sum_{j,k=1}^{n} b_{jk} \frac{dx_j}{dt} \frac{dx_k}{dt} \tag{11-26}$$

Thus, the characteristic equation (11-17), which when written in terms of $\mathbf{p} = \nabla\phi$ determines the normals **p** to the characteristic surfaces $\phi(\mathbf{x}) = 0$, can, by using the inverse matrix of the matrix of coefficients, be written as an equation for the tangents $d\mathbf{x}/dt$ to the bicharacteristic curves $\mathbf{x}(t)$. In other words, a bicharacteristic direction **q** is any solution of the equation

$$\sum_{j,k=1}^{n} b_{jk} p^{(j)} p^{(k)} = 0 \tag{11-27}$$

where $(b_{jk}) = (a_{jk})^{-1}$ and where the $p^{(j)}$'s are the components of **q**.

11-3. CHARACTERISTIC, SPACELIKE, AND TIMELIKE DIRECTIONS

If we fix a point **x**, then $Q(\mathbf{x}, \mathbf{p})$ is a function of **p**. Because Q is homogeneous and quadratic in **p**, the characteristic equation $Q = 0$ represents a cone in **p** space. This cone is real and one-sheeted if (11-1) is hyperbolic. Because $\mathbf{p} = \nabla\phi$ represents the normal to a characteristic at **x**, this cone is called the *normal cone* at **x**. Correspondingly, because $\mathbf{q} = d\mathbf{x}/dt$ represents the tangent to a bicharacteristic, and because the bicharacteristics are tangent to the characteristic surfaces, the cone given by (11-27) is called the *characteristic* *cone* at $\mathbf{x}$. Some authors call bicharacteristics "rays," so that the characteristic cone is sometimes called the *ray cone*, but, since we wish to reserve the term "rays" for something slightly different from the bicharacteristics, we shall not use the term "ray cone" here (cf. Sec. 11-4).

Thus, a direction $\mathbf{q}$ is characteristic for (11-1) if it satisfies (11-27), i.e., if it is tangent to the characteristic cone of (11-1). Notice that the coefficients in the equation of the normal cone of (11-1) are those of the principal part of (11-1), while those in the equation of the characteristic cone are the elements of the inverse matrix to the matrix of coefficients of the principal part of (11-1). This means, in practice, that it is easier to work with the normal cone because its equation can be written down immediately without any calculation.

In Chap. 9 we pointed out that the number of initial data that can be prescribed on an initial curve for a hyperbolic equation in two dimensions depends upon whether that curve is timelike or spacelike. In that chapter we also pointed out that the distinction between the independent variables, and hence between timelike and spacelike directions, is purely arbitrary, but we emphasized that this arbitrariness is limited to two dimensions. Now we wish to define spacelike and timelike directions in higher dimensions. In order to do this, we need to distinguish one variable and call it the time. The question is which one. The answer depends upon the form of the equation. Even assuming that the PDE in question is hyperbolic, we still cannot always call one variable "time" and expect it to have the same physical significance as time. For example, when the one-dimensional wave equation $u_{xx} - c^{-2}u_{tt} = 0$ is written in the form $u_{\xi\eta} = 0$, neither ξ nor η is the time t. Both are linear combinations of x and t. Hence we shall call one variable "time" only when the PDE has a special form.

In order to describe this special form, we first assume that the PDE has been transformed into canonical form at a point $\mathbf{x}$. Then its characteristic form $Q(\boldsymbol{\eta})$ has the form (11-4) there. If this is the case, then we call the variable x_n (the one which corresponds to the negative eigenvalue) "time." Once we have done this, we can permit any other form of the PDE which gives rise to a characteristic form that looks like

$$Q(\boldsymbol{\xi}) = \sum_{j,k=1}^{n-1} a_{jk}(\mathbf{x})\, \xi_j\xi_k - \lambda_n(\mathbf{x})\, \xi_n{}^2 \tag{11-28}$$

We leave it to the student as an exercise to show that if a PDE has the characteristic form (11-28) and is hyperbolic, then the matrix (a_{jk}) is positive definite, and vice versa, if (a_{jk}) is positive definite, then the PDE whose characteristic form is (11-28) is hyperbolic.

Notice that if (a_{jk}) is positive definite, then the part of the operator corresponding to it is elliptic. Hence the PDE corresponding to (11-28) can be written in the form

$$L(u) - \lambda_n(\mathbf{x})\, u_{x_n x_n} = 0 \tag{11-29}$$

where L is a second-order linear elliptic operator in $n - 1$ independent variables.

For the sake of notational convenience, we assume there are $n + 1$ independent variables instead of n and we call the $(n + 1)$th one t. We also rename λ_n and write (11-29) in the form

$$L(u) - c^{-2}(\mathbf{x}, t)\, u_{tt} = 0 \tag{11-30}$$

where L is a second-order linear elliptic operator in the n independent variables $\mathbf{x} = (x_1, \ldots, x_n)$ containing the variable t as a parameter. This means that L is permitted to contain t in its coefficients but no derivatives of u with respect to t are permitted to enter L. Actually, we could generalize a little further by permitting L to contain first derivatives with respect to t, but this would cause no alteration in the following discussion and would only necessitate using a more elaborate description of the PDE.

In any event, it is hardly necessary to point out which of the $n + 1$ independent variables in (11-30) is to be distinguished by the appelation "time." It is necessary, however, to say what spacelike and timelike directions for (11-30) are. We recall that in the two-dimensional case there are two characteristics through each point and hence four characteristic directions at each point (a positive and a negative one on each characteristic). From this fact and from the definition of spacelike and timelike directions, it follows that all directions in between any two consecutive characteristic directions at a point must be either all spacelike or all timelike. In other words, the characteristic directions at a point separate regions of spacelike directions from regions of timelike ones.

To be more specific, consider the special case of the two-dimensional wave equation $u_{xx} - c^{-2}u_{tt} = 0$. Its characteristic form is

$$Q(\xi, \tau) = \xi^2 - c^{-2}\tau^2$$

Hence the normal cone (Fig. 11-1*a*) is the pair of intersecting straight lines

$$\xi \pm c^{-1}\tau = 0 \tag{11-31}$$

in $\xi\tau$ space. The characteristic cone (Fig. 11-1*b*) is the pair of straight lines

$$\xi \pm c\tau = 0 \tag{11-32}$$

For the more general case of (11-14) the normal cone at any point $(\mathbf{x}, t)$ is the surface in $(\boldsymbol{\xi}, \tau)$ space defined by the equation $Q(\boldsymbol{\xi}, \tau) = 0$, that is, by

$$\sum_{j,k=1}^{n} a_{jk}(\mathbf{x}, t)\, \xi_j \xi_k - c^{-2}(\mathbf{x}, t)\, \tau^2 = 0 \tag{11-33}$$

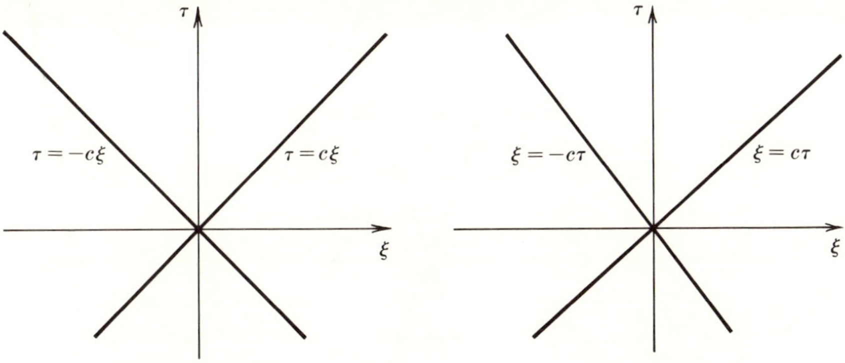

Figure 11-1

The characteristic cone is given by

$$\sum_{j,k=1}^{n} b_{jk}(\mathbf{x}, t)\, \xi_j \xi_k - c^2(\mathbf{x}, t)\, \tau^2 = 0 \tag{11-34}$$

where (b_{jk}) is the $n \times n$ inverse of (a_{jk}).

We are now ready to return to our original aim in this section, namely, the definition of spacelike and timelike directions. We remarked above that regions of spacelike directions are separated from those of timelike directions by characteristics, and we have just seen that the characteristics are the cones (11-34). However, finding the characteristic cones requires first finding the inverse of the matrix of coefficients in the principal part of the PDE. It is therefore more convenient to use the normal cone, whose equation is (11-33), rather than the characteristic cone.

Returning to the simple case of the one-dimensional wave equation for guidance, we see that the normal cone

$$Q(\xi, \tau) = \xi^2 - c^{-2}\tau^2 = 0$$

divides the $\xi\tau$ space into four regions, in two of which Q is positive and in the other two of which Q is negative (see Fig. 11-2). We also notice that all curves through the origin whose normals point into either of the regions in which $Q > 0$ are spacelike at the origin while all curves whose normals at the origin point into either of the regions in which $Q < 0$ are timelike there. Curves whose normals lie along the normal cone are, of course, characteristic at the origin.

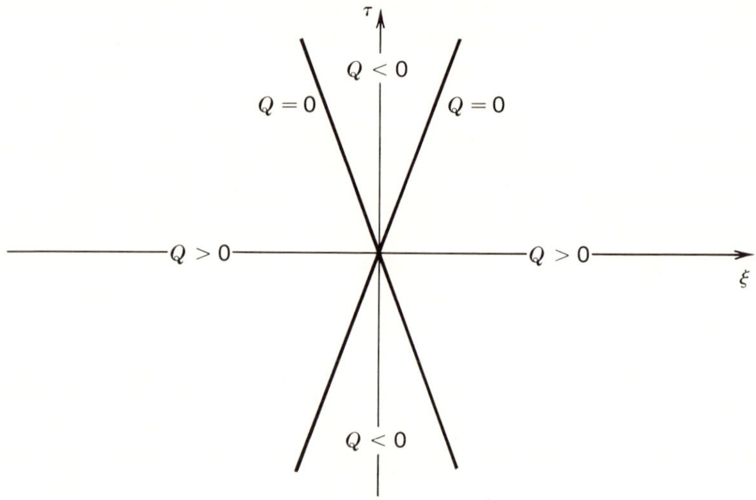

Figure 11-2

These ideas can be generalized to higher dimensions, and we proceed to do so.

DEFINITION 11-1

A surface is called *spacelike* for the hyperbolic PDE (11-30) at a point $(\mathbf{x}, t)$ if, when the components (ξ, τ) of the normal to that surface are inserted into the characteristic form $Q(\xi, \tau)$, the result is that $Q(\xi, \tau) > 0$. If, instead, $Q(\xi, \tau) < 0$, then the surface whose normal is (ξ, τ) is called *timelike* at $(\mathbf{x}, t)$. Put more qualitatively, this means that a surface is *spacelike* at a point if its normal there lies outside the normal cone at the point, *timelike* if its normal lies inside the normal cone, and characteristic if its normal lies on the normal cone.

From Definition 11-1, it can be proved that two data must be prescribed on a spacelike part of an initial surface and only one on a timelike part. We refer the interested reader to Zauderer [302] for further details.

In Chap. 9 we remarked that in the case of two independent variables the restriction on the amount of data that could be prescribed on a piece of initial curve depended on whether or not that piece of initial curve was space- or timelike. In particular, in two dimensions data could be prescribed on an everywhere timelike initial curve. It was emphasized, however, in Chap. 9 that this state of affairs is restricted to two dimensions. In three or more dimensions the Cauchy problem for a second-order linear hyperbolic equation is not well posed if the Cauchy data, i.e., function and normal derivative, are prescribed on a timelike initial manifold even if that initial manifold is *everywhere* timelike. Thus, although no characteristics could possibly carry information from another part of the initial manifold in this case, we cannot prescribe two data. However, the reason we cannot prescribe two data on a timelike initial manifold is not because the solution to such a Cauchy problem will not exist. The solution will generally exist, all right. The trouble is that it will not depend continuously on the data. Back in Chap. 3 we gave an example for the potential equation of a Cauchy problem which had a solution but whose solution blew up off the initial line when the data went to zero. The same example can be used to show that the same thing will happen for the hyperbolic equation

$$u_{xx} + u_{yy} - c^{-2}u_{tt} = 0 \tag{11-35}$$

if Cauchy data are prescribed on the timelike initial manifold $y = 0$. In fact, the solution to the example of Chap. 3 applies verbatim because it is also a solution of (11-35).

11-4. WAVEFRONTS AND RAYS

In most of what we have done so far a key idea has been that of building up integral surfaces of PDEs out of characteristic curves. Right at the outset, in Chap. 1, we formulated the Cauchy problem for a first-order linear PDE and solved it by laying characteristic curves through the given initial curve. The integral surface in that case was generated by a one-parameter family of characteristic curves, the integral surface being, at every point, tangent to a characteristic direction. Similarly, as in the present chapter for second-order hyperbolic equations, integral surfaces can be thought of as being built up out of characteristic surfaces. These characteristic surfaces, in turn, are generated by their own characteristic curves, called the bicharacteristics of the original second-order PDE.

In Chap. 2, however, and in later chapters, we saw that solutions of hyperbolic equations could be interpreted as progressing waves with characteristics as wavefronts in some sense. This wavefront-ray interpretation is important for many applications. Since it is related to, but different from, our previous interpretation of characteristics, and since this difference is not always emphasized, we wish to discuss this interpretation here.

In Chap. 2 we saw that a function of the form $\phi(x - ct)$ can be interpreted as a wave of shape $\phi(x)$ propagating with velocity c along the x axis. Notice that when we wish to interpret a function of x and t as a wave, we think of it as a function of x in which t is a parameter. We plot the function as a function of x for different times t. We do not interpret the function in the xt plane.

Similarly, if $u(\mathbf{x}, t)$ is a function of the $n + 1$ independent variables $(\mathbf{x}, t)$ and we want to think of u as a wave, we think of $u(\mathbf{x}, t)$ as a family of surfaces embedded in $\mathbf{x}$ space with t as the family parameter. Taking different values of t means taking different snapshots of the surface as it moves through $\mathbf{x}$ space. In particular, if (as we did in Sec. 11-1 with the characteristic surfaces for a second-order linear hyperbolic equation) we have a function $u(\mathbf{x}, t)$ of the form $u(\mathbf{x}, t) = v(\mathbf{x}) - ct$, then the surfaces on which $t = \text{const}$ are just the level surfaces of $v(\mathbf{x})$. Hence the surfaces $v(\mathbf{x})$ can be interpreted as wavefronts propagating through $\mathbf{x}$ space with speed c.

For example, the solutions

$$\phi(x, y, t) = \sqrt{x^2 + y^2} - ct \tag{11-36}$$

of the first-order PDE

$$\phi_x^{\,2} + \phi_y^{\,2} - c^{-2}\phi_t^{\,2} = 0 \tag{11-37}$$

are cones in space-time. Their level surfaces in the xy plane, however, are the circles

$$x^2 + y^2 = c^2t^2 \tag{11-38}$$

of radius ct, in other words, a set of circular waves spreading out from the origin with speed c.

In Sec. 11-1 we saw that the solutions of (11-37) represent the characteristic surfaces of the wave equation

$$u_{xx} + u_{yy} - c^{-2}u_{tt} = 0 \tag{11-39}$$

Thus we see that the wavefronts are the level surfaces in xy space of the characteristic cones (11-36) in xyt space, i.e., in space-time.

More generally, the characteristic surfaces of a second-order linear hyperbolic equation for a function $u(\mathbf{x}, t)$, when their equations are written in the form $\psi(\mathbf{x}) = ct$, can be interpreted as wavefronts propagating through $\mathbf{x}$ space with speed c.

Whenever there are wavefronts, there must be rays, too. The rays turn out to have an interesting connection with the characteristics. The rays for (11-38) are obviously straight lines through the origin. Notice, however, that straight lines through the origin are just the projections on the xy plane of the bicharacteristics of (11-39). This leads us to *define* the *rays* of a second-order linear hyperbolic equation to be the projections in $\mathbf{x}$ space of the bicharacteristics. Because of this definition the rays will not always be perpendicular to the wavefronts in the ordinary sense. If, as we shall do below, we associate a Riemannian metric in a natural way with a second-order linear hyperbolic equation, then the rays will always be orthogonal to the wavefronts in this metric (see Sec. 11-6).

It is important to notice the difference between the situations in space-time and in space alone. In space-time, the bicharacteristics are tangent to the characteristic surfaces of a second-order linear hyperbolic equation. In space, the rays, i.e., the projections of the bicharacteristics into the **x** space, are transverse to the wavefronts, i.e., the level surfaces of the characteristics.

The term "ray" is not always restricted to mean the projection of a bicharacteristic. Sometimes the bicharacteristics themselves, as well as their projections, are called rays. We prefer to avoid this ambiguity by reserving the term "ray" for the projection of a bicharacteristic. Nevertheless, as mentioned above, many authors use the terms "bicharacteristics" and "rays" interchangeably both for the characteristic curves in space-time and in space alone. In fact, the characteristic cones of the PDE (11-1), i.e., the conical solutions of (11-5), are frequently called the "ray cones," and their level surfaces in space for the special value $t = 1$ are called the ray surfaces.

11-5. USE OF A RIEMANNIAN METRIC IN SPACE-TIME

In Chap. 10 we found that some further insight into the nature of elliptic equations could be gained by introducing a Riemannian metric. The metric coefficients were just the elements of the inverse matrix of the matrix of the coefficients of the principal part of the PDE. More precisely, with the second-order linear elliptic PDE

$$\sum_{j,k=1}^{n} a^{jk}(\mathbf{x}) \frac{\partial^2 u}{\partial x^j \, \partial x^k} + \sum_{j=1}^{n} b^j(\mathbf{x}) \frac{\partial u}{\partial x^j} + c(\mathbf{x})u = F(\mathbf{x}) \tag{11-40}$$

we associated the Riemannian metric

$$d\sigma^2 = \sum_{j,k=1}^{n} a_{jk}(\mathbf{x}) \, dx^j \, dx^k \tag{11-41}$$

where $(a_{jk}(\mathbf{x}))$ is the inverse matrix to $(a^{jk}(\mathbf{x}))$. Because (11-40) was elliptic, (a_{jk}) was positive definite and (11-41) defined the usual type of Riemannian metric. If (11-40) is hyperbolic, however, then (a_{jk}) is indefinite and there will be curves which have zero length in the metric (11-41), namely, those whose parametric representations $x^j = x^j(\tau)$ satisfy the equation

$$\sum_{j,k=1}^{n} a_{jk}(\mathbf{x}) \frac{dx^j}{d\tau} \frac{dx^k}{d\tau} = 0 \tag{11-42}$$

However, (a_{jk}) in the present notation is just the matrix (b_{jk}) of Secs. 11-2 and 11-3. Bearing this in mind, and comparing (11-26) and (11-42), we see that (11-42) is just the equation for the bicharacteristics of (11-40). Hence the bicharacteristics are the curves of zero length in the indefinite metric (11-20).

In addition, we see from Sec. 11-2 that the $p^{(j)}$'s, as defined by (11-23), are the contravariant components of the tangent vector to a bicharacteristic, and, from (11-24), that the p_j's are the covariant components of this tangent. However, $p_j = \partial\phi/\partial x^j$. Hence the p_j's are also the covariant components of the (euclidean) normal to a characteristic surface. Since the normal to a characteristic surface should be normal (in the euclidean sense) to any direction in the surface and therefore in particular to a bicharacteristic direction, we conjecture that

$$\sum_{j=1}^{n} p_j p^j = 0$$

To prove this, note that

$$p_j = \sum_{k=1}^{n} a_{jk} p^k$$

Hence

$$\sum_{j=1}^{n} p_j p^j = \sum_{j,k=1}^{n} a_{jk} p^j p^k = 0 \qquad \text{Q.E.D.}$$

because $p^j = dx^j/dt$ and (11-42) holds.

11-6. RIEMANNIAN METRIC IN THE SPACE VARIABLES

For some purposes it is more convenient to associate a positive definite Riemannian metric with a hyperbolic PDE. For a PDE of the form (11-30) this is easy if the coefficients, which we now denote by a^{jk}, in the principal part of the elliptic operator L are independent of t and if c is a constant. Then (11-30) reads

$$\sum_{j,k=1}^{n} a^{jk}(\mathbf{x}) \frac{\partial^2 u}{\partial x^j \, \partial x^k} - c^{-2} \frac{\partial^2 u}{\partial t^2} = 0$$

where (a^{jk}) is positive definite. Using $(a_{jk}) = (a^{jk})^{-1}$, we can therefore define the Riemannian metric

$$ds^2 = \sum_{j,k=1}^{n} a_{jk}(\mathbf{x}) \, dx^j \, dx^k \tag{11-43}$$

If, as we did in Sec. 11-4, we write the equation of a characteristic surface in the form

$$\psi(\mathbf{x}) = ct \tag{11-44}$$

then the characteristic equation takes the form

$$\sum_{j,k=1}^{n} a^{jk}(\mathbf{x}) \frac{\partial \psi}{\partial x^j} \frac{\partial \psi}{\partial x^k} = 1 \tag{11-45}$$

The rays, i.e., the projections of the bicharacteristics into the **x** space, therefore satisfy

$$\frac{dx^j}{dt} = \sum_{k=1}^{n} a^{jk}(\mathbf{x}) \frac{\partial \psi}{\partial x^k} \tag{11-46}$$

which shows that they are normal to the characteristic surfaces (11-44) in the metric (11-43).

11-7. HIGHER-ORDER LINEAR HYPERBOLIC EQUATIONS IN TWO INDEPENDENT VARIABLES

Next we raise the order of the PDE but return to two independent variables. The most general linear mth-order PDE in two independent variables x and y can be written in the form

$$\sum_{j=0}^{m} a_{mj}(x, y) \frac{\partial^m u}{\partial x^j \, \partial y^{m-j}} + \sum_{j=0}^{m-1} a_{m-1,j}(x, y) \frac{\partial^{m-1} u}{\partial x^j \, \partial y^{m-j-1}} + \cdots = 0 \tag{11-47}$$

where now the superscripts on the x and y represent powers, not indices.

The characteristic form for (11-47) is

$$Q(\xi, \eta) = \sum_{j=0}^{m} a_{mj}(\mathbf{x}) \, \xi^j \eta^{m-j} \tag{11-48}$$

a form, i.e., homogeneous polynomial, of degree m. If $m > 2$, then the algebraic theory of (11-48) is complicated. However, the characteristic equation $Q(\xi, \eta) = 0$ can be reduced by the substitution

$$\zeta = \frac{\xi}{\eta} \tag{11-49}$$

to the polynomial equation

$$q(\zeta) = \eta^{-m} Q(\zeta, 1) = 0 \tag{11-50}$$

where, as indicated,

$$q(\zeta) = \sum_{j=1}^{m} a_{mj} \zeta^j \tag{11-51}$$

is an mth-degree polynomial in ζ. The PDE (11-47) is called hyperbolic if (11-50) has m real roots. It is elliptic if (11-50) has no real roots. Since we are interested in the hyperbolic case here, we will assume that (11-50) has m real roots $\zeta_j(x, y)$, $j = 1, \ldots, m$, and, for simplicity, we will assume that these roots are distinct. The case of multiple roots causes certain complications that we wish to avoid here.

If the roots are distinct, then at every point (x, y) at which (11-47) is hyperbolic, we have m different characteristic directions whose slopes are the ζ_j's. If we solve the m ODEs $dy/dx = \zeta_j(x, y)$, then we obtain m families of characteristic curves, say $C^{(j)}$, $j = 1, \ldots, m$, with one curve of each family going through each point (x, y).

Generalizing the definitions of Sec. 9-6, we choose $y = t$ as the time variable, define the positive direction on each characteristic to be that in which t increases, and then define a curve to be *spacelike* at a point if *all* m of the characteristics crossing the curve at that point cross the curve in their positive directions at that point.

Since there are now many other possibilities rather than just one, we will not define timelike or any generalization of it but rather remark on the number of conditions which can properly be prescribed on a portion of an initial curve. The number of characteristics crossing an initial curve at a point in the positive direction (on the characteristics) can vary from zero to m. If k characteristics cross an initial curve in their positive directions at a point, then precisely k conditions should be prescribed at that point, no more and no less.

EXAMPLE The equation

$$\frac{\partial}{\partial x}(u_{xx} - B_1 u_{yy}) + u_{xx} - B_2 u_{yy} = 0 \tag{11-52}$$

where B_1 and B_2 are positive constants, arises in the problem of linearized flow of a chemically relaxing gas. Its characteristic form $Q(\xi, \eta)$ is

$$Q(\xi, \eta) = \xi^3 - B_1 \xi \eta^2 \tag{11-53}$$

Setting $Q = 0$, dividing by η^3, and calling $\xi/\eta = \zeta$, we obtain the equation

$$\zeta^3 - B_1 \zeta = 0 \tag{11-54}$$

for the slopes ζ of the characteristics. From (11-54) we see that these are $\zeta = 0$ and $\zeta = \pm\sqrt{B_1}$. The characteristic curves are therefore the three families of straight lines $x = \alpha$ and $x \pm y\sqrt{B_1} = \beta_\pm$, where α and $\beta_\pm$ are the constant parameters which index the curves.

11-8. *m*TH-ORDER LINEAR EQUATIONS IN *n* INDEPENDENT VARIABLES

When we go over to the most general linear equation

$$\sum_{j_1, \ldots, j_m = 1}^{n} a_{j_1 \cdots j_m}(\mathbf{x})\, u_{x_{j_1} \cdots x_{j_m}} + \text{lower-order terms} = 0 \tag{11-55}$$

the difficulties of the two preceding cases combine. The characteristic form

$$Q(\xi) = \sum_{j_1,\ldots,j_m=1}^{n} a_{j_1\cdots j_m}\xi_{j_1}\cdots\xi_{j_m} \tag{11-56}$$

is now neither quadratic nor reducible to a function of only one variable. We can still define a characteristic surface to be a surface whose normal ξ is a root of the characteristic equation

$$Q(\xi) = 0 \tag{11-57}$$

but the surface in ξ space defined by (11-57) is no longer a cone if $m > 2$. In fact, the geometry of the characteristic surface can get quite involved in higher dimensions, so much so that it is sometimes debatable whether that geometry is an aid or a burden to the analysis of the corresponding PDE.

Although (11-57) does not represent a cone in ξ space if $m > 2$, the surface defined by (11-57) in ξ space is frequently called the *normal cone* nevertheless. Similarly, the surface generated by the bicharacteristics is frequently called the *ray cone*. These so-called "cones" can be quite complicated and frequently consist of several separate sheets. If they do consist of several sheets, and if, in order to have a wavefront-ray interpretation as in Sec. 11-4, we replace ξ by (ξ, τ), then the projections of the normal cone and ray cone from $\xi\tau$ space into ξ space may consist of several curves. If this is the case, it means that there are several different wavefronts propagating outward from a disturbance region, rather than just one, as in the case of second-order equations. Moreover, it sometimes turns out that these different wavefronts move with different speeds. This phenomenon of "fast" and "slow" waves actually arises in practice in magnetohydrodynamics. However, although the ideas involved are relatively simple, the analysis can get complicated, and we therefore refer the interested reader to [65, vol. II], [88], or [191] for details.

PROBLEMS

1. Classify and find the bicharacteristics of the PDE

$$u_{zz} - u_{xy} - u_{yz} - u_{xz} = 0$$

2. Find the characteristic form and the equation of the bicharacteristics and transform to canonical form:

$$3u_{xx} + u_{yy} + 2u_{zz} + c^{-2}u_{tt} = 0$$

3. Find the characteristics of

$$u_{xxx} + 2u_{xxy} - u_{xyy} - 2u_{yyy} = 0$$

12
First-order Systems

12-1. INTRODUCTION

It is natural to attempt to extend our results to systems of partial differential equations (PDEs). Such an extension is not limited to mere mathematical interest. More and more nowadays, physical problems give rise to systems of PDEs rather than single equations. We shall take up an important example of such a system in the next chapter.

In ordinary differential equations (ODEs) it is well known that single higher-order equations and systems of equations are equivalent; i.e., under sufficient differentiability assumptions a single nth-order ODE is equivalent to a system of n first-order ODEs. Thus, from a theoretical point of view, systems of order higher than first need not be considered for ODEs. For PDEs, however, the situation is more complicated. A system of PDEs need not always be equivalent to a single higher-order equation (see [65, vol. II] for examples). It does turn out that there is an equivalence between *initial-value problems* for systems of PDEs and for single higher-order equations but not necessarily for the equations themselves (again see [65, vol II] for the details). Since we are not developing the theory of PDEs here, we omit the fine points and take up systems directly.

We take up only first-order systems because, the foregoing pitfalls notwithstanding, most higher-order systems can be reduced to first-order ones. Moreover, many of the systems that arise in practice arise directly as first-order systems.

As in Chap. 1, we begin with a first-order linear homogeneous system in two independent variables. We assume that it is a system of n equations for the n unknown functions $u^{(1)}(x, y), \ldots, u^{(n)}(x, y)$. We use superscripts on the u's in order to save the subscript place for partial derivatives.

The assumption that the equations are linear means that $u_x^{(k)}$, $u_y^{(k)}$, and $u^{(k)}$ must all enter the equations linearly for all $k = 1, \ldots, n$. If we let $a_{jk}(x, y)$ denote the coefficient of $u_x^{(k)}$ in the jth equation, let $b_{jk}(x, y)$ denote the coefficient of $u_y^{(k)}$ in the jth equation, and let $c_{jk}(x, y)$ denote the coefficient of $u^{(k)}$ in the jth equation, we can write the most general first-order linear homogeneous system for n unknown functions of two independent variables in the form

$$\sum_{k=1}^{n} [a_{jk}(x, y)\, u_x^{(k)} + b_{jk}(x, y)\, u_y^{(k)}] = \sum_{k=1}^{n} c_{jk}(x, y)\, u^{(k)} \qquad j = 1, \ldots, n \tag{12-1}$$

In Chap. 1 we saw that the analysis of a linear equation and that of a semilinear one are nearly identical. We therefore replace (12-1) by the *semilinear system*

$$\sum_{k=1}^{n} [a_{jk}(x, y)\, u_x^{(k)} + b_{jk}(x, y)\, u_y^{(k)}] = c_j(x, y, \mathbf{u}) \qquad j = 1, \ldots, n \tag{12-2}$$

where, as usual, we use boldface type to denote vectors.

The form of (12-1) suggests the introduction of some matrix algebra. Accordingly, we let

$$U = \begin{pmatrix} u^{(1)} \\ \vdots \\ u^{(j)} \\ \vdots \\ u^{(n)} \end{pmatrix} = (u^{(j)}) \tag{12-3}$$

$$A = (a_{jk}) \qquad B = (b_{jk}) \qquad \text{and} \qquad C = (c_j) \tag{12-4}$$

Then (12-2) can be written in the abbreviated form

$$A(x, y)\, U_x + B(x, y)\, U_y = C(x, y, U) \tag{12-5}$$

12-2. CLASSIFICATION OF THE SYSTEM (12-5)

Just as we did for single first-order equations, we combine some of the derivatives in the system (12-2) into directional derivatives. If we introduce the n^2 directional derivatives

$$D_{jk} = a_{jk}\frac{\partial}{\partial x} + b_{jk}\frac{\partial}{\partial y} \tag{12-6}$$

we can write the system in the form

$$\sum_{k=1}^{n} D_{jk}u^{(k)} = c_j \qquad j = 1, \ldots, n \tag{12-6a}$$

In this form each u is differentiated in a different direction in each equation. More specifically, in the jth equation the kth u is differentiated in the direction D_{jk} given by (12-6). It would be much nicer if all the u's in a given equation were differentiated in the same direction.

We inquire whether it is possible to form a linear combination of the n equations (12-2) in such a way that all the u's in the resulting equation are differentiated in the same direction. In order to answer this question, we multiply (12-2) by arbitrary constants λ_j and sum over j. This gives

$$\sum_{j,k=1}^{n} (\lambda_j a_{jk} u_x^{(k)} + \lambda_j b_{jk} u_y^{(k)}) = \sum_{j=1}^{n} \lambda_j c_j \tag{12-7}$$

In (12-7) $u^{(k)}$ is differentiated in the direction whose direction numbers are

$$\sum_{j=1}^{n} \lambda_j a_{jk} \qquad \text{and} \qquad \sum_{j=1}^{n} \lambda_j b_{jk}$$

We wish these directions to be the same for all k. Hence their direction numbers must be proportional,

$$\sum_{j=1}^{n} \lambda_j a_{jk} = \mu \sum_{j=1}^{n} \lambda_j b_{jk} \qquad k = 1, \ldots, n \tag{12-8}$$

or

$$\sum_{j=1}^{n} \lambda_j (a_{jk} - \mu b_{jk}) = 0 \qquad k = 1, \ldots, n \tag{12-9}$$

We are looking for the λ's, the multipliers that will give us the desired PDE. Hence we need nontrivial solutions of (12-9). This means that μ must be a root of the equation

$$\det (a_{jk} - \mu b_{jk}) = 0 \tag{12-10}$$

The row vector $\Lambda = (\lambda_1, \ldots, \lambda_n)$ is then a left eigenvector of $A - \mu B$.

Equation (12-10) is an nth-degree polynomial equation for μ. Suppose it has r distinct real roots, $\mu_1, \ldots, \mu_r$. For each of these we will have an eigenvector $\Lambda_k = (\lambda_{1k}, \ldots, \lambda_{nk})$, $k = 1, \ldots, r$. Using in turn the components of these n Λ's in (12-7), we will obtain r distinct equations of the form (12-7) in each of which all u's are differentiated in the same direction. To see what a typical one of these equations looks like, use (12-8) in (12-7). Then

$$\sum_{j,k=1}^{n} \lambda_j b_{jk}(\mu u_x^{(k)} + u_y^{(k)}) = \sum_{j=1}^{n} \lambda_j c_j \tag{12-11}$$

Let

$$\sum_{j=1}^{n} \lambda_j b_{jk} = \beta_k$$

Then the left-hand side of (12-11) becomes

$$\sum_{k=1}^{n} \beta_k(\mu u_x^{(k)} + u_y^{(k)}) = \sum_{k=1}^{n} \beta_k D u^{(k)}$$

$$= D \sum_{k=1}^{n} \beta_k u^{(k)} - \sum_{k=1}^{n} (D\beta_k)\, u^{(k)}$$

where D is the directional derivative in the direction having direction numbers μ and 1, that is, whose slope is μ^{-1}. If we set

$$\hat{u} = \sum_{k=1}^{n} \beta_k u^{(k)} \qquad \text{and} \qquad \hat{C} = \sum_{k=1}^{n} [\lambda_k c_k + (D\beta_k)\, u^{(k)}]$$

then (12-11) can be written in the abbreviated form

$$D\hat{u} = \hat{C}$$

In all the above reduction μ is any one of the roots of (12-10) and the λ_j's are the components of the corresponding eigenvector. Let μ_l be a typical one of the roots of (12-10) and λ_{kl} the components of its corresponding eigenvector Λ_l. Suppose, as before, that $l = 1, \ldots, r$. Then we can form r equations such as (12-11) and eventually reduce them to the form

$$D_l \hat{u}^{(l)} = \hat{C}_l \qquad l = 1, \ldots, r \tag{12-12}$$

where

$$D_l = \mu_l \frac{\partial}{\partial x} + \frac{\partial}{\partial y} \tag{12-13}$$

$$\hat{u}^{(l)} = \sum_{k=1}^{n} \beta_{kl} u^{(k)} \tag{12-14}$$

with

$$\beta_{kl} = \sum_{j=1}^{n} \lambda_{jl} b_{jk} \tag{12-15}$$

and

$$\hat{C}_l = \sum_{j=1}^{n} [\lambda_{jl} c_j + (D_l \beta_{jl}) u^{(j)}] \tag{12-16}$$

Each equation of (12-12) contains only one of the $\hat{u}^{(l)}$'s. Equations (12-12) are therefore much simpler than the original system (12-6). However, if $r < n$, then there are not enough equations in (12-12) to replace all those in the original system (12-6) and $n - r$ equations of (12-6) have to be added to the system (12-12). Since this case is difficult to treat and rarely comes up in practice, we will not treat it here except to mention that in the extreme case, when $r = 0$, that is, when there are no real eigenvalues at all, the system is called elliptic and can be treated more easily. The classical example of an elliptic system is the pair of Cauchy-Riemann equations

$$\begin{aligned} u_x &= v_y \\ u_y &= -v_x \end{aligned} \tag{12-17}$$

Writing (12-17) in the form

$$\begin{pmatrix} 1 & 0 \\ 0 & 1 \end{pmatrix} \begin{pmatrix} u_x \\ v_x \end{pmatrix} + \begin{pmatrix} 0 & -1 \\ 1 & 0 \end{pmatrix} \begin{pmatrix} u_y \\ v_y \end{pmatrix} = 0 \tag{12-18}$$

we see that the eigenvalue equation is

$$\det \left[\begin{pmatrix} 1 & 0 \\ 0 & 1 \end{pmatrix} - \mu \begin{pmatrix} 0 & -1 \\ 1 & 0 \end{pmatrix} \right] = 0 \tag{12-19}$$

that is,

$$\begin{vmatrix} 1 & \mu \\ -\mu & 1 \end{vmatrix} = 0 \tag{12-20}$$

or

$$\mu^2 + 1 = 0 \tag{12-21}$$

Since the eigenvalues are both imaginary, the system is elliptic. This is not surprising since by elimination of either u or v from (12-17) it follows that $u_{xx} + u_{yy} = v_{xx} + v_{yy} = 0$ and the potential equation is the classical example of an elliptic equation.

12-3. HYPERBOLIC SYSTEMS; CHARACTERISTICS

If there are n real linearly independent eigenvectors Λ, the system (12-5) is called *hyperbolic*. The system (12-5) is therefore certainly hyperbolic if it has n distinct real eigenvalues, but it may also be hyperbolic if it has some repeated eigenvalues as long as each repeated eigenvalue gives rise to as many linearly independent eigenvectors as its multiplicity. When the eigenvalues are all real and distinct, (12-5) is called *totally hyperbolic*. We will deal only with totally hyperbolic systems. Of course, just as in Chap. 6, it should be borne in mind that all of the foregoing refers to the situation at a particular point (x, y). Since the coefficients in the system depend upon x and y, so do the eigenvalues and eigenvectors. Their existence, reality, and multiplicity are therefore all functions of x and y. We assume that we are treating (12-5) in a region in which it is totally hyperbolic.

This means that Eq. (12-10) has n distinct real roots μ_l, $l = 1, \ldots, n$, and that we are considering a region in the xy plane in which the μ's remain real and distinct. According to the theory developed on the previous pages, we can use the left eigenvectors that arise from the μ's to combine the original system of PDEs into n new PDEs, each of which contains derivatives of the unknowns in only one direction. The directions of these directional derivatives [see (12-13)] are μ_l^{-1}, $l = 1, \ldots, n$. Naturally, we call the directions with these slopes the *characteristic directions*. Thus, at each point (x, y) in the plane at which a semilinear system is totally hyperbolic we have a set of n distinct characteristic directions, namely, the directions whose slopes are μ_l^{-1}, $l = 1, \ldots, n$, where the μ's are the roots of (12-10). Since the μ's are functions of x and y, we can introduce the curves which have at every point (x, y) the characteristic directions μ_l^{-1}. These will be the solutions to the n ODEs

$$\frac{dy}{dx} = \frac{1}{\mu_l(x, y)} \qquad l = 1, \ldots, n \tag{12-22}$$

The PDEs then can be reduced to a canonical system in which each equation represents differentiation along one of the characteristics.

If the system is hyperbolic but not necessarily totally hyperbolic, some of the characteristic directions may be multiple. There will still be n distinct PDEs in the canonical form, but several of them may involve directional derivatives along the same characteristic.

In general, we can say that a first-order semilinear system in two independent variables will have anywhere from zero to n characteristic directions at a given point. The number of characteristic directions may vary from point to point. If there are n real characteristic directions at a point, the system is hyperbolic at that point. If there are no real characteristic directions at a point, then the system is elliptic there.

12-4. ALTERNATE INTRODUCTION OF CHARACTERISTICS

Although, as mentioned in Sec. 12-1, the equivalence between a system of n first-order ODEs and a single ODE of order n does not always carry over to PDEs, there are, nevertheless, broad analogies between results for first-order systems of n PDEs and a single PDE of order n. We have already met such an analogy in Sec. 12-2, where we saw that the first-order system consisting of the two Cauchy-Riemann equations is equivalent to the second-order potential equation.

In Sec. 11-7 we saw that a single hyperbolic equation of nth order in two independent variables has, at every point (x, y) where it is hyperbolic, n linearly independent characteristic directions, while in Sec. 12-3 we have just seen that the same situation obtains for a totally hyperbolic system of n first-order equations.

Because of these analogies it is natural to expect that the method of introducing characteristics for single higher-order equations via a solvability condition should carry over to systems. We now show that it does and thereby obtain an alternate way of arriving at the concept of characteristics for (12-5).

Suppose we give a curve $\mathscr{C}$ in the xy plane and on it prescribe the values of all the unknown functions $u^{(j)}$, $j = 1, \ldots, n$. In general, the system of differential equations (12-5) will then determine the derivatives $u_x^{(j)} u_y^{(j)}$, $j = 1, \ldots, n$, on $\mathscr{C}$. By differentiating (12-5) we can then determine the second derivatives, etc. We ask under what conditions (12-5) will fail to determine the values of the $u_x^{(j)}$'s and $u_y^{(j)}$'s on $\mathscr{C}$. We obtain the answer to this question in the same way as we did for a single equation in the previous chapter. Let $\mathscr{C}$ be given by the equation $\phi(x, y) = 0$. Introduce

$$\begin{aligned} \alpha &= \phi(x, y) \\ \beta &= \psi(x, y) \end{aligned} \tag{12-23}$$

as new curvilinear coordinates (cf. Sec. 9-2). Then

$$U_x = U_\alpha \phi_x + U_\beta \psi_x \qquad \text{and} \qquad U_y = U_\alpha \phi_y + U_\beta \psi_y$$

Hence (12-5) becomes

$$(A\phi_x + B\phi_y)\, U_\alpha + (A\psi_x + B\psi_y)\, U_\beta = C \tag{12-24}$$

The condition that this equation be solvable for the outer derivative U_α is

$$\det\,(A\phi_x + B\phi_y) \neq 0 \tag{12-25}$$

Since A and B are the given coefficient matrices in the DE, this is a condition on ϕ, that is, on $\mathscr{C}$.

If (12-25) is violated, i.e., if

$$\det\,(A\phi_x + B\phi_y) = 0 \tag{12-26}$$

then $\mathscr{C}$ is called characteristic. If we let $\mu = -\phi_y/\phi_x$ be the reciprocal of the slope of $\mathscr{C}$, then we see that for a characteristic direction

$$\det(A - \mu B) = 0 \tag{12-27}$$

which is the same condition we obtained previously in a different manner [Eq. (12-10)].

12-5. QUASI-LINEAR SYSTEMS

If we let A and B depend upon U as well as upon x and y, then we obtain the system

$$A(x, y, U)\, U_x + B(x, y, U)\, U_y = C(x, y, U) \tag{12-27a}$$

Just as for single equations, the system (12-27a) is called *quasi-linear*, and, just as for single equations, the theory of (12-27a) differs relatively little from that for the semilinear case. The major difference is that the characteristics now depend upon the unknown solution U. Thus, just as for single equations, (12-27a) has to be classified not only at a point but also for a particular solution. With this understanding, all the calculations of the present chapter remain unaltered and therefore need not be repeated. In the next chapter we analyze a special quasi-linear totally hyperbolic system in great detail.

12-6. REDUCTION OF A SINGLE FIRST-ORDER NONLINEAR EQUATION TO A QUASI-LINEAR SYSTEM

Next we should discuss fully nonlinear first-order systems. Amazingly enough, this turns out to be unnecessary from a theoretical point of view. In fact, if one is willing to deal with systems, one need never go further than quasi-linear systems. The reason for this is that a single first-order equation can always be reduced to a system of three first-order quasi-linear equations, and so a system of n first-order nonlinear equations can always be reduced to a system of $3n$ first-order quasi-linear ones.

It clearly suffices to reduce the general nonlinear first-order equation

$$F(x, y, u, p, q) = 0 \tag{12-28}$$

to a first-order quasi-linear system. To do this we first differentiate (12-28) with respect to x to obtain

$$F_x + F_u p + F_p p_x + F_q q_x = 0 \tag{12-29}$$

and then with respect to y to get

$$F_y + F_u q_y + F_p p_y + F_q q_y = 0 \tag{12-30}$$

If we use the fact that $p_y = q_x$, we can write (12-29) and (12-30) in the form

$$F_p p_x + F_q p_y = -F_x - F_u p \tag{12-31}$$

and

$$F_p q_x + F_q q_y = -F_y - F_u q \tag{12-32}$$

If we append to these the identity

$$F_p u_x + F_q u_y = F_p p + F_q q \tag{12-33}$$

then we have in Eqs. (12-31) to (12-33) a quasi-linear system of three first-order PDEs for the three unknown functions u, p, and q. These equations are already in canonical form, since the left-hand side in each case is a directional derivative in the direction whose direction numbers are F_p and F_q. In other words, the characteristics are given by

$$\frac{dx}{dt} = F_p \qquad \text{and} \qquad \frac{dy}{dt} = F_q \tag{12-34}$$

The PDEs (12-31) to (12-33) then become

$$\begin{aligned} \frac{dp}{dt} &= -F_x - F_u p \\ \frac{dq}{dt} &= -F_y - F_u q \\ \frac{du}{dt} &= pF_p + qF_q \end{aligned} \tag{12-35}$$

The equations in (12-34) and (12-35) are the five characteristic equations for (12-28) obtained previously in Chap. 8. Thus, any nonlinear equation, and hence any nonlinear system of equations, can be reduced to a quasi-linear system.

12-7. FIRST-ORDER SYSTEMS IN MORE INDEPENDENT VARIABLES

Because there were only two independent variables, we have been able to call them x and y. If we move to the general case, then to be consistent with the notation of earlier chapters we should call the number of independent variables n. If we do this, we have to change our notation for the number of equations. We therefore change our notation for the number of equations from n to m. This means that we consider a system of m equations for m unknown functions $u^{(1)}(\mathbf{x}), \ldots, u^{(m)}(\mathbf{x})$ of the n independent variables $\mathbf{x} = (x_1, \ldots, x_n)$; as before, vectors are in boldface type. Then, denoting the coefficient of $u^{(k)}_{x_l}$ in the jth equation by a $a^{(l)}_{jk}(\mathbf{x})$, we can write the most general first-order semilinear system in the form

$$\sum_{k=1}^{m} \sum_{l=1}^{n} a^{(l)}_{jk}(\mathbf{x})\, u^{(k)}_{x_l} = c_j(\mathbf{x}, \mathbf{u}) \qquad j = 1, \ldots, m \tag{12-36}$$

If we introduce the matrices

$$A^{(l)} = (a^{(l)}_{jk}) \qquad l = 1, \ldots, n$$

and the usual column vectors U and C, we can write (12-36) in the form

$$\sum_{l=1}^{n} A^{(l)} U_{x_l} = C \tag{12-37}$$

We can attempt to associate a characteristic form with (12-37) in the usual way by replacing the U_{x_l}'s by ξ_l's. The result of this replacement, however, is not a characteristic *form* but rather a *characteristic matrix*, namely,

$$A(\mathbf{x}, \boldsymbol{\xi}) = \sum_{l=1}^{n} A^{(l)}(\mathbf{x})\, \xi_l \tag{12-38}$$

Just as in the case of a system in two independent variables, the definition of characteristics by means of a solvability condition leads to a criterion in terms of the solvability of a set of algebraic equations with the matrix A defined by (12-38). The result is that a surface is characteristic for (12-37) at a point $\mathbf{x}$ if the components of its normal $\boldsymbol{\xi}$ are such that the characteristic matrix $A(\mathbf{x}, \boldsymbol{\xi})$ defined by (12-38) is singular there. However, a matrix is singular if and only if its determinant vanishes. Hence a surface is characteristic at a point $\mathbf{x}$ if and only if the components of its normal $\boldsymbol{\xi}$ satisfy

$$Q(\mathbf{x}, \boldsymbol{\xi}) = \det\,[A(\mathbf{x}, \boldsymbol{\xi})] = 0 \tag{12-39}$$

there. Equation (12-39) is therefore called the *characteristic equation* for (12-37), and $Q(\mathbf{x}, \boldsymbol{\xi}) = \det\,[A(\mathbf{x}, \boldsymbol{\xi})]$ is called the *characteristic form* of (12-37).

Most of the systems of the type (12-37) that arise in practice are hyperbolic. We therefore confine ourselves to defining hyperbolicity for (12-37). To do this, we distinguish one independent variable, calling it t, and, for notational convenience, we assume that there are n other variables, $\mathbf{x} = (x_1, \ldots, x_n)$, making a total of $n + 1$ independent variables in all. Then (12-37) can be written in the form

$$\sum_{j=1}^{n} A^{(j)} U_{x_j} - A^{(0)} U_t = C \tag{12-40}$$

If we multiply (12-40) by $(A^{(0)})^{-1}$, which we assume exists, and let $(A^{(0)})^{-1}A^{(j)} = B^{(j)}$ and $(A^{(0)})^{-1}C = B$, then we can write the resulting system in the form

$$\sum_{j=1}^{n} B^{(j)}U_{x_j} - U_t = B \tag{12-41}$$

For (12-41) the characteristic equation is

$$\det\left(\sum_{j=1}^{n} B^{(j)}\xi_j - \tau I\right) = 0 \tag{12-42}$$

where I is the identity matrix. The system (12-40) is called *totally hyperbolic* at a point $(\mathbf{x}, t)$ if for every fixed $\boldsymbol{\xi}$ (12-42) has n distinct real roots τ. For each fixed $\boldsymbol{\xi}$ the τ's are the eigenvalues of the matrix

$$B = \sum_{j=1}^{n} B^{(j)}\xi_j \tag{12-43}$$

If some of the τ's are repeated but their corresponding eigenvectors all have multiplicities equal to the multiplicity of the corresponding τ's, then (12-40) is merely called *hyperbolic*.

In practice, systems of more than two independent variables are difficult to deal with. In gas dynamics, for example, the easiest problems to treat are those of unsteady one-dimensional flow (independent variables x and t) and of steady two-dimensional flow (independent variables x and y). Unsteady two-dimensional flows (independent variables x, y, and t) are more difficult to handle and will therefore not be discussed here (see [64]).

12-8. THE HODOGRAPH TRANSFORMATION

The equations of unsteady one-dimensional gas flow and those of steady two-dimensional gas flow are two systems of considerable interest. We devote the next two chapters to the analysis of unsteady one-dimensional flow. In the present section we merely indicate some interesting features of the system of PDEs governing steady two-dimensional flow.

The Eulerian equations for steady two-dimensional irrotational isentropic flow of a compressible fluid are

$$u_y - v_x = 0 \tag{12-44}$$

and

$$[c^2(q) - u^2]\, u_x - uv(u_y + v_x) + [c^2(q) - v^2]\, v_y = 0 \tag{12-45}$$

where $u(x, y)$ and $v(x, y)$ are the velocity components of the flow at (x, y) and $q = \sqrt{u^2 + v^2}$ is the speed of the flow there.

Equations (12-44) and (12-45) constitute a first-order quasilinear system for u and v. The most interesting thing about this system is the fact that the coefficients are independent of x and y. They depend only on u and v. For single first-order PDEs we have already seen in Sec. 8-20 that when the independent variables enter a PDE in a simple way (or do not enter it at all), the Legendre transformation, which consists in introducing the derivatives as new independent variables, affords a considerable simplification. Accordingly, we let

$$\begin{aligned} \alpha &= u_x(x, y) \\ \beta &= u_y(x, y) \end{aligned} \tag{12-46}$$

and use (12-46) to interchange the roles of the pairs of variables (x, y) and (u, v). This is not exactly the same as a Legendre transformation, but it is closely related to it, as we will see below. Because u and v are the components of the velocity, (12-46) is called the *hodograph transformation.*

To reiterate, we wish to use (12-46) to transform the system (12-44) and (12-45) from a system for u and v as functions of x and y into a system for x and y as functions of u and v. To do this, we need only replace the derivatives of u and v with respect to x and y by derivatives of x and y with respect to u and v. The coefficients in (12-44) and (12-45) are already functions of the new independent variables u and v.

From (12-46) and elementary calculus we have

$$u_x = Jy_v \qquad u_y = -Jx_v \qquad v_x = -Jy_u \qquad v_y = Jx_u \tag{12-47}$$

where J is the Jacobian

$$J = \begin{vmatrix} u_x & u_y \\ v_x & v_y \end{vmatrix} \tag{12-48}$$

Since (12-44) and (12-45) are linear in u_x, u_y, v_x, and v_y, J cancels out when Eqs. (12-47) are substituted into (12-44) and (12-45) and we get the desired equations

$$x_v - y_u = 0 \tag{12-49}$$

$$(c^2 - u)_v + uv(x_v + y_u) + (c^2 - v)\, x_u = 0 \tag{12-50}$$

Notice that (12-49) and (12-50) are *linear* PDEs for $x(u, v)$ and $y(u, v)$. Notice, too, that the derivation of (12-49) and (12-50) breaks down if $J = 0$. In applied mathematics, one is usually willing to throw in such an assumption if it is needed, but in the present case it turns out that some very important physical phenomena would be missed if the case when $J = 0$ were disregarded. Nevertheless, since the analysis of this important special case would lead us too far afield, we omit it; the interested reader should consult [64] for the details.

12-9. VELOCITY POTENTIALS; CONNECTION BETWEEN LEGENDRE AND HODOGRAPH TRANSFORMATIONS

We mentioned earlier that first-order systems are usually equivalent to single higher-order equations. It is easy to derive a second-order PDE to which the system (12-44) and (12-45) is equivalent. This is done by introducing what is known in the case of the system (12-44) and (12-45) as a *velocity potential.* We notice that we can satisfy (12-44) identically if we choose any function $\phi(x, y)$ and put

$$\begin{aligned} u &= \phi_x(x, y) \\ v &= \phi_y(x, y) \end{aligned} \tag{12-51}$$

No matter what we choose for ϕ, the u and v given by (12-51) will clearly satisfy (12-44). We therefore have only to choose ϕ so as to satisfy (12-45) and the system (12-44) and (12-45) will be solved. But substituting (12-51) into (12-45), we find that ϕ must satisfy the second-order quasilinear PDE

$$[c^2(q) - \phi_x^{\,2}]\,\phi_{xx} - 2\phi_x\phi_y\phi_{xy} + [c^2(q) - \phi_y^{\,2}]\,\phi_{yy} = 0 \tag{12-52}$$

where $q = \sqrt{\phi_x^{\,2} + \phi_y^{\,2}}$

Similarly, we could reduce the hodograph equations (12-49) and (12-50) to

$$(c^2 - v^2)\Phi_{uu} + 2uv\Phi_{uv} + (c^2 - u^2)\,\Phi_{vv} = 0 \tag{12-53}$$

by introducing the potential $\Phi(u, v)$ via the equations

$$\begin{aligned} x &= \Phi_u(u, v) \\ y &= \Phi_v(u, v) \end{aligned} \tag{12-54}$$

We leave it to the reader as an instructive exercise to show that ϕ and Φ are connected by a Legendre transformation, i.e.,

$$\phi + \Phi = xu + yv \tag{12-55}$$

SUMMARY

Let $U = (u^{(j)}(x, y))$, $A = (a_{jk}(x, y))$, $B = (b_{jk}(x, y))$, and $C = (c_j(x, y))$. Then the most general first-order *semilinear* system in two independent variables can be written in the matrix form

$$AU_x + BU_y = C \tag{S12-1}$$

If A and B depend upon U in addition to x and y, the system (S12-1) is called *quasilinear*.

The *characteristic directions* of (S12-1) are, by definition, the reciprocals of the roots $\mu(x, y)$ of the equation

$$\det(A - \mu B) = 0 \tag{S12-2}$$

The curves satisfying the ODEs $dy/dx = \mu^{-1}(x, y)$ are called the *characteristic curves* of (S12-1). The system (S12-1) is called *elliptic* at a point (x, y) if none of the roots μ of (S12-2) are real there. If all the roots of (S12-2) are real and distinct at (x, y), then (S12-1) is called *totally hyperbolic* there. If some of the roots of (S12-2) are multiple, but if each such multiple eigenvalue, i.e., root of (S12-2), has as many linearly independent eigenvectors as its multiplicity, then (S12-1) is merely called *hyperbolic*. Other cases are not defined in this book.

Notice that by multiplying (S12-1) by B^{-1}, letting $y = t$, calling $B^{-1}A$ by the name A again, and $B^{-1}C$ by the name C again, we can always write (S12-1) in the form

$$U_t - AU_x = C \tag{S12-3}$$

For equations of the form (S12-3) the characteristic directions are merely the reciprocals of the eigenvalues of the matrix A.

If $\Lambda = (\lambda_j)$ is a typical eigenvector of (S12-1) corresponding to the characteristic direction μ, then, by multiplying the jth equation of the system by λ_j, that is, by taking the scalar product of (S12-3) with Λ, we obtain a single PDE

$$(\Lambda, AU_x) + (\Lambda, BU_y) = (\Lambda, C)$$

in which every $u^{(j)}$ is differentiated in the same direction, namely, the characteristic direction, μ. If (S12-1) is totally hyperbolic, this process yields a system of n distinct PDEs which are equivalent to the original system. This new system consists of n PDEs, each equation of which represents a directional derivative in a different one of the n characteristic directions. Such a system is called the *canonical form* of (S12-1).

PROBLEMS

1. Given the system:

$$u_x + 2u_t + v_x + 3v_t - u + v = 0$$
$$3u_x + u_t - 2v_x - v_t - 2v = 0$$

(*a*) Find the characteristic equation.
(*b*) Classify.
(*c*) Reduce to canonical form.

2. Find the equations of the characteristics of the following systems of PDEs and then use these characteristics to reduce the equations to canonical form. You need not solve the resulting equations.

$$u_t + u_x + v_t + 2v_x = 7u - 6v$$
$$u_t + 2u_x + v_t + v_x = 7u - 18v$$

3. Classify and reduce to canonical form:

$$\rho u_x + u\rho_x + \rho_t = -2\rho u/x$$
$$\rho uu_x + \rho u_t + c^2\rho_x = 0$$

4. Classify and transform to canonical form:

$$u_t + u_x - u + w - 1 = 0$$
$$v_t + 2u_x - w_x - 2u + w - v^2 + 2vw - w^2 = 0$$
$$w_t + 2u_x - w_x + w = 0$$

5. Consider the systems

$$(a)\quad u_y - v_x - w_x = 0 \qquad (b)\quad u_y - v_x - w_x = 0$$
$$u_x - v_y - w_y = 0 \qquad\qquad u_x - v_y - w_y = 0$$
$$v_y - w_x = 0 \qquad\qquad u_y - v_x - w_y = 2(y - x)$$

Classify them and put them into canonical form if applicable.

6. Classify and transform to canonical form:

$$4u_r + u_x + 3v_t = u - v$$
$$u_r + v_x + 2v_t = 3u + v$$

7. Classify and reduce to canonical form the following set of PDEs:

$$u_t + 4u_x + 2v_x - 2w_x + u^2 + v - w = 0$$
$$v_t - 5u_x + 3v_x + 2w_x - u^2 + v = 0$$
$$w_t - 2u_x + 4v_x + w_x - v + w = 0$$

8. Classify and change the following equations to canonical form:

$$(x + y)\, u_x + v_y = 0$$
$$(x - y)\, v_x + u_y = 0$$

9. Classify the following system of equations and reduce to canonical form:

$$u_x + u_y + u - v_x = 0$$
$$2u_y + u^2 - 1 - v_y = 0$$

10. Transform to canonical form and classify:

$$u_x + 3w_x - u_y + u^2 + v^2 - 2uv = 0$$
$$-2v_x + 6w_x + 4u_x - 2v_y - w^3 = 0$$
$$9w_x - 3w_y - 4u + 6w = 0$$

13
One-Dimensional Compressible Flow

13-1. EULERIAN EQUATIONS OF ONE-DIMENSIONAL UNSTEADY ANISENTROPIC FLOW

In section 2-11 we derived the Eulerian equations of one-dimensional isentropic flow. If the flow is *anisentropic*, then the Eulerian equations turn out to be

$$\begin{aligned} &\rho_t + (\rho u)_x = 0 \\ &u_t + uu_x + \frac{1}{\rho} p_x = 0 \\ &S_t + uS_x = 0 \end{aligned} \tag{13-1}$$

where ρ = linear density, i.e., mass per unit length
u = velocity in x direction
p = pressure
S = entropy

The first equation is the equation of continuity. It expresses the conservation of mass. The second is Newton's second law of motion, and the third says that the entropy of a given particle does not change in time. All quantities are to be considered as functions of x and t.

Since there are only three equations in four unknowns, another equation is needed. This fourth equation is provided by the assumption of an equation of state; i.e., we assume that we are given a function f such that

$$p = f(\rho, S) \tag{13-2}$$

Using (13-2), we can replace the first of Eqs. (13-1) by an equation involving u and p instead of u and ρ. To do this we note that physical considerations imply that f_ρ is positive and has the dimensions of a velocity. Hence we can put

$$c^2 = f_\rho(\rho, S) \tag{13-3}$$

Then from (13-2) we have

$$p_t = c^2\rho_t + f_S S_t$$

and

$$up_x = c^2 u\rho_x + f_S u S_x$$

Adding these two equations and using the first and third equations of (13-1), we have

$$p_t + up_x = c^2(\rho_t + u\rho_x) = -c^2\rho u_x$$

or

$$p_t + up_x + c^2\rho u_x = 0 \tag{13-4}$$

If we look upon ρ as a known function of p and S defined implicitly by (13-2), then (13-4) is a partial differential equation (PDE) for p and u as functions of x and t. If we replace the first equation of (13-1) by (13-4), we obtain the system

$$\begin{aligned} &p_t + up_x + c^2\rho u_x = 0 \\ &u_t + uu_x + \frac{1}{\rho} p_x = 0 \\ &S_t + uS_x = 0 \end{aligned} \tag{13-5}$$

These are three equations for three unknowns u, p, and S. Since ρ is, by (13-2), a function of p and S, and since c, by (13-3), is a function of ρ and S, hence of p and S, the unknowns p and S enter into (13-5) through ρ and c as well as explicitly.

13-2. CHARACTERISTICS

Writing (13-5) in matrix form, we have

$$\begin{pmatrix} u & c^2\rho & 0 \\ 1/\rho & u & 0 \\ 0 & 0 & u \end{pmatrix}\begin{pmatrix} p_x \\ u_x \\ S_x \end{pmatrix} + \begin{pmatrix} 1 & 0 & 0 \\ 0 & 1 & 0 \\ 0 & 0 & 1 \end{pmatrix}\begin{pmatrix} p_t \\ u_t \\ S_t \end{pmatrix} = 0 \tag{13-6}$$

Hence the characteristic directions μ are the roots of

$$\begin{vmatrix} u-\mu & c^2\rho & 0 \\ 1/\rho & u-\mu & 0 \\ 0 & 0 & u-\mu \end{vmatrix} = 0 \tag{13-7}$$

i.e., of

$$(u-\mu)[(u-\mu)^2 - c^2] = 0 \tag{13-8}$$

Therefore,

$$\mu = u, \qquad u \pm c \tag{13-9}$$

Since (cf. Sec. 12-3) μ is the reciprocal of the slope of a characteristic, i.e., of dt/dx, we have three families of characteristics

C_0:

$$\frac{dx}{dt} = u$$

$C^{\pm}$: (13-10)

$$\frac{dx}{dt} = u \pm c$$

In order to see the form taken by the PDEs along the characteristics, we find the eigenvectors. Let $(\lambda_1^{(0)} \quad \lambda_2^{(0)} \quad \lambda_3^{(0)})$ be the eigenvector for $\mu = u$. Then

$$(\lambda_1^{(0)} \quad \lambda_2^{(0)} \quad \lambda_3^{(0)})\begin{pmatrix} 0 & c^2\rho & 0 \\ 1/\rho & 0 & 0 \\ 0 & 0 & 0 \end{pmatrix} = 0 \tag{13-11}$$

Hence $\lambda_2^{(0)} = \lambda_1^{(0)} = 0$ and $\lambda_3^{(0)}$ is arbitrary. Therefore $(0 \quad 0 \quad 1)$ is an eigenvector. Using it to combine the PDEs (13-5), we have

$$S_t + uS_x = 0 \tag{13-12}$$

The left side of (13-12) is the particle derivative of S. Equation (13-12) tells us that it is zero. We must therefore be following the path of a specific particle. This is in agreement with the fact that the C_0 characteristics are the particle trajectories because $dx/dt = u$ on them.

Next we determine the eigenvectors $(\lambda_1^\pm \quad \lambda_2^\pm \quad \lambda_3^\pm)$ for the $C^\pm$ characteristics. These eigenvectors satisfy

$$(\lambda_1^\pm \quad \lambda_2^\pm \quad \lambda_3^\pm)\begin{pmatrix} \mp c & c^2\rho & 0 \\ 1/\rho & \mp c & 0 \\ 0 & 0 & \mp c \end{pmatrix} = 0 \tag{13-13}$$

that is,

$$c\lambda_1^\pm \mp \frac{1}{\rho}\lambda_2^\pm = 0$$
$$c^2\rho\lambda_1^\pm \mp c\lambda_2^\pm = 0$$
$$\lambda_3^\pm = 0$$

Hence

$$c\rho\lambda_1^\pm = \pm\lambda_2^\pm \qquad \lambda_3^\pm = 0$$

Therefore $(1 \quad \pm c\rho \quad 0)$ are eigenvectors. Using them, we can reduce the DEs to

$$p_t + up_x + c^2\rho u_x \pm c\rho\left(u_t + uu_x + \frac{1}{\rho}p_x\right) = 0$$

that is,

$$p_t + (u \pm c)\,p_x \pm c\rho[u_t + (u \pm c)\,u_x] = 0 \tag{13-14}$$

Just as the general theory indicates they should be, p and u are both differentiated in the same direction, namely the characteristic direction whose direction numbers are 1 and $u \pm c$. If we let $s_\pm$ be parameters along the characteristics $C^\pm$, then (13-14) takes the form

$$\frac{\partial p}{\partial s_\pm} \pm c\rho\frac{\partial u}{\partial s_\pm} = 0 \qquad \text{on } C^\pm \tag{13-15}$$

Since c depends upon S as well as ρ, (13-15) cannot be integrated along $C^\pm$ without further analysis. If c is independent of S, however, it can. This is the special case of isentropic flow.

13-3. ISENTROPIC FLOW; CHARACTERISTICS

The special case of isentropic flow can be analyzed more fully. If the entropy is a constant, the equation of state (13-2) gives p as a function of ρ

$$p = f(\rho) \tag{13-16}$$

Then

$$c^2 = f'(\rho) \tag{13-17}$$

From (13-16) and (13-17) we have

$$p_x = c^2\rho_x \tag{13-18}$$

Using (13-18), we can eliminate p from the second equation of (13-1) and get

$$u_t + uu_x + \frac{c^2}{\rho}\rho_x = 0 \tag{13-19}$$

Equation (13-19) and the first equation of (13-1) form a pair of two PDEs for the two unknown functions u and ρ as functions of x and t. Once $u(x, t)$ and $\rho(x, t)$ have been determined, $p(x, t)$ can be found from the equation of state (13-16).

We solve this system, namely,

$$\begin{aligned} \rho_t + \rho u_x + u\rho_x &= 0 \\ u_t + uu_x + \frac{c^2}{\rho}\rho_x &= 0 \end{aligned} \tag{13-20}$$

in the usual way. In matrix form (13-20) reads

$$\begin{pmatrix} u & \rho \\ c^2/\rho & u \end{pmatrix}\begin{pmatrix} \rho_x \\ u_x \end{pmatrix} + \begin{pmatrix} 1 & 0 \\ 0 & 1 \end{pmatrix}\begin{pmatrix} \rho_t \\ u_t \end{pmatrix} = 0 \tag{13-21}$$

Hence the eigenvalues μ are the roots of

$$\begin{vmatrix} u - \mu & \rho \\ c^2/\rho & u - \mu \end{vmatrix} = 0 \tag{13-22}$$

i.e., of

$$(u - \mu)^2 - c^2 = 0$$

and so

$$\mu = u \pm c \tag{13-23}$$

The eigenvectors $(\lambda_1^\pm \quad \lambda_2^\pm)$ are the solutions of

$$(\lambda_1^\pm \quad \lambda_2^\pm)\begin{pmatrix} \mp c & \rho \\ c^2/\rho & \mp c \end{pmatrix} = 0$$

i.e., of

$$c\lambda_1^\pm \mp \frac{c^2}{\rho}\lambda_2^\pm = 0$$

Hence

$$\lambda_1^\pm = \pm\frac{c}{\rho}\lambda_2^\pm$$

and so $(\pm c/\rho \quad 1)$ are eigenvectors. Using them, we obtain the DEs

$$\pm \frac{c}{\rho}(\rho_t + \rho u_x + u\rho_x) + \left(u_t + uu_x + \frac{c^2}{\rho}\rho_x\right) = 0$$

that is,

$$\pm \frac{c}{\rho}[\rho_t + (u \pm c)\rho_x] + [u_t + (u \pm c)u_x] = 0 \tag{13-24}$$

Thus, as before, we have

$$\pm \frac{c}{\rho}d\rho + du = 0 \qquad \text{on } C^\pm \tag{13-25}$$

13-4. RIEMANN INVARIANTS

In the anisentropic case, c depended on S as well as ρ, with the result that (13-15), the analog of (13-25), could not be integrated explicitly. In the present isentropic case, $c = c(\rho)$. Therefore, if we let

$$L(\rho) = \int_{\rho_0}^{\rho} \frac{c(\rho)}{\rho}\,d\rho \tag{13-26}$$

then (13-25) can be written

$$d[L(\rho) \pm u] = 0 \qquad \text{on } C^\pm \tag{13-27}$$

and integrated to give

$$L(\rho) \pm u = \text{const} \qquad \text{on } C^\pm$$

Of course, the constants will be different on C^+ and C^-. The constant value of $L(\rho) + u$ on C^+ we call $2r$, and the constant value of $L(\rho) - u$ on C^- we call $2s$. Then

$$L(\rho) + u = 2r \qquad \text{on } C^+ \tag{13-28}$$

and

$$L(\rho) - u = 2s \qquad \text{on } C^- \tag{13-29}$$

The quantities r and s are called the *Riemann invariants*. As indicated above, r is constant on each C^+ characteristic. It is not necessarily the same constant on each C^+, however. In other words, r, in general, will vary on a C^-. Similarly, s is constant on each C^- but will usually vary on each C^+. It is therefore natural to try to introduce r and s as new curvilinear coordinates. This is just another way of saying that we want to introduce the characteristics as new curvilinear coordinates. In applied mathematics, one usually assumes that a thing like this can be done, but in the present case interesting results would be missed if we failed to investigate the case in which r and s are not independent and therefore cannot be used as new independent variables. First, however, we take up the case in which r and s are independent. What happens when they are not will be investigated in Sec. 13-7.

Since r is constant on C^+, s can be taken as the parameter on C^+. Similarly, r can be used as the parameter on C^-. Then $dx = (u + c)\,dt$ on C^+ implies

$$\frac{dx}{ds} = (u + c)\frac{dt}{ds} \qquad \text{on } C^+ \tag{13-30}$$

Similarly,

$$\frac{dx}{dr} = (u - c)\frac{dt}{dr} \qquad \text{on } C^- \tag{13-31}$$

Now on C^+, r is a constant. Therefore the derivatives in (13-30) are really partial derivatives with respect to s. Similarly for the r derivatives in (13-31) and so we can write (13-30) and (13-31) as a pair of PDEs

$$x_s = (u + c)t_s \tag{13-32}$$

$$x_r = (u - c)\,t_r \tag{13-33}$$

which we now take to hold throughout some region and which then represent two first-order PDEs for the two functions $x(r, s)$ and $t(r, s)$. If we have solved them for x and t, then we can invert the solutions to obtain r and s as functions of x and t. Having done that, we solve (13-28) and (13-29) to get

$$L(\rho) = r + s \tag{13-34}$$

$$u = r - s \tag{13-35}$$

and, by substitution, obtain $u(x, t)$ and $\rho(x, t)$.

Our original problem has thus been reduced to solving the new first-order system (13-32) and (13-33), in which the coefficients u and c are known functions of the independent variables r and s by virtue of (13-34) and (13-35). [Recall that $c = c(\rho)$ is a known function of ρ, obtained directly from the given equation of state of the gas.]

Equations (13-32) and (13-33) represent a considerable simplification of the original system because they are linear. Moreover, by equating x_{rs} to x_{sr}, we can immediately eliminate t from them and obtain the single second-order equation

$$(u_r + c_r)\,t_s + (u + c)\,t_{rs} = (u - c)\,t_{rs} + (u_s - c_s)\,t_r \tag{13-36}$$

The coefficients in (13-36) are functions of r and s. To obtain these functions explicitly, we first note from (13-35) that

$$u_r = 1 \qquad u_s = -1 \tag{13-37}$$

and from (13-34) that

$$L'(\rho)\,\rho_r = L'(\rho)\,\rho_s = 1$$

From (13-26), $L'(\rho) = c/\rho$. Hence

$$\rho_r = \rho_s = \frac{\rho}{c}$$

Then

$$c_r = \frac{dc}{d\rho}\,\rho_r = \frac{\rho}{c}\frac{dc}{d\rho} \qquad \text{and} \qquad c_s = \frac{dc}{d\rho}\,\rho_s = \frac{\rho}{c}\frac{dc}{d\rho} \tag{13-38}$$

Using (13-37) and (13-38) in (13-36), we get

$$\left(1 + \frac{\rho}{c}\frac{dc}{d\rho}\right) t_s + 2ct_{rs} = -\left(1 + \frac{\rho}{c}\frac{dc}{d\rho}\right) t_r$$

or

$$t_{rs} + \frac{1 + \dfrac{\rho}{c}\dfrac{dc}{d\rho}}{2c}\,(t_r + t_s) = 0 \tag{13-39}$$

The coefficient of $t_r + t_s$ here is a function of ρ only, hence [cf. (13-34)] of $r + s$. Call this function ϕ, that is,

$$\phi(r + s) = \frac{1 + \dfrac{\rho}{c}\dfrac{dc}{d\rho}}{2c} \tag{13-40}$$

Then we can write (13-39) in the abbreviated form

$$t_{rs} + [\phi(r + s)](t_r + t_s) = 0 \tag{13-41}$$

If we have solved (13-41) for $t(r, s)$, then (13-32) and (13-33) [which are compatible from the way (13-36) was derived] yield $x(r, s)$.

13-5. INITIAL CONDITIONS

Since (13-41) is linear and hyperbolic, a natural problem to pose for it would be an initial-value problem (IVP). More precisely, we should give the values of t and an outward derivative of t along some curve Γ in the rs plane and then find the solution in a neighborhood of Γ. The original problem, however, was one for u and ρ as functions of x and t. So the natural physical conditions to give would be the initial values of u and ρ, that is, $u(x, 0)$ and $\rho(x, 0)$. Suppose, therefore, that we are given

$$u(x, 0) = \alpha(x) \tag{13-42}$$

$$\rho(x, 0) = \beta(x) \tag{13-43}$$

How do we translate these initial conditions into initial conditions for $t(r, s)$? The answer is simple. From (13-28), (13-29), (13-42), and (13-43) we have

$$r = \tfrac{1}{2}\{L[\beta(x)] + \alpha(x)\} \tag{13-44}$$

$$s = \tfrac{1}{2}\{L[\beta(x)] - \alpha(x)\} \tag{13-45}$$

Equations (13-44) and (13-45) give a parametric representation of the curve Γ in the rs plane which is the image of $t = 0$ in the xt plane. Therefore our first initial condition for t is

$$t = 0 \qquad \text{on } \Gamma$$

where Γ is the curve given parametrically by (13-44) and (13-45).

We need a second initial condition on Γ, namely, one involving the first derivatives of t there. To obtain this second condition, we differentiate (13-44) along the initial curve Γ. We get

$$dr = \tfrac{1}{2}\{L'[\beta(x)]\,\beta'(x) + \alpha'(x)\}\,dx \tag{13-46}$$

But, by (13-32) and (13-33),

$$dx = x_r\,dr + x_s\,ds = (u - c)\,t_r\,dr + (u + c)\,t_s\,ds \tag{13-47}$$

We rewrite (13-47) in the form

$$dx = (u + c)(t_r\,dr + t_s\,ds) - 2ct_r\,dr$$

because then we recognize that

$$dx = (u + c)\,dt - 2ct_r\,dr$$

But $dt = 0$ on Γ. Hence

$$dx = -2ct_r\,dr \qquad \text{on } \Gamma \tag{13-48}$$

Substituting (13-48) into (13-46) and solving for t_r, we get

$$t_r = -c^{-1}\{L'[\beta(x)]\,\beta'(x) + \alpha(x)\}^{-1} \qquad \text{on } \Gamma \tag{13-49}$$

A similar expression can be obtained for t_s along Γ. Thus the normal derivative of t along Γ is known, and we have our second initial condition. Of course, the above presumes that the denominator in (13-49) does not vanish for the given initial data.

13-6. POLYTROPIC GAS

The PDE (13-41) is too general to solve explicitly. In fact, relatively little can be done in gas dynamics when the equation of state is left in so general a form as (13-16). It is necessary to assume a more specific form for $f(\rho)$. Fortunately, a large number of actual gases have an equation of state of the form

$$p = A\rho^{\gamma} \tag{13-50}$$

where A and γ are constants.

A gas whose equation of state is given by (13-50) is called *polytropic*. At reasonable temperatures air is polytropic with $\gamma = 1.4$.

The equation of state (13-50) enables us to calculate explicitly many of the quantities introduced in the previous sections. From (13-50) and (13-17) we have

$$c = \sqrt{A\gamma}\,\rho^{(\gamma-1)/2} \tag{13-51}$$

Then (13-26) with $\rho_0 = 0$ becomes

$$L(\rho) = \frac{2\sqrt{A\gamma}}{\gamma - 1}\,\rho^{(\gamma-1)/2}$$

Using (13-51), we can write this as

$$L(\rho) = \frac{2c}{\gamma - 1} \tag{13-52}$$

To find out what the function ϕ in (13-40) looks like in the polytropic case, note that (13-51) implies

$$dc = \frac{\gamma - 1}{2}\frac{c}{\rho}\,d\rho$$

whence

$$\frac{\rho}{c}\frac{dc}{d\rho} = \frac{\gamma - 1}{2}$$

Hence [cf. (13-40)]

$$\phi(r + s) = \frac{\gamma + 1}{2(2c)}$$

Using (13-52), we can write this as

$$\phi(r + s) = \frac{\gamma + 1}{2(\gamma - 1)\,L}$$

Finally, using (13-34), we have

$$\phi(r + s) = \frac{1}{2}\frac{\gamma + 1}{\gamma - 1}\frac{1}{r + s} \tag{13-53}$$

If we let

$$\kappa = \frac{1}{2}\frac{\gamma + 1}{\gamma - 1}$$

then

$$\phi(r + s) = \frac{\kappa}{r + s}$$

and (13-41) reads

$$t_{rs} + \frac{\kappa}{r + s}(t_r + t_s) = 0 \tag{13-54}$$

We have seen in earlier chapters that first-derivative terms can be removed from equations with constant coefficients. Since r and s enter (13-54) symmetrically, it is natural to try to generalize the method to (13-54) by putting

$$t(r, s) = \psi(r + s)\,\omega(r, s) \tag{13-55}$$

where ω is the new dependent variable and ψ is to be determined so that the coefficients of ω_r and ω_s vanish.

Substituting (13-55) into (13-54) and collecting terms, we obtain

$$\psi\omega_{rs} + \left(\psi' + \frac{\kappa\psi}{r + s}\right)(\omega_r + \omega_s) + \left(\psi' + \frac{2\kappa\psi'}{r + s}\right)\omega = 0 \tag{13-56}$$

Therefore, if we choose $\psi(r + s) = (r + s)^{-\kappa}$, so that (13-55) reads

$$t(r, s) = \frac{\omega(r, s)}{(r + s)^{\kappa}}$$

then (13-56) reduces to

$$\omega_{rs} + \frac{\kappa(1 - \kappa)}{(r + s)^2}\,\omega = 0 \tag{13-57}$$

which is a telegraph equation with a variable coefficient. In Chap. 9 we saw that the Riemann function for the telegraph equation with a constant coefficient was a function of $(r - \rho)(s - \sigma)$, where ρ and σ are the coordinates of the parameter point in the Riemann function. (ρ here is not the density of the gas.) In the present case, $R(r, s, \rho, \sigma)$ must be a function of $r + s$ also, since the coefficient in (13-57) is. From these and other facts, Riemann was led to look for the Riemann function as a function of the single variable

$$z = \frac{(r - \rho)(s - \sigma)}{(r + s)(\rho + \sigma)} \tag{13-58}$$

i.e., he set

$$R(r, s, \rho, \sigma) = \psi(z) \tag{13-59}$$

where z is given by (13-58) and ψ is to be determined so that R satisfies (13-57) (which is self-adjoint) and the usual additional conditions on the Riemann function, namely (cf. Chap. 9),

$$R_s(\rho, s, \rho, \sigma) = 0 \tag{13-60}$$

that is, $R_s = 0$ when $r = \rho$,

$$R_r(r, \sigma, \rho, \sigma) = 0 \tag{13-61}$$

that is, $R_r = 0$ when $s = \sigma$,

$$R(\rho, \sigma, \rho, \sigma) = 1 \tag{13-62}$$

Substituting (13-59) into (13-57), we find that ψ satisfies the ODE

$$z(1 - z)\,\psi'' + (1 - 2z)\,\psi' - \kappa(1 - \kappa)\psi = 0 \tag{13-63}$$

Equation (13-63) is the hypergeometric equation

$$z(1 - z)\,\psi'' + [\gamma - (1 + \beta + \alpha)z]\,\psi' - \alpha\beta\psi = 0 \tag{13-64}$$

with

$$\alpha = 1 - \kappa \qquad \beta = \kappa \qquad \text{and} \qquad \gamma = 1$$

Conditions (13-60) and (13-61) are automatically satisfied by (13-59) if z is given by (13-58). Condition (13-62) requires that

$$\psi(0) = 1 \tag{13-65}$$

The solution of (13-64) which satisfies (13-65) is the hypergeometric function

$$F(\alpha, \beta, \gamma, z) = 1 + \frac{\alpha\beta}{\gamma}\,z + \frac{\alpha(\alpha + 1)\,\beta(\beta + 1)}{\gamma(\gamma + 1)}\,\frac{z^2}{2!} + \cdots \tag{13-66}$$

Hence

$$R(r, s, \rho, \sigma) = F\left[1 - \kappa, \kappa, 1; -\frac{(r - \rho)(s - \sigma)}{(r + s)(\rho + \sigma)}\right] \tag{13-67}$$

In the special case in which κ is half an integer, (13-57) can be solved directly by another method. This method consists in first using u and c

instead of r and s as independent variables. From (13-34) and (13-52) we have

$$\frac{2c}{\gamma - 1} = r + s \tag{13-68}$$

and from (13-35)

$$u = r - s \tag{13-69}$$

Using (13-68) and (13-69) to change variables in (13-54), we obtain

$$t_{uu} = \left(\frac{\gamma - 1}{2}\right)^2 \left(t_{cc} + \frac{2\kappa}{c}\, t_c\right) \tag{13-70}$$

Equation (13-70) is called the *Euler-Poisson-Darboux equation.* If

$$2\kappa = n - 1 \qquad n \text{ an integer} \tag{13-71}$$

then (13-70) reads

$$t_{uu} = \left(\frac{\gamma - 1}{2}\right)^2 \left(t_{cc} + \frac{n - 1}{c}\, t_c\right) \tag{13-72}$$

Equation (13-72) can be looked upon as a wave equation in n space variables for a problem with spherical symmetry. To see this, let

$$c = \sqrt{x_1{}^2 + \cdots + x_n{}^2} \tag{13-73}$$

let

$$\nabla^2 = \sum_{j=1}^{n} \frac{\partial^2}{\partial x_j{}^2}$$

and assume that $t = t(x_1, \ldots, x_n, u)$ is a function of c and u only,

$$t = t(c, u)$$

Then

$$\nabla^2 t = t_{cc} + \frac{n - 1}{c}\, t_c \tag{13-74}$$

and (13-72) becomes

$$t_{uu} = \left(\frac{\gamma - 1}{2}\right)^2 \nabla^2 t \tag{13-75}$$

which is a wave equation with propagation constant $(\gamma - 1)/2$.

If κ is an integer, then (13-54) can be solved by still another method. Let

$$\tau = \frac{1}{r + s}\,(t_r + t_s) \tag{13-76}$$

After a computation we find that

$$\tau_{rs} + \frac{\kappa + 1}{r + s}(\tau_r + \tau_s) = 0 \tag{13-77}$$

Equation (13-77) shows that if we have a solution to (13-54) for any value of κ, then by applying the transformation (13-76) we can obtain a solution for $\kappa + 1$. In particular, for $\kappa = 0$, (13-54) reads

$$t_{rs} = 0$$

Hence

$$t = f(r) + g(s)$$

where f and g are arbitrary. According to (13-76), the general solution to (13-77) with $\kappa = 0$, that is, of (13-54) with $\kappa = 1$, is

$$\tau = \frac{f'(r) + g'(s)}{r + s}$$

Then for $\kappa = 2$ we get

$$\tau = \frac{-2[f'(r) + g'(s)]}{(r + s)^2} + \frac{f''(r) + g''(s)}{r + s}$$

In general, if κ is an integer, the general solution to (13-54) is

$$t(r, s) = \frac{\partial^{\kappa-1}}{\partial r^{\kappa-1}} \frac{f(r)}{(r + s)^\kappa} + \frac{\partial^{\kappa-1}}{\partial s^{\kappa-1}} \frac{g(s)}{(r + s)^\kappa} \tag{13-77a}$$

where $f(r)$ and $g(s)$ are arbitrary functions of their arguments.

Formula (13-77a) is also good if κ is not an integer, provided the derivatives there are interpreted as fractional derivatives in the sense explained in Sec. 19-5. The result is that

$$t(r, s) = \frac{\partial^n}{\partial r^n} \int_0^r \frac{f(z)\, dz}{(r - s)^p (z + s)^p} + \frac{\partial^n}{\partial s^n} \int_0^s \frac{g(z)\, dz}{(r - z)^p (z + s)^p} \tag{13-77b}$$

where $\kappa = n + p$, with n a nonnegative integer and $0 \leqslant p < 1$, if $\kappa \geqslant 0$. If $\kappa < 0$, then (13-77b) holds if n is set equal to zero and $p = \kappa$.

13-7. SIMPLE WAVES

All of the preceding analysis has presupposed that r and s can be used as new independent variables. As promised previously, we now take up the case in which r and s are not independent. This means that there exists a function $F(r, s)$ with

$$F_r^2 + F_s^2 \neq 0 \tag{13-78}$$

such that

$$F(r, s) = 0 \tag{13-79}$$

and therefore

$$F_r\,dr + F_s\,ds = 0 \tag{13-80}$$

Consider (13-80) along a characteristic C^+. On C^+ r is a constant. Hence $dr = 0$. Therefore (13-80) implies that

$$F_s\,ds = 0 \qquad \text{on } C^+ \tag{13-81}$$

Similarly,

$$F_r\,dr = 0 \qquad \text{on } C^- \tag{13-82}$$

If $F_s \neq 0$, then (13-81) implies that not only r but also s is constant on C^+. Similarly, if $F_r \neq 0$, then both r and s are constant on C^-. Hence if both F_r and F_s are different from zero, r and s must be constants in the whole region covered by the characteristics. From (13-34) and (13-35) it then follows that u and ρ are constants everywhere. Hence p is, too, and we have what is called a *constant state*. From the point of view of the physical situation, the term "steady state" would be preferable because constant velocity does not mean the gas is at rest, but the term "constant state" is the one usually used and should not really cause any great confusion.

Clearly no further analysis is necessary for the case of a constant state, i.e., when both F_r and F_s are different from zero. The case in which one of them vanishes is more interesting. According to (13-78), at most one of them can vanish, and so we have only the cases

$$F_r = 0 \qquad F_s \neq 0 \tag{13-83}$$

and

$$F_r \neq 0 \qquad F_s = 0 \tag{13-84}$$

to consider.

Suppose (13-83) holds. Then both r and s are constant along the C^+ characteristics. Hence u and ρ are both constant along the C^+'s. From (13-10) it follows that the slopes of the C^+'s are constant. Hence they are straight lines. They are not in general parallel straight lines, however, because although r and s are constant along a C^+, r is a different constant on each C^+. Hence u and ρ are different constants on each C^+, and so each C^+ has a different slope.

Similarly, the case (13-84) implies that the C^- characteristics are straight lines. A flow in which one family of the characteristics consists of straight lines is called a *simple wave*. It is of interest to observe that *the flow in a region adjacent to a constant state is always a simple wave* (see Fig. 13-1). To see this,

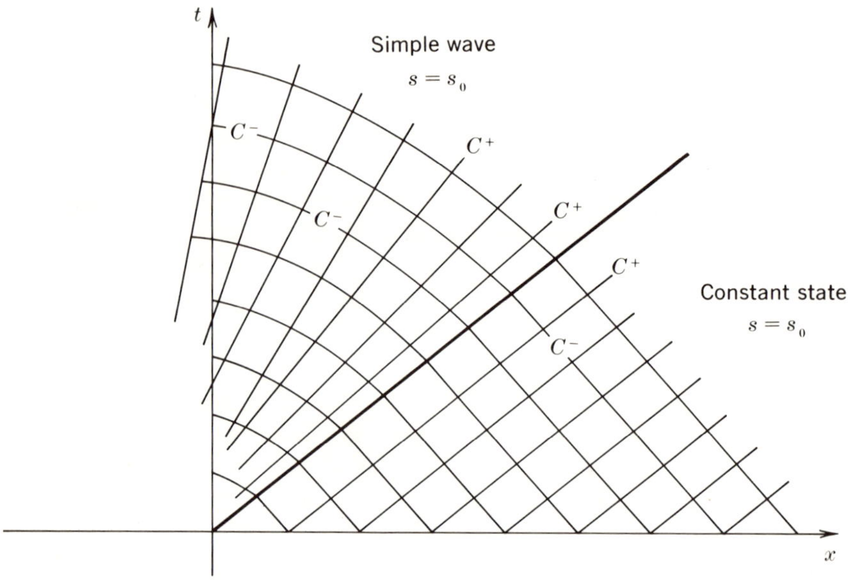

Figure 13-1

suppose that we have a region in which u and ρ are constant and an adjacent region in which u and ρ vary. Then it is not difficult to show that the curve separating the two regions must be a characteristic. Suppose it is a C^+ characteristic. Then the curves which enter the region where u and ρ vary must be C^- characteristics. Along these s is constant. Since s is constant everywhere in the region of constant state, s must be the same constant on all the C^- characteristics. Therefore s is constant in the region adjacent to the region of constant state because the C^- characteristics, so to speak, carry the constant value of s into the adjacent region. Now consider any C^+ characteristic in this adjacent region. On it r is constant because r is constant on every C^+ characteristic. But s is also constant on this C^+ because s is constant everywhere in the region. Hence both r and s are constant on every C^+ in this adjacent region. Therefore u and ρ are also constant on every C^+. Hence c is constant on every C^+ too, and so every C^+ must have constant slope, i.e., is straight. Thus the flow in the adjacent region is a simple wave.

13-8. RAREFACTION WAVES

Suppose we have a piston inserted from the left into a long tube extending to infinity in both directions and containing a gas at rest to the right of the piston. Let the piston be initially at rest at the origin, and let it start moving slowly to the left at time $t = 0$ and continue to move leftward

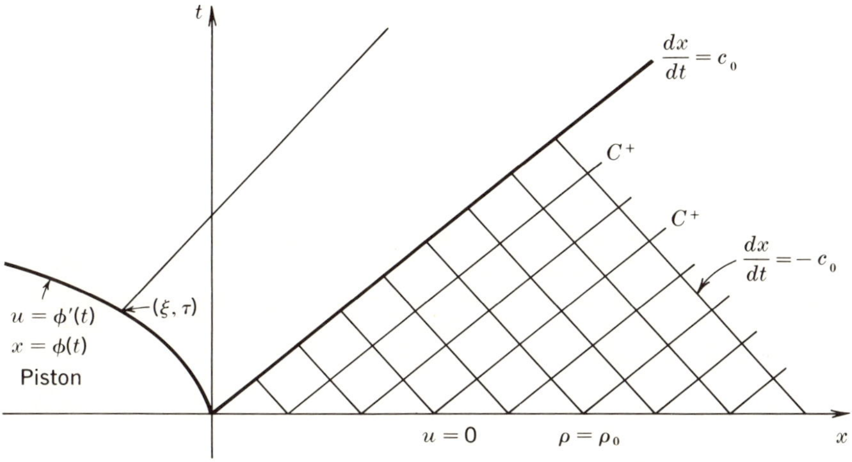

Figure 13-2

with increasing speed, its equation of motion being given by $x = \phi(t)$ with $\phi(0) = \phi'(0) = 0$ and $\phi''(t) < 0$.

Since the gas cannot penetrate the piston, its velocity u at the piston must be the same as that of the piston; that is, $u = \phi'(t)$ on the piston. This single boundary condition on the piston is sufficient because the curve in the xt plane which describes the motion of the piston is timelike.

Since the gas is initially at rest, a constant state with $u = 0$, $\rho = \rho_0$, and $c = c_0$ prevails at $t = 0$ for all $x > 0$. By a uniqueness theorem, it follows that this constant state will prevail everywhere in the st plane to the right of the C^+ characteristic through the origin, i.e., to the right of the characteristic $x = c_0 t$ (cf. Fig. 13-2).

The region to the left of this characteristic will, according to Sec.13-7, be a simple wave. For the time being we assume that this whole region is covered by the characteristics C^+. The case in which this is not true will be examined shortly.

We know that the C^+ characteristics will be straight lines in this region, their slopes being given by $dx/dt = u + c$. Since they are straight lines, dx/dt is constant along each of them. Therefore it suffices to find $u + c$ at any one point on each C^+. We pick the point (ξ, τ) on the piston curve where the particular C^+ intersects the piston curve. Then $u = \phi'(\tau)$. To get c, note that $s =$ const in the region (cf. Sec. 13-7) and therefore [cf. (13-29)]

$$L(\rho) - u = L(\rho_0) - u_0 = L(\rho_0) \qquad \text{because } u_0 = 0$$

Hence

$$L(\rho) = u + L(\rho_0) = \phi'(\tau) + L(\rho_0) \tag{13-85}$$

which determines ρ as a function of τ, the parameter on the piston curve, along C^+. From this we can determine $c = c(\rho)$ as a function of τ, and then we can determine the slope $u + c$ as a function of τ.

Thus, u, ρ, c, and the slope $m = u + c$ are all determined along a given C^+ by the value τ of t where the given C^+ hits the piston curve. If, as we will assume for the present, each C^+ intersects the piston curve only once, then each C^+ determines a unique value of τ. Hence, as already pointed out above, u, ρ, c, and m are all constant on that C^+; τ, however, will vary from one C^+ to another. It is interesting to see how the slope $m(\tau)$ of C^+ varies with τ, that is, as we move along the piston curve. Accordingly, we compute

$$m'(\tau) = u_\tau + c_\tau = \phi''(\tau) + c_\tau \tag{13-86}$$

To find c_τ we differentiate (13-85) with respect to τ, remembering $L' = c/\rho$. This gives

$$\frac{c}{\rho}\,\rho_\tau = \phi''(\tau) \tag{13-87}$$

To express ρ_τ in terms of c_τ we differentiate the definition (13-17) of c with respect to τ. This gives

$$2cc_\tau = f''(\rho)\,\rho_\tau \tag{13-88}$$

Eliminating ρ_τ between (13-87) and (13-88), solving for c_τ, and substituting the result into (13-86), we have, finally,

$$m'(\tau) = \left(1 + \frac{\rho f''}{2c^2}\right)\phi'' \tag{13-89}$$

If we assume that $f'' > 0$, a natural assumption on physical grounds, then (13-89) show us that m' has the same sign as ϕ''. Being careful to remember that we have taken the t axis as the vertical one and the x axis as the horizontal one, we see that this means that the straight C^+ characteristics fan out if the withdrawal of the piston is accelerated, while the C^+'s converge if the withdrawal of the piston is decelerated. If the piston moves with constant velocity, the C^+'s are parallel. A wave in which the characteristics fan out is called a *rarefaction wave*, while one in which they converge is called a *compression wave*.

Let us suppose for the moment that $\phi''(\tau) < 0$ on the initial portion of the piston curve. Then the C^+'s emanating from that portion form a rarefaction wave. Since $u = 0$ initially, (13-89) implies that $u < 0$ in the region of the simple wave. In other words, the gas follows the piston by flowing to the left.

According to (13-85) and (13-26),

$$u = L(\rho) - L(\rho_0) = \int_{\rho_0}^{\rho} \frac{c}{\rho}\, d\rho$$

Since $c/\rho > 0$, $L(\rho)$ decreases as ρ decreases and vice versa. Because of the vice versa, it follows that decreasing, i.e., algebraically decreasing, u means decreasing ρ. However, ρ, being a density, can never become negative. Hence if u decreases to the value u_∞ given by

$$u_\infty = \int_{\rho_0}^{0} \frac{c}{\rho}\, d\rho = -\int_{0}^{\rho_0} \frac{c}{\rho}\, d\rho$$

then any further decrease in u would result in negative values of ρ. Therefore, if the gas once reaches the state in which $u = u_\infty$ and $\rho = 0$, it remains in that state and the motion of the piston has no further influence on it as long as the velocity of the piston is not reduced. Since $u = \phi'(\tau)$, this means that once the velocity $\phi'(\tau)$ of the piston has reached the critical value $\phi'(\tau_0) = u_\infty$, the velocity u of the gas maintains the constant velocity u_∞ while the piston's speed continues to increase. In other words, the gas has become detached from the piston and continues to follow after the piston at the fastest velocity at which it (the gas) is physically capable of going, namely, u_∞. The piston outdistances the gas from the moment its velocity reaches u_∞ onward. For obvious reasons, the critical velocity u_∞ is called the *escape velocity*.

Now suppose that the piston is slowed down after a while. Then there will be a portion of the piston curve on which $\phi''(\tau) > 0$; the C^+ characteristics emanating from that portion will converge, and the resulting wave will be a compression wave. Since they are straight lines, any two C^+'s in such a compression wave will eventually intersect. At their point of intersection they will carry different values of τ and therefore give conflicting information on the values of the flow quantities at that point. This means that some sort of singularity must occur eventually in a compression wave of this kind. We shall treat this difficulty in the next chapter. Here we merely give a heuristic argument to explain what happens.

We have seen many times in past chapters that characteristics can be interpreted as wavefronts. In the present case of gas flow, the waves move with varying speeds; i.e., the characteristics have different slopes. In a rarefaction wave, the waves emitted at later times move more slowly than those emitted earlier, and the flow proceeds smoothly. In a compression wave, however, the earlier waves travel more rapidly than the later ones and therefore overtake the later ones. This piling up of waves produces a discontinuity in the flow. For a more precise analysis, see Chap. 14.

13-9. APPLICATION TO A NONLINEAR WAVE EQUATION

The methods of the present chapter can be used to solve a nonlinear wave equation of the form

$$w_{tt} - c^2(w_x)\, w_{xx} = 0 \tag{13-90}$$

where, as indicated, the velocity c is now permitted to depend upon w_x.

To reduce (13-90) to a system, put

$$w_t = u \qquad \text{and} \qquad w_x = v \tag{13-91}$$

Then (13-90) reads $u_t - c^2(v)\, v_x = 0$. Since $w_{tx} = w_{xt}$, we also have $u_x = v_t$. Thus we get the system

$$\begin{aligned} u_x - v_t &= 0 \\ u_t - c^2(v)\, v_x &= 0 \end{aligned} \tag{13-92}$$

Rearranging terms and writing (13-92) in matrix form, we have

$$\begin{pmatrix} 1 & 0 \\ 0 & -c^2(v) \end{pmatrix}\begin{pmatrix} u_x \\ v_x \end{pmatrix} + \begin{pmatrix} 0 & -1 \\ 1 & 0 \end{pmatrix}\begin{pmatrix} u_t \\ v_t \end{pmatrix} = \begin{pmatrix} 0 \\ 0 \end{pmatrix} \tag{13-93}$$

According to (12-10), the characteristic equation for the eigenvalues μ is therefore

$$\begin{vmatrix} 1 & \mu \\ -\mu & -c^2(v) \end{vmatrix} = 0 \tag{13-94}$$

Hence

$$\mu = \pm c(v) \tag{13-95}$$

The corresponding eigenvectors are $(c \mp 1)$. In keeping with our previous notation, we call the characteristics with $\mu = +c$ the C^+ characteristics and those with $\mu = -c$ the C^- characteristics.

As usual, we use the eigenvectors to write the system in canonical form. This means that we multiply the first equation of (13-92) by c and the second by -1 and add. Then we multiply the first equation of (13-92) by c and the second by $+1$ and add. The result is the following pair of PDEs:

$$\begin{aligned} u_t - cu_x + c(v_t - cv_x) &= 0 \\ u_t + cu_x - c(v_t + cv_x) &= 0 \end{aligned} \tag{13-96}$$

Equations (13-96) are the canonical form of the hyperbolic system (13-92). As the theory predicts, u and v are differentiated in the same direction in a given equation of the canonical system. In the first equation of (13-96) both u and v are differentiated in the direction with direction numbers 1 and $-c$, that is, along a C^+. In the second equation of (13-96) both u and v are differentiated in the direction of a C^-. Thus,

$$du \pm c\,dv = 0 \qquad \text{on } C^\pm \tag{13-97}$$

Since $c = c(v)$, we can introduce

$$M(v) = \int_{v_0}^{v} c(z)\,dz \tag{13-98}$$

and then write

$$c\,dv = dM(v) \tag{13-99}$$

so that (13-97) can be written

$$d[u \pm M(v)] = 0 \qquad \text{on } C^\pm \tag{13-100}$$

Equation (13-100) enables us to introduce the Riemann invariants r and s as before

$$\begin{aligned} r &= u + M(v) \\ s &= u - M(v) \end{aligned} \tag{13-101}$$

The analysis can now be continued along the lines of Sec. 13-4. The reader wishing the details should consult Zablusky [301].

A further example of the applicability of Riemann's method is provided by a pair of papers by Carrier and Greenspan [33] and by Greenspan [120]. In [33] the authors reduce the nonlinear PDEs of shallow water theory to a single hyperbolic equation of second order and thereby show that there are waves that climb a beach without breaking.

SUMMARY

One-dimensional anisotropic gas flow can be described by four functions, $u(x, t)$, $\rho(x, t)$, $p(x, t)$, and $S(x, t)$, representing respectively the velocity, density, pressure, and entropy of the gas. These four functions satisfy the four equations

$$\begin{aligned} &\rho_t + (\rho u)_x = 0 \\ &u_t + uu_x + \frac{1}{\rho} p_x = 0 \\ &S_t + uS_x = 0 \\ &p = f(\rho, S) \end{aligned} \tag{S13-1}$$

In the fourth of these equations, which is called the equation of state, $f(\rho, S)$ is a given function, with $f_\rho > 0$ and $f_{\rho\rho} > 0$.

If we introduce the *sound speed*, $c^2(\rho, S) = f'(\rho, S)$, then we can write (S13-1) in the form

$$\begin{aligned} p_t + up_x + c^2\rho u_x &= 0 \\ u_t + uu_x + \frac{1}{\rho} p_x &= 0 \\ S_t + uS_x &= 0 \end{aligned} \tag{S13-2}$$

in which form, since $c^2(\rho, S)$ is a known function of ρ and S, they represent a system of three quasi-linear PDEs for the three unknown functions $u(x, t)$, $p(x, t)$, and $S(x, t)$.

The system (S13-2) is hyperbolic. Being of third order, it has three families of characteristics. One of these families has slope, (dx/dt), equal to u; the other two have slopes $u \pm c$. The third equation of (S13-2) is already in canonical form. When the other two have been reduced to canonical form, they read

$$p_t + (u \pm c)\, p_x \pm c\rho[u_t + (u \pm c)\, u_x] = 0$$

The equations of anisotropic flow cannot be solved in general. Quite a bit can be done, however, with those of isentropic flow. Isentropic flow means that f is independent of S. Then $c^2 = f'(\rho)$, and it is easiest to reduce the equations to the pair of quasi-linear PDEs

$$\begin{aligned} \rho_t + \rho u_x + u\rho_x &= 0 \\ u_t + uu_x + \frac{c^2}{\rho} \rho_x &= 0 \end{aligned} \tag{S13-3}$$

for $u(x, t)$ and $\rho(x, t)$. In (S13-3) $c^2(\rho)$ is a known function of ρ.

The characteristics of (S13-3) have slopes, that is, dx/dt, equal to $u \pm c$. When reduced to canonical form, Eqs. (S13-3) become

$$[u_t + (u \pm c)\, u_x] \pm \frac{c}{\rho} [\rho_t + (u \pm c)\, \rho_x] = 0$$

The problem of solving the first-order quasi-linear system (S13-3) can be reduced to that of solving the second-order linear hyperbolic equation

$$t_{rs} + [\Phi(r + s)](t_r + t_s) = 0 \tag{S13-4}$$

for $t(r, s)$, where

$$\Phi(z) = (2c)^{-1} \left(1 + \frac{\rho}{c} \frac{dc}{d\rho}\right) \tag{S13-5}$$

is a function of the variable $z = r + s$ because, as will be stated precisely below, ρ is [see (S13-8)].

If we have found a function $t(r, s)$ satisfying (S13-4), then we can find a function $x(r, s)$ satisfying

$$x_r = (u - c)\, t_r$$
$$x_s = (u + c)\, t_s$$

Here again c is a function of $r + s$ because ρ is.

Having found $x(r, s)$ and $t(r, s)$, we then invert this pair of functions to find r and s as functions of x and t. Then the solution $u(x, t)$ of (S13-3) is obtained immediately from

$$u = r - s \qquad \text{(S13-6)}$$

To get $\rho(x, t)$, we first define

$$L(\rho) = \int_{\rho_0}^{\rho} \frac{c(\hat{\rho})}{\hat{\rho}}\, d\hat{\rho} \qquad \text{(S13-7)}$$

Then $\rho(x, t)$ is obtained by solving

$$L(\rho) = r + s = r(x, t) + s(x, t) \qquad \text{(S13-8)}$$

for ρ.

If the initial conditions (ICs) $u(x, 0) = \alpha(x)$ and $\rho(x, 0) = \beta(x)$ are appended to the problem (S13-3), the corresponding ICs for (S13-4) are that $t = 0$ on the curve Γ in the rs plane whose parametric equations are

$$\Gamma\colon \begin{cases} r = \frac{1}{2}\{L[\beta(x)] + \alpha(x)\} \\ s = \frac{1}{2}\{L[\beta(x)] - \alpha(x)\} \end{cases}$$

and

$$t_r = -c^{-1}\{L'[\beta(x)]\, \beta'(x) + \alpha(x)\}^{-1} \qquad \text{on } \Gamma$$
$$t_s = \ldots\ldots\ldots\ldots\ldots\ldots\ldots\ldots$$

Since (S13-4) is still too complicated for an arbitrary function Φ, a further simplifying assumption is made. The gas is called *polytropic* if its equation of state is of the form

$$p = A\rho^{\gamma} \qquad \text{(S13-9)}$$

where A and γ are constants.

With the equation of state (S13-9), i.e., for a polytropic gas, $\Phi(z)$ reduces to $\Phi(z) = \kappa/z$, where κ is a constant. Equation (S13-4) then reduces to

$$t_{rs} + \frac{\kappa}{r + s}(t_r + t_s) = 0 \qquad \text{(S13-10)}$$

and $L(\rho)$ to

$$L(\rho) = \frac{2c}{\gamma - 1} \tag{S13-11}$$

The substitution

$$t(r, s) = \frac{\omega(r, s)}{(r + s)^2}$$

converts (S13-10) into

$$\omega_{rs} + \frac{\kappa(1 - \kappa)}{(r + s)^2}\,\omega = 0$$

which has the Riemann function

$$R(r, s, \xi, \eta) = F\left[1 - \kappa, \kappa, 1; -\frac{(r - \xi)(s - \eta)}{(r + \xi)(\xi + \eta)}\right]$$

where $F(\alpha, \beta, \gamma; z)$ is the hypergeometric function, i.e., the solution to the ODE

$$z(1 - z)\,y'' + [\gamma - (1 + \beta + \alpha)\,z]\,y' - \alpha\beta y = 0$$

If we use u and c instead of r and s as independent variables, we can throw (S13-10) into the form

$$t_{uu} = \left(\frac{\gamma - 1}{2}\right)^2 \left(t_{cc} + \frac{2\kappa}{c}\,t_c\right) \tag{S13-12}$$

which is a Euler-Poisson-Darboux equation. In particular, if 2κ is an integer, say $n - 1$, then (S13-12) is a spherically symmetric wave equation in n space.

If κ itself is an integer, the general solution to (S13-10) is

$$t(r, s) = \frac{\partial^{\kappa-1}}{\partial r^{\kappa-1}}\,\frac{f(r)}{(r + s)^\kappa} + \frac{\partial^{\kappa-1}}{\partial s^{\kappa-1}}\,\frac{g(s)}{(r + s)^\kappa} \tag{S13-13}$$

where f and g are arbitrary functions.

If κ is not an integer, the general solution to (S13-10) is

$$t(r, s) = \frac{\partial^n}{\partial r^n}\int_0^r \frac{f(z)\,dz}{(r - z)^p\,(z + s)^p} + \frac{\partial^n}{\partial s^n}\int_0^s \frac{g(z)\,dz}{(r - z)^p\,(z + s)^p} \tag{S13-14}$$

when $\kappa \geqslant 0$ and $\kappa = n + p$, with n a nonnegative integer and $0 \leqslant p \leqslant 1$, and by (S13-14) with n set equal to zero and p replaced by κ if $\kappa < 0$.

14
Shocks

14-1. INTRODUCTION

In Chap. 13 we pointed out that whereas a rarefaction wave can continue indefinitely in a gas, a compression wave cannot. More precisely, if the piston considered in Chap. 13 slows down, then $\phi''(\tau) > 0$ and the C^+ characteristics will converge and eventually intersect each other. Such an intersection of two characteristics of the same family spells disaster for the solutions of the PDEs at that point. For example, suppose that the characteristics C_1^+ and C_2^+, emanating from the respective points τ_1 and τ_2 on the piston curve, intersect each other at some point D. Then, since u is constant on C_1^+, $u(D) = \phi'(\tau_1)$. On the other hand, since u is constant on C_2^+, $u(D) = \phi'(\tau_2)$, we get a contradiction.

This contradiction is not illusory. It is real, and it means that the solution to an initial-value problem (IVP) for a nonlinear partial differential equation (PDE) cannot always be extended indefinitely. If the initial values are of a certain nature, then the solution cannot exist past a certain point as a continuous function. Nevertheless, the physical problem from which the

PDEs in Chap. 13 arose must have a solution. After all, it is perfectly reasonable to slow down the piston after a while. Clearly the gas will react in some reasonable way. Therefore something must be lacking in our mathematical description of the problem.

What is lacking is the inclusion of heat conduction and viscosity in the problem. The equations of Chap. 13 neglect these two phenomena. If they are included in the analysis, then a set of equations much more complicated than those of the previous chapter can be derived. These equations are so complicated, however, that it usually pays to employ another method of attack. Instead of working with those equations of gas dynamics which include heat conduction and viscosity and seeking continuous solutions of them, we shall continue to employ the equations of Chap. 13 but now, perforce, permit their solutions to have discontinuities. Of course, we cannot permit a plethora of discontinuities. There has to be some physical motivation for the mathematical discontinuities we permit. The gas itself will not display any discontinuities. The discontinuities in the mathematical solution are to be looked at as idealizations of narrow regions in the gas in which the relevant physical quantities change very rapidly. For example, instead of requiring ρ to change from ρ_1 to ρ_2 in a region $x_1 < x < x_2$ in a time $t_1 < t < t_2$, we assign ρ the jump $\rho_2 - \rho_1$ at some point (x, t) in the indicated intervals. As in many similar situations, the problem becomes easier to deal with mathematically if we use sharp transitions, i.e., jumps, instead of gradual ones.

14-2. DISCONTINUITY CONDITIONS FOR SHOCK WAVES

The physical principles from which the sought-after discontinuity conditions will be derived will now be given. Let $\alpha_0(t)$ and $\alpha_1(t)$ be the coordinates of the end points of a "slab" of gas always consisting of the same particles. Then we shall require:

1. *Conservation of mass*:

$$\frac{d}{dt}\int_{\alpha_0(t)}^{\alpha_1(t)} \rho\, dx = 0 \tag{14-1}$$

2. *Conservation of momentum* (we assume only pressure forces act):

$$\frac{d}{dt}\int_{\alpha_0(t)}^{\alpha_1(t)} \rho u\, dx = p(\alpha_0, t) - p(\alpha_1, t) \tag{14-2}$$

3. *Conservation of energy* (the right-hand side below represents the work done by the pressure in compressing the ends of our slab of gas):

$$\frac{d}{dt}\int_{\alpha_0(t)}^{\alpha_1(t)} \rho(\tfrac{1}{2}u^2 + e)\, dx = p(\alpha_0, t)\, u(\alpha_0, t) - p(\alpha_1, t)\, u(\alpha_1, t) \tag{14-3}$$

The quantity e here is the internal energy of the gas.

4. *Nondecrease in entropy*:

$$\frac{d}{dt}\int_{\alpha_0(t)}^{\alpha_1(t)} \rho S\,dx \leqslant 0 \tag{14-4}$$

We will assume that the solutions to our problems always obey Eqs. (14-1) to (14-4) even if those solutions possess discontinuities somewhere. At any point where the solutions are continuous we can show, as we did in some special cases in Sec. 2-1, that (14-1) to (14-4) give rise to the differential equations (DEs) of Chap. 13. The question is what conditions do they give rise to at a discontinuity of the solutions? Let us suppose that there is a certain curve $\mathscr{S}$ in the xt plane given by $x = q(t)$ across which u, ρ, p, etc., may have jumps. We wish to use Eqs. (14-1) to (14-4) to find connections among the jumps across $\mathscr{S}$.

All four of our conditions involve an expression of the form dQ/dt, where

$$Q = \int_{\alpha_0(t)}^{\alpha_1(t)} \Psi(x, t)\,dx \tag{14-5}$$

and where Ψ varies with the particular condition considered. Since $x = q(t)$ is a discontinuity of Ψ, we first write

$$Q = \int_{\alpha_0(t)}^{q(t)} \Psi(x, t)\,dx + \int_{q(t)}^{\alpha_1(t)} \Psi(x, t)\,dx$$

Let

$$\Psi_0 = \lim_{x\to q-} \Psi(x, t) \qquad \Psi_1 = \lim_{x\to q+} \Psi(x, t) \tag{14-6}$$

$$\frac{dq}{dt} = U(t) = \text{velocity of the discontinuity surface } \mathscr{S} \tag{14-7}$$

and

$$u_j = \frac{d\alpha_j}{dt} = \text{velocity of the particle } \alpha_j \qquad j = 0, 1 \tag{14-8}$$

Then

$$\frac{dQ}{dt} = \Psi_0 U - \Psi(\alpha_0, t)\,u_0 + \int_{\alpha_0(t)}^{q(t)} \Psi_t\,dx + \Psi(\alpha_1, t)\,u_1 - \Psi_1 U + \int_{q(t)}^{\alpha_1(t)} \Psi_t\,dx$$

or

$$\frac{dQ}{dt} = \Psi_0 U - \Psi(\alpha_0, t)\, u_0 + \Psi(\alpha_1, t)\, u_1 - \Psi_1 U + \int_{\alpha_0(t)}^{\alpha_1(t)} \Psi_t \, dx \tag{14-9}$$

Now let $\alpha_1 \to \alpha_0$, keeping $\alpha_0 < q < \alpha_1$. Then (14-9) yields

$$\frac{dQ}{dt} = \Psi_0(U - u_0) + \Psi_1(u_1 - U)$$

or

$$\frac{dQ}{dt} = (u_1 - U)\,\Psi_1 - (u_0 - U)\,\Psi_0 \tag{14-10}$$

If we let

$$v_j = u_j - U \qquad j = 0, 1 \tag{14-11}$$

be the relative velocity of the flow with respect to the discontinuity surface $\mathscr{S}$ then (14-10) can be further abbreviated to

$$\frac{dQ}{dt} = v_1 \Psi_1 - v_0 \Psi_0 \tag{14-12}$$

Applying (14-12) to the conservation laws (14-1) to (14-4), we have:

1. Conservation of mass:

$$\rho_0 v_0 = \rho_1 v_1 = m = \text{mass flux through } \mathscr{S} \tag{14-13}$$

[m is *defined* by (14-13)].

2. Conservation of momentum:

$$(\rho_1 u_1)\, v_1 - (\rho_0 u_0)\, v_0 = p_0 - p_1 \tag{14-14}$$

or, by (14-13),

$$m u_0 + p_0 = m u_1 + p_1 \tag{14-15}$$

Equation (14-15) can also be written in terms of the relative velocities v_j as

$$\rho_0 v_0^2 + p_0 = \rho_1 v_1^2 + p_1 = p \tag{14-16}$$

where p, defined by (14-16), is the total momentum flux.

3. Conservation of energy:

$$\rho_1(e_1 + \tfrac{1}{2}u_1^2) - \rho_0(e_0 + \tfrac{1}{2}u_0^2) = p_0u_0 - p_1u_1 \tag{14-17}$$

or

$$m(e_0 + \tfrac{1}{2}u_0^2) + u_0p_0 = m(e_1 + \tfrac{1}{2}u_1^2) + u_1p_1 \tag{14-18}$$

If we let

$$\tau = 1/\rho \tag{14-19}$$

be the *specific volume* and use some of the previous results, we can write this as

$$m(e_0 + \tfrac{1}{2}v_0^2 + p_0\tau_0) = m(e_1 + \tfrac{1}{2}v_1^2 + p_1\tau_1) \tag{14-20}$$

(We leave the m in because it might be zero.)

4. Nondecreasing entropy:

$$\rho_1S_1v_1 \geqslant \rho_0S_0v_0 \tag{14-21}$$

or, by (14-13),

$$mS_0 \leqslant mS_1 \tag{14-22}$$

To summarize, we have found the following discontinuity conditions:

$$\begin{aligned} &1.\quad \rho_0v_0 = \rho_1v_1 \\ &2.\quad \rho_0v_0^2 + p_0 = \rho_1v_1^2 + p_1 \\ &3.\quad m(e_0 + \tfrac{1}{2}v_0^2 + p_0\tau_0) = m(e_1 + \tfrac{1}{2}v_1^2 = p_1\tau_1) \\ &4.\quad mS_0 \leqslant mS_1 \end{aligned} \tag{14-23}$$

Here

$$v_j = u_j - U \qquad j = 0, 1$$

and

$$m = \rho_0v_0 = \rho_1v_1$$

If $m = 0$, no gas crosses the discontinuity surface. The discontinuity surface is then called a *contact discontinuity*. If $m \neq 0$, the gas flows through the discontinuity surface, which is then called a *shock*. Note that if $m = 0$, then $v_0 = v_1 = 0$ (because $\rho \neq 0$). Hence $u_0 = u_1$. Moreover, the second of (14-23) then implies $p_0 = p_1$. Thus, both the velocity u and the pressure p are continuous across a contact discontinuity.

It was mentioned previously that the discontinuity conditions are, in part, replacements for the more complicated differential equations which would result if heat conduction and viscosity were included. It is therefore satisfying to know that under proper assumptions the discontinuity conditions can be derived from the more complicated differential equations by letting the coefficients of heat conduction and viscosity in them go to zero. For details the reader is referred to Courant and Friedrichs [64, pp. 134–138].

In the case of a shock, $m \neq 0$. If we introduce the *enthalpy* j, defined by

$$j = e + p\tau \tag{14-24}$$

where τ, as before, is the specific volume defined by $\tau = 1/\rho$, then the third shock condition of (14-23) can be written in the form

$$\tfrac{1}{2}v_0^2 + j_0 = \tfrac{1}{2}v_1^2 + j_1$$

The first three of the shock conditions of (14-23) represent three equations connecting seven quantities, namely, the velocities, densities, and pressures on either side of the shock and the velocity U of the shock. If four properly chosen ones of these seven quantities are known, then the shock conditions should determine the other three. In general the state u_0, ρ_0, p_0, on one side of the shock will be known. If in addition we know one more quantity, such as the velocity, or the pressure, or the density on the other side, or else the velocity of the shock itself, we can usually use the shock conditions to determine the unknown quantities. We will give a specific example shortly.

14-3. THE HUGONIOT FUNCTION

It is frequently advantageous to replace some of the shock conditions 1 to 3 by other ones derived from them by algebraic manipulations. For example, by eliminating v_0 and v_1 from them, one can obtain

$$(\tau_0 - \tau_1)\frac{p_0 + p_1}{2} = e_1 - e_0 \tag{14-25}$$

The internal energy e can be considered to be a function of τ and p. If we think of p_0 as fixed and $p_1 = p$ as variable and τ_0 as fixed and $\tau_1 = \tau$ as variable, and if we put

$$H(\tau, p) = e(\tau, p) - e(\tau_0, p_0) + (\tau - \tau_0)\frac{p + p_0}{2} \tag{14-26}$$

then (14-25) says

$$H(\tau, p) = 0 \tag{14-27}$$

The function $H(\tau, p)$, defined by (14-26), is called the *Hugoniot function.* The curve in the τp plane defined by (14-27) is called the *Hugoniot curve.* It gives the relationship between the values of τ and the values of p on the same side of a shock which are compatible with the shock conditions if the values of τ and p on the other side of the shock are already known.

In the case of a polytropic gas [cf. (13-50)] the internal energy e turns out to be given by

$$e = \frac{p\tau}{\gamma - 1} \tag{14-28}$$

where γ is the exponent occurring in the equation of state (13-50). If we let

$$\mu^2 = \frac{\gamma - 1}{\gamma + 1} \tag{14-29}$$

{so that $\mu^2 = \kappa/2$ [cf. the definition of κ below Eq. (13-53)]} and use (14-28) and (14-29) in (14-26), we find that in the polytropic case

$$2\mu^2 H(\tau, p) = (\tau - \mu^2\tau_0)\, p - (\tau_0 - \mu^2\tau)\, p_0 \tag{14-30}$$

The equation $H(\tau_1, p_1) = 0$ then implies

$$\frac{p_1}{p_0} = \frac{\tau_0 - \mu^2\tau_1}{\tau_1 - \mu^2\tau_0} = \frac{\rho_1 - \mu^2\rho_0}{\rho_0 - \mu^2\rho_1} \tag{14-31}$$

the second equality following from the definition of τ as $\tau = 1/\rho$.

By further manipulation, we can write (14-31) in the form

$$\frac{\tau_0}{\tau_1} = \frac{\rho_1}{\rho_0} = \frac{p_1 + \mu^2 p_0}{p_0 + \mu^2 p_1} \tag{14-32}$$

For a polytropic gas [cf. (13-50) and (13-51)]

$$\gamma p = \rho c^2 \tag{14-33}$$

Hence (14-24), (14-28), and (14-29) tell us that

$$j = \frac{\gamma}{\gamma - 1}\frac{p}{\rho} = \frac{1 - \mu^2}{2\mu^2}\, c^2 \tag{14-34}$$

Therefore (14-24) can be written in the form

$$\mu^2 v_0^2 + (1 - \mu^2)\, c_0^2 = \mu^2 v_1^2 + (1 - \mu^2)\, c_1^2 = c_*^2 \tag{14-35}$$

where c_* is defined by (14-35).

By using (14-11), (14-33), (14-35), and some of the original shock conditions, Prandtl deduced the relationship

$$v_0 v_1 = c_*^2 \tag{14-36}$$

Using (14-11), (14-35), and (14-36), it is then possible to show that

$$(1 - \mu^2)(U - u_0)^2 - (u_1 - u_0) = (1 - \mu^2)\, c_0^2 \tag{14-37}$$

EXAMPLE As an example of a problem in which a shock wave occurs, consider an infinite tube of a polytropic gas with a piston as in Chap. 13, only this time let the piston move to the right, i.e., into the gas instead of away from it. For simplicity, we suppose that the piston moves with constant velocity u_p.

Initially, we will have $u = 0$, $p = p_0$, and $\rho = \rho_0$, in other words, a constant state. If the flow were continuous, the flow adjacent to this constant state would have to be a simple wave, all of whose straight characteristics emanate from the origin. Since the characteristics fan out, the wave would be a rarefaction wave through which, as we have seen previously, the velocity of the gas decreases. The velocity starts out at zero, however, and is supposed to wind up at the positive velocity of the piston. Since this is clearly impossible in a rarefaction wave, no continuous solution to the problem can exist. Hence we look for one containing a shock.

More precisely, we seek a solution which takes the values $u = 0$, $p = p_0$, and $\rho = \rho_0$ for $x > Ut$ and the unknown values u_1, p_1, ρ_1 for $x < Ut$. The unknown values u_1, p_1, ρ_1 and the unknown shock velocity U are to be determined from u_0, p_0, ρ_0 by means of the shock conditions.

There is one fly in the ointment, however. As pointed out previously, we need to know four of seven quantities involved in the shock conditions, not just three. It is therefore necessary to determine one of the four remaining unknowns. We do this by making the plausible assumption that the gas behind the shock front moves with the piston velocity u_p. Note that the very nature of the shock as a discontinuity line in the gas automatically entails a greater velocity for the shock than for the piston. Hence the shock moves into the quiet gas at a more rapid rate than the piston. What we are assuming, therefore, is that the gas behind the shock front follows after the shock with the velocity of the piston.

If $u_0 = 0$ and $u_1 = u_p$, then (14-37) immediately determines the speed U of the shock as

$$U = \frac{u_p}{2(1-\mu^2)} + \sqrt{c_0^2 + \frac{1}{4}\left(\frac{u_p}{1-\mu^2}\right)^2} \tag{14-38}$$

Formula (14-38) for U shows that the speed U of the shock is greater than the sound speed c_0; that is, the shock moves at supersonic speed into the quiet gas. Notice that the speed of the shock is always supersonic, regardless of how slow the velocity of the piston is. It also shows that $U > u_p/(1 - \mu^2)$. For example, for air, $\gamma = 1.4$, and therefore $\mu^2 = \frac{1}{6}$. Hence in air the shock moves over 20 percent faster than the piston.

Having u_0, p_0, ρ_0, u_1, and U, we can use the various shock relations to find p_1 and ρ_1. We omit the details.

14-4. REFLECTION OF A SHOCK FROM A RIGID WALL

Suppose, instead of extending to infinity, our tube is closed at one end, $x = l$. What will happen when the shock wave hits this end? Naturally, we expect that it will be reflected. We assume that the shock hits the closed end of the tube at time $t = 0$, that the incoming shock has velocity U_+ and that the reflected shock has velocity U_-. Thus we assume that we have an initial zone of quiet for $x > l + U_+t$ and $t < 0$ ahead of the incoming shock and another zone of quiet for $x > l + U_-t$ and $t > 0$ behind the receding shock. More precisely, for $x > l + U_+ t$ and $t < 0$ we have $u_0 = 0$, $p = p_0$, and $\rho = \rho_0$, where p_0 and ρ_0 are known. For the region between the two shocks (see Fig. 14-1) we will assume that the velocity u_1 is known because, as before, it should equal the piston velocity. The pressure p_1 and density ρ_1 are unknown. Behind the reflected shock, we assume $u_2 = 0$ and p_2 and ρ_2 are unknown. In addition, the two shock velocities $U_\pm$ are unknown.

In order to solve this problem we use the middle zone as the 0 zone and the zones 1 and 2 in turn as the 1 zone in formula (14-37). Then (14-37) gives the same equation for both U_+ and U_-, U_+ being the positive root of the resulting quadratic equation and U_- the negative root. Thus, U_+ and U_- are the two roots of

$$(U - u_1)^2 + \frac{(U - u_1)\, u_1}{1 - \mu^2} - c_1^2 = 0 \tag{14-39}$$

In order to obtain some interesting information from (14-39) we introduce the *Mach numbers* $M_\pm$ defined by

$$M_\pm = \frac{u_1 - U_\pm}{c_1} \tag{14-40}$$

Note that $M_+ < 0$ and $M_- > 0$.

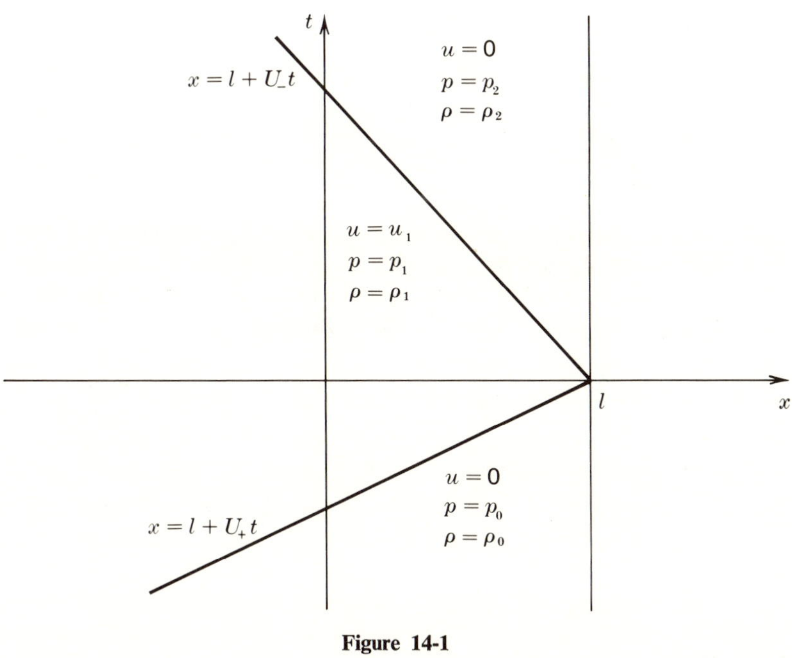

Figure 14-1

In terms of the M's, (14-39) reads

$$M^2 - \frac{Mu_1}{c_1(1 - \mu^2)} - 1 = 0 \tag{14-41}$$

Hence

$$M_+M_- = -1 \tag{14-42}$$

By manipulating the shock conditions it is possible to show that

$$\frac{p_1}{p_0} = (1 + \mu^2)\, M^2 - \mu^2 \tag{14-43}$$

Therefore in the present case

$$\frac{p_0}{p_1} = (1 + \mu^2)\, M_+{}^2 - \mu^2 \tag{14-44}$$

and

$$\frac{p_2}{p_1} = (1 + \mu^2)\, M_-{}^2 - \mu^2 \tag{14-45}$$

From (14-44) and (14-45) we obtain the ratio of the excess pressure in the reflected shock wave to that of the incoming one as

$$\frac{p_2 - p_1}{p_1 - p_0} = 1 + \frac{1 + \mu^2}{p_0/p_1 + \mu^2} \tag{14-46}$$

If the shock is strong, i.e., if p_1/p_0 is large, then p_0/p_1 is small. Neglecting it in (14-46), we find

$$\frac{p_2 - p_1}{p_1 - p_0} \approx + \frac{1}{\mu^2}$$

For $\gamma = 1.4$ (air) this yields about 8, for $\gamma = 1.2$ it gives 13, and for $\gamma = 1.1$, 23. In linear wave propagation, the reflected wave is merely doubled in intensity. This shows that the nonlinear character of shocks makes their impact on rigid walls considerably more devastating than the impact of ordinary sound waves.

15
Overdetermined Systems

15-1. INTRODUCTION

Although it is unusual for systems of equations in which the number of equations outnumbers the number of unknowns to arise in applied mathematics, it turns out that some of the concepts occurring in the treatment of such systems of partial differential equations (PDEs) are of use in some branches of applied mathematics, particularly in thermodynamics and theoretical mechanics. Therefore we shall discuss sych systems here, confining ourselves, however, to the case of systems for a single unknown function.

As an introduction to the subject, consider the system of two PDEs

$$\begin{aligned} u_x(x, y) &= f(x, y) \\ u_y(x, y) &= g(x, y) \end{aligned} \tag{15-1}$$

for the single unknown function $u(x, y)$. $f(x, y)$ and $g(x, y)$ are given functions. The system (15-1) is obviously solvable if and only if

$$f_y(x, y) = g_x(x, y) \tag{15-2}$$

If (15-2) is satisfied, then the solution to (15-1) exists and is determined to within an arbitrary constant. There are no arbitrary functions in it. If (15-2) is not satisfied, then (15-1) has no solution.

The fact that a system of equations containing more equations than unknowns is in general overdetermined is, of course, no surprise. It is interesting, however, that when we deal with such systems of PDEs, there still may be solutions even if all the equations of the system are independent. For example, the homogeneous system

$$\begin{aligned} u_x(x, y) &= 0 \\ u_y(x, y) &= 0 \end{aligned} \tag{15-3}$$

always has a solution, namely, $u(x, y) = \text{const}$. Of course from the point of view of PDEs, this is still a trivial solution, but consider the system

$$\begin{aligned} u_x(x, y, z) &= 0 \\ u_y(x, y, z) &= 0 \end{aligned} \tag{15-4}$$

of two equations for the single function $u(x, y, z)$. Its solution is $u(x, y, z) = \phi(z)$, where $\phi(z)$ is an arbitrary function of z [which, if you like, need not even be differentiable in order to satisfy (15-4)].

The nonhomogeneous form of (15-4), namely,

$$\begin{aligned} u_x(x, y, z) &= f(x, y, z) \\ u_y(x, y, z) &= g(x, y, z) \end{aligned} \tag{15-5}$$

where f and g are given differentiable functions, is solvable if and only if

$$f_y - g_x = 0 \tag{15-6}$$

and, if (15-6) is satisfied, then the general solution of (15-5) depends upon the arbitrary function of z that is the general solution of the homogeneous equation.

The extra freedom in the solution of the system (15-5), in contrast to that of (15-3), is obviously due to the extra independent variable in the system (15-5). Thus, in systems of PDEs, the property of having more *in*dependent variables than the number of equations, while not as good as having more dependent variables, may nevertheless be of some avail. We reiterate this point: in systems of PDEs, the number of *in*dependent variables as well as the number of dependent variables is important.

15-2. LINEAR HOMOGENEOUS SYSTEMS; COMPLETE SYSTEMS

Because they are the simplest type, we start with systems of linear homogeneous PDEs. We assume we have a system

$$\sum_{k=1}^{n} a_{jk}(\mathbf{x})\, u_{x_k} = 0 \qquad j = 1, \ldots, r \tag{15-7}$$

of r linear homogeneous PDEs for the single unknown function $u(\mathbf{x})$. We assume that all the equations of the system are linearly independent, i.e., that the matrix (a_{jk}) has maximal rank at every point $\mathbf{x}$. More precisely, we assume that the rank of (a_{jk}) is r if $r < n$ and n if $r \geqslant n$, but we shall show shortly that the case $r \geqslant n$ is uninteresting. So our assumption is tantamount to assuming that (a_{jk}) has rank r.

Regardless of the nature of the a_{jk}'s, (15-7) always has at least one solution, namely, $u =$ const. We call this the *trivial solution* and ask if there are any nontrivial ones.

First we show that if $r \geqslant n$ and (a_{jk}) has rank n, then (15-7) has only the trivial solution, $u =$ const. This fact follows immediately from linear algebra because at each point $\mathbf{x}$ (15-7) represents a system of r linear algebraic equations for the n unknowns u_{x_k}. If the matrix of the coefficients in these equations has rank $n \leqslant r$, then the equations have only the solution $u_{x_k} = 0$, that is, $u =$ const.

Thus, we can and shall assume that $r < n$. Then Eqs. (15-7) are algebraically compatible at each point. This algebraic compatibility by no means guarantees the compatibility of the equations of (15-7) as PDEs. We have already seen in Chap 6 that a second-order linear PDE in n independent variables can be transformed into canonical form at a given point but not globally. Similarly, Eqs. (15-7), if $r < n$, can be solved at each point $\mathbf{x}$ for the n quantities u_{x_k}, but there is no a priori reason why the values so found should fit together smoothly to form the set of partial derivatives of a single function u. The problem is to find what if any additional conditions must be satisfied in order for Eqs.(15-7), looked at as PDEs, to have nontrivial solutions.

Since, from an algebraic standpoint, the system (15-7) is complete as it is, these additional conditions must arise from analytic considerations. In particular, if the system (15-7) has a solution u, then any equations obtained from (15-7) by differentiation and algebraic combination of the equations of (15-7) must also be satisfied by this solution u. We will see that certain special combinations of derivatives of equations of (15-7) yield some additional linear homogeneous (algebraic) equations for the u_{x_k}'s. In order to facilitate the derivation of these equations, we let

$$L_j u = \sum_{k=1}^{n} a_{jk}(\mathbf{x})\, u_{x_k} \qquad j = 1, \ldots, r \tag{15-8}$$

Then the system (15-7) can be written in the abbreviated form

$$L_j u = 0 \qquad j = 1, \ldots, r \tag{15-9}$$

In general, any equations formed from (15-9) by differentiation will be of second or higher order. However, the special *commutator equations*

$$(L_j L_k - L_k L_j) u = 0 \qquad j, k = 1, \ldots, r \tag{15-10}$$

though they look at first glance as if they were second-order PDEs, are actually first-order ones, because the derivatives $u_{x_j x_k}$ cancel and the resulting equations read

$$\sum_{l=1}^{n} [L_j(a_{jl}) - L_k(a_{kl})]\, u_{x_l} = 0 \qquad j, k = 1, \ldots, r \tag{15-11}$$

where now the operators L operate only on the a's, not on u.

Thus, any function $u(\mathbf{x})$ which satisfies the system (15-7) must also satisfy all of Eqs. (15-11). Equations (15-11) therefore represent an additional set of first-order linear homogeneous PDEs which any solution of the system (15-7) must satisfy. Equations (15-11) were obtained as analytic consequences of (15-7). Some or all of them may also be algebraic consequences of (15-7); i.e., some or all of (15-11) may be linear combinations of some of the equations in (15-7). If each of the equations of (15-11) is linearly dependent on the set (15-7), then our formation of Eqs. (15-10) has not led to any new conditions on the solution u of the original system (15-7). Since Eqs. (15-10) are far from the only possible new PDEs we could form from (15-7), it is by no means clear that there might not still be some extra conditions to be satisfied by any solution u of (15-7). Nevertheless, we shall show below, by explicit construction of the solution u, that if (15-11) fails to yield any equations linearly independent of (15-7), then (15-7) has nontrivial solutions.

Before doing this, however, we first examine the situation when some of (15-11) are linearly independent of (15-7). In this case, those equations of (15-11) which are linearly independent of (15-7) represent additional linear homogeneous equations for the u_{x_k}'s, equations which must be satisfied if (15-7) is to have a solution when considered as a system of PDEs. If we adjoin to (15-7) those equations of (15-11) which are linearly independent of (15-7), we obtain a new and larger set of equations for the same n partial derivatives of u. If the number of equations in this new system is greater than or equal to n, then they have only the trivial solution. If the number is less than n, then we form all commutators of this new system. Either all commutators of the new system are linear combinations of the equations of the new system, or else some are linearly independent of them. In the latter case, we again adjoin all linearly independent commutators. Continuing in this way, we eventually obtain either a system of n or more linearly independent equations, which therefore have only the trivial solution, or else we obtain a system of less than n linearly independent equations having the property that all commutators formed from them are linearly dependent on the equations of the system.

To summarize our results so far, we introduce a definition.

DEFINITION 15-1

A system (15-7) is called *complete* if all *commutators*, i.e., expressions of the form $(L_jL_k - L_kL_j)u$, are linear combinations of the equations of the system.

We have proved the following theorem.

Theorem 15-1 Every system of the form (15-7) can, by repeated adjunction of equations of the form (15-10), be extended to a system which is either complete or has n or more independent equations.

Next we wish to show that any complete system of $r < n$ independent equations has exactly $n - r$ independent integrals. To do this we first leave it to the reader to prove the following theorem.

Theorem 15-2 A complete system remains complete under a nonsingular change of variables, $y_j = \phi_j(\mathbf{x})$, $j = 1, \ldots, n$.

Then, after laying down Definition 15-2, we ask the reader to prove Theorem 15-3, which follows that definition.

DEFINITION 15-2

Two systems of the form (15-7) are called *equivalent* if each can be obtained from the other by a nonsingular linear transformation.

Theorem 15-3 Any system equivalent to a complete system is also complete.

15-3. JACOBIAN SYSTEMS

We can take advantage of Theorem 15-3 by writing a complete system in a convenient form. One such convenient form is obtained by observing that in a system of r independent linear equations for the n unknowns u_{x_j}, $n - r$ of the unknowns, say those with subscripts $j > r$, can be chosen arbitrarily and the equations solved for the remaining r unknowns in terms of these. Doing so yields a system of the form

$$u_{x_j} + \sum_{k=r+1}^{n} b_{jk}(\mathbf{x})\, u_{x_k} = 0 \qquad j = 1, \ldots, r \tag{15-12}$$

which, according to Theorem 15-3, is complete if the original one was.

The system (15-12) has an interesting feature. If we form the commutators $(L_jL_k - L_kL_j)u$, we see that these commutators do not contain any of the u_{x_j}'s with $j \leqslant r$. But if they do not contain any of the first r u_{x_j}'s, then they cannot contain any of the other u_{x_j}'s either, because if they did, they would provide a relation among the u_{x_j}'s with $j > r$, contrary to the fact that the last $n - r$ of the u_x 's are independent.

Since the commutators contain no u_{x_j}'s at all, they must be identically zero. In other words, the special system (15-12) has the property that

$$(L_jL_k - L_kL_j)u = 0 \qquad j, k = 1, \ldots, r \tag{15-13}$$

DEFINITION 15-3

A system for which (15-13) is satisfied is called *Jacobian*.

Note that every system of the form (15-12) is Jacobian, but the converse is false.

The foregoing analysis shows that every complete system is equivalent to a Jacobian system. We now show that a Jacobian system of r equations has exactly r independent integrals. To do this, consider the first PDE of the system. Looked at as a single PDE, it has $n - 1$ independent integrals Φ_j, $j = 1, \ldots, n - 1$. It is natural to introduce new independent variables with these surfaces as new coordinate surfaces; i.e., we put $y_j = \Phi_j(\mathbf{x})$, $j = 1, \ldots, n - 1$, and pick for y_n anything independent of the other y's. Then the first PDE of the new system reduces to $u_{y_n} = 0$. Since $u_{y_n} = 0$, it can be dropped from the remaining $n - 1$ equations of the system which thus become a system of $r - 1$ equations for a function of $n - 1$ independent variables. We leave it to the reader as an exercise to check that this smaller system is also Jacobian. Having a Jacobian system of $r - 1$ equations, we can repeat the process. Eventually we obtain a single equation in $n - r + 1$ independent variables which has $n - r$ independent integrals, the desired result.

Notice that the method of proof, although computationally intricate, is constructive. It tells us how to find the general solution of a Jacobian system. The method of derivation of a Jacobian system is also constructive. Thus, in principle at least, a linear homogeneous first-order system can be solved. That the method is rather lengthy, however, is illustrated by the following example.

EXAMPLE

$$\begin{aligned} L_1u &= u_x + (y + t - 3x)\, u_z + (z + xy + xt)\, u_t = 0 \\ L_2u &= u_y + (zt - y)\, u_z + (xzt + y - xy)\, u_t = 0 \end{aligned} \tag{15-14}$$

The only commutator of this system is

$$\begin{aligned} L_1L_2 - L_2L_1 = &\left[\frac{\partial}{\partial x} + (y + t - 3x)\frac{\partial}{\partial z} + (z + xy + xt)\frac{\partial}{\partial t}\right] \\ &\times \left[\frac{\partial}{\partial y} + (zt - y)\frac{\partial}{\partial z} + (xzt + y - xy)\frac{\partial}{\partial t}\right] \\ &- \left[\frac{\partial}{\partial y} + (zt - y)\frac{\partial}{\partial z} + (xyt + y - xy)\frac{\partial}{\partial t}\right] \\ &\times \left[\frac{\partial}{\partial x} + (y + t - 3x)\frac{\partial}{\partial z} + (z + xy + xt)\frac{\partial}{\partial t}\right] \end{aligned}$$

$$= (zt - y)\frac{\partial}{\partial t} + (y + t - 3x)\left(t\frac{\partial}{\partial z} + xt\frac{\partial}{\partial t}\right)$$
$$+ (z + xy + xt)\left(z\frac{\partial}{\partial z} + xz\frac{\partial}{\partial t}\right)$$
$$- \left[\frac{\partial}{\partial z} + x\frac{\partial}{\partial t} + (zt - y)\frac{\partial}{\partial t} + (xzt + y - xy)\left(\frac{\partial}{\partial z} + x\frac{\partial}{\partial t}\right)\right]$$
$$= t(y + t - 3x)\left(\frac{\partial}{\partial z} + x\frac{\partial}{\partial t}\right) + z(z + xy + xt)\left(\frac{\partial}{\partial z} + x\frac{\partial}{\partial t}\right)$$
$$- \left(\frac{\partial}{\partial z} + x\frac{\partial}{\partial t}\right) - (xzt + y - xy)\left(\frac{\partial}{\partial z} + x\frac{\partial}{\partial t}\right)$$
$$= [t(y + t - 3x) + z(z + xy + xt) - 1 - (xzt + y - xy)]\left(\frac{\partial}{\partial z} + x\frac{\partial}{\partial t}\right)$$

Hence $(L_1L_2 - L_2L_1)u = 0$ implies

$$u_z + xu_t = 0 \tag{15-15}$$

Writing (15-14) in the form

$$\begin{aligned} &u_x - 3xu_z + zu_t + (y + t)(u_z + xu_t) = 0 \\ &u_y + yu_t + (zt - y)(u_z + xu_t) = 0 \end{aligned} \tag{15-16}$$

we see from (15-15) and (15-16) that any solution of (15-14) must also be a solution of the system

$$u_x - 3xu_z + zu_t = 0 \tag{15-17}$$
$$u_y + yu_t = 0 \tag{15-18}$$
$$u_z + xu_t = 0 \tag{15-19}$$

EXERCISE Show that the system (15-17) to (15-19) is complete.

The characteristic equations for (15-19) are

$$\frac{dx}{0} = \frac{dy}{0} = \frac{dz}{1} = \frac{dt}{x} \tag{15-20}$$

three independent integrals of which are

$$\begin{aligned} x &= \alpha \\ y &= \beta \\ \alpha z - t &= \gamma \end{aligned} \tag{15-21}$$

Hence

$$u = F(\alpha, \beta, \gamma) = F(x, y, xz - t) \tag{15-22}$$

where F is arbitrary, is the general solution of (15-19). According to the general theory expounded above, we now have to use α, β, γ as new independent variables together with any independent fourth variable. We choose

$$\begin{aligned} \alpha &= x \\ \beta &= y \\ \gamma &= xz - t \\ \tau &= t \end{aligned} \tag{15-23}$$

as new independent variables in Eqs. (15-17) to (15-19). Since

$$\frac{\partial}{\partial z} + x\frac{\partial}{\partial t} = \alpha\frac{\partial}{\partial \gamma} - \alpha\frac{\partial}{\partial \gamma} + \alpha\frac{\partial}{\partial \tau} = \alpha\frac{\partial}{\partial \tau}$$

(15-19) reduces, as the theory predicts, to

$$u_\tau = 0 \tag{15-24}$$

while (15-17) and (15-18) become

$$\begin{aligned} &u_\alpha + zu_\gamma - 3\alpha^2 u_\gamma - zu_\gamma + zu_\tau = 0 \\ &u_\beta - \beta u_\gamma + \beta u_\tau = 0 \end{aligned}$$

Canceling terms and using the fact that $u_\tau = 0$, we have

$$u_\alpha - 3\alpha^2 u_\gamma = 0 \tag{15-25}$$

$$u_\beta - \beta u_\gamma = 0 \tag{15-26}$$

a complete system for $u(\alpha, \beta, \gamma)$.

The characteristic equations of (15-26) are

$$\frac{d\alpha}{0} = \frac{d\beta}{1} = \frac{d\gamma}{-\beta} \tag{15-27}$$

two independent integrals of which are

$$\alpha = \xi \qquad \beta^2 + 2\gamma = \eta \tag{15-28}$$

where ξ and η are constants. Hence

$$u = G(\alpha, \beta^2 + 2\gamma) \tag{15-29}$$

where G is arbitrary, is the general solution of (15-26).

With new variables

$$\begin{aligned} \xi &= \alpha \\ \eta &= \beta^2 + 2\gamma \\ \zeta &= \gamma \end{aligned} \tag{15-30}$$

(15-26) becomes

$$2\beta u_\eta - \beta(2u_\eta + u_\zeta) = 0$$

that is,

$$u_\zeta = 0$$

in conformity with the general theory. Equation (15-25) becomes (remember $u_\zeta = 0$)

$$u_\xi - 3\alpha^2(2u_\eta) = 0$$

that is,

$$u_\xi - 6\xi^2 u_\eta = 0 \tag{15-31}$$

The characteristic equation of (15-31) is

$$\frac{d\xi}{1} = \frac{d\eta}{-6\xi^2}$$

Hence $2\xi^3 + \eta =$ const, and so

$$u = H(2\xi^3 + \eta) \tag{15-32}$$

where H is arbitrary, is the general solution of (15-31).

Replacing variables in (15-32) from (15-30), we have

$$u = H(2\alpha^3 + \beta^2 + 2\gamma) \tag{15-33}$$

and then using (15-23), we have

$$u = H(2x^3 + y^2 + 2xz - 2t) \tag{15-34}$$

where H is an arbitrary function of a single variable, as the general solution of the system (15-14).

15-4. DEPENDENCE ON ARBITRARY FUNCTIONS

In Sec. 8-8, we used the method for solving the Cauchy problem for a single first-order linear homogeneous PDE such as a typical one of (15-7) in order to show that the general solution of such an equation was an arbitrary function of n definite functions. When applied to a complete system of r first-order linear homogeneous PDEs, this method yields an alternate demonstration of the result of the present section that the general solution of such a system is an arbitrary function of $n - r$ definite functions. For simplicity, we illustrate this fact for the special case of two equations. The case of r equations requires more writing but entails no essential difficulties. We therefore leave it to the enterprising reader.

Let

$$\sum_{j=1}^{n} a_j(\mathbf{x})\, u_{x_j} = 0 \tag{15-35}$$

and

$$\sum_{j=1}^{n} b_j(\mathbf{x})\, u_{x_j} = 0 \tag{15-36}$$

be a complete system. We also assume that (15-35) and (15-36) are independent. Then (15-35) will have an $(n - 1)$-parameter family of characteristic curves, and (15-36) will have an $(n - 1)$-parameter family of characteristic curves that are independent of those of (15-35). Because of this independence, we can choose t_1 to be both the running parameter on the characteristics of (15-35) and also one of the arbitrary constants in the general solution to the characteristic equations of (15-36). Similarly, we can let t_2 be the running parameter on the characteristics of (15-36) and also one of the arbitrary constants in the general solution to the characteristic equations of (15-35). This means that we can write the equations of the characteristics of (15-35) in the form

$$x_j = f_j(t_1, t_2, \tau_1, \ldots, \tau_{n-2}) \qquad j = 1, \ldots, n \tag{15-37}$$

where t_1 is the running parameter on the characteristics and $t_2, \tau_1, \ldots, \tau_{n-2}$

are arbitrary constants. Similarly, the equations of the characteristics of (15-36) can be written in the form

$$x_j = g_j(t_1, t_2, \tau_1, \ldots, \tau_{n-2}) \qquad j = 1, \ldots, n \tag{15-38}$$

where t_2 is the running parameter on the characteristics and $t_1, \tau_1, \ldots, \tau_{n-2}$ are arbitrary constants.

Now the PDE (15-35) tells us that u is constant along any characteristic of the family whose equations are given by (15-37), and the PDE (15-36) tells us that u is constant along any characteristic of the family whose equations are given by (15-38). Thus,

$$u(0, t_2, \tau_1, \ldots, \tau_{n-2}) = F(t_2, \tau_1, \ldots, \tau_{n-2}) \tag{15-39}$$

and

$$u(t_1, 0, \tau_1, \ldots, \tau_{n-2}) = G(t_1, \tau_1, \ldots, \tau_{n-2}) \tag{15-40}$$

where F and G are the given initial values of u on the respective characteristics. The assumed compatibility of the PDEs implies that both (15-39) and (15-40) have solutions for arbitrary initial data F and G. Hence u must be independent of both t_1 and t_2. Thus,

$$u(t_1, t_2, \tau_1, \ldots, \tau_{n-2}) = H(\tau_1, \ldots, \tau_{n-2}) \tag{15-41}$$

where H is arbitrary.

If we solve (15-37) for the τ's in terms of the x's, then we get equations of the form

$$\tau_j = \phi(\mathbf{x}) \qquad j = 1, \ldots, n-2 \tag{15-42}$$

which, when inserted into (15-41), will give u as an arbitrary function of the $n - 2$ definite functions $\phi_j(\mathbf{x})$. If, instead of solving (15-37) for the τ's, we were to solve (15-38) for them, we would obtain a different set of functions of $\mathbf{x}$ for the τ's, but by our assumption of compatibility of the original PDEs, this set would not yield any new solutions u.

To summarize, we can say that the system of two PDEs (15-35) and (15-36), if compatible, independent, and complete, has a general solution that is an arbitrary function of $n - 2$ definite functions. A similar analysis shows that a system of r PDEs of the form (15-7) will, if compatible, independent, and complete, have a general solution which is an arbitrary function of $n - r$ definite functions.

15-5. CONNECTION WITH EXACT DIFFERENTIALS

The existence of a nontrivial solution of Eqs. (15-1) is just the condition that the differential expression $E = f(x, y)\,dx + g(x, y)\,dy$ be exact. Similarly, (15-2) is the well-known necessary and sufficient condition for E to be exact.

This suggests a possible connection between overdetermined systems of PDEs and exact differential expressions. We will investigate the possibility of such a connection.

The system (15-16) can be written in the vector form

$$\mathbf{a}_j(\mathbf{x}) \cdot \nabla u = 0 \qquad j = 1, \ldots, r \tag{15-43}$$

In this form it tells us that the vectors $\mathbf{a}_j$ are all tangent to the integral surfaces $u(\mathbf{x}) = \text{const}$ at the point $\mathbf{x}$. Since the $\mathbf{a}_j$'s are by assumption linearly independent, they form an r-dimensional vector space at $\mathbf{x}$. Suppose $\mathbf{a}_j^*$, $j = r + 1, \ldots, n$, form a basis for the orthogonal complement of this vector space. Then we have

$$\mathbf{a}_k^* \cdot \mathbf{a}_j = 0 \qquad \begin{aligned} j &= 1, \ldots, r \\ k &= r + 1, \ldots, n \end{aligned} \tag{15-44}$$

We shall show that an appropriately chosen set of $\mathbf{a}^*$'s will serve as coefficients in a set of exact DEs closely related to the system of PDEs (15-43). It is not possible to choose the $\mathbf{a}^*$'s as an arbitrary basis of the orthogonal complement of the space spanned by the $\mathbf{a}$'s. The reason for this is that the $\mathbf{a}^*$'s, like the $\mathbf{a}$'s, depend upon $\mathbf{x}$. It has to be expected that they depend upon $\mathbf{x}$ in some continuous way. Hence they cannot be chosen completely arbitrarily. We proceed to give a way of choosing appropriate $\mathbf{a}^*$'s.

In the previous section we saw that a system such as (15-43) is either trivial or else can be extended to a complete system. We shall therefore assume that (15-43) is complete. Then, as was shown in the previous section, any integral $u(\mathbf{x})$ of (15-43) is an arbitrary function of $s = n - r$ fixed functions. Hence

$$u(\mathbf{x}) = U(\phi_1, \ldots, \phi_s) \tag{15-45}$$

where $\phi_j = \phi_j(\mathbf{x})$, and where U is arbitrary.

Using (15-45) in

$$du = \sum_{j=1}^{n} u_{x_j}\, dx_j \tag{15-46}$$

we have

$$du = \sum_{j=1}^{n} \sum_{k=1}^{s} U_{\phi_k} \frac{\partial \phi_k}{\partial x_j}\, dx_j \tag{15-47}$$

or

$$du = \sum_{k=1}^{s} \left(\sum_{j=1}^{n} \frac{\partial \phi_k}{\partial x_j}\, dx_j \right) U_{\phi_k} \tag{15-48}$$

Since $du = 0$ on an integral surface, and since the U_{ϕ_k}'s are independent, (15-48) implies

$$\sum_{j=1}^{n} \frac{\partial \phi_k}{\partial x_j} dx_j = 0 \qquad k = 1, \ldots, s \tag{15-49}$$

On the other hand, writing out (15-43) as

$$\sum_{k=1}^{n} a_{jk} u_{x_k} = 0 \qquad j = 1, \ldots, r \tag{15-50}$$

and substituting (15-45) into (15-50), we find

$$\sum_{k=1}^{n} \sum_{l=1}^{s} a_{jk} U_{\phi_l} \frac{\partial \phi_l}{\partial x_k} = 0, \qquad j = 1, \ldots, r \tag{15-51}$$

or

$$\sum_{l=1}^{s} \left(\sum_{k=1}^{n} a_{jk} \frac{\partial \phi_l}{\partial x_k} \right) U_{\phi_l} = 0 \qquad j = 1, \ldots, r \tag{15-52}$$

which implies

$$\sum_{k=1}^{n} a_{jk} \frac{\partial \phi_l}{\partial x_k} = 0 \qquad \begin{array}{l} l = 1, \ldots, s \\ j = 1, \ldots, r \end{array} \tag{15-53}$$

Equations (15-53) imply that the s independent vectors with components $\partial\phi_l/\partial x_k$ are in the orthogonal complement of the space spanned by the $\mathbf{a}_j$'s. We take these vectors for our basis vectors $\mathbf{a}_j^*$, $j = 1, \ldots, s$; that is, we define $\mathbf{a}_j^*$ by

$$a_{jk}^* = \frac{\partial \phi_j}{\partial x_k} \tag{15-54}$$

Then Eqs. (15-49) read

$$\sum_{j=1}^{n} a_{kj}^*(\mathbf{x})\, dx_j = 0 \qquad k = 1, \ldots, s \tag{15-55}$$

Thus the system of PDEs (15-50) has given rise to the system of equations (15-55), which can be written in the vector form

$$\mathbf{a}_k^* \cdot d\mathbf{x} = 0 \qquad k = 1, \ldots, s \tag{15-56}$$

Moreover, (15-54) implies

$$\mathbf{a}_k^* = \nabla \phi_k \qquad k = 1, \ldots, s \tag{15-57}$$

which tells us that the equations of (15-55) are exact. Thus, the complete system of r PDEs (15-43) has given rise to the system of s exact equations (15-56). Recall that $r + s = n$.

Conversely, suppose we have a system of s exact equations of the form (15-56). Then their exactness implies the existence of s functions $\phi_k(\mathbf{x})$, $k = 1, \ldots, s$, such that Eqs. (15-57) hold. If $\mathbf{a}_j$, $j = 1, \ldots, r$, form a basis for the orthogonal complement of the space spanned by the $\mathbf{a}_k^*$'s then it follows from (15-57) that

$$\mathbf{a}_j \cdot \nabla\phi_k = 0 \qquad \begin{matrix} j = 1, \ldots, r \\ k = 1, \ldots, s \end{matrix} \tag{15-58}$$

In other words, the ϕ_k's are a set of s independent integrals of (15-43).

EXAMPLES The system

$$\begin{aligned} u_x(x, y, z) &= 0 \\ u_y(x, y, z) &= 0 \end{aligned} \tag{15-59}$$

is complete because it is Jacobian. Its only independent integral is $\phi(z)$, where ϕ is arbitrary. If ϕ is differentiable, then

$$\mathbf{a}^* = (0 \quad 0 \quad \phi'(z)) \tag{15-60}$$

and indeed, the resulting differential expression $\phi'(z)\,dz$ is exact.

As a second example, consider the case of a single PDE

$$\sum_{j=1}^{n} a_j(\mathbf{x})\, u_{x_j} = 0 \tag{15-61}$$

Equation (15-61) will give rise to $n - 1$ exact equations

$$\sum_{k=1}^{n} a_{jk}^*(\mathbf{x})\, dx_k = 0 \qquad j = 1, \ldots, n - 1 \tag{15-62}$$

Equations (15-62) can be looked at as $n - 1$ linear homogeneous algebraic equations for the n differentials dx_k, $k = 1, \ldots, n$. Because they are linearly independent, their solutions must depend upon exactly one parameter; i.e., we must have

$$dx_k = \lambda_k(\mathbf{x})\, dt \qquad k = 1, \ldots, n \tag{15-63}$$

where dt is the above-mentioned one parameter.

From (15-63) it follows that

$$du = \sum_{k=1}^{n} u_{x_k}\, dx_k = \sum_{k=1}^{n} \lambda_k(\mathbf{x})\, u_{x_k}\, dt \tag{15-64}$$

Since $du = 0$ along an integral surface, and since we have assumed that (15-61) is the only relation satisfied by u along an integral surface, it follows by comparison of (15-61) and (15-64) that

$$\lambda_k(\mathbf{x}) = a_k(\mathbf{x}) \qquad k = 1, \ldots, n \tag{15-65}$$

Thus, Eqs. (15-63) read

$$dx_k = a_k(\mathbf{x})\, dt \qquad k = 1, \ldots, n \tag{15-66}$$

which are just the characteristic equations of the PDE (15-61).

15-6. CHARACTERISTIC MANIFOLDS

If there are r PDEs instead of just one, there will be only $s = n - r$ exact differential equations and therefore r free differentials. Thus, the analog of characteristic curves for a system of r PDEs for *one* unknown function is an r-dimensional characteristic manifold. We have already seen this in the analysis of the system (15-35) and (15-36). There the solution u of a system of two PDEs was constant along two sets of curves, one with t_1 as running parameter and the other with t_2 as running parameter. Thus, u in that case was constant along the two-dimensional manifold generated by the two families of characteristic curves. This leads us to the following procedure. Given the complete system of r PDEs

$$\sum_{k=1}^{n} a_{jk}(\mathbf{x})\, u_{x_k} = 0 \qquad j = 1, \ldots, r \tag{15-67}$$

we define as *characteristic manifolds* of (15-67) those r-dimensional manifolds whose parametric representations $x_j = x_j(t_1, \ldots, t_s)$ satisfy the differential equations

$$\frac{\partial x_k}{\partial t_j} = a_{jk}(\mathbf{x}) \qquad \begin{aligned} j &= 1, \ldots, r \\ k &= 1, \ldots, n \end{aligned} \tag{15-68}$$

Along a characteristic manifold (15-68) we have

$$\frac{\partial u}{\partial t_j} = \sum_{k=1}^{n} u_{x_k} \frac{\partial x_k}{\partial t_j} = \sum_{k=1}^{n} a_{jk}(\mathbf{x})\, u_{x_k} = 0 \qquad j = 1, \ldots, r \tag{15-69}$$

by (15-67). Thus, u is constant along a characteristic manifold.

In the case of a single PDE, the characteristic equations (15-68) are ODEs, and their solutions always exist. For more than one PDE, however, (15-68) represent rn equations for only n functions. Hence some compatibility conditions have to be satisfied. These are obviously

$$\frac{\partial a_{jk}}{\partial t_l} = \frac{\partial a_{lk}}{\partial t_j} \qquad \begin{aligned} j, l &= 1, \ldots, r \\ k &= 1, \ldots, n \end{aligned} \tag{15-70}$$

that is,

$$\sum_{\mu=1}^{n} \left(\frac{\partial a_{jk}}{\partial x_\mu} \frac{\partial x_\mu}{\partial t_l} - \frac{\partial a_{lk}}{\partial x_\mu} \frac{\partial x_\mu}{\partial t_j} \right) = 0 \qquad \begin{aligned} j, l &= 1, \ldots, r \\ k &= 1, \ldots, n \end{aligned} \tag{15-71}$$

or, by (15-68),

$$\sum_{\mu=1}^{n} \left(a_{l\mu} \frac{\partial a_{jk}}{\partial x_\mu} - a_{j\mu} \frac{\partial a_{lk}}{\partial x_\mu} \right) = 0 \tag{15-72}$$

Equation (15-72) is the condition that the system (15-67) be Jacobian. So we see that characteristic r-dimensional manifolds exist if and only if the system of PDEs is Jacobian. Next we wish to derive a system of s exact equations equivalent to this Jacobian system. To do this, we first observe that since

$$dx_k = \sum_{j=1}^{s} \frac{\partial x_k}{\partial t_j} dt_j \qquad k = 1, \ldots, n \tag{15-73}$$

(15-68) tells us that if the system is Jacobian, then the n differential expressions

$$dx_k = \sum_{j=1}^{s} a_{jk}(\mathbf{x})\, dt_j \qquad k = 1, \ldots, n \tag{15-74}$$

are exact. In order to reduce these n exact differential expressions to s exact ones, it is necessary to write the Jacobian system in the special form (15-12), i.e., in the form wherein

$$a_{jk}(\mathbf{x}) = \begin{cases} \delta_{jk} & k = 1, \ldots, r \\ b_{jk}(\mathbf{x}) & k = r+1, \ldots, n \end{cases} \tag{15-75}$$

Then Eqs. (15-74) reduce to

$$dx_k = dt_k \qquad k = 1, \ldots, r \tag{15-76}$$

and

$$dx_k = \sum_{j=1}^{s} b_{jk}(\mathbf{x})\, dt_j \qquad k = r+1, \ldots, n \tag{15-77}$$

Using (15-76) in (15-77), we see that the equations

$$dx_k = \sum_{j=1}^{s} b_{jk}(\mathbf{x})\, dx_j \qquad k = r+1, \ldots, n \tag{15-78}$$

represent a system of s exact equations for the characteristic manifolds.

Thus, we have again demonstrated a connection between systems of PDEs and systems of exact equations. In the present derivation the system of exact equations appears as a system for determining the characteristic manifolds of the system of PDEs. Since any solution of the PDEs is constant on a characteristic manifold, this is tantamount to determining the solution of the PDEs.

There are, however, some important differences between the present method and that given in Sec. 15-5. In the method of those pages the set of characteristic equations, i.e., the adjoint system of exact equations, was constructed from the complete set of s independent integrals of the system of PDEs. In that method, therefore, it is necessary to solve the PDE completely in order to obtain the characteristic equations. In the present method, we can go the other way around. If we can solve the system (15-78) of exact equations, then we can solve the adjoint system of PDEs.

Couldn't we get the adjoint system to (15-43) by choosing any $\mathbf{a}^*$'s as long as they are a basis for the orthogonal complement of the space spanned by the $\mathbf{a}$'s? The answer, as indicated in Sec. 15-5, is not for the same reason that transformation of a second-order linear equation in more than three independent variables to canonical form was impossible in Chap. 6. At each point $\mathbf{x}$ we can choose a basis $\mathbf{a}_k^*$, $k = 1, \ldots, s$, for the orthogonal complement of the space of the $\mathbf{a}$'s, but there is no a priori guarantee that the bases chosen at different points $\mathbf{x}$ will fit together smoothly, just as there was no guarantee that the different linear transformations in Chap. 6 would fit together to form a global transformation. After all, if we fix a point $\mathbf{x}$, then all vectors are constants when evaluated at that point and all transformations at that point are linear. If, however, we change to another point $\hat{\mathbf{x}}$ then the vectors at $\hat{\mathbf{x}}$ will not in general be linear functions of those at $\mathbf{x}$. This is a difficulty well known to those familiar with tensor analysis and covariant derivatives.

In any event, in order to ensure the proper connection between the vector spaces at two different points, some conditions have to be satisfied. These are precisely the compatibility conditions (15-70).

Thus, we have two ways of obtaining the system of s exact equations adjoint to a system of r PDEs. One way is to find the general solution of the PDEs first, i.e., find s independent integrals. The coefficient of dx_j in the kth equation of the adjoint system of s exact equations is then just the jth component of the gradient of the kth independent integral of the system of PDEs. This method obviously requires solving the PDEs first.

The second method consists in first writing the system of PDEs in the special Jacobian form (15-12). No other Jacobian form will do for what follows. Then the adjoint system of exact equations is (15-78). Notice that in this case the same coefficients $b_{jk}(\mathbf{x})$ appear in the system of exact equations as appeared in the system of PDEs. Since each of (15-78) is exact, each has an integral, $\phi_j(\mathbf{x}) = \text{const}$, $j = 1, \ldots, s$. The general solution of the system of PDEs, (15-12), is then (cf. Sec 15-4) $u = F[\phi_1(\mathbf{x}), \ldots, \phi_s(\mathbf{x})]$, where F is arbitrary.

EXAMPLE As an example of the use of the adjoint system of exact equations in solving systems of PDEs, we solve the system (15-14) by this method. First we have to obtain a Jacobian system from (15-14). We have already done this in Sec. 15-3, where we obtained the system (15-17) to (15-19). Although Jacobian, that system is still not of the special form we need. In order to get it into that form, we solve

(15-19) for u_z and substitute the result into (15-17). Taking the resulting equation together with (15-18) and (15-19), we then have the system

$$u_x + (3x^2 + z)\, u_t = 0 \tag{15-79}$$

$$u_y + y u_t = 0 \tag{15-80}$$

$$u_z + x u_t = 0 \tag{15-81}$$

which is of the special Jacobian form we need.

Since there are three PDEs and four independent variables, we have $4 - 3 = 1$ exact equation adjoint to Eqs. (15-79) to (15-81). According the general theory developed on the previous pages, this exact equation is

$$dt = (3x^2 + z)\, dx + y\, dy + x\, dz \tag{15-82}$$

Writing (15-82) in the form

$$dt = 3x^2\, dx + y\, dy + (z\, dx + x\, dz) \tag{15-83}$$

we see immediately that its integral is

$$x^3 + \frac{y^2}{2} + xz - t = \text{const} \tag{15-84}$$

Hence the general solution of the system of PDEs (15-14) is

$$u = F\left(x^3 + \frac{y^2}{2} + xz - t\right) \tag{15-85}$$

where F is arbitrary. Since F and H are arbitrary, (15-85) agrees with the previously obtained solution (15-34).

15-7. INVOLUTORY SYSTEMS; POISSON BRACKETS

Since the device used in Sec. 8-10 converts a quasi-linear system in n independent variables into a linear system in $n + 1$ independent variables, we will not treat quasi-linear equations next but rather go on to nonlinear systems in general. These include quasi-linear ones as special cases.

For the procedures which follow it is essential that the dependent variable be absent from the systems we consider.† According to Sec. 8-10, we lose no generality in considering such systems and so we let

$$F_j(\mathbf{x}, \mathbf{p}) = 0 \qquad j = 1, \ldots, r \tag{15-86}$$

be a system of r first-order PDEs for the single unknown function $u(\mathbf{x})$. Here, as usual, $\mathbf{x} = (x_1, \ldots, x_n)$, and we assume in all of what follows that the F_j's are independent.

If $r > n$, Eqs. (15-86) are inconsistent and therefore have no solution in this case. So we can assume that $r \leqslant n$.

If we consider (15-86) purely as algebraic equations, we can choose $n - r$ of the p's as arbitrary functions of x and solve (15-86) for the other r p's in terms of them and the x's. However, there is no guarantee that the n

† For a treatment in which the dependent variable is permitted to enter the PDEs explicitly, see Caratheodory [29, Chap. 4].

p's so obtained will all be partial derivatives of a single function u. So, just as we did for linear systems, we need some compatibility conditions that take into account the fact that (15-86) are PDEs, not merely algebraic equations.

These compatibility conditions must obviously derive from the condition

$$\frac{\partial p_j}{\partial x_k} = \frac{\partial p_k}{\partial x_j} \qquad j, k = 1, \ldots, n \tag{15-87}$$

Since (15-87) are conditions on the unknowns, they have to be translated into conditions on the F's in order to be of use. To do this, it is natural to differentiate the equations of (15-86) and compare the equations that result. As in the case of linear systems, it suffices to compare two equations at a time. In order to conserve subscripts, we call the two equations to be used $F(\mathbf{x}, \mathbf{p}) = 0$ and $G(\mathbf{x}, \mathbf{p}) = 0$. Then, differentiating the first of these with respect to x_j, we have

$$\frac{\partial F}{\partial x_j} + \sum_{l=1}^{n} \frac{\partial F}{\partial p_l}\frac{\partial p_l}{\partial x_j} = 0 \qquad j = 1, \ldots, n \tag{15-88}$$

We would like to interchange j and l, but l is summed and j is not. So we multiply (15-88) by $\partial G/\partial p_j$ and sum over j. This gives

$$\sum_{j=1}^{n} \frac{\partial F}{\partial x_j}\frac{\partial G}{\partial p_j} + \sum_{j=1}^{n}\sum_{l=1}^{n} \frac{\partial F}{\partial p_l}\frac{\partial G}{\partial p_j}\frac{\partial p_l}{\partial x_j} = 0 \tag{15-89}$$

If we interchange j and l in the double sum and F and G in the entire equation here, we obtain the new equation

$$\sum_{j=1}^{n} \frac{\partial G}{\partial x_j}\frac{\partial F}{\partial p_j} + \sum_{j,l=1}^{n} \frac{\partial G}{\partial p_j}\frac{\partial F}{\partial p_l}\frac{\partial p_j}{\partial x_l} = 0 \tag{15-90}$$

Subtracting (15-90) from (15-89) and using (15-87), we find

$$\sum_{j=1}^{n} \left(\frac{\partial F}{\partial x_j}\frac{\partial G}{\partial p_j} - \frac{\partial G}{\partial x_j}\frac{\partial F}{\partial p_j}\right) = 0 \tag{15-91}$$

The expression on the left of (15-91) is just the *Poisson bracket* (F, G) introduced in Chap. 8

$$(F, G) = \sum_{j=1}^{n} \left(\frac{\partial F}{\partial x_j}\frac{\partial G}{\partial p_j} - \frac{\partial G}{\partial x_j}\frac{\partial F}{\partial p_j}\right) \tag{15-92}$$

Using (15-92), we can write the compatibility conditions (15-91) for (15-86) in the abbreviated form

$$(F_j, F_k) = 0 \qquad j, k = 1, \ldots, r \tag{15-93}$$

Recall that $(F_j, F_k) = -(F_k, F_j)$; that is, the Poisson brackets are antisymmetric. In particular, $(F, F) = 0$.

Conditions (15-93) represent the generalization of the commutator equations (15-10) to a nonlinear system. As in the case of linear systems, (15-93) are analytic consequences of (15-86). Some or all of them may also be algebraic consequences of (15-86). Those which are not have to be adjoined to (15-86) to form a larger system. Then we form all Poisson brackets of this larger system and adjoin any that are not algebraic consequences of the system whence they came. Continuing in this way, we eventually arrive either at a system of more than n independent equations, which therefore shows that the original system is incompatible, or else we obtain a system of $r \leqslant n$ equations of the form (15-86) for which all Poisson brackets vanish, either identically or as algebraic consequences of the system. We again call such a system *complete*. In the linear case it was possible to transform a complete system into a form in which all commutators were zero not merely because they were linear combinations of the equations of the system but because they were themselves identically zero. Such a special complete linear system was called Jacobian. We will now show that every complete nonlinear system can be transformed into a form in which all Poisson brackets are identically zero. Such a nonlinear system is called *involutory*.

The transformation will be to a special involutory system analogous to the special Jacobian system (15-12). The method of obtaining it will also be analogous to the method used to get (15-12) and to show that it is Jacobian. Thus, suppose we have a complete system of r independent equations of the form (15-86). Then we can solve (15-86) for r of the p's in terms of the other $n - r$ p's. We can, by renumbering the variables if necessary, assume that (15-86) can be solved for the first r p's in terms of the other $s = n - r$ p's and the x's. Thus, (15-86) can be rewritten in the form

$$p_j = f_j(\mathbf{x}, p_{r+1}, \ldots, p_n) \qquad j = 1, \ldots, r \tag{15-94}$$

To compute the Poisson brackets for (15-94), we put

$$F_j = p_j - f_j(\mathbf{x}, p_{r+1}, \ldots, p_n) \qquad j = 1, \ldots, r \tag{15-95}$$

Then

$$\frac{\partial F_j}{\partial x_k} = -\frac{\partial f_j}{\partial x_k} \tag{15-96}$$

and

$$\frac{\partial F_j}{\partial p_k} = \begin{cases} \delta_{jk} & k = 1, \ldots, r \\ -\dfrac{\partial f_j}{\partial p_k} & k = r+1, \ldots, n \end{cases} \tag{15-97}$$

From (15-96) and (15-97) we see that (F_j, F_l) contains no p_k's with $k \leqslant r$. Because the system has been assumed complete, all (F_j, F_l) vanish. They therefore cannot contain any of the s other p's either, because if they did, they would provide a relation among the s independent p's. Hence the Poisson brackets can contain no p's at all. But then they cannot contain any x's either, because the x's are independent variables and a vanishing Poisson bracket containing only x's would violate their independence. Thus we see that all Poisson brackets for the special system (15-94) vanish identically. Hence (15-94) is involutory.

Having shown that every system of the form (15-86) is either inconsistent or reducible to an involutory system, we now have to show how to solve an involutory system. It turns out that the case in which the system has n equations is easiest. If $r = n$, then (15-86) represent n equations for the n unknown p's and therefore determine the p's uniquely. It would be nice, though, if we could get at least a little arbitrariness into the solution. We can do this by observing that the system

$$F_j(\mathbf{x}, \mathbf{p}) = \alpha_j \qquad j = 1, \ldots, n \tag{15-98}$$

is just as involutory as (15-86) is and has the advantage that, when solved for the p's, it yields functions of $\mathbf{x}$ and the n parameters $\boldsymbol{\alpha}$:

$$p_j = p_j(\mathbf{x}, \boldsymbol{\alpha}) \qquad j = 1, \ldots, n \tag{15-99}$$

As pointed out at the outset of the present section, however, there is no a priori guarantee that the p's given by (15-99) represent the partial derivatives of a function u. Stated another way, this means that we need to know that the differential

$$\sum_{j=1}^{n} p_j(\mathbf{x}, \boldsymbol{\alpha})\, dx_j \tag{15-100}$$

is exact. As is to be expected, the exactness of (15-100) is a consequence of the assumption that the system (15-98) is involutory. To be precise we state this as a theorem.

Theorem 15-4 If (15-98) is involutory, i.e., if all Poisson brackets (F_j, F_k) vanish identically, then (15-100) is exact, that is, $\partial p_j/\partial x_k = \partial p_k/\partial x_j$.

Proof. We leave the proof to the reader as an instructive exercise.

From Theorem 15-4 we see that the general solution of (15-98) is a function of $n + 1$ arbitrary constants and can be obtained by merely integrating (15-100) after having obtained the p's by solving (15-98), looked at as algebraic equations, for the p's.

Finally, we take up the case of involutory systems in which $r < n$. To treat this case, Jacobi conceived the idea of adding $s = n - r$ more equations to the original r equations in such a way that the entire set of n equations is involutory. The advantage of having an involutory system of n equations has just been pointed out. Only one integration is required. The rest of the work is algebra. So the principal problem is finding s more equations that are in involution with the original r and with each other. This problem, however, turns out to reduce to one we are already very familiar with. For, suppose (15-86) is an involutory system. We want to adjoin to (15-86) a function $F_{r+1}(\mathbf{x}, \mathbf{p})$ such that the new system of $r + 1$ equations is also involutory. Clearly this requires that F_{r+1} satisfy

$$(F_j, F_{r+1}) = 0 \qquad j = 1, \ldots, r \tag{15-101}$$

From the definition (15-92) of the Poisson bracket we see that (15-101) is a system of r linear homogeneous first-order PDEs for the single function F_{r+1}. (The F_j's with $j = 1, \ldots, r$ are known functions of their arguments.) Introducing the notation

$$L_j u = (F_j, u) \qquad j = 1, \ldots, r \tag{15-102}$$

we are naturally led to hope that the system $L_j u = 0$, $j = 1, \ldots, r$ is Jacobian, i.e., that

$$(L_j L_k - L_k L_j)u = 0 \qquad j, k = 1, \ldots, n \tag{15-103}$$

Substituting (15-102) into (15-103), we see that (15-102) will be Jacobian if

$$(F_j, (F_k, u)) - (F_k, (F_j, u)) = 0 \tag{15-104}$$

To prove (15-105), Jacobi used the identity

$$(f, (g, h)) + (h, (f, g)) + (g, (h, f)) = 0 \tag{15-105}$$

Jacobi's identity (15-105) can be proved either by brute force or by checking that terms cancel appropriately. We leave the proof to the reader and proceed to employ (15-105), using it to replace the second term in (15-104). Then (15-104) becomes

$$(F_j, (F_k, u)) + (u, (F_k, F_j)) + (F_j, (u, F_k)) = 0 \tag{15-106}$$

But $(F_j, (u, F_k)) = -(F_j, (F_k, u))$. Hence the first and third terms in (15-106) cancel. The middle term is zero because the system (15-86) is involutory. Hence (15-106) is satisfied, and the system $L_j u = 0$ is indeed Jacobian. It therefore has s independent integrals. Choosing one of them, we adjoin it to the system (15-86), which thus becomes an involutory system of $r + 1$ equations. Continuing in this way, we eventually arrive at a system of n involutory equations, which can then be solved by the method already given.

Notice that the method for solving nonlinear systems yields complete integrals while that for linear homogeneous systems gives general solutions.

15-8. EXAMPLE: JACOBI'S METHOD FOR SOLVING A SINGLE FIRST-ORDER PARTIAL DIFFERENTIAL EQUATION

As an application of Jacobi's method we solve a single nonlinear equation, looked at as a system of 1 equation, namely, (8.23)

$$u = xu_x + yu_y + u_x u_y \tag{15-107}$$

First we have to get rid of u. Let $U(x, y, u) = 0$ be an implicit representation of the solution of (15-107). Then, in the usual way, (15-107) becomes

$$xU_x + yU_y + uU_u + \frac{U_x U_y}{U_u} = 0 \tag{15-108}$$

Put

$$F(x, y, u, p_1, p_2, p_3) = xp_1 + yp_2 + up_3 + \frac{p_1 p_2}{p_3} \tag{15-109}$$

Then

$$\begin{aligned} &F_x = p_1 \qquad F_y = p_2 \qquad F_u = p_3 \qquad F_{p_1} = x + \frac{p_2}{p_3} \\ &F_{p_2} = y + \frac{p_1}{p_3} \qquad F_{p_3} = u - \frac{p_1 p_2}{p_3^2} \end{aligned} \tag{15-110}$$

Now we need a function $G(x_1, x_2, x_3, p_1, p_2, p_3)$ such that $(F, G) = 0$, that is, such that

$$\begin{aligned} (F, G) &= p_1 G_{p_1} + p_2 G_{p_2} + p_3 G_{p_3} \\ &\quad - \left(x + \frac{p_2}{p_3}\right) G_x - \left(y + \frac{p_1}{p_3}\right) G_y - \left(u - \frac{p_1 p_2}{p_3^2}\right) G_u = 0 \end{aligned} \tag{15-111}$$

This, as the theory predicts, is a first-order linear PDE for G. Its characteristic equations are

$$\frac{dp_1}{p_1} = \frac{dp_2}{p_2} = \frac{dp_3}{p_3} = -\frac{dx}{x + p_2/p_3} = -\frac{dy}{y + p_1/p_3} = -\frac{du}{u - p_1 p_2/p_3^2} \tag{15-112}$$

Equations (15-112) yield the two integrals

$$p_1 = \alpha p_3 \qquad p_2 = \beta p_3 \tag{15-113}$$

or

$$\frac{p_1}{p_3} = \alpha \qquad \frac{p_2}{p_3} = \beta$$

Since in the present context, $p_j = U_{x_j}$,

$$u_x = \frac{p_1}{p_3} \qquad \text{and} \qquad u_y = \frac{p_2}{p_3}$$

Hence

$$u_x = \alpha \qquad u_y = \beta$$

and (15-107) becomes

$$u = \alpha x + \beta y + \alpha\beta$$

the same solution found in Chap. 8.

16
Variational Methods

16-1. FUNCTIONALS

A vector $\mathbf{x} = (x_1, \ldots, x_j, \ldots, x_n)$ in n-dimensional space can be looked at as a function of an integral variable j: $\mathbf{x} = (x(j))$, where j takes only the values $1, \ldots, n$. An infinite sequence $x_1, x_2, \ldots, x_n, \ldots$ can also be looked at as a function of an integral variable $x(n)$, where now the argument n runs through all positive integers. We could think of the sequence $(x_1, x_2, \ldots, x_n, \ldots)$ as a point in some infinite-dimensional space.

Carrying the above analogy one step further, we could think of a function $x(t)$, defined on the interval $0 \leqslant t \leqslant 1$, as a vector in an uncountably infinite-dimensional space in which the continuous variable t indexes the "components." In this interpretation $x(t_1)$, where t_1 is any point in the interval, is looked at as a component of the infinite-dimensional vector $x(t)$.

A scalar function of a vector $f(\mathbf{x})$ is an assignment of a scalar, i.e., a number, to the vector. In finite-dimensional space such a function is merely an ordinary function of n independent variables. In function space it is something considerably more complicated, since it is essentially a function of infinitely many variables. In order to have a name for such an entity, we call it a *functional*. A functional, then, is an assignment of numbers to functions. This is not the same as the $f[g(x)]$ concept of elementary calculus, which assigns a number to each value of g. A functional assigns a single number to an entire function, not to its individual components.

A classical example of a functional is the definite integral; $\int_a^b f(x)\,dx$ obviously assigns a single number to each integrable function $f(x)$, and this number depends upon the entire function $f(x)$ in $a \leqslant x \leqslant b$, not just on its value at one or two points. Of course, just as there exist functions which assign the same number to several different values of the independent variable, so too there exist functionals whose values depend on only a few of the values of their argument functions. An example is the famous delta functional, which assigns the value $f(0)$ to $f(x)$. Nevertheless, in general a functional will depend upon infinitely many different values of its argument function rather than those at only a finite number of points.

16-2. DOMAIN OF A FUNCTIONAL

In defining a function $f(x)$ of a real variable x, the first thing we have to say is for what x's f is defined. For example, we can define $f(x) = x^2 + 17$ for $-1 \leqslant x \leqslant 14$ and for no other x's if we wish. Then $f(2) = 21$, but $f(19)$ is not defined. Similarly, in giving a functional $\mathscr{F}[\phi(x)]$ we have to specify for what functions $\phi(x)$ $\mathscr{F}$ is defined. For example, we could define $\mathscr{F}[\phi] = \int_a^b \phi(x)\,dx$ for all functions $\phi(x)$ which are continuous on the interval $a \leqslant x \leqslant b$. On the other hand, the functional $\Delta[\phi(x)] = \phi(0)$ is clearly defined for all functions $\phi(x)$ which are themselves defined at $x = 0$. ϕ need not be continuous anywhere or even defined anywhere but at $x = 0$ in order for $\Delta[\phi]$ to exist.

We will call the set of all functions $\phi(x)$ for which a functional $\mathscr{F}[\phi]$ is defined the *domain* of $\mathscr{F}$. We will always restrict the domain of the functionals we use here to that of continuous functions. For the present we assume that these functions are continuous on a closed finite interval, $a \leqslant x \leqslant b$.

16-3. NORMS

Next we would like to define continuity for functionals. An ordinary function $f(x)$ is continuous at a point x_0 if $f(x)$ is close to $f(x_0)$ when x is close to x_0. We would like continuity for functionals to mean the same thing, i.e., that $\mathscr{F}[\phi(x)]$ is close to $\mathscr{F}[\phi_0(x)]$ when $\phi(x)$ is close to $\phi_0(x)$. The only trouble is that it is not clear what it means for two functions to be close to each other. In other words, we need a way of measuring distance in the set of functions that constitute the domain of $\mathscr{F}$.

There are a number of different definitions that can be used. For our purposes, the best one is one that ensures that whenever the "distance" between two functions is small, the pointwise difference between the two functions is small. Using the notation $\| \phi - \phi_0 \|_0$ to denote the distance between $\phi(x)$ and $\phi_0(x)$, we can achieve this aim by defining

$$\| \phi - \phi_0 \|_0 = \max_{a \leqslant x \leqslant b} | \phi(x) - \phi_0(x)| \tag{16-1}$$

(The reason for the subscript 0 on the norm sign will appear later.)

If we take $\phi_0(x) = 0$, then (16-1) reduces to

$$\| \phi \|_0 = \max_{a \leqslant x \leqslant b} | \phi(x)| \tag{16-2}$$

Relation (16-2) shows that $\| \phi \|_0$ is a functional of ϕ. By using the concept of distance (16-1) based on this special functional (16-2) we will be able to define continuity for general functionals.

From the definition (16-2) it is easy to check that $\| \phi \|_0$ enjoys all the properties that $| x |$ does for ordinary numbers x, for example, the triangle inequality. Such a functional is called a *norm*, and the particular norm (16-2) is called the *maximum norm*. The maximum norm is usually defined using the least upper bound (lub) instead of max, but since our functions are continuous on a closed bounded interval, lub $=$ max here.

As indicated in (16-1), the norm (16-2) gives us a way of defining a distance between two functions. Notice that if $\| \phi - \phi_0 \|_0 < \epsilon$, then necessarily $| \phi(x) - \phi_0(x)| < \epsilon$ for all x between a and b.

Having (16-1) we can now define continuity for functionals. We call a functional $\mathscr{F}[\phi]$ continuous in the norm (16-2) if given any $\epsilon > 0$ we can find a $\delta > 0$ such that $| \mathscr{F}[\phi] - \mathscr{F}[\phi_0]| < \epsilon$ if $\| \phi - \phi_0 \|_0 < \delta$. The functional $\mathscr{F}[\phi] = \int_a^b \phi(x)\, dx$ is an example of a functional which is continuous in the norm (16-2).

The functional considered in the classical problem of the calculus of variations is obtained as follows. Let $F(x, y, z)$ be a given continuous function of three independent variables. Choose a function $\phi(x)$ and insert $\phi(x)$ for y and $\phi'(x)$ for z in F. The result is a function of x: $F[x, \phi(x), \phi'(x)]$. If we integrate this function of x from a to b, we obtain a single number, and this number is obviously a functional of ϕ

$$\mathscr{F}[\phi] = \int_a^b F[x, \phi(x), \phi'(x)]\, dx \tag{16-3}$$

However, it is not necessarily a continuous functional of ϕ in the norm (16-2) because although a small norm for $\phi - \phi_0$ implies that ϕ and ϕ_0 are pointwise close, it implies nothing at all about their derivatives. $\| \phi - \phi_0 \|_0$ could be very small, yet $\phi'(x) - \phi_0'(x)$ might be very large. A pictorial example of this is given in Fig. 16-1.

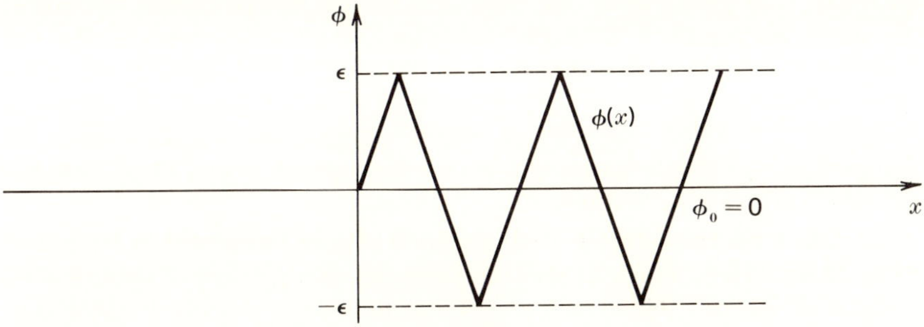

Figure 16-1

Thus, in order to make (16-3) continuous, we need a new norm. An obvious extension of (16-2) will do, namely,

$$\| \phi \|_1 = \max_{a \leqslant x \leqslant b} (| \phi(x)| + | \phi'(x)|) \tag{16-4}$$

In the norm (16-4), the functional (16-3) is continuous. However, in order to be able to use the norm (16-4) we have to further restrict the domain of $\mathscr{F}$. Now we have to require that the functions in the domain of $\mathscr{F}$ have continuous first derivatives.

16-4. MINIMA OF FUNCTIONALS

Next we wish to define what it means to say that a functional has a relative minimum at a "point" ϕ. An ordinary function, $f(x)$ has a relative minimum at a point x_0 if $f(x) \geqslant f(x_0)$ for all x in a neighborhood of x_0. A neighborhood of x_0 means the set of all points x such that $| x - x_0 | < \delta$ for some δ. For functionals we therefore say $\mathscr{F}[\phi]$ has a relative minimum at ϕ_0 if there exists a $\delta > 0$ such that $\mathscr{F}[\phi] \geqslant \mathscr{F}[\phi_0]$ for all ϕ satisfying $\| \phi - \phi_0 \| < \delta$. We have deliberately omitted the subscript on the norm sign in order to emphasize that this definition depends on the norm we use. If we use the norm (16-1), then there will be more functions in a neighborhood of ϕ_0 than if we use the norm (16-4). Therefore a functional might have a relative minimum in the norm (16-4) but fail to have one in the norm (16-1) because the competition in the latter norm is stronger. Because of this fact, a functional having a minimum in the norm (16-1) is said to have a *strong minimum*, while a functional having a minimum in the norm (16-4) is said to have a *weak minimum*. For examples of functionals having weak but not strong minima, see [232].

We are now in a position to state one of the basic problems of the calculus of variations: among all functions $\phi(x) \in C^{(1)}[a, b]$, satisfying the boundary conditions (BCs): $\phi(a) = \alpha \qquad \phi(b) = \beta$ (16-5)

where α and β are given numbers, find (if it exists) that function for which the functional (16-3) has a (weak) relative minimum. The class of competing functions ϕ is called the *class of admissible functions*. Geometrically, the problem is to find among all $C^{(1)}$ curves through (a, α) and (b, β) that one for which (16-3) is a minimum.

The question of strong relative minima is too specialized to be treated here. Moreover, as usual, we will not enter into the question of the existence of a minimum, only that of how to find it assuming that it exists. In particular, we forego a discussion of sufficient conditions for a relative minimum. In ordinary calculus, the condition $f'(x_0) = 0$ is well known to be necessary but not sufficient for the function $f(x)$ to have a relative minimum at x_0. However, if $f'(x_0) = 0$, then a sufficient condition for x_0 to yield a relative minimum is that the first nonvanishing derivative of f at x_0 be of even order and of positive sign there. Sufficiency conditions in the calculus of variations are nowhere near as simple. Indeed, in the calculus of variations, even the analog of the simplest case, namely, $f''(x_0) > 0$, is quite complicated.

Thus all we shall do here is to derive a necessary condition for a function $\phi_0(x)$ to yield a (weak) relative minimum for (16-3). Our reasoning will be as follows: suppose that $\mathscr{F}$ has a relative minimum for $\phi = \phi_0$; then what conditions must ϕ_0 satisfy? Having found some conditions that the minimizing function ϕ_0 must satisfy, i.e., necessary conditions, we leave it to the interested reader to learn from textbooks on the calculus of variations the conditions under which there actually is a ϕ_0 that satisfies these conditions.

The reader should not draw from the foregoing paragraphs the erroneous conclusion that the topics we have omitted are unimportant even from an applied point of view. They are quite important and quite interesting, but we simply do not have the space here to go into them. We must perforce restrict ourselves to those aspects of the subject which are most closely connected with differential equations.

16-5. EULER'S EQUATION

Suppose $\phi_0(x)$ minimizes (16-3) in the class of admissible functions, i.e., functions in $C^{(1)}[a, b]$ satisfying the BCs (16-5). Then surely ϕ_0 minimizes $\mathscr{F}$ in any subclass of this class. We choose a particularly convenient subclass, namely, the set of all functions $\phi(x)$ of the form

$$\phi(x) = \phi_0(x) + \mu\psi(x) \tag{16-6}$$

where $\psi(x) \in C^{(1)}[a, b]$ and

$$\psi(a) = \psi(b) = 0 \tag{16-7}$$

(in order to ensure that ϕ is admissible) and where μ is a parameter.

If $\psi(x)$ is fixed, then the right side of (16-6) is a function of the single variable μ. So therefore is $\mathscr{F}[\phi_0 + \mu\psi]$, say

$$f(\mu) = \mathscr{F}[\phi_0 + \mu\psi] \tag{16-8}$$

Since $\mathscr{F}[\phi]$ has, by assumption, a relative minimum at $\phi = \phi_0$, it follows that $f(\mu)$ has a relative minimum at $\mu = 0$. Thus, we have replaced the problem of finding a minimum for the functional (16-3) by that of finding a minimum for the function of a single real variable (16-8). Of course, the two problems are not equivalent. All we can say is that if (16-3) has a minimum, then so does (16-8), but the converse may be false. This is why the condition on ϕ_0 that we will derive in this way is only necessary for a minimum but not sufficient. Nevertheless, it is sufficient for our purposes.

If (16-8) has a minimum at $\mu = 0$, then $f'(0) = 0$. Now from (16-3) and (16-6) we have

$$f(\mu) = \mathscr{F}[\phi_0 + \mu\psi] = \int_a^b F[x, \phi_0(x) + \mu\psi(x), \phi_0'(x) + \mu\psi'(x)]\, dx \tag{16-9}$$

To compute $f'(\mu)$ from this we need a notation for the partials of F with respect to its various arguments. F is a function of three independent variables: $F(x, y, z)$. Because of the way it will be used, it is convenient to denote the third variable by y' and write $F(x, y, y')$. When we write F this way, we do not necessarily imply that the third variable is the derivative of the second one. They are still independent variables, but the notation is convenient when we want to form a functional of the form (16-3) by inserting $\phi(x)$ for y and $\phi'(x)$ for y'.

With this understanding, we shall denote the partials of F by F_x, F_y, and $F_{y'}$. Then from (16-9) we have

$$f'(0) = \int_a^b \{F_y[x, \phi_0(x), \phi_0'(x)]\, \psi(x) + F_{y'}[x, \phi_0(x), \phi_0'(x)]\, \psi'(x)\}\, dx \tag{16-10}$$

Relation (16-10) holds for arbitrary functions ψ belonging to $C^{(1)}$ and vanishing at the end points. We would like to take advantage of this fact. To do so, we change ψ' into ψ by integrating the second term in the integrand of (16-10) by parts. In order to do this we have to assume that ϕ'' exists, which further restricts the class of admissible functions. Making this additional assumption and bearing the BCs (16-7) in mind, we find

$$f'(0) = \int_a^b \left(F_y - \frac{d}{dx} F_{y'}\right) \psi(x)\, dx \tag{16-11}$$

where the arguments of F are still x, ϕ_0, and ϕ_0'. If we let

$$G(x) = F_y - \frac{d}{dx} F_{y'} \tag{16-12}$$

then we see that $f'(0) = 0$ implies that

$$\int_a^b G(x)\,\psi(x)\,dx = 0 \tag{16-13}$$

for arbitrary $\psi \in C^{(1)}$ and satisfying (16-7). From this it follows by a brief argument (which we omit) that $G(x) \equiv 0$. This result is known as the *Fundamental Lemma* of the calculus of variations. From it and (16-12) we conclude that the minimizing function $\phi_0(x)$ satisfies the ordinary differential equation (ODE)

$$F_y(x, \phi_0, \phi_0') - \frac{d}{dx} F_{y'}(x, \phi_0, \phi_0') = 0 \tag{16-14}$$

or, with the usual abbreviation, $y = \phi_0(x)$,

$$F_y(x, y, y') - \frac{d}{dx} F_{y'}(x, y, y') = 0 \tag{16-15}$$

Carrying out the differentiation in the second term of (16-15), we find

$$F_y - F_{xy'} - y'F_{yy'} - y''F_{y'y'} = 0 \tag{16-16}$$

which, unless $F_{y'y'} = 0$, is a second-order nonlinear ODE for y.

The ODE (16-15) is called *Euler's equation*. What we have shown is that a necessary condition for a function $y = \phi_0(x)$ belonging to $C^{(2)}[a, b]$ and satisfying the BCs (16-5) to minimize (16-3) is that it be a solution of Euler's equation (16-15). Since Euler's equation is second order and there are two BCs, there is a good chance of determining ϕ_0 uniquely. However, since we are interested in more general functionals than (16-3), we will not enter into a further discussion of (16-15) here. We do remark, though, that by integrating the first instead of the second term in the integrand of (16-10) by parts, it is possible to derive an integrated form of Euler's equation using only the existence of ϕ' and then to derive from this the fact that ϕ'' must exist for the minimizing function, even though it was not assumed to exist at the outset. Again we refer to standard books on the calculus of variations for details.

16-6. FREE BOUNDARY CONDITIONS

Our derivation of Euler's equation appears to depend upon the requirement (16-5) that the values of the competing functions be fixed at the end points, but this is an illusion. Euler's equation still constitutes a necessary condition for a minimizing function even if we prescribe no conditions at all at the end points. Of course, we keep the values of x at the end points but we no longer prescribe y there.

That Euler's equation holds for the minimizing function even if no BCs are prescribed in advance follows trivially from the fact that if ϕ_0 minimizes $\mathscr{F}$ in the class of functions ϕ which can take arbitrary values on the boundary, then it surely minimizes $\mathscr{F}$ in the subclass of functions which take fixed values there.

Thus Euler's equation represents a necessary condition for a minimizing function regardless of the BCs. However, it is interesting that even if we prescribe no BCs in advance, the minimizing function satisfies some anyway. To see this, note that the absence of prescribed BCs means that the function ψ in (16-6) need no longer satisfy (16-7), i.e., vanish at the end points. Consequently, when we integrate (16-10) by parts we get a contribution from the boundary, and (16-11) becomes

$$f'(0) = F_{y'}\psi\Big|_a^b + \int_a^b \left(F_y - \frac{d}{dx}F_{y'}\right)\psi\,dx \tag{16-17}$$

We have just pointed out that Euler's equation must still hold for the present case. Hence the integral in (16-17) vanishes and then the condition $f'(0) = 0$ implies

$$F_{y'}\psi\Big|_a^b = 0 \tag{16-18}$$

for arbitrary ψ. Therefore

$$F_{y'} = 0 \qquad \text{at } x = a, b \tag{16-19}$$

that is

$$F_{y'}[a, \phi_0(a), \phi_0'(a)] = F_{y'}[b, \phi_0(b), \phi_0'(b)] = 0 \tag{16-20}$$

which are two nonlinear BCs on $\phi_0(x)$. Because they arose from a problem in which no BCs were prescribed in advance, they are called *free boundary conditions*.

16-7. EULER'S EQUATIONS FOR SEVERAL FUNCTIONS

So far we have considered functionals that depend on only one function. It is natural to extend our results to functionals of several functions. To do this for (16-3) we change notation somewhat. We let $L(t, \mathbf{x}, \mathbf{y})$ be a function of the $2n + 1$ variables t, $\mathbf{x}$, $\mathbf{y}$ and form the functional

$$\mathscr{L}[\phi_1, \ldots, \phi_n] = \mathscr{L}[\boldsymbol{\phi}] = \int_a^b L[t, \boldsymbol{\phi}(t), \boldsymbol{\phi}'(t)]\,dt \tag{16-21}$$

Functionals of the type (16-21) arise in analytical mechanics with L as the Lagrangian. There the variables which we have called $\mathbf{x}$ are usually called $\mathbf{q}$, and $\mathbf{x}'$ is usually written $\dot{\mathbf{q}}$. As we shall see below, however, it will be necessary to use several different names for the variables in L. Thus, even if we were to start out calling them $\mathbf{q}$ instead of $\mathbf{x}$, we would still have to change their names to $\boldsymbol{\alpha}$ or $\boldsymbol{\beta}$ or one of the other appelations employed below.

A procedure exactly similar to that of Sec. 16-5 shows that the minimizing functions $\mathbf{x} = \boldsymbol{\phi}(t)$ for (16-21) satisfying the BCs

$$\boldsymbol{\phi}(a) = \boldsymbol{\alpha} \qquad \text{and} \qquad \boldsymbol{\phi}(b) = \boldsymbol{\beta} \tag{16-22}$$

must satisfy the n Euler equations

$$L_{x_j} - \frac{d}{dt} L_{x'_j} = 0 \qquad j = 1, \ldots, n \tag{16-23}$$

In (16-23) the arguments of L are $(t, \mathbf{x}, \mathbf{x}')$. Thus Eqs. (16-23) are a system of n second-order nonlinear ODEs for the n unknown functions $\mathbf{x}(t)$.

In mechanics, where L is the Lagrangian, Eqs. (16-23) are known as *Lagrange's equations*. Because of this fact, some authors call (16-23) the *Euler-Lagrange equations*.

16-8. EXTREMALS; THE EXTREMAL INTEGRAL

Any solution of Euler's equation is called an *extremal*. To solve a given minimum problem, then, we have to find the extremal that satisfies the appropriate BCs, but sometimes we will be interested in the extremals themselves without regard to the BCs they satisfy.

In the case of Sec. 16-7, the extremals form a $2n$-parameter family of curves $\mathbf{x}(t, \boldsymbol{\alpha}, \boldsymbol{\beta})$ in the n-dimensional $\mathbf{x}$ space. Here t is the curve parameter and $\boldsymbol{\alpha}$ and $\boldsymbol{\beta}$ are the family parameters. In the case where the extremals satisfy the BCs (16-22) the family parameters $\boldsymbol{\alpha}$ and $\boldsymbol{\beta}$ can be taken to be the same as the $\boldsymbol{\alpha}$ and $\boldsymbol{\beta}$ in (16-22), but in the general case they are any $2n$ independent parameters.

Suppose we consider a $2n$-parameter family of extremals in which the parameters $\boldsymbol{\alpha}$ and $\boldsymbol{\beta}$ are the ones appearing in the BCs (16-22). If instead of inserting various functions into (16-21) we insert only extremals, then (16-21) is no longer a functional but merely a function of the $2n$ parameters $\boldsymbol{\alpha}$ and $\boldsymbol{\beta}$ which determine the different extremals. If, in addition, we let the values of a and b vary, calling them τ_0 and τ to indicate their variability, then we can epitomize the foregoing remarks by writing

$$u(\tau_0, \boldsymbol{\alpha}, \tau, \boldsymbol{\beta}) = \int_{\tau_0}^{\tau} L[t, \boldsymbol{\phi}(t), \boldsymbol{\phi}'(t)]\, dt \tag{16-24}$$

where $\boldsymbol{\phi}(t)$ is no longer an arbitrary admissible function but now must be an extremal, in fact, the extremal that takes the values $\boldsymbol{\alpha}$ at τ_0 and $\boldsymbol{\beta}$ at τ. Thus, $\boldsymbol{\phi}(t)$ is a function of the same $2n + 2$ parameters that u is. To make this clear, we write $\boldsymbol{\phi}$ as $\boldsymbol{\phi}(t, \tau_0, \boldsymbol{\alpha}, \tau, \boldsymbol{\beta})$ and we write the BCs as

$$\boldsymbol{\phi}(\tau_0, \tau_0, \boldsymbol{\alpha}, \tau, \boldsymbol{\beta}) = \boldsymbol{\alpha} \tag{16-25}$$

and

$$\boldsymbol{\phi}(\tau, \tau_0, \boldsymbol{\alpha}, \tau, \boldsymbol{\beta}) = \boldsymbol{\beta} \tag{16-26}$$

The integral (16-24) taken over the extremals is called, naturally enough, the *extremal integral.* It can be looked upon as a sort of generalized distance between the points $(\tau_0, \boldsymbol{\alpha})$ and $(\tau, \boldsymbol{\beta})$ in the $(n + 1)$-dimensional $t\mathbf{x}$ space. If $L(t, \mathbf{x}, \mathbf{y})$ is positive homogeneous of degree one in $\mathbf{y}$, then the distance so defined is called a *Finsler metric* (see [243]). If, in addition, L is the square root of a positive definite quadratic form in $\mathbf{y}$, then the metric is the ordinary Riemann one.

In mechanics (16-24) is frequently called *Hamilton's characteristic function*, but the terminology is not universal. Some authors also call it *Hamilton's principal function*, but others make a distinction between the two. In addition, the notation for Hamilton's characteristic function varies, too. The most popular letters appear to be W and S, but since even Φ appears in at least one book ([62]), there seems to be no harm in calling it u.

If we hold τ_0 and $\boldsymbol{\alpha}$ fixed, then the u in (16-24) becomes a function of the $n + 1$ parameters τ, $\boldsymbol{\beta}$. For brevity we denote this function by u again

$$u(\tau, \boldsymbol{\beta}) = \int_{\tau_0}^{\tau} L[t, \boldsymbol{\phi}(t, \tau, \boldsymbol{\beta}), \boldsymbol{\phi}'(t, \tau, \boldsymbol{\beta})]\, dt \tag{16-27}$$

As in u, the now constant parameters τ_0 and $\boldsymbol{\alpha}$ have been suppressed in ϕ. In (16-27) we have used the notation

$$\boldsymbol{\phi}' = \partial\boldsymbol{\phi}/\partial t \tag{16-28}$$

It pays to have a special notation for the derivative of $\boldsymbol{\phi}$ with respect to t because t is on a different footing from the rest of the independent variables in $\boldsymbol{\phi}$.

16-9. THE HAMILTON-JACOBI EQUATION AGAIN

Next we return to the principal subject of the book by deriving a partial differential equation (PDE) for $u(\tau, \boldsymbol{\beta})$. To do this we naturally need the partials of u with respect to its various arguments. From (16-27) we have

$$u_\tau = L[\tau, \boldsymbol{\phi}(\tau, \tau, \boldsymbol{\beta}), \boldsymbol{\phi}'(\tau, \tau, \boldsymbol{\beta})] + \sum_{j=1}^{n} \int_{\tau_0}^{\tau} \left[L_{x_j}(t, \boldsymbol{\phi}, \boldsymbol{\phi}') \frac{\partial \phi_j}{\partial \tau} + L_{x_j'}(t, \boldsymbol{\phi}, \boldsymbol{\phi}') \frac{\partial \phi_j'}{\partial \tau}\right] dt \tag{16-29}$$

Equation (16-29) can be simplified considerably. From the BC (16-26) we have

$$\boldsymbol{\phi}(\tau, \tau, \boldsymbol{\beta}) = \boldsymbol{\beta} \tag{16-30}$$

In addition, we introduce the notation

$$\boldsymbol{\phi}'(\tau, \tau, \boldsymbol{\beta}) = \mathbf{v} \tag{16-31}$$

Since $\boldsymbol{\phi}$ is an extremal, it satisfies Euler's equations, i.e.,

$$L_{x_j}(t, \boldsymbol{\phi}, \boldsymbol{\phi}') = \frac{d}{dt} L_{x_j'}(t, \boldsymbol{\phi}, \boldsymbol{\phi}') \qquad j = 1, \ldots, n \tag{16-32}$$

The notation d/dt here leaves something to be desired because we are thinking of $\boldsymbol{\phi}$ as a function of $\boldsymbol{\beta}$ in addition to t, but $\partial/\partial t$ would not do either because this would omit the derivatives of $L_{x_j'}$ with respect to t due to the presence of t in $\boldsymbol{\phi}$ and $\boldsymbol{\phi}'$. Therefore d/dt appears to be the least objectionable notation.

In addition, we need this nebulous d/dt notation in one other term. We have

$$\boldsymbol{\phi}'_\tau = \frac{\partial \boldsymbol{\phi}'}{\partial \tau} = \frac{\partial^2 \boldsymbol{\phi}}{\partial \tau\, \partial t} = \frac{d}{dt}\frac{\partial \boldsymbol{\phi}}{\partial \tau} \tag{16-33}$$

where we have switched from the symbol $\partial/\partial t$ to d/dt used in the sense just explained.

Using relations (16-30) to (16-33) in (16-29), we have

$$u_\tau = L(\tau, \boldsymbol{\beta}, \mathbf{v}) + \sum_{j=1}^{n} \int_{\tau_0}^{\tau} \left[\frac{\partial \phi_j}{\partial \tau}\frac{d}{dt} L_{x_j'}(t, \boldsymbol{\phi}, \boldsymbol{\phi}') + L_{x_j'}(t, \boldsymbol{\phi}, \boldsymbol{\phi}')\frac{d}{dt}\frac{\partial \phi_j}{\partial \tau}\right] dt \tag{16-34}$$

or

$$u_\tau = L(\tau, \boldsymbol{\beta}, \mathbf{v}) + \sum_{j=1}^{n} \int_{\tau_0}^{\tau} \frac{d}{dt}\left(L_{x_j'}\frac{\partial \phi_j}{\partial \tau}\right) dt \tag{16-35}$$

Hence

$$u_\tau = L(\tau, \boldsymbol{\beta}, \mathbf{v}) + \sum_{j=1}^{n} L_{x_j'}(\tau, \boldsymbol{\beta}, \mathbf{v}) \frac{\partial \phi_j(\tau, \tau, \boldsymbol{\beta})}{\partial \tau} - \sum_{j=1}^{n} L_{x_j'}(\tau_0, \boldsymbol{\alpha}, \mathbf{v}_0)\frac{\partial \phi_j(\tau_0, \tau, \boldsymbol{\beta})}{\partial \tau} \tag{16-36}$$

the meaning of $\mathbf{v}_0$ being obvious.

Recalling that $\boldsymbol{\phi} = \boldsymbol{\phi}(t, \tau, \boldsymbol{\beta})$ and that $\boldsymbol{\phi}' = \partial \boldsymbol{\phi}/\partial t$, we differentiate (16-30) with respect to τ. This gives

$$\boldsymbol{\phi}'(\tau, \tau, \boldsymbol{\beta}) + \frac{\partial \boldsymbol{\phi}(\tau, \tau, \boldsymbol{\beta})}{\partial \tau} = 0$$

Hence

$$\frac{\boldsymbol{\phi}(\tau, \tau, \boldsymbol{\beta})}{\partial \tau} = -\boldsymbol{\phi}'(\tau, \tau, \boldsymbol{\beta}) = -\mathbf{v} \qquad (16\text{-}37)$$

In addition, since

$$\boldsymbol{\phi}(\tau_0, \tau, \boldsymbol{\beta}) = \boldsymbol{\alpha} \qquad (16\text{-}38)$$

we have

$$\frac{\partial \boldsymbol{\phi}(\tau_0, \tau, \boldsymbol{\beta})}{\partial \tau} = 0 \qquad (16\text{-}39)$$

Using (16-37) and (16-39) in (16-36), we find

$$u_\tau = L(\tau, \boldsymbol{\beta}, \mathbf{v}) - \sum_{j=1}^{n} L_{x_j'}(\tau, \boldsymbol{\beta}, \mathbf{v})\, v_j \qquad (16\text{-}40)$$

Notice that (16-40) expresses u_τ as a function of the $n + 1$ independent variables $(\tau, \boldsymbol{\beta})$ and n parameters $\mathbf{v}$. Since the last set of variables in $L(t, \mathbf{x}, \mathbf{y})$ in (16-40) is called $\mathbf{v}$, we rewrite (16-40) in the form

$$u_\tau = L(\tau, \boldsymbol{\beta}, \mathbf{v}) - \sum_{j=1}^{n} L_{v_j}(\tau, \boldsymbol{\beta}, \mathbf{v})\, v_j \qquad (16\text{-}41)$$

Next we find u_{β_j}. From (16-27) we have

$$u_{\beta_j} = \sum_{k=1}^{n} \int_{\tau_0}^{\tau} \left(L_{x_k} \frac{\partial \phi_k}{\partial \beta_j} + L_{x_k'} \frac{\partial \phi_k'}{\partial \beta_j} \right) dt \qquad j = 1, \ldots, n \qquad (16\text{-}42)$$

In the same way as before, we find that the integrand is a perfect differential. Integrating it, we get

$$u_{\beta_j} = \sum_{k=1}^{n} \left[L_{x_k'}(\tau, \boldsymbol{\beta}, \mathbf{v}) \frac{\partial \phi_k(\tau, \tau, \boldsymbol{\beta})}{\partial \beta_j} - L_{x_k'}(\tau_0, \boldsymbol{\alpha}, \mathbf{v}_0) \frac{\partial \phi_k(\tau_0, \tau, \boldsymbol{\beta})}{\partial \beta_j} \right] \qquad j = 1, \ldots, n \qquad (16\text{-}43)$$

Relation (16-38) implies

$$\frac{\partial \boldsymbol{\phi}(\tau_0, \tau, \boldsymbol{\beta})}{\partial \beta_j} = 0 \qquad j = 1, \ldots, n \qquad (16\text{-}44)$$

while (16-30) gives

$$\frac{\partial \phi_k}{\partial \beta_j} = \delta_{jk} \qquad j, k = 1, \ldots, n \qquad (16\text{-}45)$$

Therefore (16-43) reduces to

$$u_{\beta_j} = L_{x_j'}(\tau, \boldsymbol{\beta}, \mathbf{v}) \qquad j = 1, \ldots, n \qquad (16\text{-}46)$$

So, like u_τ, u_{β_j} is also a function of the independent variables $(\tau, \boldsymbol{\beta})$ and the n parameters $\mathbf{v}$. Again for the sake of convenience, we rewrite (16-46) as

$$u_{\beta_j} = L_{v_j}(\tau, \boldsymbol{\beta}, \mathbf{v}) \qquad j = 1, \ldots, n \tag{16-47}$$

or, with the usual notation, namely,

$$p_j = u_{\beta_j} \tag{16-48}$$

$$p_j = L_{v_j}(\tau, \boldsymbol{\beta}, \mathbf{v}) \tag{16-49}$$

Thus, in (16-41) and (16-49) we have expressed the $n + 1$ partial derivatives of u in terms of the n independent variables in u plus n parameters $\mathbf{v}$. If from these $n + 1$ equations we eliminate the n parameters $\mathbf{v}$, the result will be a single equation connecting the partial derivatives of u and the independent variables $(\tau, \boldsymbol{\beta})$, in other words, a PDE for $u(\tau, \boldsymbol{\beta})$.

In order to be able to eliminate the v's from (16-41) and (16-4) we have to assume that $\det(L_{v_jv_k}) \neq 0$. If this condition is satisfied, then we can solve (16-49) for $\mathbf{v}$ to get

$$\mathbf{v} = \mathbf{V}(\tau, \boldsymbol{\beta}, \mathbf{p}) \tag{16-50}$$

Substituting (16-50) into (16-41), we find

$$u_\tau + H(\tau, \boldsymbol{\beta}, \mathbf{p}) = 0 \tag{16-51}$$

where

$$H(\tau, \boldsymbol{\beta}, \mathbf{p}) = \sum_{j=1}^{n} v_j L_{v_j}(\tau, \boldsymbol{\beta}, \mathbf{v}) - L(\tau, \boldsymbol{\beta}, \mathbf{v}) \tag{16-52}$$

with $\mathbf{v}$ being given in terms of τ, $\boldsymbol{\beta}$, $\mathbf{p}$ by (16-50).

The PDE (16-51) is the Hamilton-Jacobi equation of Chap. 8 with t there called τ here and $\mathbf{x}$ there called $\boldsymbol{\beta}$ here. What we have therefore shown is that the extremal integral (16-27), looked upon as a function of its upper limit τ and of the values $\boldsymbol{\beta}$ taken by the extremizing functions at that upper limit, is a solution of the Hamilton-Jacobi equation. Moreover, our derivation shows that the Hamiltonian H which determines the PDE is related to the integrand L of (16-27) by means of the formula (16-52).

16-10. HAMILTON'S EQUATIONS

Note that (16-49) enables us to write (16-52) in the further abbreviated form

$$H(\tau, \boldsymbol{\beta}, \mathbf{p}) = \sum_{j=1}^{n} v_j p_j - L(\tau, \boldsymbol{\beta}, \mathbf{v}) \tag{16-53}$$

The equations (16-49) and (16-53) bear a strong resemblance to the Legendre transformation (8-192). They are not exactly the same, however, because in (16-49) and (16-53) L is a given, known function, not the unknown function. Nevertheless (16-49) and (16-53) do provide us with a change of variables from $\mathbf{v}$ to $\mathbf{p}$ and from L to H.

Before investigating this change of variables, however, we first investigate the consequence of the simpler change from $\boldsymbol{\phi}'$ to $\mathbf{v}$ made long ago. In order to do this, we have to avoid some notational pitfalls. Back in Sec. 16-7 we formulated a minimum problem by starting out with a function $L(t, \mathbf{x}, \mathbf{y})$ of $2n + 1$ independent variables. Now we have seen many times in this book that the names of variables frequently have to be changed around many times in order to obtain useful results. The present chapter is no exception. In Sec. 16-8 we introduced some variables τ and $\boldsymbol{\beta}$ which were different from t and $\mathbf{x}$ there. Nevertheless, the L in (16-49), in (16-52), and in sundry other places is still the same L we started out with. The variables in it may have different names and different meanings, but L is still a function of $2n + 1$ independent variables, regardless of whether we call them $(t, \mathbf{x}, \mathbf{y})$, $(t, \boldsymbol{\phi}, \boldsymbol{\phi}')$, $(\tau, \boldsymbol{\beta}, \mathbf{v})$, or what. Now consider Euler's equations, written in the form

$$L_{y_j}(t, \mathbf{y}, \mathbf{y}') - \frac{d}{dt} L_{y'_j}(t, \mathbf{y}, \mathbf{y}') = 0 \qquad j = 1, \ldots, n \tag{16-54}$$

As such, they represent a system of n second-order ODEs for the n unknown functions $\mathbf{y}(t)$. These functions may depend on additional parameters, but that is irrelevant here. In any event, if we let

$$\mathbf{v} = \mathbf{y}' \tag{16-55}$$

then (16-54) and (16-55) can be looked at as a system of $2n$ *first-order* ODEs for the $2n$ unknown functions $\mathbf{y}(t)$ and $\mathbf{v}(t)$

$$L_{y_j}(t, \mathbf{y}, \mathbf{v}) - \frac{d}{dt} L_{v_j}(t, \mathbf{y}, \mathbf{v}) = 0 \qquad j = 1, \ldots, n \tag{16-56}$$

$$\mathbf{y}'_j = v_j \qquad j = 1, \ldots, n \tag{16-57}$$

Now suppose we introduce a new set of n dependent variables $\mathbf{p}(t)$ in place of $\mathbf{v}(t)$ via the transformation [cf. (16-49)]

$$p_j = L_{v_j}(t, \mathbf{y}, \mathbf{v}) \qquad j = 1, \ldots, n \tag{16-58}$$

and a function $H(t, \mathbf{y}, \mathbf{p})$ in place of $L(t, \mathbf{y}, \mathbf{v})$ via [cf. (16-53)]

$$H(t, \mathbf{y}, \mathbf{p}) = \sum_{j=1}^{n} v_j p_j - L(t, \mathbf{y}, \mathbf{v}) \tag{16-59}$$

Then (16-56) becomes

$$p_j' = L_{y_j} \qquad j = 1, \ldots, n \tag{16-60}$$

where, as usual, the prime means d/dt. [Recall that $\mathbf{p} = \mathbf{p}(t)$.]

Now from (16-59) we have

$$H_{y_j} = -L_{y_j} \qquad j = 1, \ldots, n \tag{16-61}$$

so that (16-60) can be written

$$p_j' = -H_{y_j}(t, \mathbf{y}, \mathbf{p}) \qquad j = 1, \ldots, n \tag{16-62}$$

On the other hand, from (16-59) we also have

$$H_{p_j} = v_j \qquad j = 1, \ldots, n \tag{16-63}$$

and therefore (16-57) can be written

$$y_j' = H_{p_j}(t, \mathbf{y}, \mathbf{p}) \qquad j = 1, \ldots, n \tag{16-64}$$

Thus, we have replaced the first-order system (16-56) and (16-57), which is a system for the $2n$ unknowns $\mathbf{y}(t)$ and $\mathbf{v}(t)$, by the first-order system

$$y_j' = H_{p_j}(t, \mathbf{y}, \mathbf{p}) \qquad j = 1, \ldots, n \tag{16-65}$$

$$p_j' = -H_{y_j}(t, \mathbf{y}, \mathbf{p}) \qquad j = 1, \ldots, n \tag{16-66}$$

for the $2n$ unknowns $\mathbf{y}(t)$ and $\mathbf{p}(t)$.

But notice what the system (16-65) and (16-66) is. It is just Hamilton's equations, i.e., the characteristic equations for the PDE (16-51) (with τ there called t here and $\boldsymbol{\beta}$ there called $\mathbf{y}$ here). Thus we see that Hamilton's equations are just the Euler equations for the minimizing functions of (16-21) written in terms of appropriate new variables.

Once we have seen this, we also see that we could get Hamilton's equations directly by replacing the variables in the integrand of (16-21) before deriving the Euler equations. In other words, if we consider the problem of minimizing the functional

$$\mathscr{H}[\boldsymbol{\phi}, \boldsymbol{\psi}] = \int_{\tau_0}^{T} \left\{ \sum_{j=1}^{n} \phi_j'(t)\, \psi_j(t) - H[t, \boldsymbol{\phi}(t), \boldsymbol{\psi}(t)] \right\} dt \tag{16-67}$$

where $H(t, \mathbf{y}, \mathbf{p})$ is related to $L(t, \mathbf{y}, \mathbf{v})$ via (16-59), we find that the Euler equations for minimizing the functional (16-67) of $2n$ functions are precisely (16-65) and (16-66).

Since the Euler equations for the functional (16-67) and the characteristic equations for the PDE (16-51) are the same, we can say that the extremals of (16-67) and the characteristics of (16-51) are the same. We can say this provided we take the liberty we have taken in past chapters of calling the projections of the characteristics into the space of the independent variables characteristics again. Properly speaking, the characteristics of (16-51) are curves in the $(n + 2)$-dimensional $u\tau\boldsymbol{\beta}$ space, whereas the extremals of (16-67) are curves in the $(n + 1)$-dimensional $\tau\boldsymbol{\beta}$ space, and it is the projections of the characteristics of (16-51) into $\tau\boldsymbol{\beta}$ space which are the extremals of (16-67).

This fact provides us with an alternate way of finding the extremals of (16-67). If we can find the characteristics of (16-51), then their projections into $\tau\boldsymbol{\beta}$ space will give us the extremals of (16-67). But back in Chap. 8 we gave a method, Jacobi's, for finding the characteristics by forming envelopes of a complete integral of (16-51). So all we need is a complete integral of (16-51).

In the present case, we just happen to have a complete integral of (16-51) lying around, waiting to be put to use. This complete integral will not necessarily aid in actually finding the solutions of (16-65) and (16-66), but it will give a great deal of additional insight into the connection between the extremals, i.e., the (projections of the) characteristics of (16-51), and the integral surfaces of (16-51).

In deriving the Hamilton-Jacobi equation in the present chapter, we started out with the extremal integral (16-24), thinking of the point $(\tau_0, \boldsymbol{\alpha})$ as fixed and the point $(\tau, \boldsymbol{\beta})$ as variable, and we found that u, as a function of $(\tau, \boldsymbol{\beta})$, satisfied the Hamilton-Jacobi equation

$$u_\tau + H(\tau, \boldsymbol{\beta}, \nabla u) = 0 \tag{16-68}$$

This is true for any integral of the form (16-24). If we think of $(\tau_0, \boldsymbol{\alpha})$ as parameters, then the $u(\tau_0, \boldsymbol{\alpha}, \tau, \boldsymbol{\beta})$ defined by (16-24) is a solution of (16-68) depending on the $n + 1$ parameters $(\tau_0, \boldsymbol{\alpha})$, in other words, a complete integral of (16-68).

According to Chap. 8, we have to put down one arbitrary relation among the parameters and then form the envelope. But one equation connecting $n + 1$ parameters represents an n-dimensional surface lying in the $(n + 1)$-dimensional space

$\mathscr{S}$:
$$\Phi(\tau_0, \boldsymbol{\alpha}) = 0 \tag{16-69}$$

Thus the general solution u of (16-68) represents the generalized distance from points all of which lie on the surface (16-69).

16-11. WAVEFRONTS AND RAYS; HUYGENS' PRINCIPLE

Although we have mentioned only the mechanical origin of the integral (16-21), integrals of this type also arise in optics, where their values represent the time taken by a ray of light to traverse an optical path in an inhomogeneous anisotropic medium. *Fermat's principle* says that the path actually chosen by the light ray is the one which requires the least time, i.e., the one which minimizes (16-21). Thus the extremals of (16-21) in this interpretation are the paths of the light rays. The distinguished variable t is the curve parameter, which may or may not be the time at which the light ray arrives at a particular point on the extremal. In any event, this t is not the time which is minimized. The value of the integral is the total time taken by the ray in traveling from one end point to the other, and it is this time which is minimized.

More precisely, in this optical interpretation the value of $u(\tau, \boldsymbol{\beta})$, as given by (16-24), is the time it takes a light ray emanating from $(\tau_0, \boldsymbol{\alpha})$ to reach $(\tau, \boldsymbol{\beta})$. If $(\tau_0, \boldsymbol{\alpha})$ is held fixed, then the level surfaces of $u(\tau, \boldsymbol{\beta})$ constitute "spherical" (in the metric of L) wavefronts of light emanating from $(\tau_0, \boldsymbol{\alpha})$ and the extremals are the rays.

Using the Hamilton-Jacobi theory of Chap. 8, we can put these spherical waves to a very interesting use. According to Chap. 8, any complete integral of (16-51) will, upon proper envelope formation, yield the general solution of the PDE. Suppose we take the special complete integral (16-24), where the parameters $(\tau_0, \boldsymbol{\alpha})$ lie on the surface (16-69). About each point $(\tau_0, \boldsymbol{\alpha})$ we take a spherical wavefront $u = R$ of "radius" R. The envelope w of these spherical wavefronts will necessarily have $w = R$ as a level surface. Since such an envelope is a general solution of the Hamilton-Jacobi equation, it follows that the general solution can be looked at as a set of wavefronts propagating outward from the initial surface, $\Phi = 0$, and generated by a set of small spherical waves, one emanating from each point of the initial surface. This state of affairs is called *Huygens' principle.*

The foregoing analysis shows that the level surfaces of any solution of the Hamilton-Jacobi equation form a set of parallel surfaces in the metric defined by L. The converse is also true, because it merely says that the extremal integral as a function of its upper limit satisfies the Hamilton-Jacobi equation. It does because this is how we derived that equation for it.

16-12. TRANSVERSALITY IN TWO DIMENSIONS; FIELDS OF EXTREMALS

It is of interest to see how the rays intersect the wavefronts, i.e., how the extremals intersect the level surfaces of the extremal integral.

To determine a particular extremal, we have to fix its end points. Suppose we assume that $(\tau_0, \boldsymbol{\alpha})$ is fixed in the initial surface $\mathscr{S}$ given by (16-69) and $(\tau, \boldsymbol{\beta})$ is fixed in some parallel surface $\mathscr{S}_R$ given by

$$\mathscr{S}_R: \qquad \Phi(\tau, \boldsymbol{\beta}) = R \tag{16-70}$$

Since $\mathscr{S}_R$ is a level surface of the extremal integral, that integral is constant on it. Hence it does not matter which point $(\tau, \boldsymbol{\beta})$ we choose on $\mathscr{S}_R$. We can therefore choose $(\tau, \boldsymbol{\beta})$ to be fixed in space while $(\tau_0, \boldsymbol{\alpha})$ remains any point on $\mathscr{S}$. This means that a typical extremal minimizes the distance between the fixed point $(\tau, \boldsymbol{\beta})$ and the fixed surface $\mathscr{S}$. This is a new type of minimum problem. In order to see how to solve it, we first consider the simplest case, namely, that in which the functional depends on only one independent function and the fixed "surface" is therefore a curve. Moreover, since interchanging the limits on an integral merely changes the sign of the integral, we can assume that the lower limit is fixed while the upper one is the one which lies on the given curve. Then the problem is to minimize

$$\mathscr{F}[\phi] = \int_a^\xi F[x, \phi(x), \phi'(x)]\, dx \tag{16-71}$$

among all functions $\phi(x)$ satisfying the BCs

$$\phi(a) = \alpha \tag{16-72}$$

and

$$\Phi[\xi, \phi(\xi)] = 0 \tag{16-73}$$

where α is a given constant and $\Phi(x, y)$ is a given function.

Geometrically stated, the problem is to find among all curves joining the fixed point (a, α) to any point on the curve $\Phi(x, y) = 0$ that one for which (16-71) is a minimum.

If, as usual, we put

$$\phi(x) = \phi_0(x) + \mu\psi(x) \tag{16-74}$$

where ϕ_0 is the minimizing function and μ is a parameter, then the perturbation ψ must satisfy the BCs

$$\psi(a) = 0 \tag{16-75}$$

and

$$\Phi[\xi, \phi_0(\xi) + \mu\psi(\xi)] = 0 \tag{16-76}$$

The BC (16-76) shows that the upper limit ξ on the integral and the parameter μ are not independent but rather must be connected by Eq. (16-76). If we think of $\xi = \gamma(\mu)$ as being defined implicitly by (16-76), then we have, in the notation of Sec. 16-5,

$$f(\mu) = \int_a^{\gamma(\mu)} F[x, \phi_0(x) + \mu\psi(x), \phi_0'(x) + \mu\psi(x)]\, dx \tag{16-77}$$

Hence

$$f'(0) = \gamma'(0)\, F\{\gamma(0), \phi_0[\gamma(0)], \phi_0'[\gamma(0)]\} + \int_a^{\gamma(0)} (F_y\psi + F_{y'}\psi')\, dx \tag{16-78}$$

For simplicity we put

$$\gamma(0) = b \qquad \phi_0[\gamma(0)] = \phi_0(b) = \beta \qquad \phi_0'(b) = v \tag{16-79}$$

Then (16-78) reads

$$f'(0) = \gamma'(0)\, F(b, \beta, v) + \int_a^b (F_y\psi + F_{y'}\psi')\, dx \tag{16-80}$$

The usual integration by parts yields the Euler expression under the integral, and the usual arguments (cf. Secs. 16-5 and 16-6) again yield Euler's equation. Moreover, since ψ vanishes at $x = a$, we get nothing new from that end point. The interesting end point, of course, is $x = \gamma(0) = b$, the one on the given curve. Combining the contribution there with the first term in (16-78), we see that the condition there is

$$\gamma'(0)\, F(b, \beta, v) + F_{y'}(b, \beta, v)\, \psi(\xi) = 0 \tag{16-81}$$

To simplify this, we recall that $\xi = \gamma(\mu)$ is the solution to (16-76), so that (16-76) is an identity in μ if we insert $\gamma(\mu)$ for ξ in it. Differentiating this identity with respect to μ and then setting $\mu = 0$, we get [recall $\Phi = \Phi(x, y)$]

$$\Phi_x(b, \beta)\, \gamma'(0) + \Phi_y(b, \beta)[\phi_0'(b)\, \gamma'(0) + \psi(b)] = 0 \tag{16-82}$$

Hence [recall that $\phi_0'(b) = v$]

$$\gamma'(0) = \frac{-\Phi_y\psi}{\Phi_x + v\Phi_y} \tag{16-83}$$

Inserting this into (16-81), we have

$$\left(\frac{-\Phi_y F}{\Phi_x + v\Phi_y} + F_{y'}\right)\psi = 0$$

for arbitrary ψ. Hence the minimizing function $\phi_0(x)$ must satisfy

$$F_{y'}(b, \beta, v) - \frac{\Phi_y(b, \beta)\, F(b, \beta, v)}{\Phi_x(b, \beta) + v\Phi_y(b, \beta)} = 0 \tag{16-84}$$

where $\phi_0(b) = \beta$ and $\phi_0'(b) = v$.

Equation (16-84) is one equation connecting the two coordinates (b, β) of the end point of the extremal and the slope, $v = \phi_0'(b)$, of the extremal there. In addition, we also have the equation

$$\Phi(b, \beta) = 0 \tag{16-85}$$

expressing the fact that the end point (b, β) lies on the curve $\Phi = 0$.

If we think of (16-85) as solved for b in terms of β and the result inserted into (16-84), then (16-84) becomes an equation connecting β, the value of ϕ_0 at b, with v, the value of ϕ_0' there—in other words, a nonlinear BC on ϕ_0.

At the outset of Sec. 16-8 we mentioned that it is sometimes of interest to consider the extremals themselves, divorced from any minimum problems. (Recall that an extremal, by definition, is any solution of Euler's equation.) The present case is such an instance. Suppose we take a curve $\mathscr{C}$ given by (16-85). Since the extremals form a two-parameter family of curves, there will be, in general, a one-parameter family of extremals through each point (b, β) of $\mathscr{C}$. If we impose on this family of extremals the additional condition (16-84), then the resulting subfamily of extremals consists of those extremals going through (b, β) with slope v there given by (16-84). If for simplicity we assume that (16-84) has a unique solution v for each point (b, β) on $\mathscr{C}$, then there will be a unique extremal going through each point (b, β) on $\mathscr{C}$ and having the direction $v(b, \beta)$ determined from (16-84) there.

Thus, in the case where (16-84) has a unique solution v for each (b, β), there is a one-parameter family of extremals through $\mathscr{C}$, one through each point (b, β) of $\mathscr{C}$ and having a definite slope v there. Such a family of extremals is called a *field of extremals*; the direction in which a given member of the field crosses $\mathscr{C}$ is said to be *transverse* to $\mathscr{C}$, and the condition (16-84) which determines this transverse direction is called the *transversality condition.*

In the general case, (16-84) may have, for a given (b, β), more than one root v. Then there will be several transverse directions at (b, β) and therefore no unique field of extremals through $\mathscr{C}$. It is interesting to note, however, that although the transversality condition (16-84) is nonlinear in v, the slope of the extremal, it is linear in Φ_x and Φ_y and hence in the slope of the curve $\mathscr{C}$. Thus, although, given a curve $\mathscr{C}$, there may be several different transverse directions at a point on it, given an extremal $\mathscr{E}$, there can be only one transverse direction at each point on it.

The transversality condition (16-84) can obviously be written in the form

$$\frac{F_{y'}}{F} = \frac{\Phi_y}{\Phi_x + v\Phi_y} \tag{16-86}$$

or, by employing the usual elementary algebraic tricks with ratios,

$$\frac{F_{y'}}{F - vF_{y'}} = \frac{\Phi_y}{\Phi_x} \tag{16-87}$$

Note that the right side of (16-87) is just the slope of $\mathscr{C}$.

16-13. TRANSVERSALITY AND FIELDS IN *n* DIMENSIONS

In the n-dimensional case we have

$$\mathscr{L}[\phi] = \int_{\tau_0}^{\gamma(\mu)} L(t, \phi, \phi')\, dt \tag{16-88}$$

with the BCs

$$\boldsymbol{\phi}(\tau_0) = \boldsymbol{\alpha} \tag{16-89}$$

and

$$\Phi\{\gamma(\mu), \boldsymbol{\phi}[\gamma(\mu)]\} = 0 \tag{16-90}$$

where $\boldsymbol{\alpha}$ is a given constant vector and $\Phi(x, \mathbf{y})$ is a given function of $n + 1$ variables.

As in the two-dimensional case, we find

$$f'(0) = \gamma'(0)\, L(\tau, \boldsymbol{\beta}, \mathbf{v}) + \sum_{j=1}^{n} \int_{\tau_0}^{\tau} (L_{x_j}\psi_j + L_{x'_j}\psi'_j)\, dt \tag{16-91}$$

where

$$\gamma(0) = \tau \qquad \boldsymbol{\phi}[\gamma(0)] = \boldsymbol{\phi}(\tau) = \boldsymbol{\beta} \qquad \boldsymbol{\phi}'(\tau) = \mathbf{v} \tag{16-92}$$

and therefore the BC

$$\gamma'(0)\, L(\tau, \boldsymbol{\beta}, \mathbf{v}) + \sum_{j=1}^{n} L_{x'_j}\psi_j = 0 \tag{16-93}$$

Here $\gamma(\mu)$ and $\boldsymbol{\psi}(t)$ satisfy

$$\Phi\{\gamma(\mu), \boldsymbol{\phi}_0[\gamma(\mu)] + \mu\boldsymbol{\psi}[\gamma(\mu)]\} = 0 \tag{16-94}$$

Hence

$$\gamma'(0)\, \Phi_\tau(\tau, \boldsymbol{\beta}) + \sum_{j=1}^{n} \Phi_{\beta_j}(\tau, \boldsymbol{\beta})[v_j(\tau)\, \gamma'(0) + \psi_j(\tau)] = 0 \tag{16-95}$$

where, as usual, $\mathbf{v} = \boldsymbol{\phi}'(\tau)$.

From (16-95) we have

$$\gamma'(0) = -\left(\Phi_\tau + \sum_{j=1}^{n} v_j\Phi_{\beta_j}\right)^{-1} \sum_{j=1}^{n} \Phi_{\beta_j}\psi_j \tag{16-96}$$

Substituting (16-96) into (16-93) and using the fact that $\boldsymbol{\psi}(\tau)$ is arbitrary, we find the n *transversality conditions*

$$\frac{L_{v_j}(\tau, \boldsymbol{\beta}, \mathbf{v})}{L(\tau, \boldsymbol{\beta}, \mathbf{v})} = \frac{\Phi_{\beta_j}}{\Phi_\tau(\tau, \boldsymbol{\beta}) + \sum_{j=1}^{n} \Phi_{\beta_j}(\tau, \boldsymbol{\beta})\, v_j} \qquad j = 1, \ldots, n \tag{16-97}$$

By elementary algebra (16-97) can be rewritten in the form

$$\frac{L_{v_j}}{L - \sum_{j=1}^{n} v_j L_{v_j}} = \frac{\Phi_{\beta_j}}{\Phi_\tau} \qquad j = 1, \ldots, n \tag{16-98}$$

Equation (16-98) is a sort of Lagrangian formulation of the transversality conditions. If we introduce the canonical momenta, i.e., the p_j's defined by (16-58) and the Hamiltonian H defined by (16-59), then (16-98) can be written in the more compact form

$$-\frac{p_j}{H} = \frac{\Phi_{\beta_j}}{\Phi_\tau} \qquad j = 1, \ldots, n \tag{16-99}$$

which is a sort of Hamiltonian formulation of the transversality conditions.

As in two dimensions, it pays to divorce the extremals from the minimum problem from which they arose. If we pick an arbitrary point $(\tau, \boldsymbol{\beta})$ on the surface

$\mathscr{S}$:

$$\Phi(\tau, \boldsymbol{\beta}) = 0 \tag{16-100}$$

then the n transversality conditions (16-97) will determine at that point one or more directions $\mathbf{v}$ transverse to $\mathscr{S}$. If (16-97) determine a unique direction at every point of $\mathscr{S}$, then we say that the resulting one-parameter family of extremals forms a *field of extremals transverse* to $\mathscr{S}$.

Since the transversality conditions (16-97) involve only the derivatives of Φ and not Φ itself, it is natural to conjecture that a field of extremals transverse to a given surface $\Phi = 0$ is also transverse to all surfaces of the family $\Phi = \text{const}$. However, we have already seen that not all "natural" conjectures are true without qualification. This one isn't either. The reason is that the derivatives of Φ have to be evaluated on different surfaces for different members of the family and it is not a priori clear that these different values fit together coherently. In fact, we will see that they do not fit together in general but only if the family of surfaces has some special properties. Essentially, some integrability conditions are involved.

In order to gain further insight into what is involved here, we choose a family of surfaces with which we are familiar. We recall that the general solution $u(\tau, \boldsymbol{\beta})$ of the Hamilton-Jacobi equation represents the distance (in the metric of L) from an arbitrary surface, say from the surface $\mathscr{S}$ given by (16-100), and that the level surfaces

$$u(\tau, \boldsymbol{\beta}) = R \tag{16-101}$$

form a family of parallel surfaces in the metric of L. Now suppose that the level surfaces

$$\Phi(\tau, \boldsymbol{\beta}) = R \tag{16-102}$$

coincide with those of u. This means that there exists a function $\Psi(R)$ such that

$$\Phi(\tau, \boldsymbol{\beta}) = \Psi[u(\tau, \boldsymbol{\beta})] \tag{16-103}$$

Hence

$$\Phi_\tau = \Psi' u_\tau \qquad \text{and} \qquad \Phi_{\beta_j} = \Psi' u_{\beta_j} \qquad j = 1, \ldots, n \tag{16-104}$$

throughout the region of $\tau\boldsymbol{\beta}$ space containing the level surfaces under consideration. Therefore

$$\frac{\Phi_\tau}{u_\tau} = \frac{\Phi_{\beta_j}}{u_{\beta_j}} \qquad j = 1, \ldots, n \tag{16-105}$$

Since u is a solution of the Hamilton-Jacobi equation, $u_\tau = -H$ and $u_{\beta_j} = p_j$. Using these facts in (16-105), we find

$$\frac{\Phi_\tau}{-H} = \frac{\Phi_{\beta_j}}{p_j} \qquad j = 1, \ldots, n \tag{16-106}$$

which are precisely the transversality conditions (16-99). Thus, the level surfaces of any solution of the Hamilton-Jacobi equation cut all the corresponding extremals transversally. We shall show that, conversely, if all surfaces of a family of surfaces of the form (16-102) cut a field of extremals transversally, then that family is related to the level surfaces of some solution of the Hamilton-Jacobi equation as in (16-103). To see this, note that the transversality conditions (16-99), when written in the form

$$\frac{\Phi_\tau}{-H} = \frac{\Phi_{\beta_j}}{p_j} \qquad j = 1, \ldots, n \tag{16-107}$$

imply the existence of a function $\lambda(\tau, \boldsymbol{\beta})$ such that

$$\Phi_\tau = -\lambda H \qquad \Phi_{\beta_j} = \lambda p_j \qquad j = 1, \ldots, n \tag{16-108}$$

where, as usual, $p_j = u_{\beta_j}$, u being the solution of the Hamilton-Jacobi equation which generates the extremals with direction numbers $\mathbf{v}$ related to $\mathbf{p}$ via (16-58). Inserting $-H = u_\tau$ and $p_j = u_{\beta_j}$ into (16-106) and combining the resulting equations, we have

$$d\Phi = \lambda(\tau, \boldsymbol{\beta})\, du \tag{16-109}$$

If we consider (16-109) on an extremal with running parameter R, then it reads

$$\frac{d\Phi}{dR} = \lambda(R) \frac{du}{dR} \tag{16-110}$$

or

$$\frac{d\Phi}{du} = \lambda(R) \tag{16-111}$$

Since all points on the surface $u = R$ are at "distance" R from $\Phi = 0$,

(16-111) holds on any extremal, hence throughout the region. Moreover, since the values of R as we traverse an extremal obviously coincide with the values of u at corresponding points ($u = R$), we can replace R by u in (16-111) to get

$$\frac{d\Phi}{du} = \lambda(u) \tag{16-112}$$

Integration of (16-112) yields (16-103), the desired result.

16-14. COMPATIBILITY CONDITIONS FOR FIELDS

As mentioned above, an n-parameter family of extremals transverse to a given surface is called a field. In the special case where the "surface" degenerates to a point, the field is called a *central field.* Since the totality of extremals forms a $2n$-parameter family of curves, the question naturally arises whether an arbitrary n-parameter family of extremals can form a field or whether there are some conditions that have to be satisfied before a given n-parameter family of extremals can form a field.

To answer this question it is convenient to use the Hamiltonian formulation. In this formulation the extremals are described by $2n$ functions, $\mathbf{y}(t)$ and $\mathbf{p}(t)$, $\mathbf{y}(t)$ being the coordinates of a particular point on an extremal and $\mathbf{p}(t)$ being defined in terms of $\mathbf{y}(t)$ and $\mathbf{v}(t)$, the direction of the extremal, by (16-58). The extremals are determined by solving the system of ODEs (16-65) and (16-66).

Now suppose we have an n-parameter family of solutions of (16-65) and (16-66)

$$\mathbf{y} = \mathbf{y}(t, \boldsymbol{\alpha}) \tag{16-113}$$

$$\mathbf{p} = \mathbf{p}(t, \boldsymbol{\alpha}) \tag{16-114}$$

If this family forms a field, its members must all be transverse to a family of level surfaces of some solution of the Hamilton-Jacobi equation. Suppose $u(t, \mathbf{y})$ is that solution. If we insert (16-113) into u we obtain a function

$$w(t, \boldsymbol{\alpha}) = u[t, \mathbf{y}(t, \boldsymbol{\alpha})] \tag{16-115}$$

From (16-115) and the fact that $p_j = u_{y_j}$ we obtain

$$w_{\alpha_j} = \sum_{l=1}^{n} u_{y_l} \frac{\partial y_l}{\partial \alpha_j} = \sum_{l=1}^{n} p_l \frac{\partial y_l}{\partial \alpha_j} \qquad j = 1, \ldots, n \tag{16-116}$$

and therefore

$$w_{\alpha_j \alpha_k} = \sum_{l=1}^{n} \left(\frac{\partial p_l}{\partial \alpha_k} \frac{\partial y_l}{\partial \alpha_j} + p_l \frac{\partial^2 y_l}{\partial \alpha_j \, \partial \alpha_k} \right) \qquad j, k = 1, \ldots, n \tag{16-117}$$

Interchanging j and k and subtracting, we find

$$\sum_{l=1}^{n}\left(\frac{\partial y_l}{\partial \alpha_j}\frac{\partial p_l}{\partial \alpha_k}-\frac{\partial y_l}{\partial \alpha_k}\frac{\partial p_l}{\partial \alpha_j}\right)=0 \qquad j, k = 1, \ldots, n \tag{16-118}$$

that is,

$$[\alpha_j, \alpha_k] = 0 \qquad j, k = 1, \ldots, n \tag{16-119}$$

where (16-119) are the Lagrange brackets introduced in Chap. 8.

Thus we have shown that if an n-parameter family of extremals is to form a field, then the Lagrange brackets (16-119) must vanish. The converse is also true, for if Eqs. (16-119) hold, then rewriting (16-118) in the form

$$\frac{\partial}{\partial \alpha_j}\left(\sum_{l=1}^{n} p_l \frac{\partial y_l}{\partial \alpha_k}\right)=\frac{\partial}{\partial \alpha_k}\left(\sum_{l=1}^{n} p_l \frac{\partial u_l}{\partial \alpha_j}\right) \tag{16-120}$$

we see that there exists a function $\omega(t, \boldsymbol{\alpha})$ such that

$$\omega_{\alpha_j} = \sum_{l=1}^{n} p_l \frac{\partial y_l}{\partial \alpha_j} \qquad j = 1, \ldots, n \tag{16-121}$$

If we solve (16-113) for $\boldsymbol{\alpha}$ to get $\boldsymbol{\alpha} = \boldsymbol{\alpha}(t, \mathbf{y})$ and insert this into $\omega(t, \boldsymbol{\alpha})$, we find

$$\omega(t, \boldsymbol{\alpha}) = \omega[t, \boldsymbol{\alpha}(t, \mathbf{y})] = \Omega(t, \mathbf{y}) \tag{16-122}$$

(say). Therefore

$$\omega_{\alpha_j} = \sum_{k=1}^{n} \Omega_{y_k} \frac{\partial y_k}{\partial \alpha_j} \qquad j = 1, \ldots, n \tag{16-123}$$

Comparing (16-121) and (16-123), we see that

$$\sum_{k=1}^{n} p_k \frac{\partial y_k}{\partial \alpha_j} = \sum_{k=1}^{n} \Omega_{y_k} \frac{\partial y_k}{\partial \alpha_j} \qquad j = 1, \ldots, n \tag{16-124}$$

Since $\det(\partial y_k/\partial \alpha_j) \neq 0$ if the y's really contain n independent parameters $\boldsymbol{\alpha}$, (16-124) implies

$$p_k = \Omega_{y_k} \qquad k = 1, \ldots, n \tag{16-125}$$

Equations (16-125) look like the last n of the transversality conditions (16-108) with $\lambda = 1$. They therefore suggest that the level surfaces of Ω comprise the family of surfaces to which the extremals given by (16-113) and (16-114) are transverse. To prove this we have only to show that $\Omega_t + H = 0$, that is, that Ω is a solution of the Hamilton-Jacobi equation. To do this, we compute Ω_t by differentiating (16-125) with respect to t and then using

Hamilton's equations (which all extremals necessarily satisfy) to rewrite the resulting equation in a form in which it can be integrated with respect to y_k.

From (16-125) we have

$$p'_k = \Omega_{ty_k} + \sum_{j=1}^{n} \Omega_{y_j y_k} y'_j \qquad k = 1, \ldots, n \tag{16-126}$$

Transposing and using Hamilton's equations, we can write (16-126) in the form

$$\Omega_{ty_k} + H_{y_k} + \sum_{j=1}^{n} H_{p_j} \Omega_{y_j y_k} = 0 \qquad k = 1, \ldots, n \tag{16-127}$$

But since $H = H(t, \mathbf{y}, \nabla\Omega)$, where $\nabla\Omega = (\Omega_{y_1}, \ldots, \Omega_{y_n})$, (16-127) can be written in the form

$$\frac{\partial}{\partial y_k} [\Omega_t + H(t, \mathbf{y}, \nabla\Omega)] = 0 \qquad k = 1, \ldots, n \tag{16-128}$$

Integration of (16-128) gives

$$\Omega_t + H(t, \mathbf{y}, \nabla\Omega) = f(t) \tag{16-129}$$

where $f(t)$ is an arbitrary function of t. To get a Hamilton-Jacobi equation out of (16-129), we merely absorb $f(t)$ into the Hamiltonian H. This means that we have changed the integrand L of our variational problem by adding $f(t)$ to it, but the integral of $f(t)$ with respect to t is obviously independent of path and therefore influences the extremals not at all. So Ω satisfies an appropriate Hamilton-Jacobi equation, and our result is proved.

Notice, incidentally, that for $n = 1$, that is, in two dimensions (the number of dimensions is $2n$), (16-119) is vacuous. So in two dimensions any one-parameter family of extremals forms a field.

16-15. HOMOGENEOUS INTEGRANDS

In the important special case when $L(t, \mathbf{x}, \mathbf{v})$ is homogeneous of degree one in $\mathbf{v}$, we have, by Euler's relation for homogeneous functions,

$$\sum_{j=1}^{n} v_j L_{v_j} = L \tag{16-130}$$

Hence

$$\sum_{j=1}^{n} v_j L_{v_j v_k} + L_{v_k} = L_{v_k} \qquad k = 1, \ldots, n \tag{16-131}$$

that is,

$$\sum_{j=1}^{n} L_{v_j v_k} v_j = 0 \qquad k = 1, \ldots, n \tag{16-132}$$

which, since $\mathbf{v} \not\equiv 0$, implies $\det(L_{v_j v_k}) = 0$ and therewith the failure of the assumption of Sec. 16-9 that this determinant is different from zero. This means that the sort of Legendre transformation made previously cannot be used now, but that proves to be of little difficulty. The theory can still be developed and a Hamilton-Jacobi equation derived. Because of the lack of space here we omit details. The reader should consult [65, vol. II], [232], or [243] for the treatment of this homogeneous case. As mentioned above, it is not without importance. In particular, the case of geodesics in an ordinary Riemannian metric falls into this category.

16-16. GENERATING FUNCTIONS FOR CANONICAL TRANSFORMATIONS AGAIN

We have seen that a necessary condition for a functional of the form (16-3) to have a minimum is that Euler's equation (16-15) be satisfied. Euler's equation thus plays in function space a role analogous to that of the first derivative in ordinary calculus. For this reason the left side of Euler's equation is sometimes called the *functional derivative* of (16-3).

It is of interest to push the analogy a little further. If for an ordinary function $f(x)$ the derivative $f'(x)$ vanishes identically, we can conclude that that function is a constant. Analogously, suppose that the functional derivative of (16-3) vanishes identically, i.e., Euler's equation (16-15) is satisfied identically. In this case (16-15), instead of being a DE for $y(x)$, is satisfied identically by all functions $y(x)$. Reasoning by analogy with calculus, we conjecture that the functional (16-3) then does not depend on $y(x)$. This means that the integral in (16-3) must be independent of the path because $y(x)$ is the representation of the curve over which the integral is taken.

We now show that this conjecture is correct. If (16-15) vanishes identically for all *functions* $y(x)$ (notice that this is not the same as vanishing identically in the *variable* y), then from the form (16-16) we see that $F_{y'y'}$, the coefficient y'', must vanish identically. We cannot conclude that the coefficient of y' vanishes because that coefficient contains y' itself, as do the other two terms in (16-16). Nevertheless, since $y''F_{y'y'}$ is the only term containing y'', we can conclude that it vanishes identically. But if $F_{y'y'} = 0$, then F must be linear in y'

$$F(x, y, y') = B(x, y)\, y' + A(x, y) \tag{16-133}$$

From (16-133) we have

$$
\begin{aligned}
F_{y'} &= B \\
F_{xy'} &= B_x \\
F_{yy'} &= B_y \\
F_y &= B_y\, y' + A_y
\end{aligned}
$$

Inserting these results into (16-16), we find

$$A_y - B_x = 0 \tag{16-134}$$

Thus, the functional (16-3) must look like

$$\mathscr{F}[x, y, y'] = \int_a^b [B(x, y)\, y' + A(x, y)]\, dx \tag{16-135}$$

or

$$\mathscr{F}[x, y, y'] = \int_a^b [A(x, y)\, dx + B(x, y)\, dy] \tag{16-136}$$

where A and B satisfy (16-134). As conjectured, (16-134) is just the condition that (16-136) be independent of path.

Conversely, if $\mathscr{F}$ is of the form (16-136), where A and B satisfy (16-134), i.e., if the integral in (16-136) is independent of the path, then its value is obviously constant.

Just as in ordinary calculus, we can extend the foregoing results.

Theorem 16-1 If two functionals of the form (16-3) have the same extremals, then their integrands differ by an exact differential. Conversely, if their integrands differ by an exact differential, then the two functionals have the same extremals.

We leave it to the reader as an exercise to verify that Theorem 16-1 carries over to functionals of the form (16-21). This fact enables us to rediscover the generating function of a canonical transformation.

In the preceding sections we have seen that the functions $\mathbf{x}(t)$ and $\mathbf{p}(t)$ which minimize the functional

$$\mathscr{H}[\mathbf{x}, \mathbf{p}] = \int_{\tau_0}^{\tau} \left\{ \sum_{j=1}^{n} x_j' p_j - H(t, \mathbf{x}, \mathbf{p}) \right\} dt \tag{16-137}$$

are also the solutions of Hamilton's equations. As previously explained the properties of being a solution of Hamilton's equations and that of being an extremal of (16-137) are equivalent. Now suppose we transform from the variables $\mathbf{x}$ and $\mathbf{p}$ to new variables $\tilde{\mathbf{x}}$ and $\tilde{\mathbf{p}}$. The new variables $\tilde{\mathbf{x}}$ and $\tilde{\mathbf{p}}$ will describe the same mechanical motion if and only if they describe the same extremals of (16-137). According to Theorem 16-1, extended to the case of (16-137), this means that the integrands in the two coordinate systems must differ by an exact differential; i.e.,

$$\sum_{j=1}^{n} x'_j p_j - H(t, \mathbf{x}, \mathbf{p}) = \sum_{j=1}^{n} \tilde{x}'_j \tilde{p}_j - H(t, \tilde{\mathbf{x}}, \tilde{\mathbf{p}}) + d\Phi \tag{16-138}$$

Comparing (16-138) with (8-173) we see that the transformation is canonical and Φ is its generating function.

16-17. MULTIPLE INTEGRALS; PARTIAL DIFFERENTIAL EQUATIONS AS EULER EQUATIONS

Thus far all our functionals have had argument functions that were functions of a single variable. The resulting Euler equations have therefore been ODEs. As is to be expected, when we turn to functionals whose argument functions are functions of several independent variables, the resulting Euler equations will be PDEs.

For simplicity we restrict ourselves to the functional

$$\mathscr{F}[u(x, y)] = \iint_{\mathscr{R}} F[x, y, u(x, y), u_x(x, y), u_y(x, y)]\, dx\, dy \tag{16-139}$$

where $F(x, y, u, p, q)$ is a given function of five independent variables. All the essential ideas will emerge from a treatment of the functional (16-140), and so we shall not discuss functionals whose argument functions depend on more than two independent variables.

As usual, we assume that $u_0(x, y)$ minimizes (16-139) in an appropriate class of functions satisfying the BC

$$u = g(x, y) \qquad \text{on } \partial\mathscr{R} \tag{16-140}$$

and, as usual, we put

$$u(x, y) = u_0(x, y) + \mu v(x, y) \tag{16-141}$$

where

$$v = 0 \qquad \text{on } \partial\mathscr{R} \tag{16-142}$$

Dropping the subscript on u for convenience, we find

$$f'(0) = \frac{d}{d\mu}\mathscr{F}[u + \mu v]\bigg|_{\mu=0} = \iint_{\mathscr{R}} (F_u v + F_p v_x + F_q v_y)\, dx\, dy \tag{16-143}$$

where the arguments of F are the same as in (16-139), u being u_0 with the subscript omitted.

Next we set (16-143) equal to zero and integrate by parts, i.e., use Green's theorem. Taking the BC (16-142) into account, we find

$$\iint_{\mathscr{R}} \left(F_u - \frac{D}{Dx} F_p - \frac{D}{Dy} F_q\right) v \, dx \, dy = 0 \tag{16-144}$$

for arbitrary v vanishing on $\partial\mathscr{R}$. In (16-144) we have again been forced to employ a somewhat less than satisfactory notation. F contains x both explicitly and through u, u_x, and u_y. Because of the way it arose in the integration by parts, the operator D/Dx in (16-144) means the total derivative of F with respect to x, not just F_x. Similarly, D/Dy means the total derivative of F with respect to y. Most authors use the symbols $\partial/\partial x$ and $\partial/\partial y$ for our D/Dx and D/Dy, but since they do not stand for partial derivatives, this can sometimes cause confusion although it is usually clear how they are meant. On the other hand, our notation leaves something to be desired, too, since there are two independent variables, yet we have used straight D's.

From (16-144) and the arbitrariness of v we derive Euler's equation for (16-139)

$$F_u - \frac{D}{Dx} F_p - \frac{D}{Dy} F_q = 0 \tag{16-145}$$

which, regardless of notation, is a second-order (generally nonlinear) PDE for the minimizing function $u(x, y)$. Written out unambiguously, (16-145) reads

$$F_u - F_{xp} - pF_{up} - p_x F_{pp} - q_x F_{pq} - F_{yq} - qF_{uq} - p_y F_{pq} - q_y F_{qq} = 0 \tag{16-146}$$

where, as usual, $p = u_x$ and $q = u_y$ and the arguments of F are x, y, u, p, q.

16-18. EXAMPLE: THE DIRICHLET INTEGRAL

The Euler equation for the special functional

$$\mathscr{D}[u] = \iint_{\mathscr{R}} (u_x^2 + u_y^2) \, dx \, dy = \iint_{\mathscr{R}} (p^2 + q^2) \, dx \, dy \tag{16-147}$$

is

$$u_{xx} + u_{yy} = 0 \tag{16-148}$$

i.e., the potential equation. The integral in (16-148) is called the *Dirichlet integral* and in physical problems has the interpretation of kinetic energy. What we have just discovered is that the minimizing function for the Dirichlet integral is harmonic. This fact is of both theoretical and practical importance. In the calculus of variations it is proved that under proper assumptions on the class of admissible functions satisfying the BC there exists a function which minimizes the Dirichlet integral. From this it follows that the Dirichlet problem, i.e., first boundary-value problem (BVP) for the potential equation, has a solution. Unfortunately the methods involved in proving the existence of a minimizing function for the Dirichlet integral are not simple and must be omitted here. Interested readers can consult any one of numerous books, e.g., Courant and Hilbert [65, vol. II], Garabedian [115], or Pars [232].

16-19. DIRICHLET'S PRINCIPLE; RAYLEIGH-RITZ METHOD

The fact that the function that minimizes the Dirichlet integral over a region $\mathscr{R}$ is harmonic in $\mathscr{R}$ is known as *Dirichlet's principle*. We have just pointed out the theoretical importance of Dirichlet's principle. Its practical importance lies in the fact that it provides a means of approximating a harmonic function in a given region. Since the solution to the Dirichlet problem for the potential equation in a region also minimizes the Dirichlet integral over that region, it is plausible that a function that is a good approximation to the minimizing function ought to be a good approximation to the solution of the Dirichlet problem, too. A simple way of obtaining such an approximation is due to Rayleigh and Ritz, independently. There is some argument about priority, but for the sake of brevity (and no other reason) we will call it Ritz's method.

The idea of the method is somewhat similar to that behind the derivation of the Euler equation itself. Instead of choosing candidates for minimizing function from all admissible functions, we choose them from a restricted class of functions. In particular, we choose n arbitrary, fixed functions, say $\psi_1, \ldots, \psi_n$, and form the function

$$\psi = \sum_{j=1}^{n} \alpha_j \psi_j \tag{16-149}$$

where the α_j's are parameters so chosen that ψ satisfies the given BC.

Since the ψ's are fixed and there is one relation (the BC) among the n α's, (16-149) represents an $(n - 1)$-parameter family of admissible functions. Minimizing the Dirichlet integral in this class means minimizing a function of $n - 1$ independent variables or, equivalently, a function of n independent variables with one subsidiary condition. This is a standard problem in elementary calculus. If we let $\mathscr{D}[\psi] = D(\alpha_1, \ldots, \alpha_n)$, we have merely to apply Lagrange's rule for minimizing a function of n variables with a subsidiary condition. Once we have determined the α's this way, their substitution into (16-149) yields the function ψ which minimizes $\mathscr{D}[\psi]$ in the class of functions of the type (16-149) and therefore, hopefully, an approximation to the minimizing function in the wider original class of admissible functions, i.e., the harmonic function.

The Ritz method can obviously be used on functionals other than the Dirichlet integral and therefore for PDEs other than Laplace's. In some cases it can even be turned into an existence proof if a sequence of functions $\tilde{\psi}_n = \sum_{j=1}^{n} \alpha_j \psi_j$ can be found such that $\lim_{n\to\infty} \tilde{\psi}_n = \psi$ exists and is admissible.

SUMMARY

DEFINITION S16-1

A *functional* is an assignment of numbers to functions.

DEFINITION S16-2

A *linear functional* $\mathscr{L}[\phi]$ is a functional that satisfies $\mathscr{L}[\alpha\phi_1 + \beta\phi_2] = \mathscr{L}[\phi_1] + \beta\mathscr{L}[\phi_2]$ for any two functions ϕ_1, ϕ_2 in the domain of $\mathscr{L}$ and any scalars α, β.

DEFINITION S16-3

A functional $N[\phi]$ that satisfies

1. $N[\phi] > 0 \quad \text{if} \quad \phi \not\equiv 0, \qquad N[0] = 0$
2. $N[\alpha\phi] = |\alpha| N[\phi]$
3. $N[\phi_1 + \phi_2] \leqslant N[\phi_1] + N[\phi_2]$

is called a *norm* and usually denoted by $N[\phi] = \|\phi\|$.

DEFINITION S16-4

A functional $\mathscr{F}[\phi]$ is continuous at ϕ_0 in a norm if given any $\epsilon > 0$ there exists a $\delta > 0$ such that $|\mathscr{F}[\phi] - \mathscr{F}[\phi_0]| < \epsilon$ if $\|\phi - \phi_0\| < \delta$.

Theorem S16-1 A necessary condition that a function $y(x) \in C^{(2)}[a, b]$ minimize

$$\mathscr{F}[y] = \int_a^b F(x, y, y')\, dx \tag{S16-1}$$

is that it satisfy *Euler's equation*

$$F_y(x, y, y') - \frac{d}{dx} F_{y'}(x, y, y') = 0 \tag{S16-2}$$

in $[a, b]$.

DEFINITION S16-5

Any solution of Euler's equation (S16-2) is called an *extremal.*

If no BCs are imposed on the admissible functions in asking for a minimum of (S16-1), then the minimizing function satisfies some BCs anyway, namely, $F_{y'}[a, \phi(a), \phi'(a)] = F_{y'}[b, \phi(b), \phi'(b)] = 0$. Such BCs are called *free boundary conditions*. If a BC is prescribed at one end point but not at the other, there will be a free BC only at the other end point.

Theorem S16-2 The (n) *Euler equations* for the functional

$$\mathscr{L}[\mathbf{x}] = \int_a^b L(t, \mathbf{x}, \mathbf{x}')\,dt \tag{S16-3}$$

are

$$L_{x_j} - \frac{d}{dt} L_{x'_j} = 0 \qquad j = 1, \ldots, n \tag{S16-4}$$

Theorem S16-3 The function of the $n + 1$ independent variables $(\mathbf{x}, t)$ defined by

$$u(\mathbf{x}, t) = \int_{t_0}^t [\tau, \mathbf{q}(\tau), \mathbf{q}'(\tau)]\,d\tau \tag{S16-5}$$

where $\mathbf{q}(\tau)$ is an extremal such that

$$\mathbf{q}(t) = \mathbf{x} \tag{S16-6}$$

satisfies the *Hamilton-Jacobi equation*

$$u_t + H(\mathbf{x}, t, \nabla u) = 0 \tag{S16-7}$$

where $H(\mathbf{x}, t, \mathbf{p})$ is the *Hamiltonian* related to the integrand $L(t, \mathbf{x}, \mathbf{v})$ of (S16-5) by

$$H(\mathbf{x}, t, \mathbf{p}) = \sum_{j=1}^n p_j v_j - L(t, \mathbf{x}, \mathbf{v}) \tag{S16-8}$$

the variables $\mathbf{p}$ and $\mathbf{v}$ being related by

$$p_j = L_{v_j}(t, \mathbf{x}, \mathbf{v}) \tag{S16-9}$$

In mechanics the variables called $\mathbf{x}$ above are usually called the *generalized coordinates*, the variables called $\mathbf{v}$ above are usually called the *generalized velocities*, and those called $\mathbf{p}$ are usually called the *generalized momenta*. u is sometimes (but not always) called *Hamilton's characteristic function*. The integrand L is called the *Lagrangian* and Euler's equations (S16-4) are called *Lagrange's equations*.

Theorem S16-4 The function

$$u(\mathbf{x}, t, \boldsymbol{\alpha}, \tau) = \int_{t_0}^t L[s, \mathbf{q}(s), \mathbf{q}'(s)]\,ds \tag{S16-10}$$

where $\mathbf{q}(s)$ is an extremal such that

$$\mathbf{q}(t) = \mathbf{x} \tag{S16-11}$$

and

$$\mathbf{q}(\tau) = \boldsymbol{\alpha} \tag{S16-12}$$

looked upon as a function of the $n + 1$ independent variables $(\mathbf{x}, t)$ and the $n + 1$ parameters $(\tau, \boldsymbol{\alpha})$, is a complete integral of the Hamilton-Jacobi equation (S16-7).

Theorem S16-5 When (S16-8) and (S16-9) are used to replace L and $\mathbf{y}$ by H and $\mathbf{p}$, Euler's equations (S16-4) for the extremals become *Hamilton's equations*

$$\begin{aligned} x_j' &= H_{p_j} \\ p_j' &= -H_{x_j} \end{aligned} \qquad j = 1, \ldots, n \tag{S16-13}$$

DEFINITION S16-6

A surface $\Phi(\mathbf{x}, t) = 0$ is *transverse* to an extremal at a point if the n conditions

$$\frac{L_{v_j}(t, \mathbf{x}, \mathbf{v})}{L(t, \mathbf{x}, \mathbf{v})} = \frac{\Phi_{x_j}}{\Phi_t(\mathbf{x}, t) + \sum_{j=1}^n \Phi_{x_j}(\mathbf{x}, t)\, v_j} \qquad j = 1, \ldots, n \tag{S16-14}$$

are satisfied.

Theorem S16-6 The transversality conditions (S16-14) can be written in the equivalent form

$$\frac{L_{v_j}}{L - \sum_{j=1}^n v_j L_{v_j}} = \frac{\Phi_{x_j}}{\Phi_t} \qquad j = 1, \ldots, n \tag{S16-15}$$

or, in terms of the Hamiltonian variables,

$$\frac{p_j}{-H} = \frac{\Phi_{x_j}}{\Phi_t} \qquad j = 1, \ldots, n \tag{S16-16}$$

DEFINITION S16-7

An n-parameter family of extremals transverse to some surface is called a *field*.

Theorem S16-7 A necessary and sufficient condition that the n-parameter family of extremals

$$\begin{aligned} \mathbf{x} &= \mathbf{x}(t, \boldsymbol{\alpha}) \\ \mathbf{p} &= \mathbf{p}(t, \boldsymbol{\alpha}) \end{aligned} \tag{S16-17}$$

form a field is that the Lagrange brackets $[\alpha_j, \alpha_k]$ all vanish. Here

$$[\alpha_j, \alpha_k] = \sum_{l=1}^{n} \left(\frac{\partial x_l}{\partial \alpha_j} \frac{\partial p_l}{\partial \alpha_k} - \frac{\partial x_l}{\partial \alpha_k} \frac{\partial p_l}{\partial \alpha_j} \right) \tag{S16-18}$$

Theorem S16-8 The Euler equation for the functional

$$\mathscr{F}[u] = \iint_{\mathscr{R}} F(x, y, u, u_x, u_y)\, dx\, dy \tag{S16-19}$$

is

$$F_u - \frac{D}{Dx} F_p - \frac{D}{Dy} F_q = 0 \tag{S16-20}$$

where

$$\frac{D}{Dx} F(x, y, u, p, q) = F_x + F_u u_x + F_p p_x + F_q q_x \tag{S16-21}$$

and so forth.

DEFINITION S16-8

The functional

$$\mathscr{D}[u] = \iiint_{\mathscr{R}} |\, \nabla u \,|^2\, dV \tag{S16-22}$$

is known as the *Dirichlet integral.*

Dirichlet's principle states that the function that minimizes the Dirichlet integral in a region $\mathscr{R}$ is harmonic in $\mathscr{R}$.

17
Transform Methods

17-1. INTRODUCTION; LAPLACE TRANSFORM

Assuming the reader to be familiar with the Laplace transform, we summarize a few facts about it that we will need. The reader who wants more details can consult some of the many books on the subject, e.g., Churchill [46].

One of the basic properties of the Laplace transform is that it transforms the operation of taking a derivative into the operation of merely multiplying by a variable. It thus replaces some of the derivatives in a partial differential equation (PDE) (depending on which variables are involved in the transform) by mere multiples of the function, and therefore (hopefully) reduces the equation to an easier equation to solve. For example, a PDE in two independent variables will sometimes, after application of the Laplace transform, become an ODE. The Laplace transform is not capable of simplifying most PDEs, but it does help with some equations.

Basically, the idea behind any transform (not just the Laplace transform) is this. We have a difficult problem to solve. We apply a transform to it and produce an easier problem. We solve this easier problem. Then we have to transform the solution to the easier problem back into a solution of the original harder problem. This last step is the one where most of the difficulty arises. Inverting the transform may be quite difficult. For the Laplace transform it usually requires the use of contour integration in the complex plane. We will assume the reader to be familiar with the requisite techniques in the theory of functions of a complex variable and to be able to carry them out as necessary.

17-2. DEFINITION AND SOME PROPERTIES OF THE LAPLACE TRANSFORM

Let $f(t)$ be a function of the real variable t defined on the interval $0 \leqslant t < \infty$ such that† $\int_0^T |f(t)|\, dt$ exists for every $T \geqslant 0$, and let $f(t) = O(e^{s_0 t})$ for $t \to \infty$, where s_0 is a real constant (which may be positive, negative, or zero). Then (see [46])

$$\hat{f}(s) = \int_0^\infty e^{-st} f(t)\, dt \tag{17-1}$$

exists and represents an analytic function of the complex variable s in the half-plane $\text{Re}\,(s) > s_0$. $\hat{f}(s)$ is called the *Laplace transform of* $f(t)$.

If we denote the operation defined by (17-1) by

$$\hat{f} = \mathscr{L}\{f\} \tag{17-2}$$

then we can state one of the most important properties of the Laplace transform as follows (again see [46] for details):

$$\mathscr{L}\{f^{(n)}(t)\} = s^n \hat{f}(s) - \sum_{i=0}^{n-1} s^j f^{(n-j-1)}(0) \tag{17-3}$$

Here $f^{(k)}$ denotes the kth derivative of f. Notice that the coefficients of the powers of s in the sum are the values of the original function f at zero, not those of the transform $\hat{f}$.

For future reference, we write out (17-3) explicitly for the cases $n = 1$ and $n = 2$

$$\mathscr{L}\{f'(t)\} = s\hat{f}(s) - f(0) \tag{17-4}$$
$$\mathscr{L}\{f''(t)\} = s^2\hat{f}(s) - sf(0) - f'(0) \tag{17-5}$$

Also of importance is the inversion formula for (17-1). This tells us that if $\hat{f}(s)$ is obtained from $f(t)$ by means of formula (17-1), then $f(t)$ can be represented in terms of $\hat{f}(s)$ by means of the formula

$$f(t) = \frac{1}{2\pi i} \int_{s_1 - i\infty}^{s_1 + i\infty} e^{st} \hat{f}(s)\, ds \tag{17-6}$$

† $f(t) = O[g(t)]$ as $t \to \infty$ means $|f(t)/g(t)| < M$ for $t > T_0$.

where, as the limits indicate, the integration is over a line $\operatorname{Re}(s) = s_1$ parallel to the imaginary axis and where $s_1 > s_0$ (cf. the opening paragraph of this section).

The integral in (17-6) is to be interpreted as a Cauchy principal value, i.e., as $\lim_{S\to\infty} \int_{s_1-iS}^{s_1+iS}$.

Actually, (17-6) gives $f(t)$ only for $t > 0$. For $t < 0$ (17-6) gives $f \equiv 0$. This is because (17-1) is only a one-sided transform. We will clarify this point later on when we discuss one- and two-sided transforms (see Sec. 17-6).

As mentioned earlier, the method of Laplace transform sometimes affords a simpler method of solution of a problem than other methods. Sometimes, however, the transform method may be more difficult to carry through than other methods. In order to illustrate the various advantages and disadvantages of transform methods, we will illustrate how the method works when applied to a number of different problems.

Observe that unless the range of at least one variable in the PDE is from zero to infinity, the Laplace transform will not work well because the function $f(t)$ to be transformed must be defined for $0 < t < \infty$. In addition, as pointed out above, the transform (17-1) tacitly assumes that the function f being transformed vanishes identically for $t < 0$. Hence the one-sided Laplace transform (17-1) (usually called just the Laplace transform) will be of use only in problems where the functions to be transformed are defined for their arguments between 0 and ∞. If other ranges enter, different transforms must be used. Some such transforms will be discussed later in the present chapter.

Sometimes, for the sake of clarity, it is advantageous to introduce the *Heaviside unit function* $h(t)$, defined by

$$h(t) = \begin{cases} 1 & \text{if } t > 0 \\ 0 & \text{if } t < 0 \end{cases} \tag{17-7}$$

The value of $h(0)$ is usually immaterial in most applications. It can conveniently be taken to be 0, $\frac{1}{2}$, or 1, depending upon the reader's taste.

Using (17-7), we can write (17-6) in the more accurate form

$$h(t)\,f(t) = \frac{1}{2\pi i}\int_{s_1-i\infty}^{s_1+i\infty} e^{st}\hat{f}(s)\,ds \tag{17-8}$$

If $f(t)$ vanishes not only for all negative t but also for positive t up to a certain $t_0 > 0$, we can write

$$f(t - t_0) = h(t - t_0)\,f(t - t_0)$$

and, by a change of variable in

$$\hat{f}(s) = \int_0^\infty e^{-st}h(t - t_0)\,f(t - t_0)\,dt \tag{17-9}$$

we can then show that

$$\mathscr{L}\{h(t - t_0)f(t - t_0)\} = \hat{f}(s)\,e^{-st_0} \tag{17-10}$$

Note that (17-10) shows that the longer $f(t)$ stays identically zero, the more rapidly its Laplace transform dies out at infinity.

Finally, we note the important *convolution theorem for the (one-sided) Laplace transform*, which says

$$\mathscr{L}\left\{\int_0^t f(t - \tau)\,g(\tau)\,d\tau\right\} = \hat{f}(s)\,\hat{g}(s) \tag{17-11}$$

Because of the lack of symmetry between the representations for f and $\hat{f}$, (17-11) does not hold if the transform is taken in the opposite direction. Frequently the notation

$$f * g = \int_0^t f(t - \tau)\,g(\tau)\,d\tau \tag{17-12}$$

is used and called the *convolution* or *Faltung* or *fold* of f and g. Note that $f * g = g * f$.

17-3. SEMI-INFINITE STRING BY LAPLACE TRANSFORM

The Laplace transform method is not a particularly simple one to use on the finite string. In fact, in the general case of nonzero initial conditions, the Laplace transform is difficult to invert.

For the semi-infinite string, with zero initial conditions, the Laplace transform works much better. Consider the boundary-initial-value problem (BIVP):

PDE:

$$u_{xx} - c^{-2}u_{tt} = 0 \qquad \begin{matrix} 0 < x < \infty \\ 0 < t < \infty \end{matrix} \tag{17-13}$$

ICs:

$$u(x, 0) = u_t(x, 0) = 0 \qquad 0 < x < \infty \tag{17-14}$$

BCs:

$$u(0, t) = h_1(t) \qquad |u(\infty, t)| < \infty \qquad 0 < t < \infty \tag{17-15}$$

The second BC is physically reasonable and is appended to ensure the applicability of the transform.

Introduce

$$\hat{u}(x, s) = \int_0^\infty e^{-st}u(x, t)\,dt \tag{17-16}$$

Multiplying (17-13) by e^{-st}, integrating with respect to t from 0 to ∞, using (17-5) and the ICs (17-14), we obtain

$$\hat{u}_{xx} - \frac{s^2}{c^2}\hat{u} = 0 \qquad 0 < x < \infty \tag{17-17}$$

Transforming the BCs (17-15), we have the BCs

$$\hat{u}(0, s) = \hat{h}_1(s) \qquad |\hat{u}(\infty, s)| < \infty$$

The general solution to (17-17) is

$$\hat{u}(x, s) = A(s)\, e^{sx/c} + B(s)\, e^{-sx/c}$$

But $e^{sx/c} \to \infty$ for $x > 0$ and $\operatorname{Re}(s) \to \infty$. Hence the second BC implies $A(s) \equiv 0$. The first BC then implies $B(s) = \hat{h}_1(s)$, and we get

$$\hat{u}(x, s) = \hat{h}_1(s)\, e^{-sx/c}$$

From (17-10) it then follows that

$$u(x, t) = h\left(t - \frac{x}{c}\right) h_1\left(t - \frac{x}{c}\right) \tag{17-18}$$

In (17-18), h is the Heaviside unit function; h_1 is the given boundary datum. Interpretation of (17-18) in an xt plane affords the expected wave behavior with the characteristic $x = ct$ as the head of the wave (cf. Chap. 2).

Notice that the method of separation of variables does not work too well on the problem just considered because the zero ICs will not permit a nonzero solution of the equation for $\beta(t)$ if the separation $u(x, t) = \alpha(x)\,\beta(t)$ is attempted. On the other hand, temporarily omitting one of the ICs in the hope of satisfying it later leaves no condition on the separation parameter.

17-4. HEAT EQUATION BY LAPLACE TRANSFORM

Consider the flow of heat in a bar of length l. Assume that the temperature of the bar is initially zero and that one end of the bar is held at zero temperature for all time while the other end is kept at a temperature that varies in time according to the law $h_2(t)$, where $h_2(0) = 0$.

The mathematical formulation of this problem is the following:

DE:

$$u_{xx} - ku_t = 0 \qquad \begin{array}{l} 0 < x < l \\ 0 < t < \infty \end{array}$$

IC:

$$u(x, 0) = 0 \qquad 0 < x < l$$

BCs:

$$\begin{aligned} u(0, t) &= 0 \\ u(l, t) &= h_2(t) \end{aligned} \qquad 0 < t < \infty$$

the Laplace transform of which is:

DE:

$$\hat{u}_{xx} - ks\hat{u} = 0$$

BCs:

$$\begin{aligned} \hat{u}(0, s) &= 0 \\ \hat{u}(l, s) &= \hat{h}_2(s) \end{aligned}$$

Hence

$$\hat{u}(x, s) = \hat{h}_2(s) \frac{\sinh (x\sqrt{ks})}{\sinh (l\sqrt{ks})}$$

and therefore

$$u(x, t) = \frac{1}{2\pi i} \int_{s_1 - i\infty}^{s_1 + i\infty} \hat{h}_2(s)\, e^{st} \frac{\sinh (x\sqrt{ks})}{\sinh (l\sqrt{ks})} \tag{17-19}$$

Notice that because the heat equation is first order in t, rather than second order in t as the wave equation is, the Laplace transform of a solution of the heat equation contains exponentials in $\sqrt{s}$, not in s. Therefore the e^{st} in the inversion integral always dominates, and the path always gets pushed to the left. This is what makes the heat equation easier to solve by Laplace transform than the wave equation is. It is also the reason why books on the Laplace transform spend far more time on the heat equation than they do on the wave equation.

To evaluate (17-19) we observe that in spite of the square roots in it, the integrand is single-valued, with poles at the points s, where $l\sqrt{ks} = in\pi$, $n = 0, \pm 1, \pm 2, \ldots$, that is, at the points

$$s = -\frac{n^2\pi^2}{kl^2} \qquad n = 0, 1, 2, \ldots$$

The residues there are

$$\begin{aligned} &ih_2\left(-\frac{n^2\pi^2}{kl^2}\right) e^{-n^2\pi^2 t/kl^2} \frac{\sin n\pi x/l}{[(l\sqrt{k}/2in\pi)(-1)^n]/l\sqrt{k}} \\ &\quad = 2n\pi(-1)^{n+1}\, h_2\left(-\frac{n^2\pi^2}{kl^2}\right) e^{-n^2\pi^2 t/kl^2} \sin\frac{n\pi x}{l} \end{aligned}$$

Hence

$$u(x, t) = 2\pi \sum_{n=1}^{\infty} (-1)^n\, n h_2 \left(\frac{n^2\pi^2}{kl^2}\right) e^{-n^2\pi^2 t/kl^2} \sin \frac{n\pi x}{l} \tag{17-20}$$

Now

$$\hat{h}_2(s) = \int_0^\infty e^{-st} h_2(\tau)\, d\tau$$

Therefore

$$h_2\left(-\frac{n^2\pi^2}{kl^2}\right) = \int_0^\infty e^{n^2\pi^2\tau/kl^2} h_2(\tau)\, d\tau \tag{17-21}$$

Inserting (17-21) into (17-20) and interchanging summation and integration (legitimate because of the exponential and uniform convergence), we obtain, finally,

$$u(x, t) = 2\pi \int_0^\infty \sum_{n=1}^{\infty} (-1)^n\, n h_2(\tau) \sin \frac{n\pi x}{l}\, e^{n^2\pi^2(\tau - t)/kl^2}\, d\tau$$

As is to be expected, the case of a semi-infinite bar is even easier. To solve:

DE:

$$u_{xx} - ku_t = 0 \qquad \begin{matrix} 0 < x < \infty \\ 0 < t < \infty \end{matrix}$$

IC:

$$u(x, 0) = 0 \qquad 0 < x < \infty$$

BCs:

$$u(0, t) = h_1(t) \qquad |u(\infty, t)| < \infty \qquad 0 < t < \infty$$

use the Laplace transform to obtain

$$\hat{u}_{xx} - sk\hat{u} = 0$$

Then

$$\hat{u}(x, s) = A(s) \exp(x\sqrt{ks}) + B(s) \exp(-x\sqrt{ks})$$

Since $x > 0$, $x\sqrt{ks} \to \infty$ for $\operatorname{Re}(s) \to \infty$. Hence $A(s) \equiv 0$. Therefore

$$\hat{u}(x, s) = \hat{h}_1(s) \exp(-x\sqrt{ks})$$

By the convolution theorem (Sec. 17-2), we then have

$$u(x, t) = h_1(t) * \frac{x\sqrt{k}}{2\sqrt{\pi t^3}}\, e^{-kx^2/4t}$$

that is,

$$u(x, t) = \frac{x\sqrt{k}}{2\sqrt{\pi t^3}} \int_0^t e^{-kx^2/4(t-\tau)} h_1(\tau)\, d\tau \tag{17-22}$$

17-5. POISSON'S INTEGRAL FOR THE HALF-PLANE VIA FOURIER TRANSFORM

For any function $f(x)$ for which

$$\int_{-\infty}^{\infty} |f(x)|\, dx < \infty \tag{17-23}$$

the (two-sided) *Fourier transform*

$$\tilde{f}(\xi) = \int_{-\infty}^{\infty} e^{ix\xi} f(x)\, dx \tag{17-24}$$

exists and defines $\tilde{f}(\xi)$ as a function of the real variable ξ for $-\infty < \xi < \infty$.

The condition (17-23) requires that $f \to 0$ as $|x| \to \infty$ and therefore excludes many reasonable functions. In many applications it is necessary to permit f to grow exponentially at either $+\infty$ or $-\infty$ or both. Such a growth condition was permissible for the Laplace transform (see Sec. 17-2). It can be permitted for the Fourier transform, too, but then it is necessary to give up the two-sided transform (17-24) and replace it by two one-sided transforms. We do this later when it becomes necessary (see Sec. 17-6). For the present we restrict ourselves to the two-sided transform (17-24) and therefore require the growth condition (17-23).

The most important results we need for (17-24) are its inversion

$$f(x) = \frac{1}{2\pi} \int_{-\infty}^{\infty} e^{-ix\xi} \tilde{f}(\xi)\, d\xi$$

its effect on a derivative: the transform gets multiplied by $i\xi$ with no boundary or initial conditions entering; and the convolution formula, which now is symmetric in the functions and their transforms and in which the convolution integral now goes from $-\infty$ to ∞.

As an example of the use of the Fourier transform, we solve the Dirichlet problem for the potential equation for the upper half-plane under the reasonable assumption that the boundary data tend to zero at infinity. More precisely we solve the following BVP:

PDE:

$$\nabla^2 u = 0 \qquad \begin{array}{l} -\infty < x < \infty \\ 0 < y < \infty \end{array}$$

BCs:

$$u(x, 0) = f(x) \qquad -\infty < x < \infty \qquad \int_{-\infty}^{\infty} |f(x)|\,dx < \infty$$

$$u(x, y) \to 0 \qquad \text{at } \infty$$

We do this by putting

$$\tilde{u}(\xi, y) = \int_{-\infty}^{\infty} e^{-ix\xi} u(x, y)\,dx \tag{17-25}$$

The problem then becomes:

DE:

$$\tilde{u}_{yy} - \xi^2 \tilde{u} = 0 \tag{17-26}$$

BCs:

$$\tilde{u}(x, 0) = \tilde{f}(\xi) \qquad \tilde{u} \to 0 \qquad \text{as } y \to \infty$$

The solution to this transformed problem is

$$\tilde{u}(\xi, y) = \tilde{f}(\xi) e^{-|\xi| y} \tag{17-27}$$

Note that (17-27) is differentiable with respect to y. The jump in its ξ derivative is permissible since all derivatives in the DE (17-26) are with respect to y, not ξ.

Since the right-hand side of (17-27) is a product, we can use the convolution to invert it if we first find the inverse transform of the exponential in (17-27). Denoting this inverse transform by $g(x, y)$, we have

$$\begin{aligned} g(x, y) &= \frac{1}{2\pi} \int_{-\infty}^{\infty} e^{-|\xi| y - i\xi x}\,d\xi \\ &= \frac{1}{2\pi} \int_{-\infty}^{0} e^{\xi(y - ix)}\,d\xi + \frac{1}{2\pi} \int_{0}^{\infty} e^{-\xi(y + ix)}\,d\xi \\ &= \frac{1}{2\pi} \left(\frac{1}{y - ix} + \frac{1}{y + ix} \right) = \frac{1}{\pi} \frac{y}{x^2 + y^2} \end{aligned}$$

Thus,

$$u(x, y) = f(x) * \frac{y}{\pi(x^2 + y^2)} = \frac{y}{\pi} \int_{-\infty}^{\infty} \frac{f(\xi)}{(\xi - x)^2 + y^2}\,d\xi$$

17-6. ONE-SIDED FOURIER TRANSFORMS

As mentioned above, the two-sided Fourier transform exists only if the function being transformed dies out at infinity. Suppose $f(x) = O(e^{cx})$ as $|x| \to \infty$. If $c > 0$, and if ξ is real, then $\int_{-\infty}^{\infty} e^{ix\xi} f(x)\,dx$ will converge at

$x = -\infty$ but diverge at $x = +\infty$, and vice versa, if $c < 0$, this two-sided Fourier transform will converge at $x = +\infty$ but diverge at $x = -\infty$. A way out of this dilemma is to introduce two one-sided Fourier transforms

$$f_+(\xi) = \int_0^\infty e^{ix\xi} f(x)\,dx$$

and

$$f_-(\xi) = \int_{-\infty}^0 e^{ix\xi} f(x)\,dx$$

Then f_+ will exist for real ξ if f dies out at $+\infty$, and f_- will exist if f dies out at $-\infty$. Still, we would like f_+ and f_- to exist even if f grows exponentially at infinity. To achieve this aim we have only to let ξ be complex. To avoid confusing real and complex variables, we will let $\zeta = \xi + i\eta$ be the complex variable that replaces the real variable ξ. Then we define the one-sided complex Fourier transforms as follows:

$$f_+(\zeta) = \int_0^\infty e^{ix\zeta} f(x)\,dx \tag{17-28}$$

and

$$f_-(\zeta) = \int_{-\infty}^0 e^{ix\zeta} f(x)\,dx \tag{17-29}$$

Suppose

$$f(x) = O(e^{\eta_\pm x}) \qquad \text{as } x \to \pm\infty \tag{17-30}$$

that is, $f = O(e^{\eta_+ x})$ as $x \to +\infty$ and $f = O(e^{\eta_- x})$ as $x \to -\infty$, where $\eta_\pm$ are constants which may be positive, negative, or zero. Then substituting $\zeta = \xi + i\eta$ into (17-28) and (17-29), we see that (17-28) converges uniformly and absolutely for all ζ such that $\eta = \mathrm{Im}\,(\zeta) > \eta_+$, while (17-29) converges uniformly and absolutely for all ζ with $\eta < \eta_-$.

$f_+(\zeta)$ is thus an analytic function of ζ in the half-plane $\mathrm{Im}\,(\zeta) > \eta_+$, while $f_-(\zeta)$ is an analytic function of ζ in the half-plane $\mathrm{Im}\,(\zeta) < \eta_-$. If $\eta_- < \eta_+$, these two half-planes have no points in common and the two-sided Fourier transform

$$\tilde{f}(\zeta) = \int_{-\infty}^\infty e^{ix\zeta} f(x)\,dx \tag{17-31}$$

exists for no complex ζ whatsoever.

If $\eta_+ < \eta_-$, then (17-31) exists in the strip $\eta_+ < \mathrm{Im}\,(\zeta) < \eta_-$. In the special case where $\eta_+ = \eta_-$, the two-sided transform exists only on the line $\mathrm{Im}\,(\zeta) = \eta_+$, and then additional conditions on f are needed to ensure uniform convergence. The previous section is an example of this case with $\eta_\pm = 0$.

The inversion formulas for (17-28) and (17-29) are

$$h(x)\,f(x) = \frac{1}{2\pi}\int_{-\infty+i\eta^+}^{\infty+i\eta^+} e^{-ix\zeta} f_+(\zeta)\,d\zeta \tag{17-32}$$

$$h(-x)\,f(x) = \frac{1}{2\pi}\int_{-\infty+i\eta^-}^{\infty+i\eta^-} e^{-ix\zeta} f_-(\zeta)\,d\zeta \tag{17-33}$$

where $\eta^+ > \eta_+$ and $\eta^- < \eta_-$.

The Laplace transform usually used is a one-sided one, while the Fourier transform usually used is two-sided. If complex variables are needed in the analysis, however, there is absolutely no difference whether $f_\pm$ or $\hat{f}_\pm$ are used. To say that there is a difference between the Laplace transform as ordinarily used and the Fourier transform as ordinarily used is to say nothing more profound than that there is a difference between the one- and two-sided Fourier transforms. To be sure, there is a difference between one- and two-sided transforms, but this difference is obvious and not baffling.

17-7. MELLIN TRANSFORMS

Another useful transform, the *Mellin transform*, can be obtained by a change of variables in the Fourier or Laplace transform. To obtain the Mellin transform, first assume that the two-sided Laplace transform

$$\hat{F}(s) = \int_{-\infty}^{\infty} e^{-st} F(t)\,dt \tag{17-34}$$

exists in some strip

$$s_- < \operatorname{Re}(s) < s_+ \tag{17-35}$$

and let

$$e^{-t} = \sigma \tag{17-36}$$

and $F(t) = f(\sigma)$. Then

$$\hat{F}(s) = \int_0^\infty \sigma^{s-1} f(\sigma)\,d\sigma$$

Renaming $\hat{F}(s)$, we define the Mellin transform $\mathscr{F}(s)$, to be

$$\mathscr{F}(s) = \int_0^\infty \sigma^{s-1} f(\sigma)\,d\sigma \tag{17-37}$$

The existence of (17-34) in the strip (17-35) is equivalent to $F(t) = O(e^{s_\pm t})$ as $t \to \pm\infty$. This means

$$f(\sigma) = \begin{cases} O(\sigma^{-s_-}) & \text{as } \sigma \to 0 \\ O(\sigma^{-s_+}) & \text{as } \sigma \to \infty \end{cases} \tag{17-38}$$

These are clearly the correct conditions for the convergence of (17-60) at zero and infinity.

Notice that the *two*-sided Laplace transform leads to what looks like a *one*-sided Mellin transform. The terms one-sided and two-sided do not apply well to the Mellin transform, however. This fact will become even more apparent below. Accordingly we call (17-37) the Mellin transform with no qualifying adjectives. As mentioned above, it exists in the strip (17-35) whenever $f(\sigma)$ satisfies the growth conditions (17-38).

From the inversion formula for the Laplace transform we deduce the inversion formula for (17-37)

$$f(\sigma) = \frac{1}{2\pi i}\int_{s_0-i\infty}^{s_0+i\infty} \sigma^{-s}\mathscr{F}(s)\, ds \tag{17-39}$$

where s_0 lies in the strip (17-35).

The Mellin analogs of the one-sided Laplace transforms are

$$\mathscr{F}_0(s) = \int_0^1 \sigma^{s-1}f(\sigma)\, d\sigma \tag{17-40}$$

and

$$\mathscr{F}_1(s) = \int_1^\infty \sigma^{s-1}f(\sigma)\, d\sigma \tag{17-41}$$

If $f(\sigma) = O(\sigma^{-s_-})$ as $\sigma \to 0$, then $\mathscr{F}_0(s)$ is analytic for $\text{Re}\,(s) > s_-$. If $f(\sigma) = O(\sigma^{-s_+})$ as $\sigma \to \infty$, then $\mathscr{F}_1(s)$ is analytic for $\text{Re}\,(s) < s_+$. Accordingly, the inversion formulas for (17-63) and (17-64) are

$$f(\sigma) = \frac{1}{2\pi i}\int_{s_1-i\infty}^{s_1+i\infty} \sigma^{-s}\mathscr{F}_0(s)\, ds \tag{17-42}$$

where $s_1 > s_-$, and

$$f(\sigma) = \frac{1}{2\pi i}\int_{s_2-i\infty}^{s_2+i\infty} \sigma^{-s}\mathscr{F}_1(s)\, ds \tag{17-43}$$

where $s_2 < s_+$.

For a slight further generalization of (17-40) and (17-41), see Naylor [222, 223].

Notice that the Mellin transform (17-37), even though it contains an integral from zero to infinity rather than from $-\infty$ to $+\infty$, still exists only in a strip, not in a half-plane. This is because, as explained above, (17-37) is the analog of a two-sided Fourier or Laplace transform, not of a one-sided one. The analogs of one-sided Fourier and Laplace transforms are (17-40) and (17-41).

The convolution theorem for Mellin transforms also is different.

Theorem 17-1 If $\mathscr{F}(s)$ and $\mathscr{G}(s)$ are the Mellin transforms of $f(\sigma)$ and $g(\sigma)$, then the Mellin transform of the product $f(\sigma)\,g(\sigma)$ is $\mathscr{F}(s)\,\mathscr{G}(1-s)$.

17-8. HANKEL TRANSFORMS

If we change to polar coordinates in the two-dimensional Fourier transform and assume radial symmetry, then we obtain another useful one-dimensional transform, called the *Hankel* (*or Bessel*) *transform*. The two-dimensional Fourier transform

$$\tilde{f}(\xi, \eta) = \frac{1}{2\pi} \iint\limits_{-\infty}^{\infty} e^{i(x\xi + y\eta)} f(x, y)\, dx\, dy \tag{17-44}$$

possesses the inversion formula

$$f(x, y) = \frac{1}{2\pi} \iint\limits_{-\infty}^{\infty} e^{-i(x\xi + y\eta)} \tilde{f}(\xi, \eta)\, d\xi\, d\eta \tag{17-45}$$

In (17-44), let $x = r\cos\theta$, $y = r\sin\theta$, $\xi = \rho\cos\phi$, and $\eta = \rho\sin\phi$, and assume that $f(x, y) = f(r)$. Then (17-44) becomes

$$\tilde{f}(\rho, \phi) = \frac{1}{2\pi} \int_0^\infty f(r)\, r\, dr \int_0^{2\pi} e^{i\rho r\cos(\theta - \phi)}\, d\theta \tag{17-46}$$

However,

$$\int_0^{2\pi} e^{i\rho r\cos(\theta-\phi)}\, d\theta = \int_{-\phi}^{2\pi-\phi} e^{i\rho r\cos\psi}\, d\psi = \int_0^{2\pi} e^{i\rho r\cos\psi}\, d\psi \tag{17-47}$$

the first step following from the change of variable $\psi = \theta - \phi$ and the second by virtue of the periodicity of the integrand. From a known formula for the Bessel functions, the last integral in (17-47) is just 2π times $J_0(\rho r)$. Thus,

$$\tilde{f}(\rho, \phi) = \int_0^\infty J_0(\rho r) f(r)\, r\, dr \tag{17-48}$$

Formula (17-48) shows that $\tilde{f}(\rho, \phi)$ is independent of ϕ. Hence we can define the Hankel transform $F(\rho)$ as

$$F(\rho) = \int_0^\infty J_0(\rho r) f(r)\, r\, dr \tag{17-49}$$

The inversion formula for (17-73) turns out to be

$$f(r) = \int_0^\infty J_0(\rho r)\, F(\rho)\, \rho\, d\rho \tag{17-50}$$

The formula for the Hankel transform and its inverse are thus perfectly symmetric.

17-9. OTHER INTEGRAL TRANSFORMS

All the transforms of the previous sections are special cases of the transform

$$f(x) = \int_a^b K(x, y)\, \phi(y)\, dy \tag{17-51}$$

where either a or b or both may be $\pm\infty$, and where $K(x, y)$, called the *kernel*, is continuous in (x, y) on the given interval. For whatever functions $\phi(y)$ (17-51) exists, it defines a linear transformation which maps those functions $\phi(y)$ into new functions $f(x)$.

A natural question to ask is how to decide what to choose for the kernel of the transform (17-51). The answer depends upon the type of problem the transform is to be used on. The Laplace and Fourier transforms convert the operator d/dy into x or ix, respectively. The Mellin and Hankel transforms *do not*. We reemphasize this important point. *The Mellin and Hankel transforms do not convert the operation of taking the first derivative into multiplication by a variable.* What do they do then? They simplify certain differential operators. Each transform simplifies its own particular differential operator. The Laplace and Fourier transforms happen to simplify the simplest of all possible differential operators, namely, d/dx. The Hankel transform turns out, not surprisingly, to simplify Bessel's differential operator and also operators similar to Bessel's operator. For example, we leave it to the reader as an instructive exercise to show that apart from the BCs, the PDE

$$\phi_{rr} + \frac{1}{r}\phi_r + \phi_{zz} = 0 \tag{17-52}$$

(which is the three-dimensional potential equation in cylindrical coordinates with ϕ assumed independent of θ) becomes, upon application of the Hankel transform,

$$\Phi(\rho, z) = \int_0^\infty J_0(\rho r)\, \phi(r)\, r\, dr \tag{17-53}$$

$$\Phi_{zz} - \rho^2\Phi = 0$$

Hint: Multiply (17-52) by $J_0(\rho r)\, r$, integrate from 0 to ∞, and then use integration by parts to shift derivatives from ϕ to J_0 so that the ODE satisfied by J_0 can be utilized.

There are many papers on how to choose appropriate transforms for various problems as well as many which explore interconnections among transforms. We cite a number of these papers below. First, however, we remark that it is not enough that a transform simplify a given operator. It is also necessary to be able to invert the transform successfully. Some of the papers cited below attack this problem, too. Sneddon [252] answers a special case of the following interesting question: If

$$\phi(y) = \int_{\alpha}^{\beta} H(xy) f(x)\, dx \tag{17-54}$$

is the inversion formula for the transform (17-51), with $K(x, y) = K(xy)$, what must be the relationship between the kernels $K(xy)$ and $H(xy)$? In the case where $a = \alpha = 0$ and $b = \beta = \infty$, Sneddon shows that it is necessary that

$$\mathscr{K}(s)\, \mathscr{H}(1 - s) = 1$$

where $\mathscr{K}$ and $\mathscr{H}$ are the Mellin transforms of K and H, respectively.

As an incomplete list of references for further reading we mention Sneddon [252], Churchill [44–46, 49], Naylor [222, 223], and Mercer [207, 208].

EXERCISE Solve by Hankel transform the time-dependent wave equation in polar coordinates in two space dimensions with solution independent of θ:

PDE:

$$\phi_{rr} + \frac{1}{r}\phi_r = \frac{1}{c^2}\phi_{tt} \qquad \begin{matrix} 0 < r < \infty \\ 0 < t < \infty \end{matrix}$$

ICs:

$$\phi(r, 0) = \frac{k}{\sqrt{1 + r^2}} \qquad \phi_t(r, 0) = 0$$

BCs:

$$r^2\varphi_r \to 0 \qquad \begin{matrix} \text{as } r \to 0 \\ \text{and as } r \to \infty \end{matrix}$$

Hint:

$$\int_0^{\infty} e^{-\alpha t} J_0(\rho t)\, dt = \frac{1}{\sqrt{\rho^2 + \alpha^2}}$$

18
Integral Equations in Boundary-value Problems

18-1. INTRODUCTION

There are some standard techniques for converting differential equations (DEs) into integral equations. Such conversions are important both in theoretical and practical applications. The reader may recall that the proof of the standard existence theorem of ordinary differential equations (ODEs) relies on such a conversion into an integral equation. Moreover, most of the Langer related-equation techniques for attacking transition-point problems for ODEs rely on conversion into an integral equation.

Because of the lack of space, we concentrate on only a few ideas and, for the sake of definiteness, we will begin with the partial differential equation (PDE)

$$\nabla^2 u + \lambda u = 0 \tag{18-1}$$

which, as we have already seen, is an excellent prototype of an elliptic equation. Moreover, to save writing, we shall confine ourselves to two dimensions. Most of our results carry over to higher dimensions, though.

18-2. THE THIRD BOUNDARY-VALUE PROBLEM

We have already considered the Dirichlet and Neumann problems of (18-1), i.e., the problems in which a solution of (18-1) is sought in some region $\mathscr{R}$, satisfying either the first or second boundary condition (BC) on $\partial\mathscr{R}$. In order to generalize a bit we consider a BC which includes the first and second BCs as special cases. More precisely, we consider the third boundary-value problem (BVP) (sometimes called the *Robbin problem* or the *Churchill problem*):

DE:

$$\nabla^2 u + \lambda u = 0 \qquad \text{in } \mathscr{R} \tag{18-2}$$

BC:

$$\frac{\partial u}{\partial n} + k(\sigma)\, u = f(\sigma) \qquad \text{on } \partial\mathscr{R} \tag{18-3}$$

where σ is a parameter on $\partial\mathscr{R}$ and $k(\sigma)$ and $f(\sigma)$ are given functions on $\partial\mathscr{R}$. The Neumann problem is the special case in which $k(\sigma) \equiv 0$, while the Dirichlet problem emerges if we put $k(\sigma) = \mu k(\sigma)$ and $f(\sigma) = \mu f(\sigma)$, where μ is a parameter, and then let $\mu \to \infty$.

By using Green's identity in the usual way, it is easy to prove that the solution of the third BVP, if it exists, is unique except for certain eigenvalues λ and also except for certain eigenvalues k. Physically, the third BVP arises in problems in electromagnetics where the boundary is an imperfect conductor, in heat-conduction problems where the boundary absorbs some heat, being neither perfectly insulated nor perfectly conducting, and in the linearized theory of water waves at the free surface of the water. In general, the third BC typifies physical situations wherein the boundary absorbs some but not all of the energy, heat, etc., being transmitted through it. We have already met the third BC for the one-dimensional time-dependent wave equation in Eq. (2-29).

Because it is more general than the first and second BVPs, the third BVP is harder to solve. For example, the number of domains in which a Green's function can be found by reflection for the third BVP is very small owing to the fact that the image of the source point in a straight boundary has to be taken as a whole line of sources of exponentially decaying strength distributed along a line running from the mirror image of the source to infinity in a direction perpendicular to the boundary. For example, the Green's function for the upper half-plane with third BC with k a positive constant on the x axis is (see Keller [154] for details)

$$g(x, y, \xi, \eta) = g^*(x, y, \xi, \eta) + g^*(x, y, \xi, -\eta) + 2k \int_{-\infty}^{-\eta} e^{k(\eta+\tau)}\, g^*(x, y, \xi, \tau)\, d\tau$$

where g^* is the free-space Green's function.

18-3. MIXED BOUNDARY-VALUE PROBLEMS

Another important type of BVP is a *mixed BVP*. A typical mixed BVP consists in giving one type of boundary data, say u, on one part $\partial\mathscr{R}_1$ of the boundary and a different type of data, say $\partial u/\partial n$, on the rest of the boundary $\partial\mathscr{R}_2$. Such a mixed BVP can be solved if a Green's function that vanishes on $\partial\mathscr{R}_1$ and whose normal derivative vanishes on $\partial\mathscr{R}_2$ can be found, but in practice it is rare that such a Green's function can be found. It may be, however, that the Green's function for the Dirichlet or Neumann problem for (18-2) can be found. If the Green's function for the Neumann problem (sometimes called the *Neumann function*) is known, then the BVP described above can be reduced to an integral equation. Strangely enough, the mixed BVP *cannot* be reduced to an integral equation by using the Green's function for the *Dirichlet* problem. We point out why below.

Consider, now, the mixed BVP

PDE:

$$\nabla^2 u + \lambda u = 0 \qquad \text{in } \mathscr{R} \tag{18-4}$$

BCs:

$$u = f_1(\sigma) \qquad \text{on } \partial\mathscr{R}_1 \tag{18-5}$$

$$\frac{\partial u}{\partial n} = f_2(\sigma) \qquad \text{on } \partial\mathscr{R}_2 \tag{18-6}$$

If we knew u on $\partial\mathscr{R}_2$ as well as on $\partial\mathscr{R}_1$, the Green's function would give us the solution immediately. Similarly, if we knew $\partial u/\partial n$ on $\partial\mathscr{R}_1$, the Neumann function would yield an immediate solution. Thus, the problem is to find the missing boundary values. As mentioned above, it turns out to be necessary to look for the missing values of $\partial u/\partial n$ rather than those of u. If we denote the unknown values of $\partial u/\partial n$ on $\partial\mathscr{R}_1$ by $\psi(\sigma)$ and let $g(x, y, \xi, \eta)$ be the Neumann function, i.e., the Green's function for the Neumann problem, for (18-4), then Green's identity tells us that

$$u(\xi, \eta) = \frac{1}{2\pi}\int_{\partial\mathscr{R}_1} g(x, y, \xi, \eta)\, \psi(\sigma)\, d\sigma + \frac{1}{2\pi}\int_{\partial\mathscr{R}_2} g(x, y, \xi, \eta) f_2(\sigma)\, d\sigma \tag{18-7}$$

In (18-7) (ξ, η) is a point in $\mathscr{R}$, and (x, y) is a point on $\partial\mathscr{R}$. The second integral in (18-7) is a known function of (ξ, η). Call it $F(\xi, \eta)$. Then (18-7) reads

$$u(\xi, \eta) = \frac{1}{2\pi}\int_{\partial\mathscr{R}_1} g(x, y, \xi, \eta)\, \psi(\sigma)\, d\sigma + F(\xi, \eta) \tag{18-8}$$

In (18-8) let (ξ, η) approach a point on $\partial\mathscr{R}_1$ whose σ coordinate is s. Using an obvious notation, we then get the integral equation

$$f_1(s) = \frac{1}{2\pi}\int_{\partial\mathscr{R}_1} g(s, \sigma)\,\psi(\sigma)\,d\sigma + F(s) \qquad s \text{ on } \partial\mathscr{R}_1 \tag{18-9}$$

which is an integral equation of the first kind for $\psi(\sigma)$ (see Chap. 19). Once (18-9) has been solved for ψ, (18-7) then gives the solution u to the original mixed BVP.

It is interesting to see what would happen if we tried to use the Green's function for the Dirichlet problem instead of that for the Neumann problem. Suppose $G(x, y, \xi, \eta)$ is the Green's function for the Dirichlet problem, and this time let $\phi(\sigma)$ be the unknown boundary values of u on $\partial\mathscr{R}_2$. Then

$$u(\xi, \eta) = -\frac{1}{2\pi}\int_{\partial\mathscr{R}_1} G_n(x, y, \xi, \eta)\,f_1(\sigma)\,d\sigma - \frac{1}{2\pi}\int_{\partial\mathscr{R}_2} G_n(x, y, \xi, \eta)\,\phi(\sigma)\,d\sigma \tag{18-10}$$

where G_n means $\partial G/\partial n$. This time the first integral is a known function of (ξ, η), say $F_1(\xi, \eta)$, and now letting (ξ, η) approach a point s on $\partial\mathscr{R}_2$, we obtain

$$\phi(s) = F_1(s) - \frac{1}{2\pi}\int_{\partial\mathscr{R}_2} G_n(s, \sigma)\,\phi(\sigma)\,d\sigma \tag{18-11}$$

which looks like an integral equation of the second kind for ϕ. Something is wrong, however. In deriving (18-11) we never used (18-6). Clearly without (18-6) the solution to the mixed BVP cannot be unique. So something must be wrong with (18-11). What is wrong with it is due to the behavior of G_n as (ξ, η) approaches the boundary. We have already seen (Chap. 3) that G_n acts like a delta function as its parameter point approaches the boundary. The result is that a boundary integral of G_n times something picks out the value of the "something" at the boundary point approached. Since the point s being approached lies on $\partial\mathscr{R}_2$, the first integral in (18-10) approaches zero and the second approaches $\phi(s)$. Therefore a more accurate statement of (18-11) is

$$\phi(s) = \phi(s)$$

which is hardly useful for determining ϕ.

Basically, the reason why the Green's function for the Neumann problem works in the above analysis whereas the Green's function for the Dirichlet problem fails is that it is the normal derivative of the Green's function that is the delta function, not the Green's function itself. In (18-7) only the Green's function itself enters, not its normal derivative. In (18-10), however, G_n enters, and its singularity is too strong for the present purposes.

The basic idea in deriving (18-9) was to use a "wrong" Green's function, namely, that for the Neumann problem, because the right Green's function, namely, that for the mixed BVP, is too difficult to find. The same idea could have been introduced in Chap. 3 to reduce the solution of the Neumann problem to an integral equation. In Sec. 3-7 we noted that $1/r$, which is the free-space Green's function for the potential equation in three dimensions and therefore the wrong Green's function for a bounded domain, failed to remove all the unknown boundary data from Green's identity. Suppose, however, that we wish to solve the Neumann problem for $\nabla^2 u = 0$ in $\mathscr{R}$, and suppose that we denote the unknown boundary values of u on $\partial\mathscr{R}$ by $\phi(S)$, where S is a point on $\partial\mathscr{R}$. Then Green's identity applied to u and $1/r$ in the usual way yields

$$u(Q) = -\frac{1}{4\pi}\iint_{\partial\mathscr{R}} \phi(S)\left(\frac{1}{r}\right)_n dS + \frac{1}{4\pi}\iint_{\partial\mathscr{R}} \frac{1}{r} f(S)\, dS \tag{18-12}$$

where $f(S)$ are the given boundary values of $\partial u/\partial n$.

The last integral in (18-12) is a known function $F(Q)$. If we let Q approach the boundary, we obtain the integral equation

$$\phi(Q) = -\frac{1}{4\pi}\iint_{\partial\mathscr{R}} \left(\frac{1}{r}\right)_n \phi(S)\, dS + F(Q) \tag{18-13}$$

for ϕ. Having ϕ, we find u from (18-12).

For a number of applications of the ideas behind (18-13) to problems for (18-1) in three dimensions, the reader is referred to Hochstadt [140].

EXAMPLE: Water waves in the presence of an infinite dock As an example, we reduce a mixed BVP involving the third BC to an integral equation. Consider the problem:

PDE:

$$\nabla^2 u = 0 \qquad \begin{matrix} -\infty < x < \infty \\ 0 < y < \infty \end{matrix} \tag{18-14}$$

BCs:

$$u_y(x, 0) = 0 \qquad -\infty < x < 0 \tag{18-15}$$

$$u_y(x, 0) - ku(x, 0) = 0 \qquad \begin{matrix} 0 < x < \infty \\ k = \text{const} \end{matrix} \tag{18-16}$$

This problem (see Fig. 18-1) occurs in the linearized theory of water waves. The Neumann condition represents a rigid barrier which the water cannot penetrate, while the third BC is satisfied on the free surface of the water.

The Green's function for the Neumann problem for (18-14) for the upper half-plane is

$$g(x, y, \xi, \eta) = -\log\sqrt{(x-\xi)^2 + (y-\eta)^2} - \log\sqrt{(x-\xi)^2 + (y+\eta)^2} \tag{18-17}$$

Hence

$$g(x, 0, \xi, \eta) = -\log\,[(x-\xi)^2 + \eta^2] \tag{18-18}$$

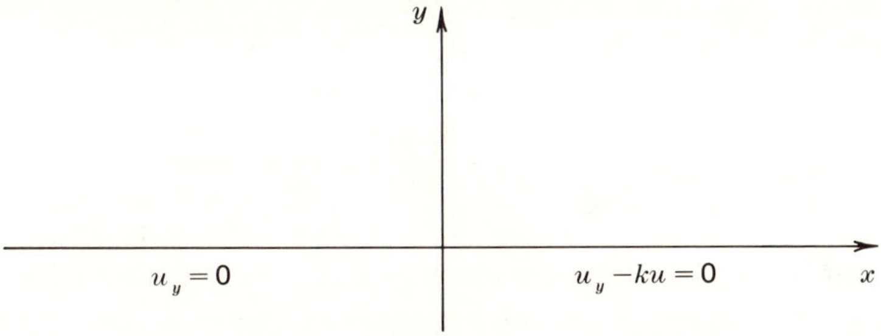

Figure 18-1

Using (18-15) to (18-17) in the Green's formula for the solution to the Neumann problem, we obtain

$$u(\xi, \eta) = \frac{k}{2\pi} \int_0^\infty u(x, 0) \log [(x - \xi)^2 + \eta^2]\, dx \tag{18-19}$$

In (18-19) let $u(x, 0) = \phi(x)$. Then

$$u(\xi, \eta) = \frac{k}{2\pi} \int_0^\infty \phi(x) \log [(x - \xi)^2 + \eta^2]\, dx \tag{18-20}$$

If we let $\eta \to 0$ in (18-20), we obtain the integral equation

$$\phi(\xi) = \frac{k}{\pi} \int_0^\infty \phi(x) \log | x - \xi |\, dx \tag{18-21}$$

for ϕ. Once ϕ has been found, (18-20) yields $u(\xi, \eta)$.

EXERCISE Use the above method to derive an integral equation for the following mixed BVP for a strip (see Fig. 18-2):

PDE:

$$\nabla^2 u + k^2 u = 0 \qquad \begin{array}{l} -\infty < x < \infty \\ 0 < y < b \end{array}$$

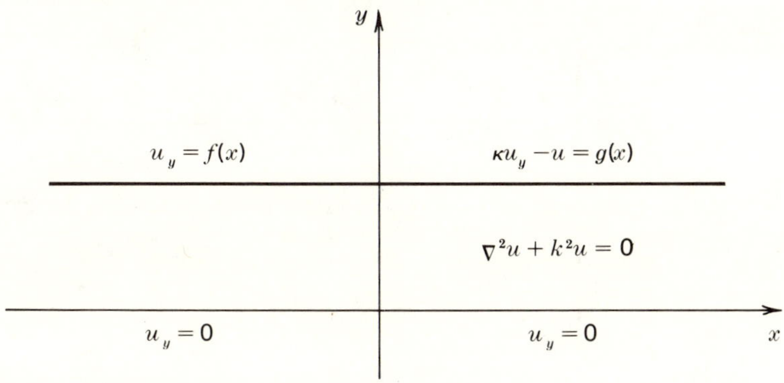

Figure 18-2

BCs:

$$u_y(x, 0) = 0 \qquad -\infty < x < \infty$$
$$u_y(x, b) = f(x) \qquad -\infty < x < 0$$
$$\kappa u_y(x, b) - u(x, b) = g(x) \qquad 0 < x < \infty$$

κ is a constant and $f(x)$ and $g(x)$ are given functions.

18-4. DUAL INTEGRAL EQUATIONS

Consider the following mixed BVP for the potential equation in a quarter-plane (see Fig. 18-3):

PDE:

$$\nabla^2 u(x, y) = 0 \qquad \begin{matrix} 0 < x < \infty \\ -\infty < y < 0 \end{matrix} \tag{18-22}$$

BCs:

$$u_x(0, y) = 0 \qquad -\infty < y < 0 \tag{18-23}$$

$$u(x, 0) = 1 \qquad 0 < x < 1 \tag{18-24}$$

$$u_y(x, 0) = 0 \qquad 1 < x < \infty \tag{18-25}$$

$$u,\ u_x \to 0 \qquad \text{at } \infty \tag{18-26}$$

One of the physical problems it represents is the steady flow of heat into the fourth quadrant through the segment $0 < x < 1$ of the x axis, the rest of the boundary being perfectly insulated. Another problem it represents is the steady irrotational flow of an incompressible fluid into the fourth quadrant through the segment $0 < x < 1$ of the x axis, the rest of the x axis and the entire y axis being rigid walls. This BVP can be reduced to a pair of integral equations called *dual integral equations*. To obtain them we first observe that the BC $u_x(0, y) = 0$ implies that u can be continued into the

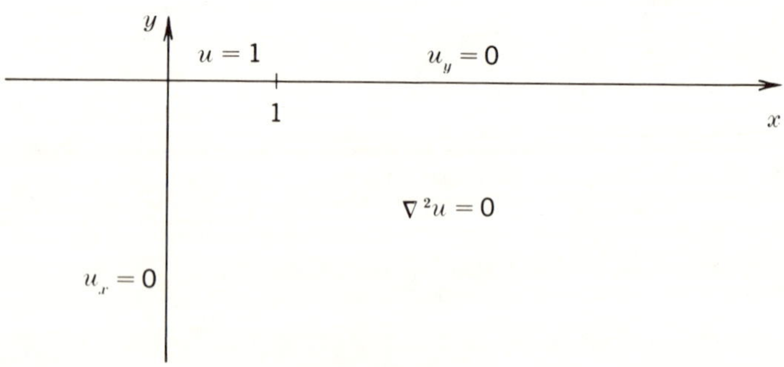

Figure 18-3

third quadrant as an even function of x. This suggests that we use the Fourier cosine transform

$$\tilde{u}(\xi, y) = \int_0^\infty \cos \xi x \, u(x, y) \, dx \tag{18-27}$$

Then integrating by parts twice and using the BCs (18-23) and (18-26), we find

$$\int_0^\infty \cos \xi x \, u_{xx}(x, y) \, dx = -\xi^2 \tilde{u}(\xi, y)$$

Hence the PDE (18-22) becomes

$$\tilde{u}_{yy} - \xi^2 \tilde{u} = 0 \tag{18-28}$$

the general solution to which is

$$\tilde{u}(\xi, y) = \alpha(\xi) \, e^{\xi y} + \beta(\xi) \, e^{-\xi y} \tag{18-29}$$

For $y < 0$ (the region in which the DE holds), $e^{-\xi y} \to \infty$ as $\xi \to +\infty$. Therefore, in order to be able to use a real transform (18-27) we take $\beta(\xi) \equiv 0$ in (18-29). This leaves

$$\tilde{u}(\xi, y) = \alpha(\xi) \, e^{\xi y} \tag{18-30}$$

Using (18-30) in the inversion formula for (18-27), we have

$$u(x, y) = \frac{2}{\pi} \int_0^\infty \alpha(\xi) \, e^{\xi y} \cos x\xi \, d\xi \tag{18-31}$$

The function u given by (18-31) satisfies the PDE (18-22) and the BCs (18-23) and (18-26). It remains to choose the function $\alpha(\xi)$ in (18-31) so as to satisfy the BCs (18-24) and (18-25). The latter two conditions require α to be a solution of the dual integral equations

$$\frac{2}{\pi} \int_0^\infty \alpha(\xi) \cos x\xi \, d\xi = 1 \qquad 0 < x < 1 \tag{18-32}$$

$$\int_0^\infty \xi \, \alpha(\xi) \cos x\xi \, d\xi = 0 \qquad 1 < x < \infty \tag{18-33}$$

Equations (18-32) and (18-33) represent two equations for the single unknown function $\alpha(\xi)$. They hold in different regions, however. So there is no danger of overdetermination. If they have once been solved for α, then (18-31) gives the solution u of the original problem (18-22) to (18-26).

Another pair of dual integral equations arises when a problem similar to the previous one is considered. Instead of two-dimensional flow, we consider an irrotational flow in three dimensions with cylindrical symmetry. We assume that an incompressible liquid lies in the lower half-space and that a steady flow of liquid is fed into the lower half-space through a circular hole of radius 1 about the origin in the xy plane. Using cylindrical coordinates, we can formulate this problem as follows:

PDE:

$$u_{rr} + \frac{1}{r} u_r + u_{zz} = 0 \qquad \begin{array}{c} 0 < r < \infty \\ -\infty < z < 0 \end{array} \tag{18-34}$$

BCs:

$$u(r, 0) = 1 \qquad 0 < r < 1 \tag{18-35}$$

$$u_z(r, 0) = 0 \qquad 1 < r < \infty \tag{18-36}$$

$$\lim_{z \to -\infty} u(r, z) = 0 \qquad \lim_{r \to 0} u(r, z) < \infty \tag{18-37}$$

In Sec. 17-9 we asked the reader to show that the Hankel transform

$$U(\rho, z) = \int_0^\infty ru(r, z)\, J_0(\rho r)\, dr \tag{18-38}$$

reduces (18-34) to

$$U_{zz} - \rho^2 U = 0 \tag{18-39}$$

The bounded solution of (18-39) for negative z and positive ρ is

$$U(\rho, z) = \alpha(\rho)\, e^{\rho z} \tag{18-40}$$

The Hankel inversion formula then yields

$$u(r, z) = \int_0^\infty \rho\alpha(\rho)\, e^{\rho z} J_0(\rho r)\, d\rho \tag{18-41}$$

as the solution to (18-34) which satisfies (18-37). In order for (18-41) to satisfy (18-35) and (18-36), α must be a solution of the dual integral equations

$$\int^\infty \rho\alpha(\rho)\, J_0(\rho r)\, d\rho = 0 \qquad 0 < r < 1 \tag{18-42}$$

and

$$\int_0^\infty \rho^2\alpha(\rho)\, J_0'(\rho r)\, d\rho = 0 \qquad 1 < r < \infty \tag{18-43}$$

If we introduce the new unknown function β by $\beta(\rho) = \rho\alpha(\rho)$, we can write (18-42) and (18-43) as

$$\int_0^\infty \beta(\rho)\, J_0(\rho r)\, d\rho = 1 \qquad 0 < r < 1 \tag{18-44}$$

and

$$\int_0^\infty \rho\beta(\rho)\, J_0'(\rho r)\, d\rho = 0 \qquad 1 < r < \infty \tag{18-45}$$

Although we have used transforms to reduce the two previous mixed BVPs to dual integral equations, the same pair of dual integral equations can also be obtained by attempting to solve the given problems by the less sophisticated method of separation of variables. We leave it to the reader as an exercise to obtain (18-32) and (18-33) and also (18-42) and (18-43) by separating variables in the PDE and then integrating over the separation parameter, calling it ξ in the first case and ρ in the second. In addition we mention that the use of a Laplace transform in z in the second case again leads to (18-42) and (18-43) when the BCs are applied. In using the Laplace transform in this case, however, it is necessary to use it in the proper direction. We have to write

$$u(r, z) = \int_0^\infty e^{\rho z}\hat{u}(r, \rho)\, d\rho$$

rather than the other way around. This means we think of u as being a Laplace transform rather than taking the Laplace transform of the PDE. In other words, we use the inverse Laplace transform directly rather than the Laplace transform itself.

19
Techniques in Solving Integral Equations

19-1. LINEAR INTEGRAL EQUATIONS

DEFINITION 19-1

The equation

$$\phi(x) = f(x) + \int_a^b K(x, y)\, \phi(y)\, dy \tag{19-1}$$

in which f and K are given and ϕ is sought, is called a *linear Fredholm integral equation of the second kind*. K is called the *kernel* (in some books the *nucleus*) of the integral equation. If $f(x) \equiv 0$, (19-1) is called *homogeneous*. Otherwise it is called *nonhomogeneous*.

DEFINITION 19-2

The equation

$$f(x) = \int_a^b K(x, y)\, \phi(y)\, dy \tag{19-2}$$

in which f and K are given and ϕ sought, is called a *linear Fredholm integral equation of the first kind.* If $f \equiv 0$, it is called *homogeneous*, otherwise *nonhomogeneous.* K is again called its *kernel.*

The difference between Fredholm integral equations of the first and second kinds is the occurrence of the unknown function outside as well as inside the integral in the equation of the second kind. Strangely enough, this makes equations of the second kind easier to handle than those of the first kind. To see this, suppose that a and b are finite and that f is continuous on the closed interval $[a, b]$ and K on the closed square $a \leqslant x \leqslant b, a \leqslant y \leqslant b$. Then (19-1) implies that ϕ is continuous on $[a, b]$, too, while (19-2) implies no such thing. ϕ might be highly discontinuous yet give rise to a continuous f when inserted into (19-2) because x enters the right side of (19-2) through K, not through ϕ. In (19-1), on the other hand, the right side is continuous if f and K are (and ϕ is such that the integral exists). Hence the left side ϕ must also be continuous.

DEFINITION 19-3

The equation

$$\phi(x) = f(x) + \int_a^x K(x, y)\, \phi(y)\, dy \tag{19-3}$$

is called a *linear Volterra integral equation of the second kind.*

It is clear from Definitions 19-1 and 19-2 what the definitions of homogeneous, nonhomogeneous, and *Volterra equation of the first kind* are.

A Volterra equation can be looked at as a special case of a Fredholm equation in which $K(x, y) \equiv 0$ if $y > x$, but it usually does not pay to do so because Fredholm and Volterra equations are fundamentally different and are better treated separately, just as boundary-value problems (BVPs) for ODEs differ from initial-value problems (IVPs) and have to be treated separately.

If both a and b are finite and if certain integrals of K exist, then there is a theory, naturally called the *Fredholm theory* (see, for example, Mikhlin [213] or Courant and Hilbert [65, vol. I]), that gives copious information on the solutions of (19-1). In the case of an infinite interval or a nonintegrable K, however, the ordinary Fredholm theory breaks down and other methods must be applied. It is some of these other methods in which we are interested in the present chapter.

19-2. INTEGRAL EQUATIONS OF CONVOLUTION TYPE

Most of the methods discussed in the present chapter involve the use of integral transforms. In order for them to work, it is necessary to have a special type of kernel. Henceforth, unless the contrary is stated, we assume that

$$K(x, y) = k(x - y)$$

This means that the kernel depends only on the *difference* $x - y$. Such a kernel, not surprisingly, is called a *difference kernel.* Assuming that all functions involved possess a two-sided Fourier transform, we can solve one particular integral equation with a difference kernel with consummate ease. This is the equation

$$\phi(x) = f(x) + \int_{-\infty}^{\infty} k(x - y)\, \phi(y)\, dy \tag{19-4}$$

Using the definition of convolution for the Fourier transform [cf. (17-12) and the remark (Sec. 17-5) that the convolution theorem for the Fourier transform uses an integral from $-\infty$ to ∞], we can write (19-4) in the form

$$\phi = f + k * \phi \tag{19-5}$$

Taking the Fourier transform of (19-5), we have

$$\tilde{\phi} = \tilde{f} + \tilde{k}\tilde{\phi} \tag{19-6}$$

Hence

$$\tilde{\phi} = \frac{\tilde{f}}{1 - \tilde{k}} \tag{19-7}$$

and so

$$\phi(x) = \frac{1}{2\pi} \int_{-\infty}^{\infty} \frac{e^{-ix\xi}\tilde{f}(\xi)}{1 - \tilde{k}(\xi)}\, d\xi \tag{19-8}$$

For obvious reasons, (19-4) is called an *integral equation of convolution type.* There is an analogous one for the Laplace transform, namely, the special Volterra equation

$$\phi(x) = f(x) + \int_{0}^{x} k(x - y)\, \phi(y)\, dy \tag{19-9}$$

19-3. THE HOMOGENEOUS WIENER-HOPF INTEGRAL EQUATION

The homogeneous Wiener-Hopf integral equation is

$$\phi(x) = \int_{0}^{\infty} k(x - y)\, \phi(y)\, dy \qquad 0 < x < \infty \tag{19-10}$$

It differs from the homogeneous form of (19-4) in that it holds only for positive x instead of all x and the integral goes only from 0 to ∞ instead of from $-\infty$ to ∞.

In order to use the convolution theorem for the Fourier transform, we need an integral from $-\infty$ to ∞. One way to get it would seem to be merely to define ϕ to be identically zero when its argument is negative. Then the integral in (19-10) could be written as an integral from $-\infty$ to ∞, and the convolution theorem would apply. There is only one thing wrong with this idea. It would extend ϕ so that the left-hand side of (19-10) would be identically zero for $x < 0$. The right-hand side of (19-10), however, is *not* identically zero for $x < 0$, even if $\phi(y) \equiv 0$ for $y < 0$. The reason for this is the following. In (19-10), as originally set up, both x and y run from 0 to ∞. The argument of k, namely, $x - y$, therefore runs from $-\infty$ to ∞. In other words, k is already defined for all values of its argument. If we insert a negative value of x into $k(x - y)$, we can carry out the integration on the right-hand side of (19-10) and obtain a value for the integral using only the values of ϕ for positive argument (of ϕ). Thus, we have no right to define the value of the integral in (19-10) to be identically zero for $x < 0$ because that integral is already defined for $x < 0$ and will not in general be zero for those values of x.

At this point many authors define ϕ to be identically zero for negative x anyway but then introduce $\psi(x)$ to denote the values of the integral in (19-10) for negative x. This seems unnecessary, and we will proceed in a slightly simpler way. The first few steps in the analysis below are equivalent to the usual procedure but may look a little different from some treatments. Once the transforms have been taken, the analysis below will then look much like the usual Wiener-Hopf development.

Having convinced ourselves that we cannot extend the values of ϕ arbitrarily, we let the integral equation itself extend them. Using the Heaviside unit function $h(x)$, we write

$$\phi(x) = \int_{-\infty}^{\infty} k(x - y)\, h(y)\, \phi(y)\, dy \qquad -\infty < x < \infty \tag{19-11}$$

Since the integral now goes from $-\infty$ to ∞, the convolution theorem will now apply, provided k has a two-sided transform. In order to guarantee that it does, we *assume*

$$k(t) = O(e^{k_{\pm} t}) \qquad \begin{array}{l}\text{as } t \to \pm\infty \\ \text{where } k_+ < k_-\end{array} \tag{19-12}$$

Since we are interested in ϕ only for positive x, we do not assume that ϕ has a two-sided transform. We merely assume that it has a one-sided one. More precisely, we *assume*

$$\phi(x) = O(e^{cx}) \qquad \text{for } x \to \infty \tag{19-13}$$

c to be specified more precisely below. If we fix a value of $x > 0$, then the integral in (19-11) will converge only if the integrand goes to zero strongly enough as $y \to \infty$. For large y the behavior of the integrand in (19-11) is

$$\sim e^{k_-(x-y)} e^{cy} = e^{k_- x} e^{(c-k_-)y}$$

Therefore, for this to go to zero as $y \to \infty$, we have to add the restriction

$$c < k_- \tag{19-14}$$

The same computation that produced (19-14) also shows that

$$\phi(x) = O(e^{k_- x}) \qquad \text{as } x \to -\infty \tag{19-15}$$

Condition (19-13) implies that the Fourier transform $\phi_+(\zeta)$ exists for $\text{Im}\,\zeta > c$ and (19-15) implies that ϕ_- exists for $\text{Im}\,\zeta < k_-$. Condition (19-14) then implies that the two-sided transform $\tilde{\phi}$ exists in the strip $c < \text{Im}\,\zeta < k_-$. Thus, even though we assumed only a one-sided transform for ϕ, it turns out that the two-sided one exists too. Nevertheless, we shall see below that it will be necessary to split the two-sided transform into two one-sided ones.

The transform of the convolution integral in (19-11) will exist only where both $\tilde{k}$ and ϕ_+ exist, i.e., where $\text{Im}\,\zeta > \max(k_+, c)$. So we might as well add $c > k_+$ to our list of assumptions.

To summarize, we assume that k satisfies (19-12), and we look for a solution of (19-11) which satisfies (19-13), where

$$k_+ < c < k_- \tag{19-16}$$

To find it, we apply the Fourier transform to (19-11) to obtain

$$\tilde{\phi}(\zeta) = \tilde{k}(\zeta)\, \phi_+(\zeta) \qquad k_+ < c < \text{Im}\,(\zeta) < k_- \tag{19-17}$$

The only difference between (19-17) and the homogeneous form of (19-6) is that ϕ_+ occurs instead of $\tilde{\phi}$ on the right. ϕ_+ and $\tilde{\phi}$ are not the same unless ϕ vanishes identically for negative x, and we have taken pains to show that ϕ very definitely does not vanish for negative x. So $\tilde{\phi}$ and ϕ_+ are different. However, it follows trivially from their definitions that

$$\tilde{\phi} = \phi_+ + \phi_- \tag{19-18}$$

Inserting (19-18) into (19-17), we have

$$\phi_+(\zeta) + \phi_-(\zeta) = \tilde{k}(\zeta)\, \phi_+(\zeta) \qquad k_+ < c < \text{Im}\,(\zeta) < k_- \tag{19-19}$$

or

$$[\tilde{k}(\zeta) - 1]\, \phi_+(\zeta) = \phi_-(\zeta) \qquad k_+ < c < \text{Im}\,(\zeta) < k_- \tag{19-20}$$

As indicated, (19-20) holds in the strip $c < \text{Im}(\zeta) < k_-$. Now comes the fundamental idea of Wiener and Hopf. The right side of (19-20) is analytic in the lower half-plane $\text{Im}(\zeta) < k_-$. The ϕ_+ on the left side of (19-20) is analytic in the overlapping upper half-plane $\text{Im}(\zeta) > c$. Suppose we could write the coefficient of ϕ_+, namely, $\tilde{k}(\zeta) - 1$, as the quotient of two functions, one of which is analytic in an upper half-plane and the other of which is analytic in an overlapping lower half-plane with the strip of overlap lying in the strip in which (19-20) holds. More precisely, suppose we could find two functions $\psi_\pm(\zeta)$ such that ψ_+ is analytic for $\text{Im}(\zeta) > \tilde{l}_+$ and ψ_- is analytic for $\text{Im}(\zeta) < \tilde{l}_-$, where the strips $c < \text{Im}(\zeta) < k_-$ and $\tilde{l}_+ < \text{Im}(\zeta) < \tilde{l}_-$ have a strip $l_+ < \text{Im}(\zeta) < l_-$ in common and such that

$$\tilde{k}(\zeta) - 1 = \frac{\psi_+(\zeta)}{\psi_-(\zeta)} \qquad l_+ < \text{Im}(\zeta) < l_- \tag{19-21}$$

Inserting (19-21) into (19-20), we would have

$$\psi_+(\zeta)\,\phi_+(\zeta) = \psi_-(\zeta)\,\phi_-(\zeta) \qquad l_+ < \text{Im}(\zeta) < l_- \tag{19-22}$$

The right side of (19-22) is analytic for $\text{Im}(\zeta) < l_-$, and the left side is analytic for $\text{Im}\,\zeta > l_+$. Since the two sides are equal in the strip $l_+ < \text{Im}\,\zeta < l_-$, each side is the analytic continuation of the other into its respective half-plane, and each side is therefore just a different representation in a different region for the same analytic function. This analytic function obviously has no singularities in the finite part of the plane, hence is entire. If we could determine its behavior at infinity, we would know it exactly. Below we will show how to do this. First, however, we have to find the functions $\psi_\pm$ in the decomposition (19-21). We will do this in several steps, first choosing functions that have some singularities and then gradually removing the singularities.

For brevity, we set

$$1 - \tilde{k}(\zeta) = l(\zeta) \tag{19-23}$$

Our goal is to find $\psi_\pm(\zeta)$ as described previously so that

$$l(\zeta) = \frac{\psi_+(\zeta)}{\psi_-(\zeta)} \tag{19-24}$$

The next few steps will be inductive rather than deductive. When we have finally found the desired $\psi_\pm$ we will say so.

Take the log of (19-24)

$$\log l(\zeta) = \log \psi_+(\zeta) - \log \psi_-(\zeta) \tag{19-25}$$

Then, we have the problem of writing a given analytic function, namely, $\log l(\zeta)$, as the *difference* of two functions, one analytic in an upper half-plane and the other analytic in an overlapping lower half-plane. Of course, $\log l(\zeta)$ will not generally be analytic in the strip $l_+ < \text{Im}(\zeta) < l_-$ even though $l(\zeta)$ is. Nevertheless, let us temporarily consider the problem of writing a function $f(\zeta)$ analytic in the strip $l_+ < \text{Im}(\zeta) < l_-$ as the difference of two functions $g_\pm(\zeta)$, where g_+ is analytic in the upper half-plane $\text{Im}\,\zeta > l_+$ and g_- is analytic in the lower half-plane $\text{Im}\,\zeta < l_-$. If $f \to 0$ in the strip when $|\text{Re}(\zeta)| \to \infty$, then the solution to this problem can be written down immediately. By Cauchy's theorem, we have

$$f(\zeta) = \frac{1}{2\pi i}\int_{-\infty+i\eta_+}^{\infty+i\eta_+} \frac{f(z)}{z-\zeta}\,dz - \frac{1}{2\pi i}\int_{-\infty+i\eta_-}^{\infty+i\eta_-} \frac{f(z)}{z-\zeta}\,dz \tag{19-26}$$

where (see Fig. 19-1),

$$l_+ < \eta_+ < \eta_- < l_- \tag{19-27}$$

Thus, if we knew that $\log l(\zeta)$ were analytic in the strip $l_+ < \text{Im}(\zeta) < l_-$ and that it went to zero as $|\text{Re}(\zeta)| \to \infty$, we could replace $f(\zeta)$ in (19-26) by $\log l(\zeta)$ and we would have the decomposition (19-25). In general, however, $\log l(\zeta)$ will not satisfy these requirements. We therefore proceed to concoct a function from it that will. We do this by first examining how $\log l(\zeta)$ can fail to satisfy the desiderata and then eliminating these shortcomings.

First we state a theorem on Fourier transforms.

Theorem 19-1 If $\tilde{k}(\zeta)$ is a Fourier transform analytic in a strip $k_+ < \text{Im}(\zeta) < k_-$, and if ζ is a point in this strip, then $\lim_{|\text{Re}(\zeta)|\to\infty} \tilde{k}(\zeta) = 0$.

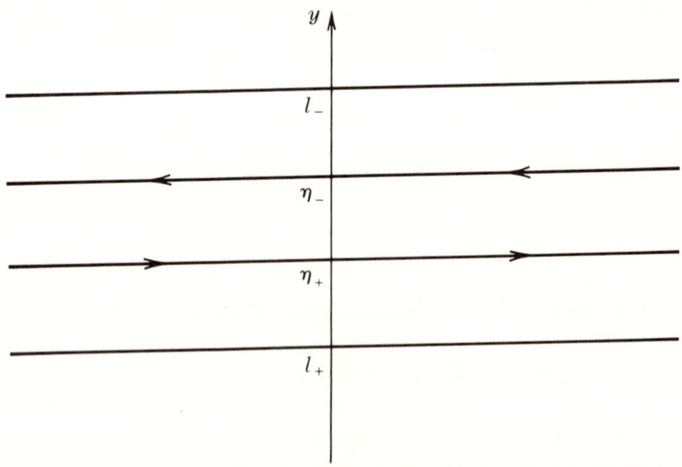

Figure 19-1

From this theorem and (19-23), it follows that $\log l(\zeta) \to 0$ in the strip as $|\operatorname{Re}(\zeta)| \to \infty$. So the only trouble in applying (19-26) to $\log l(\zeta)$ must come from points where $\log l(\zeta)$ has singularities. Since $\tilde{k}$ is analytic in the strip, the only singularities of $\log l(\zeta)$ must be branch points, and since $\tilde{k} \to 0$ at infinity in the strip, the only branch points of $\log l(\zeta)$ in the strip must be those points at which $l(\zeta)$ vanishes. These are the points where $\tilde{k}(\zeta) = 1$. There can be at most a finite number of such points, because the zeros of an analytic function can have no finite point of accumulation within its domain of analyticity. $l(\zeta)$ is analytic in the strip. Hence its zeros could have only infinity as a point of accumulation. They cannot have that as a point of accumulation either, however, because we have just pointed out that $\tilde{k}$, being a Fourier transform, tends to zero at infinity. Hence $l \to 1$, not zero, at infinity. This proves that there are at most a finite number of roots of the equation $\tilde{k}(\zeta) = 1$. Let $\zeta_1, \ldots, \zeta_n$ be the complete set of these roots; i.e., let

$$\tilde{k}(\zeta_j) = 1 \qquad j = 1, \ldots, n \tag{19-28}$$

in the strip. Then the function

$$\frac{l(\zeta)}{\prod_{j=1}^{n} (\zeta - \zeta_j)}$$

is analytic in the strip and different from zero therein. The only trouble is that it does not approach 1 at infinity as $l(\zeta)$ did. We remedy this difficulty by multiplying the numerator by a polynomial of degree n with no zeros in the strip. There are many possible choices. For example, $(\zeta - l_-)^n$ will do. Then the function

$$\alpha(\zeta) = \frac{l(\zeta)(\zeta - l_-)^n}{\prod_{j=1}^{n} (\zeta - \zeta_j)} \tag{19-29}$$

is analytic in the strip, different from zero therein, and tends to 1 at infinity. Hence its logarithm has no branch points in the strip. Its log might still have branch points outside the strip, however, and we know from function theory that such branch points could cause the argument of the logarithm to increase by a multiple of 2π when the strip is traversed from $-\infty$ to $+\infty$. Since $\alpha \to 1$ as $\operatorname{Re}(\zeta) \to \pm\infty$, the change in the argument of $\log \alpha$ in going from $-\infty$ to $+\infty$ in the strip must be an *integral* multiple of 2π. Suppose this integer is m. If we multiply (19-29) by a function whose argument increases by $-2\pi m$ in traversing the strip from $-\infty$ to $+\infty$, the resulting function will have a single-valued argument in the strip, *provided* we do not disturb any of the other properties of α that we have worked so hard to achieve, namely, that α is analytic in the strip, different from 0 therein, and tends to 1 at infinity. Again there are many functions we can use. One such is

$$\frac{(\zeta - l_-)^m}{(\zeta - l_+)^m}$$

Multiplying (19-29) by this, we obtain the function

$$\beta(\zeta) = \frac{l(\zeta)(\zeta - l_-)^{n+m}}{(\zeta - l_+)^m \prod_{j=1}^n (\zeta - \zeta_j)} \tag{19-30}$$

Because of the way in which it was constructed, $\beta(\zeta)$ has the property that its logarithm is single-valued and analytic in the strip $l_+ < \operatorname{Im}(\zeta) < l_-$ and $\log \beta(\zeta)$ goes to zero as $|\operatorname{Re}(\zeta)| \to \infty$ in the strip. Hence (19-27) applies to $\log \beta$ [obtaining a suitable function to which we could apply (19-26) was our goal], and we have

$$\log \beta(\zeta) = g_+(\zeta) - g_-(\zeta) \tag{19-31}$$

where g_+ is analytic for $\operatorname{Im}(\zeta) > l_+$ and g_- is analytic for $\operatorname{Im}(\zeta) < l_-$. For the present the fact that $g_\pm$ can be realized concretely as Cauchy integrals is immaterial but will be used shortly.

From (19-31) we deduce that

$$\beta(\zeta) = \frac{\beta_+(\zeta)}{\beta_-(\zeta)} \tag{19-32}$$

where β_+ is analytic for $\operatorname{Im}(\zeta) > l_+$ and β_- is analytic for $\operatorname{Im}(\zeta) < l_-$. From (19-32) and (19-30) we then get

$$l(\zeta) = \frac{(\zeta - l_+)^m \beta_+(\zeta) \prod_{j=1}^n (\zeta - \zeta_j)}{(\zeta - l_-)^{n+m} \beta_-(\zeta)} \tag{19-33}$$

as the long-sought decomposition (19-24).

It remains to determine the behavior of $\psi_\pm$ [cf. (19-24) and (19-33)] at infinity. As remarked above, the $g_\pm$ in (19-31) are Cauchy integrals. As such they tend to zero at infinity. Hence the $\beta_\pm$ tend to 1 at infinity. Therefore both numerator and denominator in (19-33) behave like ζ^{n+m} at infinity. Hence the analytic function of which both numerator and denominator are elements behaves like ζ^{n+m} at infinity. This implies that it is a polynomial of degree $n + m$. From this and the fact that ϕ_+, being a Fourier transform, goes to zero at infinity, it follows from (19-22) that

$$\phi_+(\zeta) = \frac{P(\zeta)}{\psi_+(\zeta)} \tag{19-34}$$

where $P(\zeta)$ is an arbitrary polynomial of degree $n + m$ in ζ and $\psi_+(\zeta)$ is the numerator of (19-33)

$$\psi_+(\zeta) = (\zeta - l_+)^m \beta_+(\zeta) \prod_{j=1}^{n} (\zeta - \zeta_j) \tag{19-35}$$

Having ϕ_+, we can then find ϕ for positive x by the Fourier inversion formula. Since we only wanted ϕ for positive values of x in the first place, this solves the problem. Notice that the solution is not unique but contains an arbitrary polynomial whose degree depends on the nature of the Fourier transform of k.

In explaining the Wiener-Hopf technique we have provided a constructive way of obtaining the decomposition (19-21). In practice, however, the method given above is often difficult to carry out. It is frequently simpler to use some properties of the particular k at hand in order to obtain the desired decomposition. The kernels that arise in practice often arise in the form of an infinite series or an infinite product. In such cases it may be possible to spot a decomposition of their transforms after a little manipulation of their representations. Nevertheless, considerable ingenuity or hard work or both are frequently required. Examples using the Wiener-Hopf technique in the literature are legion. Most are far more involved than the simple illustration considered below. The paper by Heins [127] contains a large number of references. Since its appearance, the list has naturally grown; see, for example, Bailin [6], Baldwin and Heins [9], and numerous papers by Heins and his coworkers.

EXAMPLE Lalesco's equation

$$\phi(x) = \lambda \int_0^\infty e^{-|x-y|}\phi(y)\,dy \qquad 0 < x < \infty \tag{19-36}$$

Here

$$k(t) = \lambda e^{-|t|} \tag{19-37}$$

So $k_\pm = \mp 1$.

By a trivial computation we find

$$\tilde{k}(\zeta) = \frac{2\lambda}{1+\zeta^2} \tag{19-38}$$

As the theory predicts, $\tilde{k}(\zeta)$ is analytic in the strip $-1 < \operatorname{Im}(\zeta) < 1$. The function to be decomposed is

$$\tilde{k}(\zeta) - 1 = \frac{2\lambda - 1 - \zeta^2}{1+\zeta^2} \tag{19-39}$$

and we do not need any high-powered theorems to decompose it. It has two zeros $\zeta_\pm$, where

$$\zeta_\pm = \pm\sqrt{2\lambda - 1} \tag{19-40}$$

If $\lambda < \frac{1}{2}$, the $\pm$ sign in front of the radical is to be interpreted as merely meaning that $\zeta_\pm$ are the two different complex roots of the radical. In order to avoid unnecessary complications, we assume that $0 < \lambda < 1$ so that both $\zeta_\pm$ lie in the strip $-1 < \text{Im}(\zeta) < 1$. Then we write (19-39) as

$$\tilde{k}(\zeta) - 1 = \frac{(\zeta - \zeta_-)(\zeta - \zeta_+)}{(\zeta - i)(\zeta + i)} \tag{19-41}$$

Rather than attempt any decomposition of (19-41), we substitute it into (19-20) and then separate the two sides appropriately

$$\frac{(\zeta - \zeta_-)(\zeta - \zeta_+)\,\phi_+(\zeta)}{\zeta + i} = (\zeta - i)\,\phi_-(\zeta) \tag{19-42}$$

Why, the reader may ask, that particular decomposition? First of all, the $\zeta - i$ had to be put on the right because otherwise the left side would be analytic only for $\text{Im}(\zeta) > 1$, that is, above the desired strip. Why not put the $\zeta + i$ over there with it then? The reason for leaving $\zeta + i$ where it is will be indicated below. We have already mentioned (Theorem 19-1) that a Fourier transform goes to zero at infinity. Now we mention and employ a slight strengthening of this theorem. A Fourier transform goes to zero at least like $1/\zeta$ at infinity. This means that both sides of (19-42) are bounded at infinity. By Liouville's theorem, it follows that the entire function of which they are different representations is a constant; i.e.,

$$\varphi_+(\zeta) = \frac{c(\zeta + i)}{(\zeta - \zeta_-)(\zeta - \zeta_+)} \tag{19-43}$$

where c is a complex constant. We leave the details of inverting (19-43) to the reader as an exercise in contour integration. Lalesco's equation is rather easy to solve by the Wiener-Hopf technique. It is even easier to solve by more elementary methods after one has done the following exercise.

EXERCISE Show that Lalesco's equation (19-36) is equivalent to the following linear ordinary differential equation (ODE) with constant coefficients

$$\phi''(x) + 2\lambda\phi'(x) - \phi(x) = 0$$

Just above we mentioned that it is preferable to leave the factor $\zeta + i$ where it is in (19-42). We are now able to say why. If $\zeta + i$ were put on the right of (19-42), we could conclude only that each side of the resulting equation was a linear function of ζ, not necessarily a constant. Since this would introduce a second constant, namely, the coefficient of ζ, into the solution, it therefore leads to the suspicion that something has been overlooked. What has been overlooked is this: with $\zeta + i$ on the right-hand side of (19-42) we conclude, among other things, that

$$\phi_-(\zeta) = \frac{A\zeta + B}{\zeta^2 + 1} \tag{19-44}$$

which, as mentioned above, seems to have two arbitrary complex constants instead of the single constant c found from the previous decomposition. However, ϕ_- is a Fourier transform and must therefore be analytic in a lower half-plane. The factor $\zeta + i$ in the denominator of (19-44) will give ϕ_-

a pole at $-i$ unless this factor is canceled by a zero of the numerator. In other words, A and B are not arbitrary but rather must satisfy $B = iA$. Then $\zeta + i$ cancels, and we are back to the form $c/(\zeta - i)$ for ϕ_- and the previous form for ϕ_+.

19-4. THE NONHOMOGENEOUS WIENER-HOPF INTEGRAL EQUATION

It remains to treat the nonhomogeneous equation

$$\phi(x) = f(x) + \int_0^\infty k(x-y)\,\phi(y)\,dy \qquad 0 < x < \infty \tag{19-45}$$

Naturally we have to assume that f has a Fourier transform and that this transform is analytic in the strip in which we have been working. We therefore assume

$$f(x) = O(e^{ax}) \qquad a > k_+$$

Then we can take the Fourier transform of (19-45) and obtain

$$\phi_+(\zeta) + \phi_-(\zeta) = f_+(\zeta) + \tilde{k}(\zeta)\,\phi_+(\zeta)$$

Rearranging and decomposing, we can write

$$\psi_+(\zeta)\,\phi_+(\zeta) + \psi_-(\zeta)f_+(\zeta) = \psi_-(\zeta)\,\phi_-(\zeta) \tag{19-46}$$

and we see that the term $\psi_- f_+$ is analytic only in a strip instead of in a half-plane. This difficulty is easily remedied, however. Given any function analytic in a strip, we have already seen that it can be written as the difference of two functions, one analytic in an upper half-plane and the other analytic in an overlapping lower half-plane. Thus, we can write

$$\psi_-(\zeta)f_+(\zeta) = \omega_+(\zeta) - \omega_-(\zeta) \tag{19-47}$$

the meaning of the notation being obvious.

Substituting (19-47) into (19-46) and separating terms appropriately, we have, finally,

$$\psi_+(\zeta)\,\phi_+(\zeta) + \omega_+(\zeta) = \psi_-(\zeta)\,\phi_-(\zeta) + \omega_-(\zeta)$$

and the rest of the procedure is now standard.

19-5. FRACTIONAL INTEGRATION; ABEL'S INTEGRAL EQUATION

As indicated in Chap. 17, dual integral equations frequently arise in practice. The literature on them has been growing apace, and we cite a number of papers later on. They were solved by Titchmarsh, using Mellin transforms and the ideas of the Wiener-Hopf method.

In the last few years a number of somewhat simpler methods have been developed to solve dual integral equations. Since many of them use fractional integrals in one form or another, and since such integrals are of use in their own right, we discuss these topics in the present section and then apply them in later sections.

Let $f(x)$ be continuous on $[0, \infty)$, and let I denote the (linear) operation of taking the indefinite integral of f

$$If = \int_0^x f(t)\,dt \tag{19-48}$$

Note that If is a function of x. As such it can be integrated from zero to x. Then

$$I^2f = I(If) = \int_0^x du \int_0^u f(t)\,dt \tag{19-49}$$

By interchanging the order of integration in the double integral in (19-49) we can easily show that

$$I^2f = \int_0^x (x - t)f(t)\,dt \tag{19-50}$$

and, by repeating the process, that

$$I^nf = \frac{1}{(n-1)!}\int_0^x (x-t)^{n-1}f(t)\,dt \tag{19-51}$$

for any positive integer n.

Noting that $(n-1)! = \Gamma(n)$, we can write (19-51) in the form

$$I^nf = \frac{1}{\Gamma(n)}\int_0^x (x-t)^{n-1}f(t)\,dt \tag{19-52}$$

But now look at (19-52). The right side of it exists for every real number, $n > 0$, not just for integers. As such, it defines an integral operator I^n. For example, if $n = \frac{1}{2}$, then

$$I^{1/2}f = \frac{1}{\sqrt{\pi}}\int_0^x \frac{f(t)\,dt}{\sqrt{x-t}} \tag{19-53}$$

where we have used the fact that $\Gamma(\frac{1}{2}) = \sqrt{\pi}$.

To summarize, we have the following definition.

DEFINITION 19-4

If p is any positive real number, the integral operator I^p, called a *fractional integral* or the *Riemann-Liouville integral*, is defined to be

$$I^p f = \frac{1}{\Gamma(p)} \int_0^x (x-t)^{p-1} f(t)\, dt \tag{19-54}$$

$I^p f$ clearly exists for any continuous f.

Having defined I^p for positive p, we would naturally like to define it for negative p, too. The definition (19-54) is no good for this purpose because the integral in it diverges. So we have to use a more roundabout method of defining I^p for negative p.

We start by asking ourselves what I^p should be for negative *integral* p. Clearly I^{-m}, for m an integer, should mean D^m, where $D^m = d^m/dx^m$. Also, it is natural to hope that $I^{p+q} = I^p I^q$. This leads to another definition.

DEFINITION 19-5

Let q be any nonpositive real number. Write $q = p - n$, where $0 < p \leqslant 1$ and n is a positive integer. (It is clear that this can always be done.) Then $I^q f$ is defined to be

$$I^q f = \frac{d^n}{dx^n} I^p f \tag{19-55}$$

where $I^p f$ is given by (19-54).

EXERCISE Prove that $I^\alpha I^\beta = I^{\alpha+\beta}$, for any real numbers α and β.

The notion of fractional integral leads to an immediate solution of *Abel's integral equation*. Abel's equation is

$$\int_0^x \frac{\phi(t)\, dt}{(x-t)^\alpha} = f(x) \qquad 0 < \alpha < 1 \tag{19-56}$$

In (19-56) f is given; ϕ is sought.

Using (19-54), we can write (19-56) in the operational form

$$\Gamma(1-\alpha)\, I^{1-\alpha} \phi = f \tag{19-57}$$

the solution to which is obviously

$$\phi = \frac{I^{\alpha-1} f}{\Gamma(1-\alpha)} = \frac{I^{-1} I^\alpha f}{\Gamma(1-\alpha)} = \frac{1}{\Gamma(\alpha)\,\Gamma(1-\alpha)} \frac{d}{dx} \int_0^x (x-t)^{\alpha-1} f(t)\, dt \tag{19-58}$$

Since the Γ function satisfies

$$\Gamma(\alpha)\,\Gamma(1-\alpha) = \frac{\pi}{\sin \pi\alpha} \tag{19-59}$$

we have

$$\phi(x) = \frac{\sin \pi\alpha}{\pi} \frac{d}{dx} \int_0^x (x - t)^{\alpha-1} f(t)\, dt \tag{19-60}$$

The differentiation in (19-60) cannot be carried out under the integral sign because doing so would create a divergent integral. We could differentiate under the integral sign if we could somehow build up the power of $(x - t)^{\alpha-1}$. Since this factor is nearly symmetric in x and t, we can accomplish this by integrating by parts. This gives

$$\begin{aligned} \phi(x) &= \frac{\sin \pi\alpha}{\pi} \frac{d}{dx} \left[\frac{(x-t)^\alpha}{\alpha} f(t) \bigg|_0^x + \frac{1}{\alpha} \int_0^x (x-t)^\alpha f'(t)\, dt \right] \\ &= \frac{\sin \pi\alpha}{\pi} \frac{d}{dx} \left[\frac{x^\alpha}{\alpha} f(0) + \frac{1}{\alpha} \int_0^x (x-t)^\alpha f'(t)\, dt \right] \end{aligned} \tag{19-61}$$

Thus,

$$\phi(x) = \frac{\sin \pi\alpha}{\pi} f(0)\, x^{\alpha-1} + \int_0^x (x-t)^{\alpha-1} f'(t)\, dt \tag{19-62}$$

Either (19-60) or (19-62) gives the solution to (19-56).

19-6. THE ELEMENTARY SOLUTION OF DUAL INTEGRAL EQUATIONS

In 1959 Sneddon [254] gave a way of solving a pair of dual integral equations without using the function-theoretic techniques of Titchmarsh. To solve

$$\int_0^\infty f(\rho)\, J_0(\rho r)\, d\rho = g(r) \qquad 0 < r < 1 \tag{19-63}$$

$$\int_0^\infty \rho f(\rho)\, J_0(\rho r)\, d\rho = 0 \qquad 1 < r < \infty \tag{19-64}$$

Sneddon observed† that it is known from the theory of Bessel functions that

$$\int_0^\infty \sin \rho t\, J_0(\rho r)\, d\rho = 0 \qquad \text{if } r > t \tag{19-65}$$

If we differentiate (19-65) with respect to t, we find

$$\int_0^\infty \rho \cos \rho t\, J_0(\rho r)\, d\rho = 0 \qquad \text{if } r > t \tag{19-66}$$

Equation (19-66) shows that $f(\rho) = \cos \rho t$ is a solution of (19-64) for all $t < 1$. If we multiply this solution by an arbitrary function $\phi(t)$ and integrate with respect to t from zero to 1, the resulting function is still a solution of (19-64) and the hope is that we can determine the function ϕ so that the

† Our development does not follow Sneddon verbatim but is essentially the same.

resulting f is also a solution of (19-63). More precisely, we try to find $\phi(t)$ so that

$$f(\rho) = \int_0^1 \phi(t) \cos \rho t \, dt \tag{19-67}$$

is a solution of (19-63) and (19-64).

As already pointed out, (19-67) automatically satisfies (19-64) for arbitrary ϕ because of (19-66). It remains to determine ϕ so that (19-67) satisfies (19-63). Hence we insert (19-67) into (19-63) and interchange the order of integration in the resulting iterated integral. This gives

$$\int_0^1 \phi(t) \, dt \int_0^\infty \cos \rho t \, J_0(\rho r) \, d\rho = g(r) \qquad 0 < r < 1 \tag{19-68}$$

Again we use a formula from the theory of Bessel functions, this time

$$\int_0^\infty \cos \rho t \, J_0(\rho r) \, d\rho = \begin{cases} 0 & \text{if } 0 < r < t \\ \dfrac{1}{\sqrt{r^2 - t^2}} & \text{if } r > t \end{cases} \tag{19-69}$$

Use of (19-69) in (19-68) gives

$$\int_0^r \frac{\phi(t) \, dt}{\sqrt{r^2 - t^2}} = g(r) \qquad 0 < r < 1 \tag{19-70}$$

Equation (19-70) looks similar to Abel's equation and can easily be reduced to it by putting $t^2 = \tau$ and $r^2 = x$. Then (19-70) becomes

$$\int_0^x \frac{\phi(\sqrt{\tau})}{2\sqrt{\tau}} \frac{d\tau}{\sqrt{x - \tau}} = g(\sqrt{x}) \tag{19-71}$$

which is Abel's equation for $\phi(\sqrt{\tau})/2\sqrt{\tau}$. According to (19-60), its solution is

$$\frac{\phi(\sqrt{\tau})}{2\sqrt{\tau}} = \frac{1}{\pi} \frac{d}{d\tau} \int_0^\tau \frac{g(\sqrt{x}) \, dx}{\sqrt{\tau - x}} \tag{19-72}$$

Putting back $\tau = t^2$ and $x = r^2$ in (19-72), we obtain

$$\phi(t) = \frac{2}{\pi} \frac{d}{dt} \int_0^t \frac{r g(r) \, dr}{\sqrt{t^2 - r^2}} \tag{19-73}$$

as the solution to (19-70).

Just as we did for Abel's equation, we can integrate by parts in (19-73) and then carry out the differentiation. The result is

$$\phi(t) = \frac{2}{\pi} \left[g(0) + t \int_0^t \frac{g'(r) \, dr}{\sqrt{t^2 - r^2}} \right] \tag{19-74}$$

Inserting (19-74) into (19-67), we have the solution of our dual integral equations

$$f(\rho) = \frac{2}{\pi\rho} g(0) \sin \rho + \frac{2}{\pi} \int_0^1 \int_0^t \frac{t \cos \rho t \, g'(r)}{\sqrt{t^2 - r^2}} \, dr \, dt \tag{19-75}$$

In the special case of Chap. 18 where $g(r) = 1$, $0 < r < 1$, (19-75) reduces to

$$f(\rho) = \frac{2}{\pi} \frac{\sin \rho}{\rho} \tag{19-76}$$

Referring back to (18-42) to (18-45), we see that $f(\rho) = \beta(\rho) = \rho\alpha(\rho)$. Therefore substituting (19-76) into (18-41), we have

$$u(r, z) = \frac{2}{\pi} \int_0^\infty \frac{\sin \rho}{\rho} e^{\rho z} J_0(\rho r) \, d\rho \tag{19-77}$$

as the solution to the mixed BVP in Eqs. (18-34) to (18-37). That (19-77) satisfies the PDE (18-34) is obvious because it is a superposition of products of functions of r and of z that do. That it satisfies (18-35) and (18-36) follows from a property of J_0, namely (see Sneddon [252, p. 528])

$$\frac{2}{\pi} \int_0^\infty \frac{\sin \rho}{\rho} J_0(\rho r) \, d\rho = 1 \qquad \text{if } 0 < r < 1 \tag{19-78}$$

and

$$\int_0^\infty \sin \rho \, J_0(\rho r) \, d\rho = 0 \qquad \text{if } r > 1 \tag{19-79}$$

which is (19-65) with $t = 1$.

As the reader must have noticed, Sneddon's method would not work if the zero on the right side of (19-64) were replaced by $h(r)$ because the method relies on satisfying (19-64) automatically by an integral of the form (19-67). However, we know that superposition would enable us to solve the general case if we could solve

$$\int_0^\infty f(\rho) \, J_0(r\rho) \, d\rho = 0 \qquad 0 < r < 1 \tag{19-80}$$

$$\int_0^\infty \rho f(\rho) \, J_0(r\rho) \, d\rho = h(r) \qquad 1 < r < \infty \tag{19-81}$$

in addition to (19-63) and (19-64). Equation (19-80) can be satisfied automatically, however, by putting

$$f(\rho) = \int_1^\infty \phi(t) \cos \rho t \, dt \tag{19-82}$$

because

$$\int_0^\infty \cos \rho t \, J_0(\rho t)\, d\rho = 0 \qquad \text{if } t > \rho \tag{19-83}$$

Sneddon's method then applies again, and we get an integral equation for $\phi(t)$ similar to the previous one. The solution to it (see Lowengrub and Sneddon [184]) turns out to be

$$\phi(t) = \frac{2}{\pi} \int_t^\infty \frac{rh(r)\, dr}{\sqrt{r^2 - t^2}} \tag{19-84}$$

19-7. MORE GENERAL DUAL INTEGRAL EQUATIONS

In 1960, Copson generalized Sneddon's method to the more general pair of dual integral equations

$$\int_0^\infty f(\rho)\, J_\nu(\rho r)\, d\rho = g(r) \qquad 0 < r < 1 \tag{19-85}$$

$$\int_0^\infty \rho^{2\gamma} f(\rho)\, J_\nu(\rho r)\, d\rho = 0 \qquad 1 < r < \infty \tag{19-86}$$

Copson recalled that

$$J_{1/2}(z) = \sqrt{\frac{2}{\pi}} \frac{\sin z}{\sqrt{z}}$$

Thus the identity (19-65) could be written entirely in terms of Bessel functions. In fact, its generalization is known to be (cf. Copson [61])

$$\int_0^\infty J_\lambda(at)\, J_\mu(bt)\, t^{1+\mu-\lambda}\, dt = 0 \qquad \text{if } 0 < a < b \tag{19-87}$$

This leads to generalizing (19-67) to

$$f(\rho) = \rho^{1-\gamma} \int_0^1 \phi(t)\, J_{\nu-\gamma}(\rho r)\, dt \tag{19-88}$$

The function $f(\rho)$ given by (19-88) then satisfies (19-86) identically, and (19-85) yields an integral equation similar to the previous one for ϕ. This time the solution turns out to be (see Copson [61])

$$\phi(t) = \frac{2^\gamma t^{\gamma-\nu}}{\Gamma(1-\gamma)} \frac{d}{dt} \int_0^t \frac{r^{\nu+1} g(r)}{(t^2 - r^2)^\gamma}\, dr \tag{19-89}$$

19-8. MORE GENERAL FRACTIONAL INTEGRALS AND THEIR APPLICATION TO DUAL INTEGRAL EQUATIONS

After various authors had treated a number of different problems in dual integral equations, Erdélyi and Sneddon [98] noticed that the sort of generalized Abel integral equations to which many such problems eventually reduce can be used in much the same manner as fractional integrals were in Sec. 19-5 to solve Abel's equation. To exploit this fact they introduced the integral operators

$$I_{\eta,\alpha}f(x) = \frac{x^{-\eta-\alpha}}{\Gamma(\alpha)} \int_0^x (x-y)^{\alpha-1} y^\eta f(y)\, dy \tag{19-90}$$

$$K_{\eta,\alpha}f(x) = \frac{x^\eta}{\Gamma(\alpha)} \int_x^\infty (x-y)^{\alpha-1} y^{-\eta-\alpha} f(y)\, dy \qquad \begin{matrix}\alpha > 0 \\ \eta > -\frac{1}{2}\end{matrix} \tag{19-91}$$

$$S_{\eta,\alpha}f(x) = x^{-\alpha/2} \int_0^\infty y^{-\alpha/2} J_{2\eta+\alpha}(2\sqrt{xy})\, f(y)\, dy \qquad \alpha \geqslant 0 \tag{19-92}$$

The operator $I_{\eta,\alpha}$ in (19-90) generalizes (19-54) and, apart from the factor $x^{-\alpha}$, reduces to it when $\eta = 0$. The operator $K_{\eta,\alpha}$ in (19-91) is obviously needed for cases when the Abel-type equation looks like (19-84). The operator $S_{\eta,\alpha}$ in (19-92) is a variant of the Hankel transform. For values of the parameters other than the indicated ones, it is necessary to do things analogous to Definition 19-5 by pulling out a certain number of derivatives. For example, if $\alpha < 0$, we define

$$I_{\eta,\alpha}f(x) = x^{-\eta-\alpha} \frac{d^n}{dx^n} x^{\eta+\alpha+n} I_{\eta,\alpha+n} f(x) \qquad n \text{ a positive integer such that } \alpha + n \geqslant 0 \tag{19-93}$$

If we assume that the dual integral equations to be solved are of the form

$$x^{-\alpha} \int_0^\infty y^{-\alpha} \psi(y)\, J_\nu(2\sqrt{xy})\, dy = f(x) \qquad 0 < x < 1 \tag{19-94}$$

$$x^{-\beta} \int_0^\infty y^{-\beta} \psi(y)\, J_\nu(2\sqrt{xy})\, dy = g(x) \qquad 1 < x < \infty \tag{19-95}$$

then we can write them in the operational form

$$S_{\nu/2-\alpha,2\alpha}\psi = f \qquad \text{and} \qquad S_{\nu/2-\beta,2\beta}\psi = g \tag{19-96}$$

It can be shown that

$$S_{\eta,\alpha}^{-1} = S_{\eta+\alpha,-\alpha} \tag{19-97}$$

where S^{-1} means S inverse, the inverse transformation. In addition, it turns out that

$$I_{\eta+\alpha,\beta} S_{\eta,\alpha} = S_{\eta,\alpha+\beta} \tag{19-98}$$

and

$$K_{\eta,\alpha} S_{\eta+\alpha,\beta} = S_{\eta,\alpha+\beta} \tag{19-99}$$

The first equation in (19-96) holds only for $0 < x < 1$ and the second only for $1 < x < \infty$. In order to use (19-97) directly on (19-96) we would have to know f and g outside the intervals in which they are given. This is an old difficulty. We have already seen the troubles that arise when we introduce unknown extensions for the given data. The idea of [98], therefore, is to apply appropriate I's and K's to both equations in (19-96) so as to produce through (19-98) and (19-89) the same S on the left sides of both equations. Then (19-97) can be used. To carry out this idea, apply $I_{\frac{1}{2}\nu+\alpha,\beta-\alpha}$ to the first equation in (19-97) and $K_{\frac{1}{2}\nu-\alpha,\alpha-\beta}$ to the second. Using (19-98) and (19-99), we obtain

$$S_{\nu/2-\alpha,\alpha+\beta}\psi = \begin{cases} I_{1/2+\alpha,\beta-\alpha}f & 0 < x < 1 \qquad (19\text{-}100) \\ K_{1/2-\alpha,\alpha-\beta}\,g & 1 < x < \infty \qquad (19\text{-}101) \end{cases}$$

Since (19-100) and (19-101) have the same S on the left, we can define

$$h(x) = \begin{cases} I_{1/2+\alpha,\beta-\alpha}f & 0 < x < 1 \\ K_{1/2-\alpha,\alpha-\beta}\,g & 1 < x < \infty \end{cases}$$

and combine them into the one equation

$$S_{\nu/2-\alpha,\alpha+\beta}\psi = h \qquad 0 < x < \infty \tag{19-102}$$

Equation (19-102) can be solved immediately by (19-97)

$$\psi = S_{\nu/2+\beta,-\alpha-\beta}h \tag{19-103}$$

Equation (19-103) can then be written out explicitly by using (19-92). The special case considered in a previous section in which $\nu = 0$, $\alpha = 0$, $\beta = -1$, is left as an exercise.

EXERCISE Find $\psi = S_{-1,1}h$.

Still more general dual integral equations have been treated by Burlak [24], who uses Sonine's integral for Bessel functions, as Copson used the simpler identity (19-87), to solve

$$\int_k^\infty u^{-\mu-\nu}(u^2 - k^2)^\alpha\, \psi(u)\, J_\mu(xu)\, du = f(x) \qquad 0 \leqslant x \leqslant 1$$

$$\int_0^\infty \psi(u)\, J_\nu(xu)\, du = g(x) \qquad 1 < x < \infty$$

where $k \geqslant 0$, μ, and α are real constants.

In addition, Erdélyi and Sneddon [98], Love [183], and Schmeltzer and Lewin [245] have considered more general equations which they are able to reduce to Fredholm-type ones.

19-9. THE SOLUTION OF DUAL INTEGRAL EQUATIONS WITH TRIGONOMETRIC KERNELS

All the dual integral equations of the previous sections have had Bessel function kernels. Work has also been done on dual integral equations with trigonometric kernels. In this section we merely mention how to solve the dual integral equations (18-32) and (18-33). Use (cf. [254]) the integral identities

$$\int_0^\infty \cos x\xi \, J_1(\xi) \, d\xi = \begin{cases} 1 & \text{if } 0 < x < 1 \\ 0 & \text{if } x > 1 \end{cases}$$

The solution to (18-32) and (18-33) is therefore $\alpha(\xi) = (\pi/2) \, J_1(\xi)$.

References

1. Abbott, Michael B.: "An Introduction to the Method of Characteristics," American Elsevier, New York, 1966.
2. Alblas, J. B.: On the Diffraction of Sound Waves in a Heat-conducting Viscous Medium, *Koninkl. Nedl. Akad. Wetenschap. Proc Ser. B*, **64**:351–367 (1961).
3. Ames, W. F.: "Nonlinear Partial Differential Equations in Engineering," Academic, New York, 1965.
4. Ames, W. F., and S. E. Jones: Integrated Lagrange Expansions for a Monge-Ampere Equation, *J. Math. Anal. Appl.*, **21**:479–848 (1968).
5. Avila, Geraldo S. S., and Calvin R. Wilcox: The Near-field Behavior of the Green's Matrix in Anisotropic Wave Motion, *J. Math. Mech.*, **16**:867–884 (1967).
6. Bailin, Louis L.: An Analysis of the Effect of the Discontinuity in a Bifurcated Circular Guide upon Plane Longitudinal Waves, *J. Res. Natl. Bur. Std.*, **47**:315–335 (1951).
7. Baker, Bevan B., and Edward T. Copson: "The Mathematical Theory of Huygens' Principle," Oxford University Press, Fair Lawn, N.J., 1949.
8. Balazs, Nandor L.: On the Solution of the Wave Equation with Moving Boundaries, *J. Math. Anal. Appl.*, **3**:472–484 (1961).
9. Baldwin, George L., and Albert E. Heins: On the Diffraction of a Plane Wave by an Infinite Plane Grating, *Math. Scand.*, **2**:103–118 (1954).
10. Bartlett, C. C., and B. Noble: A Derivation of Certain Variational Principles for Mixed Boundary Value Problems in Potential Theory, *Proc. Edinburgh Math. Soc.*, **12**:113–117 (1961).
11. Bateman, Harry: "Partial Differential Equations of Mathematical Physics," Cambridge University Press, London, 1959.
12. Bers, Lipman: "Mathematical Aspects of Subsonic and Transonic Gas Dynamics," Wiley, New York, 1958.
13. Biot, M. A.: Generalized Variational Principles for Convective Heat Transfer and Irreversible Thermodynamics, *J. Math. Mech.*, **15**:177–186 (1966).
14. Bluman, George W., and Julian D. Cole: The General Similarity Solution of the Heat Equation, *J. Math. Mech.*, **18**:1025–1042 (1969).
15. Bobisud, Larry: Laplace's Equation in a Quarter Space as the Limit of the Wave Equation, *J. Differential Equations*, **4**:309–313 (1968).
16. Boyer, Robert H.: On Some Solutions of a Non-linear Diffusion Equation, *J. Math. Phys.*, **40**:41–45 (1961).
17. Bragg, L. R., and John W. Dettman: An Operator Calculus for Related Partial Differential Equations, *J. Math. Anal. Appl.*, **22**:261–271 (1968).
18. Bragg, L. R., and John W. Dettman: Related Partial Differential Equations and Their Applications, *J. Soc. Indus. Appl. Math.*, **16**:459–467 (1968).
19. Bragg, L. R., and John W. Dettman: Related Problems in Partial Differential Equations, *Bull. Am. Math. Soc.*, **74**:375–378 (1968).
20. Burckhardt, C. B.: Diffraction of a Plane Wave at a Sinusoidally Stratified Dielectric Grating, *J. Opt. Soc. Am.*, **56**:1502–1509 (1966).
21. Bürger, W.: A Note on the Breaking of Waves on Non-uniformly Sloping Beaches, *J. Math. Mech.*, **16**:1131–1142 (1967).
22. Burke, J. E.: Scattering of Surface Waves on an Infinitely Deep Fluid, *J. Math. Phys.*, **5**:805–819 (1964).

23. Burlak, J.: A Pair of Dual Integral Equations Occurring in Diffraction Theory, *Proc. Edinburgh Math. Soc.*, **13**:179–187 (1962).
24. Burlak, J.: On the Solution of Certain Dual Integral Equations, *Proc. Glasgow Math. Assoc.*, **6**:39–44 (1963).
25. Burlak, J.: On Gordon's Method of Solving Dual Integral Equations, *Proc. Glasgow Math. Assoc.*, **6**:117–122 (1964).
26. Burlak, J.: A Further Note on Certain Integral Equations of Abel Type, *Proc. Edinburgh Math. Soc.*, **14**:255–256 (1965).
27. Burns, J. C.: The Iterated Equation of Generalized Axially Symmetric Potential Theory, *J. Australian Math. Soc.*, **7**:263–300 (1967).
28. Callen, Herbert B.: "Thermodynamics," Wiley, New York, 1960.
29. Caratheodory, Constantin, "Calculus of Variations and Partial Differential Equations of the First Order," Holden-Day, San Francisco, Vol. I, 1968, Vol. II, 1967.
30. Carrier, George F.: On the Nonlinear Vibration Problem of the Elastic String, *Quart. Appl. Math.*, **3**:157–165 (1945).
31. Carrier, George F.: A Note on the Vibrating String, *Quart. Appl. Math.*, **7**:97–101 (1949).
32. Carrier, George F.: Useful Approximations in Wiener-Hopf Problems, *J. Appl. Phys.*, **30**:1769–1774 (1959).
33. Carrier, George F., and H. P. Greenspan: Water Waves of Finite Amplitude on a Sloping Beach, *J. Fluid Mech.*, **4**:97–109 (1958).
34. Carrier, George F., Max Krook, and Carl E. Pearson: "Functions of a Complex Variable: Theory and Technique," McGraw-Hill, New York, 1966.
35. Case, K.: An Approximation Method for Diffraction Problems, *Rev. Mod. Phys.*, **36**:669–679 (1964).
36. Chadwick, P., and G. E. Tulpholme: Generation of an Acoustic Pulse by a Baffled Circular Piston, *Proc. Edinburgh Math. Soc.*, **15**:263–267 (1967).
37. Chambers, Ll. G.: Derivation of Solutions of the Klein-Gordon Equation from Solutions of the Wave Equation, *Proc. Edinburgh Math. Soc.*, **15**:125–129 (1966).
38. Chang, C. C., and T. S. Lundgren: Airfoil in a Sonic Shear Flow Jet: A Mixed BVP for the Generalized Tricomi Equation, *Quart. Appl. Math.*, **17**:375–392 (1960).
39. Chong, Frederick: Solution by Dual Integral Equations of a Plane-strain Boussinesq Problem for an Orthotropic Medium, *Iowa State Coll. J. Sci.*, **27**:321–334 (1953).
40. Chu, Chong-Wei: A Class of Reducible Systems of Quasi-linear Partial Differential Equations, *Quart. Appl. Math.*, **23**:275–278 (1954).
41. Chu, S. C., and J. B. Diaz: The Goursat Problem for the Partial Differential Equation $u_{xyz} = f$: A Mirage, *J. Math. Mech.*, **16**:709–714 (1967).
42. Church, Alonzo: Remarks on the Elementary Theory of Differential Equations as Area of Research, *Colloq. Acad. Internat. Phil. Sci. Brussels*, September, 1962.
43. Church, Alonzo: A Generalization of Laplace's Transformation, *Ann. Acad. Sci. Fennicae Ser. A* I, **377**:1–34 (1966).
44. Churchill, Ruel V.: Integral Transforms and Boundary Value Problems, *Am. Math. Monthly*, **59**:149–155 (1952).
45. Churchill, Ruel V.: Extensions of Operational Mathematics, *Proc. Conf. Differential Equations, Univ. of Maryland*, March, 1955.
46. Churchill, Ruel V.: "Operational Mathematics," 2d ed., McGraw-Hill, New York, 1958.
47. Churchill, Ruel V.: "Complex Variables and Applications," 2d ed., McGraw-Hill Book Company, New York, 1960.
48. Churchill, Ruel V.: "Fourier Series and Boundary Value Problems," 2d ed., McGraw-Hill, New York, 1963.

49. Churchill, Ruel V.: Integral Transforms Associated with Boundary Conditions of the Third Type, in J. P. LaSalle and J. B. Diaz (eds.), "Contributions to Differential Equations," vol. 3, Wiley, New York, 1964.
50. Cohen, Donald S.: Separation of Variables and Alternative Representations for Non-selfadjoint Boundary Value Problems, *Commun. Pure Appl. Math.*, **17**:1–22 (1964).
51. Cohen, Donald S.: Eigenfunction Expansions and Non-selfadjoint Boundary Value Problems, *Commun. Pure Appl. Math.*, **17**:23–34 (1964).
52. Cohen, Donald S.: New Eigenfunction Expansions and Alternative Representations for the Reduced Wave Equation, *J. Math. Mech.*, **14**:403–412 (1965).
53. Collin, Robert E.: "Field Theory of Guided Waves," McGraw-Hill, New York, 1960.
54. Cooke, J. C.: Triple Integral Equations, *Quart. J. Mech. Appl. Math.*, **26**:193–203 (1963).
55. Cooke, J. C.: Some Further Triple Integral Equation Solutions, *Proc. Edinburgh Math. Soc.*, **13**:303–316 (1963).
56. Cooke, J. C.: The Solution of Triple Integral Equations in Operational Form, *Quart. J. Mech. Appl. Math.*, **18**:72–57 (1965).
57. Copson, Edward T.: On the Problem of the Electrified Disc, *Proc. Edinburgh Math. Soc.*, **8**:14–19 (1947).
58. Copson, Edward T.: On Sound Waves of Finite Amplitude, *Proc. Roy. Soc. (London), Ser. A.*, **216**:539–547 (1953).
59. Copson, Edward T.: On a Singular Boundary Value Problem for an Equation of Hyperbolic Type, *Arch. Rational Mech. Anal.*, **1**:349–356 (1958).
60. Copson, Edward T.: On the Riemann-Green Function, *Arch. Rational Mech. Anal.*, **1**:324–348 (1958).
61. Copson, Edward T.: On Certain Dual Integral Equations, *Proc. Glasgow Math. Assoc.*, **5**:21–24 (1961).
62. Corben, H. C., and Philip Stehle: "Classical Mechanics," 2d ed., Wiley, New York, 1960.
63. Coulson, C. A.: "Waves," 7th ed., Interscience, New York, 1955.
64. Courant, Richard, and Kurt O. Friedrichs: "Supersonic Flow and Shock Waves," Interscience, New York, 1948.
65. Courant, Richard, and David Hilbert: "Methods of Mathematical Physics," Interscience, New York, vol. I, 1953, vol. II, 1962.
66. Courant, Richard, and Peter Lax: On Non-linear Partial Differential Equations with Two Independent Variables, *Commun. Pure Appl. Math.*, **2**:255–273 (1949).
67. Crank, J.: "The Mathematics of Diffusion," Oxford University Press, Fair Lawn, N.J., 1956.
68. Crease, J.: Long Waves on a Rotating Earth in the Presence of a Semi-infinite Barrier, *J. Fluid Mech.*, **1**:86–96 (1956).
69. Crease, J.: The Propagation of Long Waves into a Semi-infinite Channel in a Rotating System, *J. Fluid Mech.*, **4**:306–320 (1958).
70. Darling, D. A.: Some Relations between Potential Theory and the Wave Equation, *Univ. Mich. Rad. Lab. Rept.* 2871-5-7, December, 1960.
71. Davis, Harry F.: "Fourier Series and Orthogonal Functions," Allyn and Bacon, Boston, 1963.
72. Dennemeyer, Rene: "Introduction to Partial Differential Equations and Boundary Value Problems," McGraw-Hill, New York, 1968.
73. Dettman, John W.: Initial-boundary Value Problems Related through the Stieltjes Transform, *J. Math. Anal. Appl.*, **25**:341–349 (1969).
74. Dettman, John W.: "Mathematical Methods in Physics and Engineering," 2d ed., McGraw-Hill, New York, 1969.

75. Dhaliwal, R. S.: An Axisymmetric Mixed Boundary Value Problem for a Thick Slab, *J. Soc. Ind. Appl. Math.*, **15**:98–106 (1967).
76. Diaz, J. B.: On Cauchy's Problem and Fundamental Solutions, *Ann. Math. Studies*, **33**:235–247 (1954).
77. Diaz, J. B.: On Cauchy's Problem and Fundamental Solutions, L. Bers, S. Bochner, and F. John (eds.), "Contributions to Theory of Partial Differential Equations," Princeton University Press, Princeton, N.J., 1954.
78. Diaz, J. B., and Abdul R. Kiwan: A Remark on the Singular Cauchy Problem, for All Values of the Time, for the Euler-Poisson-Darboux Equation, *J. Math. Mech.*, **16**:197–202 (1966).
79. Diaz, J. B., and G. S. S. Ludford: A Transonic Approximation, *Proc. 2d U.S. Natl. Congr. Appl. Mech., Univ. Mich.*, 1954, 651–658.
80. Diaz, J. B., and G. S. S. Ludford: On Two Methods of Generating Solutions of Linear Partial Differential Equations by Means of Definite Integrals, *Quart. Appl. Math.*, **12**:422–427 (1955).
81. Diaz, J. B., and G. S. S. Ludford: On the Singular Cauchy Problem for a Generalization of the Euler-Poisson-Darboux Equation in Two Space Variables, *Ann. Math.*, (4) **38**:33–50 (1955).
82. Diaz, J. B., and G. S. S. Ludford, On a Theorem of Le Roux, *Can. J. Math.*, **8**:82–85 (1956).
83. Diaz, J. B., and M. H. Martin: A Generalization of Riemann's Method for Partial Differential Equations, *Ann. Math.*, **36**:335–359 (1954).
84. Diprima, R. C.: On the Diffusion of Tides into Permeable Rock of Finite Depth, *Quart. Appl. Math.*, **15**:329–339 (1958).
85. Douglis, Avron: On Linear, Hyperbolic Equations of Second Order, *Univ. Maryland Tech. Note* BN-139, June, 1958.
86. Duff, G. F. D.: "Partial Differential Equations," University of Toronto Press, Toronto, 1956.
87. Duff, G. F. D.: Mixed Problems for Linear Systems of First Order Equations, *Can. J. Math.*, **10**:127–160 (1958).
88. Duff, G. F. D.: The Cauchy Problem for Elastic Waves in an Anisotropic Medium, *Phil. Trans. Roy. Soc. (London) Ser. A.*, **252**:249–273 (1960).
89. Duff, G. F. D., and Derek Naylor: "Differential Equations of Applied Mathematics," Wiley, New York, 1966.
90. Duff, G. F. D., and R. A. Ross: Indefinite Green's Functions and Elementary Solutions, *Can. Math. Bull.*, **6**:71–103 (1963).
91. Eichler, Martin: Allgemeine Integration linearer partieller Differentialgleichungen von elliptischem Typ bei zwei Grundvariablen, *Abhandl. Math. Seminar Univ. Hamburg*, **15**:179–210 (1947).
92. Eichler, Martin: On the Differential Equation $u_{xx} + u_{yy} + N(x)u = 0$, *Trans. Am. Math. Soc.*, **65**:259–278 (1949).
93. Eichler, Martin: Eine Modifikation der Riemannschen Integrationsmethode bei partiellen Differentialgleichungen vom hyperbolischen Typ, *Math. Z.*, **53**:1–10 (1950).
94. Elsgolc, L. E.: "Calculus of Variations," Pergamon, London, 1962.
95. Epstein, Bernard: "Partial Differential Equations," McGraw-Hill, New York, 1962.
96. Erdélyi, Arthur: Some Applications of Fractional Integral Equations, *Boeing Sci. Res. Lab. Rept.*, June, 1963.
97. Erdélyi, Arthur: An Application of Fractional Integrals, *J. Anal. Math.*, **14**:113–126 (1965).
98. Erdélyi, Arthur, and Ian N. Sneddon: Fractional Integration and Dual Integral Equations, *Can. J. Math.*, **14**:685–693 (1962).

99. Erdogan, F., and Leon Y. Bahar: On the Solution of Simultaneous Dual Integral Equations, *J. Soc. Indus. Appl. Math.*, **12**:666–675 (1964).
100. Faulkner, T. R.: Diffraction of Singular Fields by a Wedge, *Arch. Rational Mech. Anal.*, **18**:196–204 (1965).
101. Faulkner, T. R.: Diffraction of an Electromagnetic Plane-wave by a Metallic Strip, *J. Inst. Math. Appl.*, **1**:149–163 (1965).
102. Faulkner, T. R.: Diffraction by a Perfectly Conducting Wedge in an Anisotropic Plasma, *Proc. Cambridge Phil. Soc.*, **61**:767–776 (1965).
103. Felsen, Leo, and Lawrence Levey: A Relation between a Class of Boundary Value Problems in a Homogeneous and an Inhomogeneous Region, *IEEE Trans. Antennas Propagation*, **AP-14**:308–317 (1966).
104. Ficken, F. A.: An Unusual Parabolic Problem, *Commun. Pure Appl. Math.*, **14**:295–307 (1961).
105. Flatto, Leopold, D. J. Newman, and H. S. Shapiro: The Level Curves of Harmonic Functions, *Trans. Am. Math. Soc.*, **123**:425–436 (1966).
106. Fleishman, Bernard A.: Progressing Waves in an Infinite Nonlinear String, *Proc. Am. Math. Soc.*, **10**:329–334 (1959).
107. Fleishman, Bernard A.: Wave Propagation in Non-simple Media, in "Nonlinear Differential Equations and Nonlinear Mechanics," pp. 211–217, Academic, New York, 1963.
108. Ford, Lester R.: "Differential Equations," 2d ed., McGraw-Hill, New York, 1955.
109. Fox, Phyllis: Perturbation Theory of Wave Propagation Based on the Method of Characteristics, *J. Math. Phys.*, **34**:133–151 (1955).
110. Foy, Linus R.: Steady State Solutions of Hyperbolic Systems of Conservation Laws with Viscosity Terms, *Commun. Pure Appl. Math.*, **17**:177–188 (1964).
111. Fredricks, R. W.: Solution of a Pair of Integral Equations from Elastostatics, *Proc. Natl. Acad. Sci.*, **44**:309–312 (1958).
112. Friedlander, F. G.: On an Improperly Posed Characteristic Initial Value Problem, *J. Math. Mech.*, **16**:907–916 (1967).
113. Friedman, Bernard: "Principles and Techniques of Applied Mathematics," Wiley, New York, 1956.
114. Funk, Paul: "Variationsrechnung und ihre Anwendung in Physik und Technik," Springer-Verlag, Berlin, 1962.
115. Garabedian, Paul R.: "Partial Differential Equations," Wiley, New York, 1964.
116. Gelfand, I. M., and S. V. Fomin: "Calculus of Variations," Prentice-Hall, Englewood Cliffs, N.J., 1963.
117. Germain, P.: Remarks on Transforms and Boundary Value Problems, *J. Rational Mech. Anal.* **4**:925–941 (1955).
118. Gordon, A.: Dual Integral Equations, *J. London Math. Soc.*, **29**:360–363 (1954).
119. Goursat, Edouard: "A Course in Mathematical Analysis," vol. II, pt. 2, transl. E. R. Hedrick and Otto Dunkel, Dover, New York, 1945.
120. Greenspan, H. P.: On the Breaking of Water Waves of Finite Amplitude on a Sloping Beach, *J. Fluid Mech.*, **4**:339–334 (1958).
121. Greenspan, H. P.: A String Problem, *J. Math. Anal. Appl.*, **6**:339–348 (1963).
122. Guderly, Karl G.: "The Theory of Transonic Flow," Addison-Wesley, Reading, Mass., 1962.
123. Heins, Albert E.: Water Waves over a Channel of Finite Depth with a Dock, *Am. J. Math.*, **70**:730–748 (1948).
124. Heins, Albert E.: Water Waves over a Channel of Finite Depth, *Can. J. Math.*, **2**:210–222 (1950).

125. Heins, Albert E.: Water Waves over a Channel of Finite Depth with a Submerged Plane Barrier, *Can. J. Math.*, **2**:210–222 (1950).

126. Heins, Albert E.: Some Remarks on the Coupling of Two Ducts, *J. Math. Phys.*, **30**:164–169 (1951).

127. Heins, Albert E.: The Scope and Limitations of the Method of Wiener and Hopf, *Commun. Pure Appl. Math.*, **9**:447–466 (1956).

128. Heins, Albert E.: The Green's Function for Periodic Structures in Diffraction Theory with an Application to Parallel Plate Media, II, *J. Math. Mech.*, **6**:629–640 (1957).

129. Heins, Albert E.: Function-theoretic Aspects of Diffraction Theory, in R. E. Langer (ed.), "Electromagnetic Waves," University of Wisconsin Press, Madison, 1962.

130. Heins, Albert E.: Axially Symmetric Boundary Value Problems, *Bull. Am. Math. Soc.*, **71**:787–808 (1965).

131. Heins, Albert E.: A Boundary Value Problem Associated with the Tricomi Equation, *Ann. Del. Scuola Norm. Sup. Pisa*, **19**:465–479 (1965).

132. Heins, Albert E.: Some Remarks on Axially-symmetric Potential Theory, *J. Math. Mech.*, **16**:203–212 (1966).

133. Heins, Albert E., and Richard C. MacCamy: A Function Theoretic Solution of Certain Integral Equations, I and II, *Quart. J. Math. Oxford Ser.*, **9**:132–143 (1958); **10**:280–293 (1959).

134. Heins, Albert E., and Richard C. MacCamy: On Mixed Boundary Value Problems for Axially Symmetric Potentials, *J. Math. Anal. Appl.*, **1**:331–333 (1960).

135. Helliwell, J. B.: A Flow Pattern at High Subsonic Speeds Past a Wedge at Incidence in a Free Stream and a Choked Channel, *J. Math. Phys.*, **40**:1–22 (1961).

136. Herglotz, Gustav: Die Greensche Funktion der Wellengleichung für eine keilformige Begrenzung, *Math. Ann.* **124**:219–234 (1962).

137. Heyda, James F.: A Note Concerning Hyperbolic Equations with Constant Coefficients, *Quart. Appl. Math.*, **18**:299–300 (1960).

138. Heyda, James F.: Green's Function for Laplace's Equation in a Circular Ring with Radiation Type Boundary Conditions, *Z. Angew. Math. Phys.*, **12**:322–328 (1961).

139. Hildebrand, Francis B.: "Advanced Calculus for Applications," Prentice-Hall, Englewood Cliffs, N.J., 1962.

140. Hochstadt, Harry: Some Diffraction by Convex Bodies, *Arch. Rational Mech. Anal.*, **3**:422–438 (1959).

141. Hopf, Eberhard: The Partial Differential Equation $u_t + uu_x = \mu u_{xx}$, *Commun. Pure Appl. Math.*, **3**:201–230 (1950).

142. Hutson, V.: Asymptotic Solutions of Integral Equations with Convolution Kernels, *Proc. Edinburgh Math. Soc.*, **14**:5–19 (1964).

143. Jardetzky, W. S.: On the General Solution of a Wave Equation, *Trans. N.Y. Acad. Sci.*, **15**:297–301 (1953).

144. Jeffrey, Alan: The Development of Jump Discontinuities in Nonlinear Hyperbolic Systems of Equations in Two Independent Variables, *Arch. Rational Mech. Anal.*, **14**:27–37 (1963).

145. Jeffrey, Alan: The Breaking of Waves on a Sloping Beach, *Z. Angew. Math. Phys.*, **15**:97–106 (1964).

146. Jeffrey, Alan: The Development of Singularities of Solutions of Nonlinear Hyperbolic Equations of Order Greater than Unity, *J. Math. Mech.*, **15**:585–598 (1966).

147. Jeffrey, Alan, and T. Taniuti, "Nonlinear Wave Propagation with Applications to Physics and Magnetohydrodynamics," Academic, New York, 1964.

148. John, Fritz: Partial Differential Equations, mimeographed lecture notes, New York University, 1952–1953.

149. John, Fritz: "Plane Waves and Spherical Means," Interscience, New York, 1955.

150. Johnson, J. L., and J. A. Smoller: Global Solutions for Certain Systems of Quasilinear Hyperbolic Equations, *J. Math. Mech.*, **17**:561–576 (1967).
151. Johnson, J. L., and J. A. Smoller: Global Systems of Certain Hyperbolic Systems of Quasilinear Equations, *Bull. Am. Math. Soc.*, **73**:666–667 (1967).
152. Jones, Douglas S.: The Unsteady Motion of a Thin Aerofoil in an Incompressible Fluid, *Commun. Pure Appl. Math.*, **10**:1–21 (1957).
153. Kane, Julius: The Superposition Principle and Diffraction in Sectors, *J. Math. Mech.*, **15**:207–220 (1966).
154. Keller, Joseph B.: The Scope of the Image Method, *Commun. Pure Appl. Math.*, **6**:505–512 (1953).
155. Keller, Joseph B.: Finite Amplitude Sound Waves, *J. Acoust. Soc. Am.*, **25**:212–216 (1953).
156. Keller, Joseph B.: On Solutions of a Nonlinear Wave Equation, *NYU Rept.* MME-3, April, 1957.
157. Keller, Joseph B.: Large Amplitude Motion of a String, *Am. J. Phys.*, **27**:584–586 (1959).
158. Keller, Joseph B., and Clifford S. Gardner: The Field of a Pulsed Dipole in an Interface, *NYU Rept.* EM-177, April, 1963.
159. Kellogg, Oliver Dimon: "Foundations of Potential Theory," Dover, New York, 1953.
160. Klamkin, Murray S., and Donald J. Newman: The Philosophy and Applications of Transform Theory, *Soc. Ind. Appl. Math. Rev.*, **3**:10–36 (1961).
161. Kleinman, Ralph Ellis: The Dirichlet Problem for the Helmholtz Equation, *Arch. Rational Mech. Anal.*, **18**:205–229 (1965).
162. Kleinman, Ralph Ellis, and R. Tinman: Studies in Radar Cross Sections XLIV: Integral Representations of Solutions of the Helmholtz Equation with Application to Diffraction by a Strip, *Univ. Mich. Rad. Lab. Rept.* 3648-3-T, February, 1961.
163. Kline, Morris, and Irwin W. Kay: "Electromagnetic Theory and Geometrical Optics," Interscience, New York, 1965.
164. Kneschke, A.: "Differentialgleichungen und Randwertproblemen," 2d ed., 3 vols., VEB Verlag Technik, Berlin, 1960.
165. Kranzer, Herbert C.: Asymptotic Factorization in Nondissipative Wiener-Hopf Problems, *J. Math. Mech.*, **17**:577–600 (1967).
166. Kranzer, Herbert C., and James Radlow: Asymptotic Factorization for Perturbed Wiener-Hopf Problems, *J. Math. Anal. Appl.*, **4**:240–256 (1962).
167. Kranzer, Herbert C., and James Radlow: An Asymptotic Method for Solving Perturbed Wiener-Hopf Problems, *J. Math. Mech.*, **14**:41–60 (1965).
168. Kruskal, Martin D., and Norman J. Zabusky: Stroboscopic Perturbation Procedure for Treating a Class of Nonlinear Wave Equations, *J. Mathematical Phys.*, **5**:231–244 (1964).
169. Latter, Richard: Approximate Solutions for a Class of Integral Equations, *Quart. Appl. Math.*, **16**:21–31 (1958).
170. Lax, Peter D.: Partial Differential Equations, mimeographed lecture notes by J. Berkowitz and A. and P. Lax, New York University, 1950–1951.
171. Lax, Peter D.: Hyperbolic Systems of Conservation Laws, II, *Commun. Pure Appl. Math.*, **10**:539–566 (1957).
172. Lax, Peter D.: Nonlinear Hyperbolic Systems of Conservation Laws, in R. E. Langer (ed.), "Nonlinear Problems," University of Wisconsin Press, Madison, 1963.
173. Lax, Peter D.: Development of Singularities of Solutions of Nonlinear Hyperbolic Partial Differential Equations, *J. Mathematical Phys.*, **5**:611–613 (1964).
174. Lax, Peter D.: Integrals of Nonlinear Equations of Evolution and Solitary Waves, *Commun. Pure Appl. Math.*, **21**:467–490 (1968).

175. Lebedev, N. N., and Ya. S. Uflyand, The Axisymmetric Contact Problem for an Elastic Layer, *Prikl. Math. Mech.*, **22**:320–326 (1958).
176. Leiter, Erich: Ein Beitrag zur Charakteristikentheorie der instationären ebenen und achsensymmetrischen Strömungen, *Z. Angew. Math. Mech.*, **47**:175–190; 229–237 (1967).
177. Levi-Civita, Tullio: "The Absolute Differential Calculus," Blackie, London, 1926.
178. Levine, Harold: A Corner Effect in Plane Diffusion Theory, *Appl. Sci. Res., Sec. B*, **8**:105–127 (1960).
179. Levine, Lawrence E., and A. G. Mackie, Two-dimensional Unsteady Motion of a Medium with a Linear Pressure-Density Relationship, *J. Math. Mech.*, **18**:1075–1086 (1969).
180. Levine, Leo: A Uniqueness Theorem for the Reduced Wave Equation, *Commun. Pure. Appl. Math.*, **17**:147–176 (1964).
181. Levinson, Norman, N. Bogert, and R. M. Redheffer: Separation of Laplace's Equation, *Quart. Appl. Math.*, **7**:241–262 (1949).
182. Lewis, Robert M.: The Progressing Wave Formalism, *Polytechnic Institute Brooklyn Symp. Quasi-optics*, June, 1964, 71–103.
183. Love, E. R.: Dual Integral Equations, *Can. J. Math.*, **15**:631–640 (1963).
184. Lowengrub, M., and Ian N. Sneddon: An Axisymmetric Boundary Value Problem of Mixed Type for a Half-space, *Proc. Edinburgh Math. Soc.*, **13**:39–46 (1962).
185. Ludford, G. S. S.: The Boundary Layer Nature of Shock Transition in a Real Fluid, *Quart. Appl. Math.*, **10**:1–16 (1952).
186. Ludford, G. S. S.: Riemann's Method of Integration: Its Extensions with an Application, *Seminar Math. Barcelone*, *Collectanea Math.*, **6**:3–33 (1953).
187. Ludford, G. S. S.: Extensions in the Applicability of Riemann's Formula, *J. Rational Mech. Anal.*, **3**:77–88 (1954).
188. Ludford, G. S. S.: Generalized Riemann Invariants, *Pacific J. Math.*, **5**:441–450 (1955).
189. Ludford, G. S. S., and M. H. Martin: One Dimensional Anisentropic Flows, *Commun. Pure Appl. Math.*, **7**:45–63 (1954).
190. Ludford, G. S. S., J. Martinek, and G. C. K. Yeh: The Sphere Theorem in Potential Theory, *Proc. Cambridge Phil. Soc.*, **51**:389–393 (1954).
191. Ludwig, Donald: The Singularities of the Riemann Function, *AEC Computing Appl. Math. Center NYU Rept.* 9351, 1960.
192. Ludwig, Donald, and Barry Granoff: Propagation of Singularities along Characteristics with Nonuniform Multiplicity, *J. Math. Anal. Appl.*, **21**:556–574 (1968).
193. Luré, K. A.: On the Propagation of Disturbances in Systems with Nonlinear Boundary Conditions (in Russian), *J. Appl. Math. Mech.*, **26**:749–760 (1962).
194. MacCamy, Richard C.: On Babinet's Principle, *Can. J. Math.*, **10**:632–640 (1958).
195. MacCamy, Richard C.: On the Scattering of Water Waves by a Circular Disk, *Arch. Rational Mech. Anal.*, **8**:120–138 (1961).
196. MacCamy, Richard C.: On an Axially Symmetric Singular Integral Equation, *J. Math. Mech.*, **15**:435–454 (1966).
197. Mackie, A. G.: One-dimensional Unsteady Motion of a Gas Initially at Rest and the Dam-break Problem, *Proc. Cambridge Phil. Soc.*, **50**:131–138 (1954).
198. Mackie, A. G.: Contour Integral Solutions of a Class of Differential Equations, *J. Rational Mech. Anal.*, **4**:733–750 (1955).
199. Mackie, A. G.: Initial Value Problems in Water Wave Theory, *J. Australian Math. Soc.*, **3**:340–350 (1963).
200. Mackie, A. G.: On Riemann's Method and a Variation by Martin, *Am. J. Math.*, **86**:728–734 (1964).
201. Mackie, A. G.: "Boundary Value Problems," Oliver & Boyd, London, 1965.

202. Mackie, A. G.: Green's Functions and Riemann's Method, *Proc. Edinburgh Math. Soc.*, **14**:293–302 (1965).
203. Mackie, A. G.: The Nonlinear Oscillations of a String, *Quart. Appl. Math.*, **25**:468–469 (1968).
204. MacMillan, William Duncan: "The Theory of the Potential," Dover, New York, 1958.
205. Martin, Monroe H.: The Propagation of a Plane Shock into a Quiet Atmosphere, *Can. J. Math.*, **5**:37–39 (1953).
206. Martinek, Johann, and Henry R. Thielman: New Solutions of the Laplace Equation in Spherical Coordinates, *J. Math. Mech.*, **16**:1177–1182 (1967).
207. Mercer, A. McD.: On Integral Transform Pairs Arising from Second Order Differential Equations, *Proc. Edinburgh Math. Soc.*, **13**:63–68 (1962).
208. Mercer, A. McD.: On Integral Transform Pairs Arising from Differential Equations of Any Even Order, *Quart. J. Math. Oxford*, (2) **14**:9–15 (1963).
209. Meyer, Norman G.: An Example of Non-uniqueness in the Theory of Quasi-linear Elliptic Equations of the Second Order, *Arch. Rational Mech. Anal.*, **14**:177–179 (1963).
210. Meyer, R. E.: Uniformization of a Quasi-linear Hyperbolic Equation, I and II, *J. Math. Mech.*, **16**:257–286 (1966).
211. Migdal, David, and Vito D. Agosta: A Source Flow Model for Continuum Gas-particle Flow, *J. Appl. Mech.*, **1967**:860–865.
212. Migdal, David, and Vito D. Agosta: Supersonic Gas-particle Flow with Chemical Reactions, *Phys. Fluids*, **10**:880–884 (1967).
213. Mikhlin, S. G. (ed.): "Linear Equations of Mathematical Physics," Holt, New York, 1967.
214. Miles, John W.: A Note on Riemann's Method Applied to the Diffusion Equation, *Quart. Appl. Math.*, **8**:95–101 (1950).
215. Miles, John W.: The Potential Theory of Unsteady Supersonic Flow, Cambridge University Press, London, 1959.
216. Millar, Robert F.: On a Non-linear Integral Equation Occurring in Diffraction Theory, *Proc. Cambridge Phil. Soc.*, **62**:249–261 (1966).
217. Mises, Richard von: "Mathematical Theory of Compressible Fluid Flow," Academic, New York, 1958.
218. Morse, Phillip McCord, and Herman Feshback: "Methods of Theoretical Physics," McGraw-Hill, New York, 1953.
219. Murray, J. D.: Singular Perturbations of a Class of Nonlinear Hyperbolic and Parabolic Equations, *J. Math. Phys.*, **47**:111–133 (1968).
220. Narain, Prem: A Note on an Asymmetric Mixed Boundary Value Problem for a Half Space with a Cylindrical Cavity, *Proc. Glasgow Math. Assoc.*, **7**:45–47 (1965).
221. Naylor, Derek: Unsteady Rectilinear Gas Flow, *J. Math. Mech.*, **9**:1-18 (1960).
222. Naylor, Derek: On a Mellin Type Integral Transform, *J. Math. Mech.*, **12**:265–274 (1963).
223. Naylor, Derek: On a Finite Lebedev Transform, *J. Math. Mech.*, **12**:375–384 (1963).
224. Naylor, Derek: On the Direct Solution of Certain Half-plane Mixed Boundary-value Problems, *Proc. Cambridge Phil. Soc.*, **62**:753–759 (1966).
225. Noble, Ben: "Methods Based on the Wiener-Hopf Technique for the Solution of Partial Differential Equations," Pergamon, New York, 1958.
226. Noble, Ben: The Solution of Bessel Function Dual Integral Equations by a Multiplying-factor Method, *Proc. Cambridge Phil. Soc.*, **59**:351–362 (1963).
227. Noble, Ben, and Arthur S. Peters: A Multiplying Factor Method for the Solution of Wiener-Hopf Integral Equations, *NYU Rept.* EM-137, July, 1959.

228. Odeh, Farouk M.: Uniqueness Theorems for the Helmholtz Equation in Domains with Infinite Boundaries, *J. Math. Mech.*, **12**:857–868 (1963).
229. O'Keeffe, J.: The Initial Value Problem for the Wave Equation in the Distributions of Schwartz, *Quart. J. Mech. Appl. Math.*, **8**:422–434 (1955).
230. Oswatitsch, Klaus: "Gas Dynamics," English version by G. Kuerti, Academic, New York, 1956.
231. Papoulis, Athanasios: "The Fourier Integral and Its Applications," McGraw-Hill, New York, 1962.
232. Pars, L. A.: "An Introduction to the Calculus of Variations," Heinemann, London, 1962.
233. Peters, Arthur S.: Certain Dual Integral Equations and Sonine's Integrals, *NYU Rept.* IMM-285, August, 1961.
234. Peters, Arthur S.: Abel's Equation and the Cauchy Integral Equation of the Second Kind, *Commun. Pure Appl. Math.*, **21**:51–66 (1968).
235. Peters, Arthur S.: Residue Expansions for Certain Green's Functions and Resolvent Kernels, *NYU Rept.* IMM-366, April, 1968.
236. Peters, Arthur S.: Some Generalized Eigenfunction Expansions and Uniqueness Theorems, *NYU Rept.* IMM-368, June, 1968.
237. Petrovskii, I. G.: "Lectures on Partial Differential Equations," transl. Abe Schenitzer, Interscience, New York, 1954.
238. Pounder, J. R., and J. L. Synge: Note on the Initial Value Problem for the Wave Equation in *N* Dimensions, *Proc. Royal Irish Acad.*, **57**:151–159 (1955).
239. Ranger, K. B.: On the Construction of Some Integral Operators for Generalized Axially Symmetric Harmonics and Stream Functions, *J. Math. Mech.*, **14**:383–402 (1965).
240. Ranger, K. B.: Note on Some Integral Transformations for Generalized Axially Symmetric Potentials, *J. Math. Mech.*, **16**:1377–1380 (1967).
241. Rosen, Gerald: Formal Equivalence of the Nonlinear String and One-dimensional Fluid Flow, *Quart. Appl. Math.*, **23**:286–287 (1965).
242. Rühs, F.: Über gewisse singuläre Volterrasche Integralgleichungen vom Faltungstypus, *Z. Angew. Math. Mech.*, **43**:361–367 (1963).
243. Rund, Hanno: "The Hamilton-Jacobi Theory in the Calculus of Variations," Van Nostrand, Princeton, N.J., 1966.
244. Sauer, Robert: Einfache Wellen in der Charakteristikentheorie von Systemen quasilinearer partieller Differentialgleichungen, *Z. Angew. Math. Mech.*, **44**:203–209 (1964).
245. Schmeltzer, Robert A., and Myrna Lewin: Function-theoretic Solution to a Class of Dual Integral Equations and an Application to Diffraction Theory, *Quart. Appl. Math.*, **21**:269–283 (1964).
246. Schubert, Hans: Über das dritte Randwertproblem der Potentialtheorie für den beiderseits unendlich langen Kreiszylinder, *Z. Angew. Math. Mech.*, **38**:194–199 (1958).
247. Segel, Lee A.: Application of Conformal Mapping to Boundary Perturbation Problems for the Membrane Equation, *Arch. Rational Mech. Anal.*, **8**:228–237 (1961).
248. Senior, T. B. A.: Diffraction by a Semi-infinite Metallic Sheet, *Proc. Roy. Soc.* (*London*), *Ser. A*, **213**:436–458 (1952).
249. Shinbrot, Marvin: A Generalization of Latta's Method for the Solution of Integral Equations, *Quart. Appl. Math.*, **16**:415–421 (1959).
250. Smirnov, V. I.: "A Course of Higher Mathematics," vol. IV, Addison-Wesley, Reading, Mass., 1964.
251. Smith, P.: An Example of a Plane Shock of Variable Strength, *Proc. Edinburgh Math. Soc.*, **13**:297–302 (1963).

252. Sneddon, Ian N.: "Fourier Transforms," McGraw-Hill, New York, 1951.

253. Sneddon, Ian N.: "Elements of Partial Differential Equations," McGraw-Hill, New York, 1957.

254. Sneddon, Ian N.: The Elementary Solution of Dual Integral Equations, *Proc. Glasgow Math. Assoc.*, **4**:108–110 (1960).

255. Sneddon, Ian N.: A Note on Some Relations between Fourier and Hankel Transforms, *Bull. Acad. Polon. Sci., Ser. Math.*, **9**:799–806 (1961).

256. Sneddon, Ian N.: A Procedure for Deriving Inversion Formulae for Integral Transform Pairs of a General Kind, *Glasgow Math. J.*, **9**:67–77 (1968).

257. Sobolev, S. L.: "Partial Differential Equations of Mathematical Physics," transl. E. R. Dawson, Addison-Wesley, Reading, Mass., 1964.

258. Sokolnikoff, Ivan S.: "Mathematical Theory of Elasticity," 2d ed., McGraw-Hill, New York, 1956.

259. Somerfeld, Arnold: "Partial Differential Equations in Physics," Academic, New York, 1949.

260. Sreedharan, V. P.: Function-theoretic Solutions of Certain Boundary Value Problems, *J. Math. Mech.*, **14**:211–230 (1965).

261. Srivastav, R. P.: A Note on Certain Integral Equations of Abel-type, *Proc. Edinburgh Math. Soc.*, **13**:271–272 (1963).

262. Srivastav, R. P.: An Axisymmetric Mixed Boundary Value Problem for a Half-space with a Cylindrical Cavity, *J. Math. Mech.*, **13**:385–393 (1964).

263. Srivastav, R. P.: Certain Two-dimensional Mixed Boundary Value Problems for Wedge-shaped Regions and Dual Integral Equations, *Proc. Edinburgh Math. Soc.*, **14**:321–332 (1964/1965).

264. Srivastav, R. P., and Prem Narain: Some Mixed Boundary Value Problems for Regions with Spherical Boundaries, *J. Math. Mech.*, **14**:613–627 (1965).

265. Srivastav, R. P.: On Certain Integral Equations of Convolution Type with Bessel Function Kernels, *Proc. Edinburgh Math. Soc.*, **15**:111–116 (1966).

266. Stakgold, Ivar: "Boundary Value Problems of Mathematical Physics," Macmillan, New York, vol. I, 1967, vol. II, 1968.

267. Stallybrass, M. P.: On a Pointwise Variational Principle for the Approximate Solution of Linear Boundary Value Problems, *J. Math. Mech.*, **16**:1247–1286 (1967).

268. Stoker, James J.: "Water Waves," Interscience, New York, 1957.

269. Stoker, James J., and Arthur S. Peters: A Uniqueness Theorem and a New Solution for Sommerfeld's and Other Diffraction Problems, *Commun. Pure Appl. Math.*, **7**:565–585 (1954).

270. Stuart, I. M.: String Vibrating at a Finite Amplitude, *J. Australian Math. Soc.*, **6**:369–382 (1966).

271. Synge, John L.: The Hamiltonian Method and Its Application to Water Waves, *Proc. Roy. Irish Acad., Ser. A*, **63**:1–34 (1963).

272. Tikhonov, A. N., and A. A. Samarskii: "Equations of Mathematical Physics," transl. A. R. M. Robson and P. Basu, Pergamon, New York, 1963.

273. Tralli, Nunzio: "Classical Electromagnetic Theory," McGraw-Hill, New York, 1963.

274. Trantner, C. J.: Some Triple Integral Equations, *Proc. Glasgow Math. Assoc.*, **4**:200–203 (1960).

275. Tricomi, Francesco G.: "Differential Equations," Blackie, London, 1961.

276. Tupholme, G. E.: Generation of an Axisymmetrical Acoustic Pulse by a Deformable Sphere, *Proc. Cambridge Phil. Soc.*, **63**:1285–1308 (1967).

277. Tychonov, A. N., and A. A. Samarskii: "Partial Differential Equations of Mathematical Physics," transl. S. Radding, Holden-Day, San Francisco, vol. I, 1964, vol. II, 1968.

278. Varley, Eric: A Class of Nonlinear Partial Differential Equations, *Commun. Pure Appl. Math.*, **15**:91–94 (1962).

279. Webster, Arthur Gordon: "Partial Differential Equations of Mathematical Physics," Dover, New York, 1955.

280. Weinberger, Hans F.: "A First Course in Partial Differential Equations," Blaisdell, New York, 1965.

281. Weinstein, Alexander: The Generalized Radiation Problem and the Euler-Poisson-Darboux Equation, *Summa Brasiliensis Math.*, **3**:125–147 (1955).

282. Weir, D. G.: A Family of Exact Solutions of One-dimensional Anisentropic Flow, *Proc. Cambridge Phil. Soc.*, **57**:89–894 (1961).

283. Weir, D. G.: The Propagation of a Constant Strength Shock through a Simple Wave, *Proc. Edinburgh Math. Soc.*, **13**:277–284 (1963).

283*a*. Weitz, Mortimer, and Joseph B. Keller: Reflection of Water Waves from Floating Ice in Water of Finite Depth, *Commun. Pure Appl. Math.*, **3**:305–318 (1950).

284. Wilcox, Calvin H.: Spherical Means and Radiation Conditions, *Arch. Rational Mech. Anal.*, **3**:133–148 (1959).

285. Wilcox, Calvin H.: Wave Operators and Asymptotic Solution of Wave Propagation Problems of Classical Type, *Arch. Rational Mech. Anal.*, **22**:37–78 (1966).

286. Wilcox, Calvin H.: Steady-state Wave Propagation in Homogeneous Anisotropic Media, *Arch. Rational Mech. Anal.*, **25**:201–242 (1967).

287. Williams, W. Elwyn: Diffraction of an E-polarized Plane Wave by an Imperfectly Conducting Wedge, *Proc. Roy. Soc.* (*London*), *Ser. A*, **252**:376–393 (1959).

288. Williams, W. Elwyn: Diffraction of an Electromagnetic Plane Wave by a Metallic Sheet, *Proc. Roy. Soc.* (*London*), *Ser. A*, **257**:413–419 (1960).

289. Williams, W. Elwyn: Waves on a Sloping Beach, *Proc. Cambridge Phil. Soc.*, **57**:160–165 (1961).

290. Williams, W. Elwyn: Vertex Generated Waves outside Metallic Wedges, *Proc. Cambridge Phil. Soc.*, **57**:393–400 (1961).

291. Williams, W. Elwyn: The Reduction of Boundary Value Problems to Fredholm Integral Equations of the Second Kind, *Z. Angew. Math. Phys.*, **13**:133–152 (1962).

292. Williams, W. Elwyn: A Class of Integral Equations, *Proc. Cambridge Phil. Soc.*, **59**:589–597 (1963).

293. Williams, W. Elwyn: Note on the Electrostatic Problem for a Circular Annulus, *Quart. Mech. Appl. Math.*, **16**:205–207 (1963).

294. Williams, W. Elwyn: Note on the Scattering of Long Waves in a Rotating System, *J. Fluid Mech.*, **20**:115–119 (1964).

295. Williams, W. Elwyn: Two Dimensional Mixed Boundary Value Problems in a Wedge-shaped Domain, *Proc. Cambridge Phil. Soc.*, **64**:503–505 (1968).

296. Wolfersdorf, Lothar von: Abelsche Integralgleichungen und Randwertprobleme für die verallgemeinerte Tricomi-Gleichung, *Math. Nachr.*, **29**:161–178 (1965).

297. Wu, Tai Te, and Tai Tsun Wu: Iterative Solutions of Wiener-Hopf Integral Equations, *Quart. Appl. Math.*, **20**:341–352 (1963).

298. Yeh, G. C. K., J. Martinek, and G. S. S. Ludford: The Potentials due to Certain Singularities in the Presence of a Fixed Sphere, *J. Soc. Ind. Appl. Math.*, **3**:142–152 (1955).

299. Yeh, G. C. K., J. Martinek, and G. S. S. Ludford: A General Sphere Theorem for Hydrodynamics, Heat, Magnetism and Electrostatics, *Z. Angew. Math. Mech.*, **36**:111–116 (1956).

300. Yu, Yi-Yuan: Generalized Hamilton's Principle and Variational Equation of Motion in Nonlinear Elasticity Theory with Application to Plate Theory, *J. Acoust. Soc. Am.*, **36**:111–120 (1964).

301. Zabusky, Norman J.: Exact Solution for the Vibrations of a Nonlinear Continuous Model String, *J. Math. Phys.*, **3**:1–12 (1962).
302. Zauderer, Erich: Domains of Dependence for Mixed Problems for Wave Equations, *Arch. Rational Mech. Anal.*, **15**:69–78 (1964).
303. Zuckerberg, Hyam I.: Solution of the Dirichlet Problem in Parallel Slit Domains, *J. Math. Phys.*, **45**:77–94 (1966).

INDEX

INDEX